## Symmetrization in Analysis

Symmetrization is a rich area of mathematical analysis whose history reaches back to antiquity. This book presents many aspects of the theory, including symmetric decreasing rearrangement and circular and Steiner symmetrization in Euclidean spaces, spheres, and hyperbolic spaces. Many energies, frequencies, capacities, eigenvalues, perimeters, and function norms are shown to either decrease or increase under symmetrization.

The book begins by focusing on Euclidean space, building up from two-point polarization with respect to hyperplanes. Background material in geometric measure theory and analysis is carefully developed, yielding self-contained proofs of all the major theorems. This leads to the analysis of functions defined on spheres and hyperbolic spaces, and then to convolutions, multiple integrals, and hypercontractivity of the Poisson semigroup. The author's star function, which preserves subharmonicity, is developed with applications to semilinear partial differential equations. The book concludes with a thorough self-contained account of the star function's role in complex analysis, covering value distribution theory, conformal mapping, and the hyperbolic metric.

Albert Baernstein II was a professor in the Department of Mathematics at Washington University in St. Louis until his death in 2014. He gained international renown for innovative solutions to extremal problems in complex and harmonic analysis. His invention of the "star function" method in the 1970s prompted an invitation to the International Congress of Mathematicians held in Helsinki in 1978, and during the 1980s and 1990s he substantially extended the breadth and applications of this method.

NEW MATHEMATICAL MONOGRAPHS



All the titles listed below can be obtained from good booksellers or from Cambridge University Press. For a complete series listing visit www.cambridge.org/mathematics.

1. M. Cabanes and M. Enguehard *Representation Theory of Finite Reductive Groups*
2. J. B. Garnett and D. E. Marshall *Harmonic Measure*
3. P. Cohn *Free Ideal Rings and Localization in General Rings*
4. E. Bombieri and W. Gubler *Heights in Diophantine Geometry*
5. Y. J. Ionin and M. S. Shrikhande *Combinatorics of Symmetric Designs*
6. S. Berhanu, P. D. Cordaro and J. Hounie *An Introduction to Involutive Structures*
7. A. Shlapentokh *Hilbert's Tenth Problem*
8. G. Michler *Theory of Finite Simple Groups I*
9. A. Baker and G. Wüstholz *Logarithmic Forms and Diophantine Geometry*
10. P. Kronheimer and T. Mrowka *Monopoles and Three-Manifolds*
11. B. Bekka, P. de la Harpe and A. Valette *Kazhdan's Property (T)*
12. J. Neisendorfer *Algebraic Methods in Unstable Homotopy Theory*
13. M. Grandis *Directed Algebraic Topology*
14. G. Michler *Theory of Finite Simple Groups II*
15. R. Schertz *Complex Multiplication*
16. S. Bloch *Lectures on Algebraic Cycles (2nd Edition)*
17. B. Conrad, O. Gabber and G. Prasad *Pseudo-reductive Groups*
18. T. Downarowicz *Entropy in Dynamical Systems*
19. C. Simpson *Homotopy Theory of Higher Categories*
20. E. Fricain and J. Mashreghi *The Theory of H(b) Spaces I*
21. E. Fricain and J. Mashreghi *The Theory of H(b) Spaces II*
22. J. Goubault-Larrecq *Non-Hausdorff Topology and Domain Theory*
23. J. Śniatycki *Differential Geometry of Singular Spaces and Reduction of Symmetry*
24. E. Riehl *Categorical Homotopy Theory*
25. B. A. Munson and I. Volić *Cubical Homotopy Theory*
26. B. Conrad, O. Gabber and G. Prasad *Pseudo-reductive Groups (2nd Edition)*
27. J. Heinonen, P. Koskela, N. Shanmugalingam and J. T. Tyson *Sobolev Spaces on Metric Measure Spaces*
28. Y.-G. Oh *Symplectic Topology and Floer Homology I*
29. Y.-G. Oh *Symplectic Topology and Floer Homology II*
30. A. Bobrowski *Convergence of One-Parameter Operator Semigroups*
31. K. Costello and O. Gwilliam *Factorization Algebras in Quantum Field Theory I*
32. J.-H. Evertse and K. Győry *Discriminant Equations in Diophantine Number Theory*
33. G. Friedman *Singular Intersection Homology*
34. S. Schwede *Global Homotopy Theory*
35. M. Dickmann, N. Schwartz and M. Tressl *Spectral Spaces*

# Symmetrization in Analysis

ALBERT BAERNSTEIN II
*Washington University in St Louis*

with

DAVID DRASIN
*Purdue University*

RICHARD S. LAUGESEN
*University of Illinois, Urbana–Champaign*

CAMBRIDGE
UNIVERSITY PRESS

**CAMBRIDGE**
**UNIVERSITY PRESS**

University Printing House, Cambridge CB2 8BS, United Kingdom

One Liberty Plaza, 20th Floor, New York, NY 10006, USA

477 Williamstown Road, Port Melbourne, VIC 3207, Australia

314–321, 3rd Floor, Plot 3, Splendor Forum, Jasola District Centre, New Delhi – 110025, India

79 Anson Road, #06–04/06, Singapore 079906

Cambridge University Press is part of the University of Cambridge.

It furthers the University's mission by disseminating knowledge in the pursuit of education, learning and research at the highest international levels of excellence.

www.cambridge.org
Information on this title: www.cambridge.org/9780521830478
DOI: 10.1017/9781139020244



First published 2019

Printed and bound in Great Britain by Clays Ltd, Elcograf S.p.A.

*A catalogue record for this publication is available from the British Library.*

*Library of Congress Cataloging-in-Publication Data*
Names: Baernstein, Albert, II, 1941–2014, author. | Drasin, D. (David), 1940– author. | Laugesen, Richard, 1967– author.
Title: Symmetrization in analysis / Albert Baernstein II (Washington University in St Louis) ; with David Drasin (Purdue University, Indiana), Richard Laugesen (University of Illinois, Urbana-Champaign).
Description: Cambridge ; New York, NY : Cambridge University Press, 2019. | Series: New mathematical monographs | Includes bibliographical references and index.
Identifiers: LCCN 2018043277 | ISBN 9780521830478 (hardback)
Subjects: LCSH: Mathematical analysis. | Symmetric functions. |
Classification: LCC QA300 .B1685 2019 | DDC 515/.22–dc23
LC record available at https://lccn.loc.gov/2018043277

ISBN 978-0-521-83047-8 Hardback

# Contents

# Notation

$|A|$, the operator norm of a matrix $A$, §7.8
$|A|$, the linear measure of a subset $A$ of $\mathbb{R}$ or $\mathbb{T}$, §11.1
$\alpha_n$, the volume of the unit $n$-ball, §1.4
$AL$ and $SAL$ classes, §2.1
$AL_0$ and $SAL_0$ classes, §2.2
$A(R_1, R_2)$, open spherical shell, §9.2
$\beta_{n-1}$, the surface measure of the unit $(n-1)$-sphere, §4.5
$\mathcal{B}$, the Borel $\sigma$-algebra, §1.3
$\mathcal{B}_c$, the set of Borel sets contained in some compact set, §9.2
$\mathbb{B}^n(r)$, the open $n$-ball of radius $r$ centered at the origin, §1.4
$\mathbb{B}^n(a, r)$, the open $n$-ball of radius $r$ centered at $a$, §1.5
$B_t$, Brownian motion, §5.7
$BV$, the space of functions of bounded variation, §4.2
$\widehat{\mathbb{C}}$, extended complex numbers $\mathbb{C} \cup \{\infty\}$, §11.4
$C_c$, continuous functions with compact support, §2.2
$c(E)$, the center of mass of $E$, §2.6
$C^\gamma$, Hölder spaces, §4.8
$\operatorname{Cap} K$, the Newtonian capacity of $K$, §5.5
$\operatorname{Cap}_p K$, the variational $p$-capacity of $K$, §5.6
$C_\alpha K$, the Riesz $\alpha$-capacity of $K$, §5.6
$\mathbb{D}$, open unit disk, §11.1
$\mathbb{D}(r)$, open disk of radius $r$, §11.1
$\operatorname{diam} E$, the diameter of $E$, §1.7
$d(x, E)$, the distance from point $x$ to set $E$, §2.4
$\partial^* E$, the reduced boundary of $E$, §4.3
$\partial_i f$, the partial derivatives of $f$, §3.4
$\partial_v f$, the derivative of $f$ in direction $v$, §3.1
$D^\alpha f$, multiindex notation for derivatives, §4.8

$\Delta$, the Laplace operator, §5.1
$d(x, y)$, spherical distance in §7.1; hyperbolic distance in §7.6
$\Delta_s$, spherical Laplacian, §7.2
$\Delta^\star$ and related elliptic operators, §9.6, 9.7, 9.8, 9.10, 9.11
$\Delta^{\star}$, a variant of the $\Delta^\star$ operator, §10.6
$\nabla_s$, spherical gradient, §7.2
$\nabla_h$, hyperbolic gradient, §7.6
$E^\#$, rearrangement of a set $E$, §1.6 and later. The particular type of rearrangement (e.g., symmetric decreasing, Steiner, spherical, cap) depends on the context.
$E(\delta)$, the $\delta$-collar of $E$, §4.4
$E(-\delta)$, the $\delta$-core of $E$, §4.4
$E_x$, expected value with respect to the Brownian motion starting at $x$, §5.7
$\tilde{E}$, the Gauss symmetrization of set $E$, §7.7
$f^*$, the decreasing rearrangement of $f$, §1.2
$f^+ = \max(f, 0)$ and $f^- = \max(-f, 0)$, positive and negative parts of $f$, §1.3
$f^\#$, rearrangement of a function $f$, §1.6 and later. The particular type of rearrangement (e.g., symmetric decreasing, Steiner, spherical, cap) depends on the context.
$f^\star$, the $\star$-function, for various types of rearrangement, §9.1, 9.2, 9.6, 9.7, 9.8, 9.10, 9.11
$f^{\star}$, a variant of the $\star$-function, §11.1
$f_H$, the polarization of $f$ with respect to $H$, §1.7
$\tilde{f}$, the Gauss symmetrization of function $f$, §7.7
$f_\#\mu$, the pushforward of $\mu$ by $f$, §1.3
$G(x, y, \Omega)$, Green's function, §9.4
$G(x)$, Green's function of unit ball with pole at 0, §9.4
$G(\mathbb{R}^n)$, the conformal group of $\mathbb{R}^n$, §7.6
$\gamma_n$, the Gauss measure, §7.7
$\mathbb{H}^n$, the $n$-dimensional hyperbolic space, §1.6
$\mathbb{H}$, the upper halfplane, §11.1
$\mathcal{H}(\mathbb{R}^n)$, the set of affine hyperplanes in $\mathbb{R}^n$, §1.7
$\mathcal{H}(\mathbb{S}^n)$, the intersections of linear hyperplanes in $\mathbb{R}^{n+1}$ with $\mathbb{S}^n$, §7.1
$H^+$ and $H^-$, two halfspaces determined by affine hyperplane $H$, §1.7
$\mathcal{H}^s$, Hausdorff measure, §4.1
$\mathcal{H}(f)$, the Riesz energy of $f$, §8.7
$\mathcal{H}^{k,n}$, spherical harmonics on $\mathbb{S}^n$ of degree $k$, §8.8
$J_f$, the Jacobian determinant of $f$, §4.1
$J(f, g, h)$, triple convolution evaluated at zero, §8.1. More general version in §8.5.

$J$-operator for definite integration over balls, spherical caps, and so on, depending on the type of rearrangement, §9.6, 9.7, 9.10, 9.11
$\mathcal{K}(\theta)$, open spherical cap on $\mathbb{S}^n$ centered at $e_1$, §7.1
$K(x, y, t)$ the Dirichlet heat kernel, §8.6
$K_\lambda(x) = |x|^{-\lambda}$, the Riesz kernel, §8.7
$\mathcal{L}$, the Lebesgue measure on $\mathbb{R}$, §1.2
$\mathcal{L}^n$, the Lebesgue measure on $\mathbb{R}^n$, §1.4
$\lambda_f$, distribution function, §1.1
$\lambda(t+)$ and $\lambda(t-)$, one-sided limits, §1.1
Lip, the Lipschitz class, §3.1
$\lambda_1(\Omega)$, the principal Dirichlet eigenvalue of $\Omega$, §5.3
Lcap $K$, the logarithmic capacity of $K$, §5.6
$\mathcal{M}^s$, $\mathcal{M}^s_*$, Minkowski content, §4.4
Mod $\Omega$, the conformal modulus of $\Omega$, §5.6 (extended to dimensions $n > 2$ in §7.8)
$M(K)$, the space of finite measures on $K$, §9.2
$M_{\mathrm{loc}}(X)$, the space of locally finite measures on $X$, §9.2
$\mu^\#$, rearrangement of a measure, for various types of rearrangement, §§9.5–9.9, 9.11
$\mu^\star$, star operation applied to a measure, for various types of rearrangement, §9.6, 9.7, 9.8, 9.10, 9.11
$\nu_n$, normalized spherical measure, §8.8
$O(n)$, orthogonal group, §7.1
$\omega(t,f)$, the modulus of continuity of $f$, §1.7
$p^*$, the Sobolev conjugate exponent of $p$, §4.6
$P(E)$, the perimeter of set $E$, §4.3
$\mathbb{R}^+ = [0, \infty)$, the set of nonnegative real numbers
$R_G$ and $R_T$, the Grötszsch and Teichmüller rings, §7.8
$\rho_H$, reflection in hyperplane $H$, §1.7
$\mathbb{S}^n$, the $n$-dimensional unit sphere, §1.6
$\mathbb{S}(r)$, the circle of radius $r$ centered at the origin, §11.1
$\sigma_{n-1}$, the restriction of $\mathcal{H}^{n-1}$ to $\mathbb{S}^{n-1}$, §4.5
$\star$-function, see entries above for $f^\star$ and $\mu^\star$
$\star$-function, see entry above for $f^{\star}$
$T(\Omega)$, the torsional rigidity of $\Omega$, §5.7
$T^{\star}$, the $\star$-function of the Nevanlinna characteristic $T$, §11.2
$\tau(E)$, the canonical measure on hyperbolic space, §7.6
$\mathrm{Tr}(t, \Omega)$, the trace of the heat kernel, §8.6
$u_K$, the equilibrium potential of $K$, §5.5
$V(f)$, the total variation of $f$, §4.2

$W^{1,p}$, $W_0^{1,p}$, Sobolev spaces, §3.4
$\mathcal{W}(\Omega)$, the space of functions whose weak Laplacian is a measure, §9.2
$(X, \mathcal{M}, \mu)$, measure space, §1.1
$\chi_A$, characteristic function of set $A$, §1.1
$Y_k$, a spherical harmonic of degree $k$, §8.8

# Foreword

At the 1973 Symposium at Canterbury on Complex Analysis there were two stars. One was the star function and the other was its inventor, Al Baernstein.

Suppose $u(z)$ is subharmonic in the annulus $\{re^{i\theta} : r_1 < r < r_2\}$, and define

$$u^{\star}(re^{i\theta}) = \sup_E \int_E u(re^{it})\, dt,$$

where the supremum is taken over all sets $E$ of measure $2\theta$ in $[0, 2\pi]$. Then $u^{\star}$ is the star function of $u$.

**Theorem I** *With the above hypotheses, $u^{\star}$ is subharmonic in the semiannulus $\{re^{i\theta} : r_1 < r < r_2, 0 < \theta < \pi\}$.*

(See Corollary 9.10 and the Chapter 9 Notes, with $u^{\star} = u^{\star}/r$.)

It is amazing how many consequences have been deduced by Baernstein and others from this innocent-looking theorem, and one purpose of this volume is to provide a coherent account of what have been the most important. There are generalizations to higher dimensional Euclidean space but some of the most interesting results occur when $u(z) = \log |f(z)|$ and $f$ is analytic or meromorphic. These are described in Chapter 11 of this book.

In a short foreword it is impossible to do justice to the full portfolio of results in this comprehensive book on symmetrization. So I would like to concentrate on two results in Chapter 11, which were first announced at the Canterbury conference.

1. Let $\mathcal{S}$ be the class of functions

$$f(z) = z + a_2 z^2 + \cdots$$

analytic and univalent in the unit disk $|z| < 1$. If $\Phi$ is a convex increasing function we have

$$\int_0^{2\pi} \Phi\big(\log|f(re^{i\theta})|\big)\,d\theta \leq \int_0^{2\pi} \Phi\big(\log|k(re^{i\theta})|\big)\,d\theta,$$

where $k(z) = z/(1-z)^2$ is the Koebe function, with strict inequality unless $f$ equals the Koebe function or one of its rotates $(e^{-it}k(ze^{it}))$.

As a corollary, we have for $0 < p < \infty$ and $0 < r < 1$ that

$$M_p(r,f) \leq M_p(r,k),$$

where

$$M_p(r,f) = \left(\int_0^{2\pi} |f(re^{i\theta})|^p\,d\theta\right)^{1/p}.$$

2. Baernstein originally introduced the star function in order to prove A. Edrei's spread conjecture. This is more complicated to formulate and needs a few definitions. Suppose that $u(z)$ is $\delta$-subharmonic in the plane, that is, $u(z) = u_1(z) - u_2(z)$ where $u_1$ and $u_2$ are subharmonic. We define $u^+(z) = \max(u(z), 0)$, and

$$m(r) = m(r,u) = \frac{1}{2\pi}\int_0^{2\pi} u^+(re^{it})\,dt$$

for $r > 0$. Let

$$n(r) = \frac{1}{2\pi}\int_{\{|z|\leq r\}} \Delta u_2\,dxdy$$

and

$$N(r) = N(r,u) = \int_0^r t^{-1}\,(n(t) - n(0))\,dt + n(0)\log r.$$

Then

$$T(r) = T(r,u) = m(r,u) + N(r,u)$$

is the Nevanlinna characteristic of $u$ (the designation is in honor of Rolf Nevanlinna, the founder of this theory). The deficiency of $u$ is defined by

$$\delta(u) = \liminf_{r\to\infty} \frac{m(r,u)}{T(r,u)}$$

(so that $0 \leq \delta(u) \leq 1$), and the lower order of $u$ is

$$\mu = \liminf_{r\to\infty} \frac{\log T(r,u)}{\log r}.$$

Then Baernstein's spread theorem states:

**Theorem II** *If $\delta(u) > 0$ then for every positive $\eta$ there is a positive $\varepsilon$ and a sequence of radii $r_n$ and sets $E_n$ of measure at least $2\beta - \eta$ in $[0, 2\pi]$ so that $r_n \to \infty$ and*

$$u(r_n e^{i\theta}) > \varepsilon T(r_n) \text{ on } E_n,$$

*where*

$$\beta = \min\left(\pi, (2/\mu)\sin^{-1}\sqrt{\delta/2}\right).$$

See Proposition 11.7, which handles the case of $u = \log|f|$ with $f$ meromorphic and is phrased in terms of a more general growth measure than $\mu$.

This value of $\beta$ is sharp. When $f$ is meromorphic in the plane and $u = \log|f|$, we deduce Edrei's spread conjecture on the size of the set where $|f| > 1$; in fact on this set $\log|f|$ is comparable to $T(r)$. Applying the result to $u = -\log|f|$ we obtain sharp bounds for the sum of the deficiencies of $f$. These were known for $\mu \leq 1/2$ and conjectured by Edrei when $1/2 < \mu < 1$. Very little is known when $\mu > 1$ other than for certain isolated values, as discussed in Note 3 at the end of Chapter 11.

—Walter Hayman

# Preface

Albert Baernstein passed away June 10, 2014, a great loss to mathematics and to his many friends, colleagues, and students. As recounted in Walter Hayman's foreword, Al's early discovery of the star function, along with its immediate applications to classical complex analysis, gave him international prominence, and provided the foundation for a long and productive career in mathematical analysis. An obituary, including a mathematical sketch and bibliography, has appeared in the *Notices of the American Mathematical Society* (Drasin, 2015).

Al worked on this symmetrization book for many years. Toward the end of his life, when it became clear he might not finish the work, a group of friends and former students committed to completing the project in the manner he envisaged. Al gave his blessing, and shared his files. He had planned eleven chapters. Eight and a half were essentially complete, and he left a rough outline of his goals for the remainder.

Richard Laugesen[1] and David Drasin were the general coordinators, aided by Juan Manfredi. In the early stages, Leonid Kovalev provided an essential service by editing Al's original LaTeX files to label the results, create an index file, put references into .bib format, and make the notation list. Juan Manfredi wrote most of the Introduction, and he and Almut Burchard drew the figures for the book. Burchard provided invaluable help revising Chapter 8, and Laugesen revised and added material to Chapter 9. Al had prepared the first section of Chapter 10, and Laugesen and Jeffrey Langford wrote the remainder. David Drasin and Allen Weitsman wrote Chapter 11, using Al's earlier account (Baernstein, 2002) as a guide.

The extended period required to finish the book was warranted, we feel, by the mathematical vision embedded in Al's manuscript. We hope his monograph

[1] Supported by the Simons Foundation (#429422 to Richard Laugesen).

will serve as a foundation for further research in symmetrization and its applications.

Cambridge University Press has always been patient and supportive. Special thanks go to Juan Manfredi, who contributed to and encouraged the group effort at every stage. We warmly thank Judy Baernstein for decades of friendship, and for the hospitality she showed to all mathematicians making pilgrimages to Washington University in St. Louis.

— David Drasin (Purdue University),
Richard S. Laugesen (University of Illinois),
June 2018

# Introduction

This is a book primarily about symmetrization of real-valued functions and sets. Many extremal problems in mathematics and physics have symmetric solutions, the quintessential example being the isoperimetric inequality (see Chapter 4) that among all sets with given volume, the ball possesses minimal surface area. The book develops and applies symmetrization techniques for problems in geometry, partial differential equations, and complex analysis.

Other treatments of symmetrization with applications to analysis and partial differential equations can be found in the works of Bandle (1980), Bennett and Sharpley (1988), Kawohl (1985), Kesavan (2006), Lieb and Loss (1997), and Pólya and Szegő (1951). For applications to complex analysis see Duren (1983), Hayman and Kennedy (1976), Hayman (1989, 1994), and Dubinin (2014). For applications to Fourier analysis and hyperbolic geometry, one may consult Beckner (1995).

Each chapter ends with Notes that contain historical remarks and additional information.

**Chapter 1** presents the theory of rearrangements of functions, where one compares a real-valued function $f$ on a measure space $(X, \mathcal{M}, \mu)$ with another function $g$, defined on a possibly different measure space, such that $f$ and $g$ have the same "size." The notion of size corresponds to the distribution function $\lambda_f(t) = \mu(f > t)$. To avoid technical difficulties with infinity, we always assume that $\lambda_f(t) < \infty$, for every $t > \operatorname{ess\,inf} f$. We consider $f$ and $g$ to have the same size if they have the same distribution function, in which case $f$ and $g$ are called rearrangements of each other. We would like to find a rearrangement $g$ that has "more symmetry" than $f$.

The simplest case (§1.2) is the decreasing rearrangement of $f$, denoted $f^*$, which is a decreasing one-variable function defined on the interval $[0, \mu(X)]$.

Next in simplicity is the symmetric decreasing rearrangement on $\mathbb{R}^n$ (§1.6), written $f^\#(x)$. It has the property that $(f^\# > t)$ is a ball centered at the origin. Before studying $f^{\#}$ prerequisites in measure theory are covered (§§1.3–1.4) in order to present a general version of Ryff's factorization theorem (1970). Ryff's theorem asserts that if $(X, \mathcal{M}, \mu)$ is a nonatomic measure space with $\mu(X) < \infty$ and $f: X \to \mathbb{R}$ is $\mathcal{M}$ measurable, then a measure preserving transformation $T: X \to [0, \mu(X)]$ exists such that $f = f^* \circ T$ for almost every $x \in X$. Note that if $T$ is measure preserving, then $f$ and $f \circ T$ have the same distribution function. A particular case is when $T(x) = \alpha_n |x|^n$, with $\alpha_n$ the volume of the unit ball in $\mathbb{R}^n$. In that case $f^{\#} = f^* \circ T$ (see §1.6), which connects the symmetric decreasing rearrangement to the decreasing rearrangement through the change of variable $T$.

Another type of rearrangement central to this book is the polarization of $f$ with respect to an affine hyperplane $H \subset \mathbb{R}^n$, denoted by $f_H$ (§1.7). Polarization involves moving the larger values of $f$ preferentially to one side of the hyperplane. Polarization with respect to all hyperplanes that do not contain the origin yields the symmetric decreasing rearrangement $f^{\#}$.

The chapter ends with convergence theorems for $f^*$ and $f^{\#}$, covering the cases of almost everywhere convergence and convergence in measure.

Examples and graphs are included throughout the chapter, in line with the author's pedagogical intentions. Some new notions are introduced first in the discrete case, where functions are just finite sequences and all calculations can be carried out explicitly.

**Chapter 2** covers the foundational inequalities for integrals of functions on $\mathbb{R}^n$. In Baernstein's approach a key notion is that of an AL function $\Psi(x, y)$, which generalizes the condition of nonnegative mixed partials $\Psi_{xy} \geq 0$. The two key results in this chapter are that symmetric decreasing rearrangement of a continuous function decreases its modulus of continuity, and that certain integral expressions increase when functions are replaced by their symmetric decreasing rearrangements.

The proof presented for the decrease of modulus of continuity (Theorem 2.12) is based on elementary polarization inequalities and the Arzelà–Ascoli theorem, and does not rely on other inequalities such as the isoperimetric or Brunn–Minkowski type inequalities.

Given nonnegative functions $f, g$, along with a nonnegative kernel $K$ and an AL function $\Psi: \mathbb{R}^+ \times \mathbb{R}^+ \to \mathbb{R}^+$, the basic inequality in Theorem 2.15 says that a certain integral expression increases under symmetrization:

$$\int_{\mathbb{R}^{2n}} \Psi(f(x), g(y)) K(|x-y|)\, dx\, dy \leq \int_{\mathbb{R}^{2n}} \Psi(f^{\#}(x), g^{\#}(y)) K(|x-y|)\, dx\, dy.$$

The proof is presented in stages. First, an analogous inequality is proved in the simple case of a space consisting of two points (Theorem 2.8), and then for the case of polarization with respect to an affine subspace (Theorem 2.9), and finally for the symmetric decreasing rearrangement (Theorem 2.15). This structured approach permits easy modification later (Chapter 7) to spheres and hyperbolic spaces. The proof is done first in the case of continuous $\Psi$, which is the most important case for applications, and completed in §§2.8–2.9 for general AL functions.

In §2.7 many direct consequences of Theorem 2.15 are presented, including the classical Hardy–Littlewood inequality

$$\int_{\mathbb{R}^n} fg\,dx \le \int_{\mathbb{R}^n} f^\# g^\#\,dx, = \int_{\mathbb{R}^+} f^* g^*\,dx,$$

as well as the contractivity of rearrangement in the $L^\infty$-norm (Corollary 2.23).

**Chapter 3** develops the basic Dirichlet integral inequalities for symmetric decreasing rearrangement. The main result is the inequality

$$\int_{\mathbb{R}^n} |\nabla f^\#|^p\,dx \le \int_{\mathbb{R}^n} |\nabla f|^p\,dx, \quad 1 \le p < \infty,$$

for $f \in \mathrm{Lip}(\mathbb{R}^n, \mathbb{R})$ satisfying $\lambda_f(t) < \infty$ for all $t > \inf f$ (Theorem 3.7) and its extension to $f \in W^{1,p}(\mathbb{R}^n, \mathbb{R}^+)$ (Theorem 3.20). The inequality when $p = \infty$ is easier, $\|\nabla f^\#\|_{L^\infty(\mathbb{R}^n)} \le \|\nabla f\|_{L^\infty(\mathbb{R}^n)}$, and follows from the monotonicity of the modulus of continuity (Theorem 3.6). Background on Lipschitz functions is given in §3.1. The proof of Theorem 3.7 (the Lipschitz case) is in §3.2, ultimately based on the basic inequality in Theorem 2.15. Various comments are made on the equality case. This section also includes a version valid for nonnegative functions on a domain $\Omega \subset \mathbb{R}^n$ (Corollary 3.9), assuming the function vanishes on the boundary.

Section 3.3 presents a more general inequality for $\Phi$-Dirichlet integrals

$$\int_{\mathbb{R}^n} \Phi(|\nabla f^\#|)\,dx \le \int_{\mathbb{R}^n} \Phi(|\nabla f|)\,dx,$$

where $\Phi\colon \mathbb{R}^+ \to \mathbb{R}^+$ is convex and increasing with $\Phi(0) = 0$. The proof is again based on Theorem 2.15. Another approach due to Dubinin based on polarization is included too.

Sections 3.4 and 3.5 include background material on Sobolev spaces and functional analysis needed to extend the Dirichlet integral inequality to functions in the Sobolev space $W^{1,p}(\mathbb{R}^n, \mathbb{R}^+)$. The extension is presented in §3.6. The chapter ends with §3.7, discussing the continuity of the rearrangement operator $f \mapsto f^\#$ in various situations. The operator is continuous in $L^p(\mathbb{R}^n, \mathbb{R}^+)$, continuous at the zero function in $W^{1,p}(\mathbb{R}^n, \mathbb{R}^+)$, and continuous everywhere in $W^{1,p}(\mathbb{R}, \mathbb{R}^+)$ (dimension $n = 1$), but is discontinuous at a general Sobolev function when $n \ge 2$. The condition for continuity at $f$, the coarea regularity condition discovered by Almgren and Lieb, is presented in this section.

**Chapter 4** is devoted to the isoperimetric inequality and sharp Sobolev inequalities. It begins with a review of geometric measure theory tools (Hausdorff measures, area formula, and Gauss–Green theorem) used in this and later chapters. The convention of Evans and Gariepy (1992) is followed in this chapter: "measure" means "outer measure."

Three isoperimetric inequalities are presented: for perimeters (Theorem 4.10), for Hausdorff measures (Corollary 4.13), and for Minkowski content (Theorem 4.16). If $E \subset \mathbb{R}^n$ with finite perimeter, finite measure, or finite Minkowski content, one has

$$\begin{aligned} P(E) &\geq P(E^\#), \\ \mathcal{H}^{n-1}(\partial E) &\geq \mathcal{H}^{n-1}(\partial(E^\#)), \\ \mathcal{M}_*^{n-1}(\partial E) &\geq \mathcal{M}^{n-1}(\partial(E^\#)), \end{aligned}$$

where $P(E^\#) = \mathcal{H}^{n-1}(\partial(E^\#)) = \mathcal{M}^{n-1}(\partial(E^\#)) = n\alpha_n^{1/n}\mathcal{L}^n(E)^{\frac{n-1}{n}}$, and $E^\#$ is a ball of the same volume as $E$. (Here $\mathcal{L}^n$ is the $n$-dimensional Lebesgue measure.) All three isoperimetric inequalities are deduced from the fact that symmetrization decreases the Dirichlet integral (Theorem 3.7) or the variation of a function (Theorem 4.8).

Additional facts from geometric measure theory (the coarea formula and polar coordinates) are stated in §4.5. This section also shows that the coarea formula and the isoperimetric inequality for perimeter together imply decrease of the Dirichlet integral under symmetrization.

Section 4.6 presents the proof of the sharp Sobolev embedding inequalities for $f \in BV(\mathbb{R}^n)$, $n \geq 2$, which is

$$\|f\|_{\frac{n}{n-1}} \leq n^{-1}\alpha_n^{-1/n}V(f).$$

Equality holds when $f = \chi_B$ for some ball $B \subset \mathbb{R}^n$. The proof is reduced to the radial case by symmetrization. Another proof based on the isoperimetric inequality and the coarea formula is also included. This shows that the sharp Sobolev inequality in $BV(\mathbb{R}^n)$ is indeed equivalent to the sharp isoperimetric inequality. Section 4.7 gives the corresponding sharp result for $W^{1,p}(\mathbb{R}^n)$ when $1 < p < n$, $n \geq 2$:

$$\|f\|_{p^*} \leq (n\alpha_n^{1/n})^{-1}(p^*/p')^{1/p'}\left(\frac{p'}{n}\frac{\Gamma(n)}{\Gamma(n/p)\Gamma(n/p')}\right)^{1/n}\|\nabla f\|_p,$$

where $p^* = np/(n-p)$ is the Sobolev conjugate of $p$, and $p' = p/(p-1)$ is the Hölder conjugate, and $\Gamma$ is the Gamma function. Equality holds for $g_{n,p}(x) = (1 + |x|^{p'})^{-n/p^*}$. The proof of this inequality starts with a symmetrization to reduce to radial functions, and then follows a constructive version of the strategy of the proof by Cordero-Erausquin, Nazaret and Villani (2004) based on Monge–Kantorovich mass transportation ideas. The point is that in this proof, the transport map is explicitly constructed.

The last part of the chapter, §4.8, deals with the cases $p = n$ (Moser's theorem) and $p > n$ (Morrey's embedding theorem). Sharp inequalities are not known in the latter case, while partial results are available in the former.

**Chapter 5** covers three classical topics in symmetrization, and includes historical remarks as well as the needed background in physics to guide the reader. The first result is that symmetrizing a fixed membrane into a disk of the same area decreases its principal frequency (the first eigenvalue of the Laplacian with Dirichlet boundary conditions), as conjectured by Rayleigh in 1877 and proved independently by Faber (1923) and Krahn (1925). The second result is that symmetrization increases the torsional rigidity of a planar domain, as conjectured by St Venant in 1856 and proved by Pólya (1948). Lastly, a closed ball in $\mathbb{R}^3$ is shown to have the smallest Newtonian capacity among all compact sets with the same volume. This conjecture was raised by Poincaré in 1887 and proved by Szegő (1930). The proofs depend on the decrease of the Dirichlet integral under symmetric decreasing rearrangement of the function.

Background on weak solutions and spectral theory for the Laplace operator is presented in §§5.1–5.2 with all details carefully presented. In §5.3 we reach the proof of the Faber–Krahn theorem: when $\Omega$ is a bounded open set in $\mathbb{R}^n$ and $\Omega^\#$ is a ball of the same volume, the first eigenvalue $\lambda_1$ of the Laplacian is smallest for the ball: $\lambda_1(\Omega) \geq \lambda_1(\Omega^\#)$. The proof relies on expressing the first eigenvalue as the minimum value of the Rayleigh quotient, by

$$\lambda_1(\Omega) = \min_u \frac{\int_\Omega |\nabla u|^2\, dx}{\int_\Omega u^2\, dx},$$

where the minimum is over all $u \in W_0^{1,2}(\Omega)$ with $u \not\equiv 0$.

Two useful domain approximation lemmas are proved in §5.4, and then the Newtonian capacity of a compact set is developed from Coulomb's inverse square law in electrostatics, in §5.5. Szegő's Theorem's follows from the variational characterization of Newtonian capacity in terms of Dirichlet integrals:

$$\mathrm{Cap}(K) = \inf\left\{\frac{1}{4\pi}\int_{\mathbb{R}^3} |\nabla v|^2\, dx\colon\ v \in \mathcal{A}(K)\right\}$$

where the class of admissible functions is

$$\mathcal{A}(K) = \{v \in \mathrm{Lip}(\mathbb{R}^3)\colon\ 0 \leq v \leq 1 \text{ in } \mathbb{R}^3,\ v = 1 \text{ on } K,\ \lim_{|x|\to\infty} v(x) = 0\}.$$

The key point is that if $v$ is admissible for $K$ then $v^\#$ is admissible for the symmetrized set $K^\#$. Extensions to variational $p$-capacities, Riesz $\alpha$-capacities, and logarithmic capacities are considered in §5.6.

The torsional rigidity of a bounded open set $\Omega \subset \mathbb{R}^n$ is the quantity

$$T(\Omega) = 2 \int_\Omega u(x)\, dx,$$

where $u$ satisfies $\Delta u = -2$ in $\Omega$ with $u = 0$ in $\partial\Omega$. It turns out that $u(x)$ can also be interpreted as the expected lifetime of Brownian motion starting at $x \in \Omega$ and that $T(\Omega)/2|\Omega|$ equals the average lifetime of a particle born somewhere in $\Omega$. The key result of §5.7 is that symmetrization increases both quantities, that is,

$$T(\Omega) \le T(\Omega^\#).$$

**Chapter 6** discusses Steiner symmetrization. The Steiner symmetrization of a set or function on $\mathbb{R}^n = \mathbb{R}^k \times \mathbb{R}^m$ is obtained by performing symmetric decreasing rearrangement on the $k$-dimensional slice $\mathbb{R}^k \times \{z\}$, for each $z \in \mathbb{R}^m$.

Basic properties of symmetric decreasing rearrangement that were developed in Chapter 1 are adapted to Steiner symmetrization in §6.2, and properties of polarization are adapted in §6.3. Then Theorem 6.8 is an analogue of the main inequality (Theorem 2.15), taking the form

$$\int \Psi(f(x), g(\bar{x}))K(|x - \bar{x}|)\, dx\, d\bar{x} \le \int \Psi(f^\#(x), g^\#(\bar{x}))K(|x - \bar{x}|)\, dx\, d\bar{x}.$$

In §6.5 we see Steiner symmetrization decreases the modulus of continuity (Theorem 6.10) and the diameter (Theorem 6.12), and acts contractively on $L^\infty(X)$ (Theorem 6.14).

When considering the effect of Steiner symmetrization on Dirichlet integrals (§6.6), one first splits the gradient as

$$\nabla f(x) = \big(\nabla_y f(x), \nabla_z f(x)\big)$$

where $x = (y, z)$. Applying on each slice the result for symmetric decreasing rearrangement from Chapter 3, we find under suitable conditions on $f$ that

$$\int \Phi(|\nabla_y f^\#(y, z)|)\, dy \le \int \Phi(|\nabla_y f(y, z)|)\, dy$$

for each $z$. Integrating over $z$ gives

$$\int \Phi(|\nabla_y f^\#(x)|)\, dx \le \int \Phi(|\nabla_y f(x)|)\, dx.$$

The corresponding inequalities for the transverse gradient $\nabla_z f$ and full gradient $\nabla f$ are obtained in Theorem 6.16:

$$\int \Phi(|\nabla_z f^{\#}(y,z)|)\,dy \le \int \Phi(|\nabla_z f(y,z)|)\,dy$$

and

$$\int \Phi(|\nabla f^{\#}(y,z)|)\,dy \le \int \Phi(|\nabla f(y,z)|)\,dy$$

for each $z$. Once again, integrating over $z$ yields inequalities on all of $\mathbb{R}^n$.

While the above statements are simple, the proofs requires a technical lemma postponed to §6.7. In §6.8 the case of $p$-Dirichlet integrals is considered for Sobolev functions (Theorem 6.19). The case $p > 1$ follows from Theorem 6.16, but the case $p = 1$ needs additional work.

Steiner symmetrization decreases perimeter and Minkowski content, but in general it is not known whether it decreases the $(n-1)$-dimensional Hausdorff measure (§6.9). Steiner symmetrization also decreases the principal frequency and various capacities, and increases the torsional rigidity and mean lifetime of a Brownian particle (§6.10).

**Chapter 7** covers symmetrization in the sphere $\mathbb{S}^n$, hyperbolic space $\mathbb{H}^n$, and Gauss space, and includes as an application a landmark theorem of Gehring on quasiconformal mappings.

Spheres and hyperbolic spaces have a canonical distance and measure, and possess rich isometry groups of measure preserving mappings. There are plenty of hyperplanes in which to polarize, and so most of the theory from Chapters 2–6 can be extended.

Sections 7.1 and 7.2 introduce the distance and measure on the sphere. The distance $d(x, y)$ is the length of the shortest circular arc joining points $x$ and $y$, and so $0 \le d(x, y) \le \pi$. The measure $\sigma_n$ is the restriction of the $n$-dimensional Hausdorff measure $\mathcal{H}^n$ to $\mathbb{S}^n$. The unit vector $e_1$ plays the role of origin, in the sphere, and the metric balls centered at this origin are the open spherical caps

$$K(\theta) = \{x \in \mathbb{S}^n : d(x, e_1) < \theta\}, \qquad \theta \le \pi.$$

Hyperplanes in $\mathbb{S}^n$ are given by the intersection of the sphere with hyperplanes in $\mathbb{R}^{n+1}$ that pass through the origin. Hence the polarization theory from §1.7 carries over to the sphere. Symmetric decreasing rearrangement for sets and functions extends to the sphere also, using spherical caps rather than Euclidean balls.

Spherical analogs of inequalities from Chapters 1 and 2 are developed in §7.3. The basic polarization inequality is Theorem 7.2, and the foundational inequality for integrals of functions on $\mathbb{S}^n$ under symmetric decreasing rearrangement is Theorem 7.3. The proofs are somewhat simpler than in Euclidean space, due to compactness of the sphere. In §7.4 one finds the decrease of spherical Dirichlet integrals under symmetric decreasing rearrangement on the sphere (Theorem 7.4) and the spherical isoperimetric inequality for Minkowski content (Theorem 7.5).

Cap symmetrization on $\mathbb{R}^n$ is presented in §7.5, where spherical $(k, n)$-cap symmetrization corresponds to $(k, n)$-Steiner symmetrization except now rearranging on $k$-spheres rather than $k$-planes. For example, circular symmetrization in the complex plane is exactly $(1, 2)$-cap symmetrization, with the function made symmetric decreasing about the positive real axis, on each circle centered at the origin.

Section 7.6 is devoted to symmetrization in the hyperbolic space $\mathbb{H}^n$, which is modeled by the unit ball $\mathbb{B}^n$ endowed with the hyperbolic metric

$$ds = \frac{2}{1 - |x|^2}\,|dx|,$$

where $|dx|$ is the Euclidean length element. The corresponding hyperbolic measure has density $2^n(1 - |x|^2)^{-n}$. Polarization is defined in terms of hyperbolic hyperplanes, and hyperbolic symmetric decreasing rearrangement is constructed in terms of balls centered at the origin, but with respect to the hyperbolic measure rather than Euclidean measure. The majority of the symmetrization results from Chapters 1–5 are shown to hold for hyperbolic symmetric decreasing rearrangement.

Section 7.7 presents a brief discussion of symmetrization in the Gauss space $(\mathbb{R}^n, d\mu)$, where $d\mu = (2\pi)^{-n/2}e^{-|x|^2/2}\,dx$. Here sets and functions are rearranged with respect to the measure $d\mu$. The lack of appropriate hyperplanes makes the theory quite different from Euclidean, spherical, or hyperbolic symmetrization. A version of the isoperimetric inequality for the Gaussian Minkowski content can be proved by using the fact that Gauss space is the limit of spheres of increasing radius and dimension going to infinity; see Corollary 7.12.

In the final section, §7.8, the basic theory of quasiconformal mappings in $\mathbb{R}^n$ is discussed, including the equivalence of the analytic and geometric definitions of quasiconformality (Theorem 7.15). The sharp Hölder continuity exponent $1/K$ for $K$-quasiconformal mappings is obtained by using $(n-1, n)$-cap symmetrization, in Theorem 7.16 and Corollary 7.17. This is a celebrated theorem of Gehring (1962).

**Chapter 8** studies symmetrization and convolution. The Riesz–Sobolev convolution theorem for nonnegative functions $f, g, h$ on $\mathbb{R}^n$ asserts that the triple convolution

$$\int_{\mathbb{R}^n}\int_{\mathbb{R}^n} f(-x)g(y)h(x-y)\,dxdy = f * g * h(0)$$

increases when $f, g, h$ are replaced by their symmetric decreasing rearrangements. The theorem is proved for functions on the circle $\mathbb{S}^1$ in §8.1, using ideas suggested by the star function in Chapter 9. The version on the circle implies the version on the real line (§8.2), which in turn implies the version in $\mathbb{R}^n$ (§8.3) for symmetric decreasing rearrangement and $(k, n)$-Steiner symmetrization. The Brunn–Minkowski inequality is proved in §8.4 as an application of Riesz–Sobolev.

A significant extension of the Riesz–Sobolev inequality, valid for multiple integrals with arbitrarily many functions, is the Brascamp–Lieb–Luttinger inequality proved in §8.5. It implies that the Dirichlet heat kernel increases under symmetrization (§8.6). On a bounded open set $\Omega \subset \mathbb{R}^n$ the Dirichlet Laplacian has eigenvalues $0 < \lambda_1 \leq \lambda_2 \leq \ldots$, and by writing the heat kernel $K(x, y, t)$ as an eigenfunction series one arrives at the heat trace

$$\mathrm{Tr}(t, \Omega) = \int_{\Omega} K(x, x, t)\,dx = \sum_{j=1}^{\infty} e^{-\lambda_j t}.$$

Luttinger's Theorem 8.9 says the heat trace increases under rearrangement:

$$\mathrm{Tr}(t, \Omega) \leq \mathrm{Tr}(t, \Omega^{\#}),$$

where $\Omega^{\#}$ denotes the symmetric decreasing rearrangement or $(k, n)$-Steiner symmetrization of the domain.

The Hardy–Littlewood–Sobolev inequality

$$\int_{\mathbb{R}^{2n}} f(x)g(y)|x-y|^{-\lambda}\,dxdy \leq C\|f\|_p\|g\|_q$$

holds when $p > 1, q > 1, 0 < \lambda < n$. Section 8.7 presents a result of Lieb that determines the sharp constants for certain special values of the parameters. A key ingredient is to observe the conformal invariance of the integral (Proposition 8.12). Theorem 8.15 presents the sharp version (best constant) of the Hardy–Littlewood–Sobolev inequality for $1 < p < 2$ and $\lambda = 2n/p'$. In this case the extremals are constant multiples of $(a^2 + |x - v|^2)^{-n/p}$, where $a > 0$ and $v \in \mathbb{R}^n$.

In §8.8 and §8.10 the endpoint cases $\lambda \to n$ and $\lambda \to 0$ are investigated, following ideas of Beckner. The first case yields Gross's logarithmic Sobolev inequality (8.62, 8.63), as an infinite dimensional version of Beckner's logarithmic Sobolev inequality in $\mathbb{S}^n$ (Theorem 8.17). The second case gives sharp inequalities for exponential integrals known as the Lebedev–Milin inequality (8.69) and Onofri's inequality (8.70). In §8.9 Beckner's logarithmic Sobolev inequality is used to establish the hypercontractivity of the Poisson semigroup in a sharp range.

**Chapter 9** marks the debut of the star function in the book. Each type of rearrangement $u^\#$ has an associated star function $u^\star$, which is an indefinite integral of $u^\#$. This chapter proves "subharmonicity" theorems for the star function, expressing the fact that if $u$ satisfies a Poisson-type partial differential equation then $u^\star$ satisfies a related differential inequality. In the simplest case of a function $u$ in the plane, subject to circular symmetrization, the result says that if $u$ is subharmonic then so is $u^\star/r$. Subharmonicity yields comparison theorems for solutions of partial differential equations (Chapter 10), and extremal results in complex analysis (Chapter 11). Recall the complex plane with circular symmetrization is where the star function first made an impact.

Section 9.1 defines the star function in terms of the decreasing rearrangement on a general measure space, by

$$u^\star(x) = \int_0^x u^*(s)\,ds = \sup\left\{\int_E u\,d\mu : \mu(E) = x\right\}.$$

This formula motivates the star function definition for each of the specific geometries considered later in the chapter: spherical shells, spheres, Euclidean domains, and so on. Section 9.2 provides an overview of the chapter, and the next section establishes some facts on measurability.

The Laplacian is usually regarded as a differential operator, but it is more convenient in §9.4 to formulate the Laplacian as a limit of integral operators, so that later we can apply rearrangement results for convolutions. Specifically, the Laplacian at a point equals the difference between the average value of the function over a small neighborhood and its value actually at the point, as made precise by Lemma 9.5 for functions and Lemma 9.6 for measures.

The theory of the star function is easiest to grasp in the case of $(n-1,n)$-cap symmetrization on a spherical shell, because no boundary conditions need be imposed. Accordingly, we start with that case in §9.5. Given a measure with decomposition

$$d\mu = f\,d\mathcal{L}^n + d\tau - d\eta$$

where the function $f$ is locally integrable, $\mathcal{L}^n$ is Lebesgue measure and $\tau$ and $\eta$ are nonnegative measures, the cap symmetrization of $\mu$ is defined by

$$d\mu^\# = f^\# d\mathcal{L}^n + d\tau^\# - d\eta_\#.$$

Here $f^\#$ is symmetric decreasing on each sphere centered at the origin, $\tau^\#$ is the measure obtained by sweeping the mass of $\tau$ on each sphere to the positive $x_1$-axis, and $\eta^\#$ is obtained by spherically sweeping the mass of $\eta$

to the negative $x_1$-axis. The "pre-subharmonicity" result in Theorem 9.7 says that if $u$ satisfies a semilinear Poisson equation then its cap symmetrization $u^\#$ satisfies a corresponding differential inequality:

$$-\Delta u = \phi(r,u) + \mu \quad \Longrightarrow \quad -\Delta(u^\#) \leq \phi(r,u^\#) + \mu^\#,$$

although only with respect to test functions that are nonnegative and cap-symmetric decreasing. The Riesz rearrangement inequality for convolutions is the key step in the proof of pre-subharmonicity, because it leads to an integral inequality relating $\Delta(u^\#)$ and $\Delta u$, in the weak sense.

Section 9.6 defines the star function for cap symmetrization, which is

$$u^\star(r,\theta) = \int_{\mathcal{K}(\theta)} u^\#(rs)\, r^{n-1}\, d\sigma_{n-1}(s),$$

where $\mathcal{K}(\theta) \subset \mathbb{S}^{n-1}$ is the spherical cap of aperture $\theta$ centered on the positive $x_1$-axis. Also defined is an operator $\Delta^\star$ that is a modified Laplacian in the $r$ and $\theta$ variables. In 2 dimensions $\Delta^\star = r\Delta r^{-1}$. The desired subharmonicity property of the star function (Theorem 9.9) says that

$$-\Delta u = \phi(r,u) + \mu \quad \Longrightarrow \quad -\Delta^\star u^\star \leq J\phi(r,u^\#) + \mu^\star,$$

where the $J$-operator involves integrating over spherical caps, and $\mu^\star$ is defined by $d\mu^\star = f^\star\, dr d\theta + d\tau^\star$ (with $\eta$ being discarded).

A subharmonicity property for the star function in the complex plane follows as a special case (Corollary 9.10):

$$\Delta u \geq 0 \quad \Longrightarrow \quad \Delta(u^\star/r) \geq 0,$$

where this star function relates to circular symmetrization, that is, symmetric decreasing rearrangement on each circle centered at the origin. (Note the star function definition in this book differs from the original one in the literature by a factor of $r^{n-1}$ – hence the occurrence of $u^\star/r$ in the last formula.) In other words, if $u$ is subharmonic then so is $u^\star/r$.

The spherical shell results are used in §9.7 to deduce a subharmonicity theorem on the sphere, and in §9.8 to arrive at a subharmonicity theorem for $(k,n)$-cap symmetrization on ring-type domains such as domains in $\mathbb{R}^3$ that are rotationally symmetric around the vertical axis.

Sections 9.9 and 9.10 develop the star function and subharmonicity result for the situation most widely used in applications: symmetric decreasing rearrangement on a domain in $\mathbb{R}^n$ (Theorem 9.20). The function $u$ is now required to be nonnegative and satisfy a zero Dirichlet boundary condition. Once more, the Riesz rearrangement inequality plays the key role in the proof.

Finally, subharmonicity for $(k, n)$-Steiner symmetrization is proved in §9.11, assuming $u \geq 0$ and $u \to 0$ at "horizontal" boundary points.

**Chapter 10** establishes comparison principles for solutions of partial differential equations. The prototypical result says that the solution of Poisson's equation increases in an integral sense when the data in the equation is rearranged: if $-\Delta u = f$ on $\Omega$ and $-\Delta v = f^\#$ on $\Omega^\#$, with $u$ and $v$ nonnegative and satisfying Dirichlet boundary conditions, then $\|u\|_{L^p(\Omega)} \leq \|v\|_{L^p(\Omega^\#)}$ for each $p \geq 1$. Such comparisons have been used in the literature for deriving sharp bounds on certain eigenvalues, obtaining a priori bounds on solutions, and comparing Green functions, among other uses.

Integral norm comparisons follow from star function comparisons. The theory of majorization in §10.1 implies that $\int \Phi(u)\, d\mathcal{L}^n \leq \int \Phi(v)\, d\mathcal{L}^n$ for all convex increasing $\Phi$ if and only if $u^\star \leq v^\star$. Thus the task is to prove that rearranging the data in Poisson's equation increases the star function of the solution.

Maximum principles for distributional solutions are proved in §10.2, as a preliminary step. Section 10.3 contains comparison principles under symmetric decreasing rearrangement on a Euclidean domain, firstly for linear equations of the form $-\Delta u + cu = \mu$ where $u \geq 0$ satisfies a zero Dirichlet boundary condition (Theorem 10.10), and then for the semilinear equation $-\Delta u = \phi(u) + \mu$ where $\phi$ is convex decreasing (Theorem 10.12). The equation $-\Delta u = e^{-u}$, for example, relates to the hyperbolic metric, as explained later in Chapter 11. The comparison results are adapted to Steiner symmetrization in §10.4.

Underlying each result is a maximum principle argument applied to $u^\star - Jv$. Chapter 9 says that $u^\star$ is a subsolution of a certain differential inequality and $Jv$ (which behaves like $v^\star$) is a solution of the corresponding differential equation. Thus in the linear case, $u^\star - Jv$ is a subsolution and so attains its maximum on the boundary. We can exploit the boundary conditions on $u$ and $v$ to complete the proof that $u^\star - Jv \leq 0$. An additional argument is needed for the nonlinear terms $\phi(u)$ and $\phi(v)$, to rule out interior maximum points.

Symmetric decreasing rearrangement on the sphere is handled in §10.5. If $-\Delta_s u = \phi(u) + \mu$ and $-\Delta_s v = \phi(v) + \mu^\#$, where $\Delta_s$ is the spherical Laplacian, and if $u$ and $v$ have the same mean value over the sphere, then $v$ has larger $L^p$-norms than $u$, and larger oscillation (Theorem 10.16).

Cap symmetrization on spherical shells is the subject of §10.6. Boundary conditions must be imposed on the inner and outer boundary spheres. Options include a Dirichlet condition $u \leq v$ (Theorem 10.18) or a Neumann condition $\partial u/\partial n = \partial v/\partial n = 0$ (Theorem 10.20). Results on balls follow as a special case, by setting the inner radius to zero.

The final chapter, **Chapter 11**, presents the original context in which the star function appeared. As the subject developed and extended beyond one complex variable some definitions and formulas were slightly modified, and differ in this book by a factor of $2\pi$ from the original papers, as discussed before Proposition 11.2.

The star function was created to solve the long-standing "spread conjecture" in Rolf Nevanlinna's classical theory of meromorphic functions. Being able to continue Nevanlinna's growth function $T(r)$ into the upper halfplane as a subharmonic function, and using it as a vehicle to resolve the conjecture, drew immediate attention from a broad mathematical community, as indicated in Professor Hayman's Foreword.

The first three sections of Chapter 11 provide background in Nevanlinna theory. For example, the elegant Phragmén–Lindelöf theory of the indicator, §11.3, is a powerful tool in the study of entire functions. We provide a full account of the theory, based on geometric properties of trigonometric convexity, and use it to obtain several applications exploiting this viewpoint.

The first application, Proposition 11.7 in §11.4, is the spread relation. An immediate corollary is a striking improvement of Nevanlinna's celebrated defect relation for general meromorphic functions. Nevanlinna had proved nearly a century ago that if $f$ is meromorphic in the plane, then

$$\sum_a \delta(a,f) \leq 2,$$

as described in §11.2. Corollary 11.8 shows that the upper bound 2 in the defect relation can be diminished if we know in addition that the *order* of $f$ satisfies $0 < \lambda < 1$. In fact, $f$ need only have Pólya peaks of that order (see Definition 11.3).

Proposition 11.11 gives another refinement of the defect relation for functions of order $\lambda < 1$. Few sharp results are known for $\lambda > 1$, other than isolated values, with Proposition 11.10 a rare exception. Further refinements, dealing with the "spread" of the value $a$, are presented too, and a complete proof is provided for a lemma due to Fuchs (1963) on which a key property of the indicator depends. The elegant Lemma 11.17 of H. Cartan is proved along the way; it yields remarkably effective lower bounds on the modulus of a polynomial $P(z)$ outside a small exceptional set containing the zeros.

After an interlude on subordination in §11.5, the next section, §11.6, presents applications to functions $f(z)$ analytic and univalent in the unit disk. The main result, Theorem 11.22, is that if $\Phi\colon \mathbb{R} \to \mathbb{R}$ is convex, then the integral mean

$$\int_0^{2\pi} \Phi(\log|f(re^{i\theta})|)\, d\theta$$

is maximized when the range of $f$ is symmetric with respect to a ray, such as for the Koebe function. The primary interest is for $L^p$-norms, obtained when $\Phi(x) = e^{px}, p > 0$. Since the result involves the range of $f$, there are interpretations using the Green function (Corollary 11.24) and harmonic measure (Theorem 11.27).

A feature available only in 2 dimensions is that on a simply connected domain such as the disk, to any harmonic function $u$ in $\mathbb{D}$ may be associated a *conjugate* harmonic function $v$ such that $f = u + iv$ is analytic. This is the subject of §11.8, which also presents historical background. The relation between $u$ and $v$ has long been a theme in Fourier analysis. Viewing the situation from complex analysis allows the geometry to appear in a natural way, in §11.8. The hypotheses about $u$ or $f$ are usually phrased in terms of "boundary values." We suppose that

$$f(z) = u + iv = \int_{-\pi}^{\pi} \frac{e^{i\varphi} + z}{e^{i\varphi} - z} d\mu(\varphi),$$

with total variation $\|\mu\| < \infty$. For $f$ in a suitable Hardy space, the measure $\mu$ relates to the boundary values of $u$ by $d\mu(\varphi) = u(e^{i\varphi})\, d\varphi/2\pi$.

When $f$ has this representation, the $L^p$ norms

$$\int_{-\pi}^{\pi} |f(re^{i\theta})|^p\, d\theta$$

are maximal for measures $\mu$ that concentrate in a particular manner at $\theta = 0, \pi$, and similarly for the $L^p$ norms of $u$ and $v$; see Theorem 11.28.

If the boundary function $u(e^{i\theta})$ belongs to $L^p$ then so does $v(e^{i\theta})$, and there is a norm bound

$$\|v\|_p \leq C_p\|u\|_p, \qquad 1 < p < \infty,$$

due to M. Riesz. The best constant $C_p$ was obtained later by Pichorides (1972). For $p = 1$ the inequality fails, although a weak type inequality still holds. The main application of this section is to obtain for $0 < p < 1$, by using the star function (and thus complex analysis), the inequality

$$\|v\|_p \leq C_p\|u\|_1, \qquad 0 < p < 1,$$

also identifying the sharp constant $C_p$. This objective requires significant additional apparatus, where star functions of the real and imaginary parts of the boundary functions themselves appear. The key step is Proposition 11.30, whose formulation uses the counting function $N(r,f)$ from Nevanlinna theory.

The final section, §11.9, considers the effect of symmetrization on the hyperbolic (or Poincaré) metric $\rho_\Omega$ for simply or multiply connected domains $\Omega$ in the plane. This metric may be constructed (Proposition 11.31) on any domain whose complement has at least two points, as we show by developing the theory of the universal cover.

A result due to Weitsman (Theorem 11.33) compares the hyperbolic metric on $\Omega$ to the metric on the circularly symmetrized domain $\Omega^\#$, and shows that the integral means of $\log 1/\rho_\Omega$ are smaller than those of $\log 1/\rho_{\Omega^\#}$. Corollary 11.35 deduces that the hyperbolic metric achieves smaller values on $\Omega^\#$ than on $\Omega$, meaning in essence that the symmetrized domain contains points "farther from the boundary." Another consequence says that if $f\colon \mathbb{D} \to \Omega$ is holomorphic (not necessarily univalent) then $|f'(0)| \leq |\psi'(0)|$, where $\psi$ is the universal cover map of the unit disk $\mathbb{D}$ to $\Omega^\#$. This last inequality generalizes to the maximum modulus of the two functions, and concludes the book.

# 1

# Rearrangements

In the theory of symmetrization one compares a given real-valued function $f$ on a measure space $(X, \mathcal{M}, \mu)$ with another function $g$, defined on a possibly different measure space, such that $f$ and $g$ have the same "size" but $g$ perhaps has more symmetry. In this chapter we quantify the notion "size of a function" by introducing the distribution function $\lambda_f(t) = \mu(f > t)$ of $f$. We consider $f$ and $g$ to have the same size if they have the same distribution function, in which case $f$ and $g$ are called rearrangements of each other. The decreasing rearrangement of $f$, denoted $f^*$, which is defined on the interval $[0, \mu(X)] \subset \mathbb{R}$, is of particular interest. When $X = \mathbb{R}^n$ and $\mu$ is $n$-dimensional Lebesgue measure, an important role is played by the symmetric decreasing rearrangement $f^\#$ of $f$. Other notions introduced in this chapter include measure preserving transformations and polarization of functions. The latter is a very simple rearrangement process defined for functions on $\mathbb{R}^n$.

## 1.1 The Distribution Function

Let $(X, \mathcal{M}, \mu)$ be a measure space and $f: X \to \mathbb{R}$ be a measurable function. For $t \in \mathbb{R}$, we'll write

$$(f > t) = \{x \in X: f(x) > t\}.$$

The *distribution function* $\lambda: \mathbb{R} \to [0, \infty]$ of $f$, with respect to $\mu$, is defined as follows:

**Definition 1.1** For $t \in \mathbb{R}$,

$$\lambda(t) = \lambda_f(t) = \mu(f > t).$$

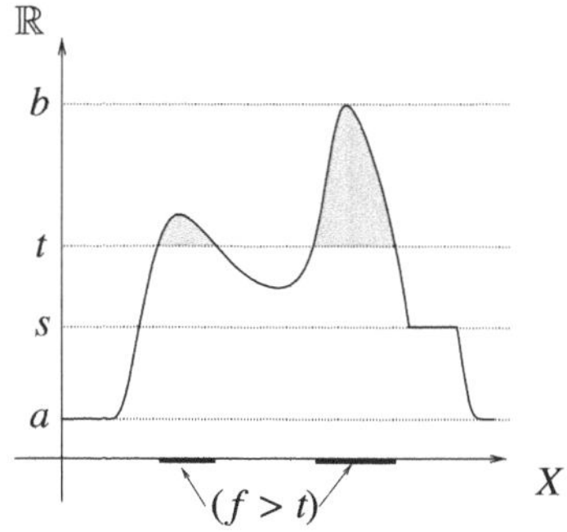


Figure 1.1 Level sets of a real-valued function $f$ on a measure space $X$. The essential infimum of $f$ is denoted by $a$, and its essential supremum by $b$. All level sets at heights $t > a$ have finite measure, in accordance with Condition (1.1) below. The example shown has a flat spot at height $s$.

Our definition differs from that of many authors, who define $\lambda(t) = \mu(|f|>t)$. If it is necessary to specify $\mu$, we'll write $\lambda_{f,\mu}$.

Here are some examples.

**Example 1.2** Let $f = \chi_A$, the characteristic function of a set $A \in \mathcal{M}$. Then $\lambda(t) = 0$ for $t \geq 1$, $\lambda(t) = \mu(A)$ for $0 \leq t < 1$, and $\lambda(t) = \mu(X)$ for $t < 0$.

More generally, let $f: X \to \mathbb{R}$ be a simple function. Then $f = \sum_{i=1}^{n} \alpha_i \chi_{E_i}$, where the $\alpha_i$ are distinct real numbers, and the $E_i$ form a measurable partition of $X$; that is, each $E_i \in \mathcal{M}$, the $E_i$ are disjoint, and their union is $X$. By relabeling, if necessary, we may assume that the $\alpha_i$ are arranged in descending order: $\alpha_1 > \alpha_2 > \cdots > \alpha_n$. Then

$$(f > t) = \bigcup_{i=1}^{k} E_i$$

for $t \in [\alpha_{k+1}, \alpha_k)$ when $1 \leq k \leq n-1$. Also, $(f > t)$ is empty when $t \geq \alpha_1$, $(f > t) = X$ when $t < \alpha_n$. Thus, $\lambda$ is a step function on $\mathbb{R}$ with jumps $\mu(E_k)$ at the points $t = \alpha_k$. It is given by the formula

$$\lambda(t) = \sum_{i=1}^{k} \mu(E_i), \quad \alpha_{k+1} \leq t < \alpha_k, \quad k = 0, \cdots n,$$

where $\alpha_0 \equiv \infty, \alpha_{n+1} \equiv -\infty$, and the empty sum is interpreted to be zero.

**Example 1.3** Let $X = [-1, 2]$ and $f(x) = x^2$. To find $(f > t)$, think of $t$ as starting at $\infty$ then moving downward. We see that the level set $(f > t)$ is empty for $t \geq 4$ and $(f > t) = [-1, 2]$ for $t < 0$. For $1 \leq t < 4$, one calculates that $(f > t)$ is the single interval $(t^{1/2}, 2]$, and for $0 \le t < 1$, $(f > t)$ is the union of two intervals $[-1, -t^{1/2}) \cup (t^{1/2}, 2]$. Thus,

$$\begin{aligned} \lambda(t) &= 0, \quad t \ge 4 \\ &= 2 - t^{\frac{1}{2}}, \quad 1 \le t < 4 \\ &= 3 - 2t^{\frac{1}{2}}, \quad 0 \le t < 1 \\ &= 3, \quad t < 0. \end{aligned}$$

**Example 1.4** Let $X = [0, c]$, where $0 < c < \infty$, and $f(x) = c^2 - x^2$. If $0 \le t < c^2$, then $(f > t)$ is the set of $x$ such that $c^2 - x^2 > t$. Thus $(f > t) = \{x \in [0, c] : x^2 < c^2 - t\} = [0, (c^2 - t)^{1/2})$, so that

$$\lambda(t) = (c^2 - t)^{\frac{1}{2}}, \quad t \in [0, c^2].$$

Also, $\lambda(t) = 0$ for $t \ge c^2$, $\lambda(t) = c$ for $t < 0$.

Note that $f$ is a strictly decreasing function on $[0, c]$, and that $\lambda$ restricted to the range of $f$ is the inverse function of $f$. This foreshadows the fact to be shown in the next section that in general the distribution function of a function $f$ is the inverse, properly interpreted, of the decreasing rearrangement of $f$.

We prove now some simple properties of distribution functions.

**Proposition 1.5** *Let $f : (X, \mathcal{M}, \mu) \to \mathbb{R}$ be measurable. Then*

*(a)* $\lambda(t) \searrow$ *as* $t \nearrow$ *on* $\mathbb{R}$.
*(b)* $\lambda(t+) = \lambda(t), \quad t \in \mathbb{R}$.
*(c) If* $\lambda(t-) < \infty$, *then* $\lambda(t-) = \mu(f \ge t) = \lambda(t) + \mu(f = t)$.
*(d) If* $t \in (\operatorname{ess\,inf} f, \operatorname{ess\,sup} f)$, *then* $0 < \lambda(t) < \mu(X)$.
*(e) If* $t \ge \operatorname{ess\,sup} f$ *then* $\lambda(t) = 0$. *If* $t < \operatorname{ess\,inf} f$ *then* $\lambda(t) = \mu(X)$.
*(f) If* $f_n : (X, \mathcal{M}, \mu) \to \mathbb{R}$ *is measurable for* $n \ge 1$ *and* $f_n \nearrow f$ *pointwise on* $X$, *then* $\lambda_{f_n} \nearrow \lambda_f$ *pointwise on* $\mathbb{R}$.

We remind the reader that $\lambda(t+)$ and $\lambda(t-)$ denote limits from the right and left, respectively. In (f), pointwise convergence on $X$ means that $f_n(x) \to f(x)$ for every $x \in X$. The notations $\operatorname{ess\,sup} f$ and $\operatorname{ess\,inf} f$ denote respectively the essential supremum and essential infimum of $f$ on $X$.

Parts (a) and (b) of Proposition 1.5 say that $\lambda$ is decreasing (= nonincreasing) and right continuous on $\mathbb{R}$. Part (c) tells us that $\lambda$ has jumps of size $\mu(f = t)$ at values of $t$ which $f$ assumes on sets of positive measure, provided $\lambda(t-)$ is finite. The function $f : [0, \infty) \to \mathbb{R}^+$ defined by $f(x) = x$ for $x \in [0, 1]$, $f(x) = -\frac{1}{x}$ for $1 < x < \infty$, has $\lambda(0-) = \infty$, $\lambda(0) + \mu(f = 0) = \mu(f \ge 0) = 1$. Thus, (c) can fail when $\lambda(t-) = \infty$.

*Proof of Proposition 1.5* (a) Let $-\infty < t_1 < t_2 < \infty$. Then $(f > t_2) \subset (f > t_1)$. Hence $\lambda(t_2) \leq \lambda(t_1)$.
(b) Since $\lambda$ is decreasing,

$$\lambda(t+) = \lim_{n\to\infty} \lambda\left(t + \frac{1}{n}\right) = \lim_{n\to\infty} \mu\left(f > t + \frac{1}{n}\right).$$

Also, $(f > t)$ is the increasing union of the sets $(f > t + \frac{1}{n})$. Thus,

$$\lim_{n\to\infty} \mu\left(f > t + \frac{1}{n}\right) = \mu(f > t) = \lambda(t).$$

(c) Since $\lambda(t-) < \infty$, there exists $N$ such that $\lambda(t - \frac{1}{n}) < \infty$ for $n \geq N$. Moreover, $(f \geq t)$ is the decreasing intersection of the sets $(f > t - \frac{1}{n})$. Thus,

$$\mu(f \geq t) = \lim_{n\to\infty} \mu\left(f > t - \frac{1}{n}\right) = \lim_{n\to\infty} \lambda\left(t - \frac{1}{n}\right) = \lambda(t-).$$

This proves the first identity in (c). The second identity follows from $\mu(f \geq t) = \mu(f > t) + \mu(f = t) = \lambda(t) + \mu(f = t)$.
(d) and (e) These are exercises with the definitions of ess inf and ess sup which are left to the reader. Note that when $t = \operatorname{ess\,inf} f$, either $\lambda(t) < \mu(X)$ or $\lambda(t) = \mu(X)$ can occur, depending on whether or not $\lambda$ has a jump at $t = \operatorname{ess\,inf}$.
(f) For each $t$ we have $(f > t) = \bigcup_{n=1}^{\infty}(f_n > t)$ and $(f_n > t) \subset (f_{n+1} > t)$ for each $n$. It follows that $\lambda_{f_n}(t) = \mu(f_n > t) \nearrow \mu(f > t) = \lambda_f(t)$. □

In the proofs of (b) and (f) we used the facts that if $\{A_n\}$ is an increasing sequence of measurable sets and $A = \bigcup_{n=1}^{\infty} A_n$, then $\{\mu(A_n)\}$ is an increasing sequence with limit $\mu(A)$. In (c), we used the corresponding fact about decreasing sequences of sets.

We introduce now one of the principal themes of this book: the concept of *rearrangement* of functions. To motivate, we examine first the simpler case of sequences. Two finite real sequences $\{a_i\}_{i=1}^n$ and $\{b_i\}_{i=1}^n$ of the same length are said to be rearrangements of each other if they consist of the same numbers but differ perhaps in the order in which the terms are listed. More formally, the two sequences are rearrangements of each other if for each $t \in \mathbb{R}$, $|\{i \in \{1,\ldots,n\}\colon a_i = t\}| = |\{i \in \{1,\ldots,n\}\colon b_i = t\}|$, where $|S|$ denotes the number of elements in the set $S$. For example, the sequences $(a_1, a_2, a_3, a_4, a_5) = (5, 7, 2, 5, 6)$ and $(b_1, b_2, b_3, b_4, b_5) = (6, 7, 5, 2, 5)$ are rearrangements of each other.

It is easy to see that $\{a_i\}_{i=1}^n$ and $\{b_i\}_{i=1}^n$ are rearrangements of each other if and only if the sequences have the same distribution functions. That is,

$$|\{i \in \{1,\dots,n\}\colon a_i > t\}| = |\{i \in \{1,\dots,n\}\colon b_i > t\}|, \quad t \in \mathbb{R}.$$

Now let $f$ and $g$ be real valued measurable functions defined on possibly different measure spaces which have the same total measure. $f$ and $g$ will be called rearrangements of each other if they have the same size, in the sense of measures. The distribution functions permit us to make this notion precise.

**Definition 1.6** Measurable functions $f\colon (X,\mathcal{M},\mu) \to \mathbb{R}$ and $g\colon (Y,\mathcal{N},\nu) \to \mathbb{R}$ are said to be rearrangements of each other if $\mu(X) = \nu(Y)$ and $\lambda_f(t) = \lambda_g(t)$ for every $t \in \mathbb{R}$. We shall also say in this case that $f$ and $g$ are equidistributed.

For example, if $A \in \mathcal{M}$, $B \in \mathcal{N}$, then $\chi_A$ and $\chi_B$ are equidistributed if and only if $\mu(X) = \nu(Y)$ and $\mu(A) = \nu(B)$. Note also that the concept of rearrangement of sequences is in fact the special case of Definition 1.3 when $X = Y = \{1,..,n\}$ and $\mu$ and $\nu$ are counting measures.

## 1.2 The Decreasing Rearrangement

Suppose that $\{a_i\}_{i=1}^n$ is a finite sequence in $\mathbb{R}$. Then one can form a possibly new sequence by listing the terms $\{a_i\}$ in decreasing order, with the understanding that if there are $k$ values of $i$ for which $a_i$ equals some number $t$, then $t$ will be listed $k$ times in the new sequence. The terms of the new sequence will be denoted $a_i^*$. The new sequence $\{a_i^*\}_{i=1}^n$ is a rearrangement of the original sequence, in the sense of §1.1. We call $\{a_i^*\}_{i=1}^n$ the *decreasing rearrangement* of $\{a_i\}_{i=1}^n$. For example, if $(a_1,a_2,a_3,a_4) = (5,7,2,6)$, then $(a_1^*,a_2^*,a_3^*,a_4^*) = (7,6,5,2)$, while if $(a_1,a_2,a_3,a_4,a_5) = (5,7,2,5,6)$, then $(a_1^*,a_2^*,a_3^*,a_4^*,a_5^*) = (7,6,5,5,2)$.

Now let $(X,\mathcal{M},\mu)$ be a measure space and let $f\colon X \to \mathbb{R}$ be a measurable function. We would like to devise a new function $f^*$ which is related to $f$ in the same way that $\{a_i^*\}_{i=1}^n$ is related to $\{a_i\}_{i=1}^n$. Thus, $f^*$ should take on the same values as $f$, but should do so in "decreasing order." In general, there is no ordering of points on $X$, so for "decreasing order" to make sense the domain of $f^*$ might have to be some space other than $X$. We shall take the domain of $f^*$ to be the closed interval $[0,\mu(X)]$ of the extended real line, equipped with the $\sigma$-algebra of Lebesgue measurable sets and the 1-dimensional Lebesgue measure $\mathcal{L}$.

To make precise the phrase "takes on the same values," we shall interpret it to mean that $f$ and $f^*$ have the same distribution function. Thus, we aspire to define a Lebesgue measurable function $f^*\colon [0, \mu(X)] \to [-\infty, \infty]$ such that $x \le y \implies f^*(y) \le f^*(x)$, and $\lambda_f = \lambda_{f^*}$ on $\mathbb{R}$, where $\lambda_{f^*}(t) = \mathcal{L}(f^* > t)$. If $f$ is a constant $c$ we can take $f^* \equiv c$. In the discussion below we shall always assume that $f$ is nonconstant.

The theory of $f^*$ becomes more awkward and less useful when $\lambda_f(t) = \infty$ for some $t > \operatorname{ess\,inf} f$. Accordingly, we shall only carry out the construction when $f$ satisfies the following condition:

$$\lambda_f(t) < \infty, \qquad \text{for every } t > \operatorname{ess\,inf} f. \tag{1.1}$$

Condition (1.1) will appear in the hypotheses of many, many results throughout this book. It is automatically fulfilled when $\mu(X) < \infty$.

Suppose, then, that $f\colon (X, \mathcal{M}, \mu) \to \mathbb{R}$ is measurable and nonconstant, and that $f$ satisfies (1.1). An example of such a function is shown in Figure 1.1. Draw the graph of $x = \lambda(t)$, with $t$ being on the horizontal axis, as in Figure 1.2. For aesthetic reasons, write

$$a = \operatorname{ess\,inf} f, \quad b = \operatorname{ess\,sup} f.$$

Let us temporarily add the assumptions that $\lambda$ is continuous and strictly decreasing on $[a, b]$ and that $\lambda(a) = \lambda(a-)$. Then $\lambda(a) = \mu(X)$ and $\lambda(b) = 0$. Take $x \in (0, \mu(X))$. The horizontal line through $x$ intersects the graph of $\lambda$ in exactly one point. Call this point $(t, x)$. Then $a < t < b$. Define $f^*(x) = t$. Define also $f^*(x) = a$ for $x = \mu(X)$ and $f^*(0) = b$. To verify that $f^*$ and $f$ are equidistributed, take $t \in (a, b)$, and let $x$ be the corresponding point on the graph. Then

$$(f^* > t) = [0, x).$$

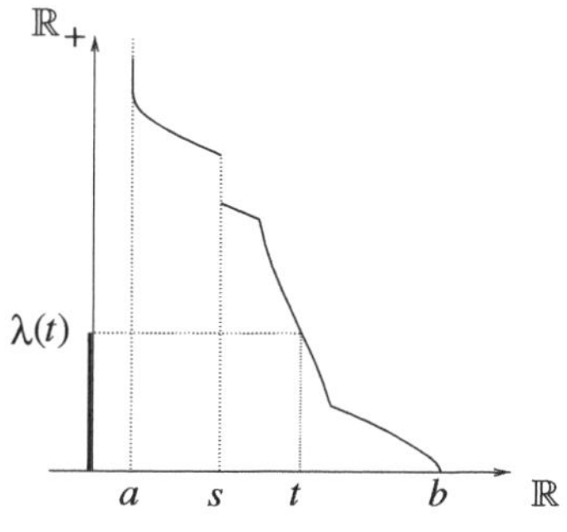


Figure 1.2 The distribution function $\lambda$ of the function $f$ from Figure 1.1. By construction, $\lambda$ is decreasing and right continuous. Its value at $t$ equals the measure of the level set $(f > t)$. The flat spot of $f$ gives rise to a jump of $\lambda$ at $s$, whose height equals the measure of $f^{-1}(s)$.

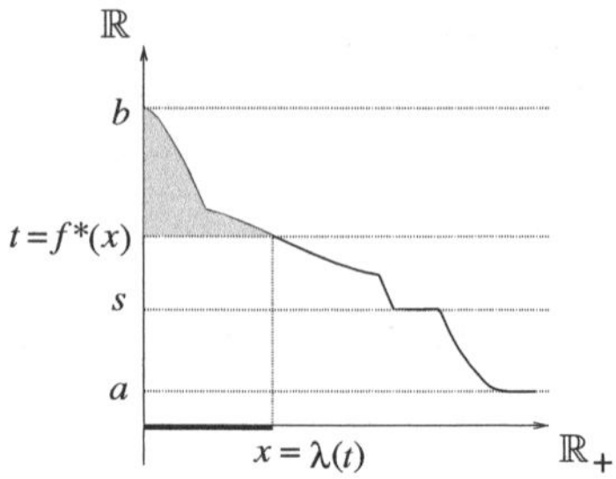


Figure 1.3 The decreasing rearrangement $f^*$ of the function $f$ from Figure 1.1. By definition, $f^*$ is decreasing, right continuous, and equidistributed with $f$. It is constructed as a generalized inverse of the distribution function $\lambda$, shown in Figure 1.2. The flat spot of $f^*$ at height $s$ has the same measure as the flat spot $(f = s)$.

Thus, $\lambda_{f^*}(t) = \mathcal{L}([0, x)) = x$. But also $\lambda_f(t) = x$, since $(t, x)$ is on the graph of $\lambda_f$. We conclude that $\lambda_{f^*}(t) = \lambda_f(t)$ for all $t \in (a, b)$. It is easy to see that both distribution functions equal $\mu(X)$ for $t \leq a$ and both equal zero for $t \geq b$. Thus, $f$ and $f^*$ are equidistributed. Moreover, $f^*$ is strictly decreasing and continuous on $[0, \mu(X)]$.

We have achieved our goal when $\lambda_f$ is continuous and strictly decreasing on $(a, b)$ and $\lambda(a) = \lambda(a-)$. Of course, in this case our construction shows that the function $f^*\colon [0, \mu(X)] \to \mathbb{R}$ is just the inverse of $\lambda_f\colon [a, b] \to \mathbb{R}$.

Now we define $f^*$ in the more general case when $\lambda$ might have jumps and/or intervals of constancy. Take $x \in (0, \mu(X))$. Set

$$E(x) = \{t \geq 0\colon \lambda(t) \leq x\}.$$

Since $\lambda$ is decreasing and right continuous, $E(x)$ is a closed interval of the form $[u, \infty)$. Using Proposition 1.5, one can show that $a \leq u \leq b$. Define $f^*(x) = u$. Define also $f^*(0) = b$ and $f^*(\mu(X)) = a$.

Pictorially, for $x \in (0, \mu(X))$, the horizontal line at height $x$ either hits the graph of $\lambda$ or passes through a vertical gate created by a jump of $\lambda$: see Figure 1.2. In the first case, $f^*(x)$ is the smallest value of $t$ at which the line meets the graph. In the second case, $f^*(x)$ is the value of $t$ over which the gate sits: see Figure 1.3.

The reader may verify that $f^*\colon [0, \mu(X)] \to [-\infty, \infty]$ is given by the following formula:

$$f^*(x) = \begin{cases} \inf\{t \geq 0\colon \lambda_f(t) \leq x\}, & 0 < x < \mu(X), \\ b = \operatorname{ess\,sup} f, & x = 0, \\ a = \operatorname{ess\,inf} f, & x = \mu(X). \end{cases} \tag{1.2}$$

**Proposition 1.7** *Let $f\colon (X, \mathcal{M}, \mu) \to \mathbb{R}$ be measurable and satisfy* (1.1). *Then*

*(a) The function $f^*$ is decreasing and right continuous on $[0, \mu(X))$, and $f^*(\mu(X)) = f^*(\mu(X)-)$.*
*(b) $f^*$ is Borel measurable and satisfies $\lambda_{f^*} = \lambda_f$ on $\mathbb{R}$.*
*(c) If $g\colon [0, \mu(X)] \to [-\infty, \infty]$ satisfies the conditions in (a) and (b), then $g = f^*$ on $[0, \mu(X)]$.*

*Proof* (a) If $0 < x < y < \mu(X)$ then $E(x) \subset E(y)$, so $f^*(x) \geq f^*(y)$. It is also clear that $f^*(0) \geq f^*(x)$ and that $f^*(\mu(X)) \leq f^*(y)$. Thus, $f^*$ is decreasing on $[0, \mu(X)]$.

To prove right continuity at $x \in (0, \mu(X))$, write $u = f^*(x)$. Then $u = \inf E(x)$. Given small $\epsilon > 0$, let $y = \lambda(u-\epsilon)$. Then $y \geq x$, since $\lambda$ is decreasing. If $y = x$ then $u - \epsilon \in E(x)$, so $f^*(x) \leq u - \epsilon$, which contradicts $f^*(x) = u$. So $y > x$. Also, if $x < z < y$ then $\lambda(u - \epsilon) > z$, so $u - \epsilon \notin E(z)$. Since $E(z)$ is a semi-infinite interval, we have $u - \epsilon \leq \inf E(z) = f^*(z)$. Since $f^*$ is decreasing and $u = f^*(x)$, we have

$$f^*(x) - \epsilon \leq f^*(z) \leq f^*(x)$$

when $x < z < y$. Thus, $f^*$ is right-continuous at $x$. We leave it to the reader to check that $f^*$ is right continuous at 0 and is left continuous at $\mu(X)$.
(b) For monotonic functions $g$ on an interval of $\mathbb{R}$ the sets $(g > t)$ are intervals. Thus, such $g$ are Borel measurable. In particular, $f^*$ is Borel measurable on $[0, \mu(X)]$.

To show $\lambda_f = \lambda_{f^*}$, take $t \in (\operatorname{ess\,inf} f, \operatorname{ess\,sup} f)$ and $x \in (0, \mu(X))$. If $f^*(x) > t$ then (1.2) implies $\lambda(t) > x$. Conversely, if $\lambda(t) > x$ then $t \notin E(x)$, so $t < \inf E(x) = f^*(x)$, where the strict inequality follows from $E(x)$ being closed. Thus, $\{x \in (0, \mu(X)) \colon f^*(x) > t)\} = \{x \in (0, \mu(X)) \colon 0 < x < \lambda(t)\}$, and hence $\lambda_{f^*}(t) = \lambda(t)$. The verification of this equation for $t \in \mathbb{R} \setminus (\operatorname{ess\,inf} f, \operatorname{ess\,sup} f)$ is left to the reader.
(c) Take $x_0 \in (0, \mu(X))$. Let $x_1 = \inf\{y \in [0, \mu(X)] \colon f^*(y) = f^*(x_0)\}$ Then $x_1 \leq x_0$, $f^*(x_1) = f^*(x_0)$, and $\lambda_{f^*}(f^*(x_0)) = x_1$. Since $f$, $f^*$, and $g$ all have the same distribution function, it follows that $\lambda_g(f^*(x_0)) = x_1$. If $\lambda_g(t) = x_1$ for some $t$, then the right continuity and monotone decrease of $g$ implies $g(x_1) \leq t$. Choosing $t = f^*(x_0)$. we obtain $g(x_1) \leq f^*(x_0)$. Also, $g(x_1) \geq g(x_0)$, since $g$ is decreasing. Hence, $g(x_0) \leq f^*(x_0)$. If we interchange $f^*$ and $g$, the argument works again, to produce $f^*(x_0) \leq g(x_0)$ and hence $f^*(x_0) = g(x_0)$. Thus, $f^* = g$ on $(0, \mu(X))$. The equality still holds at the endpoints because, by part (a), each function is right continuous at 0, left continuous at $\mu(X)$. □

Let $f\colon (X,\mathcal{M},\mu) \to \mathbb{R}$ and $g\colon (Y,\mathcal{N},\nu) \to \mathbb{R}$ be measurable functions satisfying (1.1). Then $f$ and $g$ are equidistributed if and only if $\mu(X) = \nu(Y)$ and $f^* = g^*$ on $[0,\mu(X)]$. The "only if" statement follows from Definition 1.6 and formula (1.2). The "if" part follows from the fact that if $f^* = g^*$ on $[0,\mu(X)] = [0,\nu(Y)]$, then $\lambda_{f^*} = \lambda_{g^*}$ on $\mathbb{R}$, which we combine with the equations $\lambda_f = \lambda_{f^*}$ on $\mathbb{R}$, $\lambda_g = \lambda_{g^*}$ on $\mathbb{R}$.

We describe now the decreasing rearrangements of the functions whose distribution functions were computed in Examples 1.2–1.4. In each case, the reader may check that the function asserted to be $f^*$ satisfies the conditions in (a) and (b) of Proposition 1.7, and hence, by (c), really is $f^*$.

**Example 1.8** $f\colon X \to \mathbb{R}$ is given by $f = \sum_{i=1}^{n} \alpha_i \chi_{E_i}$, where the $\alpha_i$ are distinct real numbers with $\alpha_1 > \cdots > \alpha_n$ and the $E_i$ form a measurable partition of a measure space $(X,\mathcal{M},\mu)$. To avoid trivialities, we shall assume that each $E_i$ has positive measure. Then $\operatorname{ess\,inf} f = \alpha_n$. To satisfy condition (1.1), we assume also that each $E_i$ has finite measure, except perhaps for $E_n$. Define positive numbers $m_1,\ldots,m_n$ by

$$m_1 = \mu(E_1), \quad m_2 = m_1 + \mu(E_2) = \mu(E_1 \cup E_2), \ldots$$
$$\ldots, m_n = m_{n-1} + \mu(E_n) = \mu(X).$$

Define $m_0 = 0$. Then $f^*\colon [0,\mu(X)] \to \mathbb{R}$ is a decreasing step function taking values $\alpha_i$ on the intervals $[m_{i-1}, m_i)$ :

$$f^* = \sum_{i=1}^{n-1} \alpha_i \chi_{[m_{i-1},m_i)} + \alpha_n \chi_{[m_{n-1},m_n]} = \sum_{i=1}^{n-1} (\alpha_i - \alpha_{i+1}) \chi_{[0,m_i)} + \alpha_n \chi_{[0,m_n]}. \quad (1.3)$$

The second formula, which follows from the first by summation by parts, expresses $f^*$ as a positive linear combination of *decreasing* characteristic functions, plus a constant function. It is an example of a *layer cake representation.* We shall visit this subject again in §1.6.

**Example 1.9** $X = [-1,2]$, $f(x) = x^2$. Then $\operatorname{ess\,inf} f = 0$, $\operatorname{ess\,sup} f = 4$, and $\mu(X) = 3$. Here $\lambda$ is continuous and strictly decreasing on $[0,4]$, so $f^*$ on $[0,3]$ equals $(\lambda|_{[0,4]})^{-1}$. Calculation gives

$$f^*(x) = (2-x)^2, \quad 0 \le x \le 1,$$
$$= \frac{1}{4}(3-x)^2, \quad 1 \le x \le 3.$$

**Example 1.10** $X = [0,a]$, where $0 < a < \infty$, and $f(x) = a^2 - x^2$. Then $\mu(X) = a$ and $f^* = f$.

As noted in Example 1.9, we observed earlier in this section that if $\lambda_f$ is continuous and strictly decreasing on $I = [\text{ess inf} f, \text{ess sup} f)$, then the inverse function of $\lambda|_I$ equals $f^*$. The next proposition tells us in what respects $f^*$ and $\lambda$ behave like inverse functions when $\lambda$ has jumps and/or intervals of constancy.

**Proposition 1.11** *Let $f\colon (X, \mathcal{M}, \mu) \to \mathbb{R}$, be measurable and satisfy* (1.1). *Set $a = \text{ess inf} f$, $b = \text{ess sup} f$. Then*

*(a) $f^*(\lambda(t)) \le t, \quad t \in (a, b)$, with equality when $\lambda$ is $1-1$ in a neighborhood of $t$.*
*(b) $\lambda(f^*(x)) \le x, \quad x \in (0, \mu(X))$, with equality when $\lambda$ is continuous at $f^*(x)$.*
*(c) $f^*(\lambda(t)) = t$ when $t = a$ or $t = b$.*
*(d) $\lambda(f^*(x)) = x$ when $x = 0$ or $x = \mu(X)$.*

The proofs are exercises, which can be expeditiously done by first drawing pictures to see how the formal arguments should go. In the same vein, but with less formality, we record some rules of thumb:

If $\lambda$ jumps downward from $x_2$ to $x_1$ at $t_0$, then $f^*$ has constant value $t_0$ on the interval $[x_1, x_2)$.

If $\lambda$ is a constant $x_0$ on $[t_1, t_2)$, then $f^*$ jumps from $t_2$ down to $t_1$ at $x_0$.

The converse statements are also true.

## 1.3 Induced Measures

**Definition 1.12** Let $(X, \mathcal{M})$ and $(Y, \mathcal{N})$ be measurable spaces, $\mu$ be a measure on $\mathcal{M}$, and $f\colon X \to Y$ be $(\mathcal{M}, \mathcal{N})$ measurable. Then $f_\#\mu$ is the measure on $\mathcal{N}$ defined by

$$(f_\#\mu)(E) = \mu(f^{-1}E), \quad E \in \mathcal{N}.$$

The measure $f_\#\mu$ is called the pushforward of $\mu$ by $f$, and also, when $\mu$ is clear from context, the measure induced by $f$ on $\mathcal{N}$ or on $Y$.

We will say that an integral $\int_X f \, d\mu$ exists if at least one of the integrals $\int_X f^+ \, d\mu$ or $\int_X f^- \, d\mu$ is finite.

**Proposition 1.13** *Let $f\colon X \to Y$ be as in Definition 1.12, and let $g\colon Y \to \mathbb{R}$ be $\mathcal{N}$ measurable. Then, whenever the integrals exist,*

$$\int_Y g \, d(f_\#\mu) = \int_X g \circ f \, d\mu. \tag{1.4}$$

*Proof* For $g = \chi_E, E \in \mathcal{N}$, (1.4) reduces to Definition 1.12. By linearity, (1.4) holds when $g$ is simple and nonnegative. If $g$ is nonnegative then $g$

is the increasing pointwise limit of nonnegative simple functions (Folland, 1999, p. 47), so (1.4) follows by the monotone convergence theorem. For general $g: Y \to \mathbb{R}$, write $g = g^+ - g^-$ and use the definition $\int_X h\,d\mu = \int_X h^+\,d\mu - \int_X h^-\,d\mu$. □

Suppose now that $f: (X, \mathcal{M}, \mu) \to \mathbb{R}$ is measurable and satisfies (1.1). Write

$$a = a(f) = \operatorname{ess\,inf} f.$$

Then $-\infty \leq a < \infty$, and $f_\#\mu$ is a Borel measure on $\mathbb{R}$ with

$$(f_\#\mu)((t,\infty)) = \lambda(t), \quad t \in \mathbb{R},$$

where $\lambda = \lambda_f$ is the distribution function of $f$. Suppose that $a = -\infty$ or that $\mu(X) < \infty$. Then $\lambda(t) < \infty$ for every $t \in \mathbb{R}$. Let $\nu = \nu_f$ be the Lebesgue–Stieltjes measure generated by $\lambda$. Then $\nu$ is the unique Borel measure on $\mathbb{R}$ such that $\nu((t,\infty)) = \lambda(t)$ for every $t$. To construct $\nu$ one may follow the method of Folland (1999, Theorem 1.16) with small changes. For example, one should begin the construction by defining $\nu((a,b]) = \lambda(a) - \lambda(b)$. The uniqueness implies that $\nu = f_\#\mu$.

If $a > -\infty$ and $\mu(X) = \infty$, then $\lambda(t) < \infty$ for $t > a$ and $\lambda(t) = \infty$ for $t < a$. Construct the associated Lebesgue–Stieltjes measure $\nu$ on $(a,\infty)$ as above, then extend $\nu$ to $\mathbb{R}$ by specifying that $\nu(\{a\}) = \mu(f = a)$ and that $\nu((-\infty,a)) = 0$. It is easily checked that again $\nu = f_\#\mu$ as measures on $\mathbb{R}$. We summarize the discussion with

**Proposition 1.14** *Let $f: (X, \mathcal{M}, \mu) \to \mathbb{R}$ be measurable and satisfy* (1.1). *Then the Borel measures $f_\#\mu$ and $\nu_f$ coincide on $\mathbb{R}$, where $\nu_f$ is the Lebesgue–Stieltjes measure generated by $\lambda_f$. Moreover, $\nu_f([t,\infty)) < \infty$ for each $t > a$.*

In brief: The measure induced by $f$ on $\mathbb{R}$ equals the Lebesgue–Stieltjes measure generated by the function $\lambda_f$, suitably extended.

For functions $g$ and sets $E \subset \mathbb{R}$, we shall often write $-\int_E g\,d\lambda$ instead of $\int_E g\,d\nu$, when the integral exists. When $a > -\infty$ and $\mu(X) = \infty$ one must bear in mind that according to the definitions

$$\int_E g\,d\lambda \equiv \int_{E\cap(a,\infty)} g\,d\lambda - g(a)\mu(f = a)\chi_E(a).$$

This identity holds also when $a > -\infty$ and $\mu(X) < \infty$. The integral on the right is an ordinary Lebesgue–Stieltjes integral to which the usual rules of calculus apply.

Let us revisit Examples 1.2 and 1.8. There $f$ is simple, $f = \sum_{i=1}^{n} \alpha_i \chi_{E_i}$, where the $\alpha_i$ are strictly decreasing real numbers and the $E_i$ form a measurable partition of $X$. Then $\lambda$ is the decreasing step function with jumps of size $\mu(E_i)$ at the $\alpha_i$. The associated Lebesgue–Stieltjes measure is the sum of point masses at the $\alpha_i$, with mass $\mu(E_i)$ at $\alpha_i$. Thus,

$$\nu_f = \sum_{i=1}^{n} \mu(E_i)\delta_{\alpha_i}.$$

Also, for suitable functions $g$,

$$\int_{\mathbb{R}} g\,d\lambda = -\sum_{i=1}^{n} g(\alpha_i)\mu(E_i).$$

Returning now to general $f$, let $g$ be a real-valued Borel measurable function defined on $[a,\infty)$ when $a > -\infty$ and on $\mathbb{R}$ when $a = -\infty$. Set $g = 0$ on $(-\infty, a)$. Combining Propositions 1.13 and 1.14 and making use, if necessary, of the convention $0 \cdot \infty = 0$, we obtain

$$\int_X g \circ f\,d\mu = -\int_{[a(f),\infty)} g\,d\lambda_f, \tag{1.5}$$

when either integral exists, where we interpret $[-\infty,\infty)$ to be $\mathbb{R}$. Here is a consequence of (1.5), useful in practice:

**Corollary 1.15** *If $f\colon (X, \mathcal{M}, \mu) \to \mathbb{R}$ is measurable and nonnegative, and $g$ is nonnegative and Borel measurable on $[0,\infty)$, then*

$$\int_X g \circ f\,d\mu = -\int_{(0,\infty)} g\,d\lambda_f + g(0)\nu_f(\{0\}). \tag{1.6}$$

*Proof* Note that for nonnegative $f$ we have $a(f) \geq 0$. If $a = 0$, then (1.5) gives

$$\int_X g \circ f\,d\mu = \int_{[0,\infty)} g\,d\nu = \int_{(0,\infty)} g\,d\nu + g(0)\nu(\{0\}).$$

Suppose $a > 0$. Since $\nu$ vanishes on $(-\infty, a)$, (1.5) gives

$$\int_X g \circ f\,d\mu = \int_{[a,\infty)} g\,d\nu = \int_{(0,\infty)} g\,d\nu + g(0)\nu(\{a\}).$$

Either way, (1.6) is proved. □

Under favorable conditions on $f$ and $g$, (1.5) and (1.6) can be integrated by parts to obtain useful formulas. Here is one such result.

**Proposition 1.16** *Let $f$ be as in Proposition 1.14, and be nonnegative. Suppose that $g\colon [0,\infty) \to \mathbb{R}$ is increasing and continuous, that $g$ is absolutely continuous on each interval $[a,b]$, $0 < a < b < \infty$, and that $g(0) = 0$. Then*

$$\int_X g \circ f\,d\mu = \int_0^\infty g'(t)\,\lambda_f(t)\,dt.$$

*Proof* The hypotheses imply that $g$ is nonnegative. Since $g$ is monotonic it is Borel measurable. Identity (1.6) gives

$$\int_X g \circ f \, d\mu = -\int_{(0,\infty)} g \, d\lambda = \lim_{a \to 0+, b \to \infty} -\int_{(a,b]} g \, d\lambda. \tag{1.7}$$

Assume for the moment that $f$ is bounded and that $\lambda(0) = \mu(f > 0) < \infty$. Then $\lambda(t) = 0$ for all sufficiently large $t$, and $\lambda$ is bounded on $(0, \infty)$. For $0 < a < b < \infty$, right continuity of $\lambda$ together with the argument in Folland (1999, Th 3.36) and the result in Folland (1999, Cor 3.33) give

$$-\int_{(a,b]} g \, d\lambda = \int_{(a,b]} g'(t)\lambda(t) \, dt - g(b)\lambda(b) + g(a)\lambda(a).$$

As $a \to 0$ and $b \to \infty$ the boundary terms on the right tend to zero. Since $g' \geq 0$ and $\lambda \geq 0$, the integral on the right tends to the corresponding integral over $(0, \infty)$. Using (1.7), we see that (1.16) is true when $f$ is bounded and $\mu(f > 0) < \infty$.

Next, suppose that $f$ satisfies the hypotheses of the proposition and also satisfies $\lambda(t) < \infty$ for each $t > 0$. Put $f_n = (\min(f, n) - \frac{1}{n})^+$. Each $f_n$ is bounded, and $\lambda_{f_n}(0) \leq \mu(f > 1/n) < \infty$. Thus, for each $n \geq 1$,

$$\int_X g \circ f_n \, d\mu = \int_0^\infty g'(t) \, \lambda_{f_n}(t) \, dt. \tag{1.8}$$

The sequence $\{f_n\}$ is increasing and converges pointwise to $f$ on $X$. Since $g$ is continuous and increasing, the sequence $\{g \circ f_n\}$ is increasing and converges pointwise to $g \circ f$. By Proposition 1.5(f), the sequence $\{\lambda_{f_n}\}$ is increasing and converges pointwise to $\lambda_f$ on $\mathbb{R}$. Also, $g' \geq 0$. Thus we can apply the monotone convergence theorem on both sides of (1.8) and conclude (1.16).

Let $f$ be a general function satisfying the hypotheses of the proposition. Then $a = \operatorname{ess\,inf} f \geq 0$ since $f$ is nonnegative, and $\lambda_{f-a}(t) = \lambda_f(t + a) < \infty$ for all $t > 0$, since $f$ satisfies (1.1), Set $G(x) = g(x + a) - g(a)$. Then $g \circ f(x) = G(f(x) - a) + g(a)$. Now (1.16) is valid for $f - a$ and $G$, and $\lambda(t) = \mu(X)$ for $t < a$, so

$$\begin{aligned}\int_X g \circ f \, d\mu &= \int_0^\infty G'(t)\lambda_{f-a}(t) \, dt + g(a)\mu(X) \\ &= \int_0^\infty g'(t + a)\lambda_f(t + a) \, dt + g(a)\mu(X) = \int_0^\infty g'(t)\lambda(t) \, dt. \qquad \square\end{aligned}$$

The choice $g(t) = t^p$, $p > 0$, is of particular interest.

**Corollary 1.17** *Let $f\colon (X, \mathcal{M}, \mu) \to \mathbb{R}^+$ be measurable, and $0 < p < \infty$. Then*

$$\int_0^\infty f^p \, d\mu = \int_0^\infty p t^{p-1} \lambda_f(t) \, dt.$$

*In particular,*

$$\int_0^\infty f\,d\mu = \int_0^\infty \lambda_f(t)\,dt.$$

Proposition 1.16 admits many variants. For example, if $\mu(X) < \infty$, then (1.16) is true when $g$ is increasing and not necessarily nonnegative, provided the term $g(0)\mu(X)$ is added to the right-hand side. This may be obtained from Proposition 1.16 by writing $g(x) = g(x) - g(0) + g(0)$. More generally, if $g$ has locally bounded variation, then by expressing $g$ as the difference of two increasing functions we can obtain formulas like (1.16) when the appropriate integrals exist.

The proof of the following proposition is left to the reader.

**Proposition 1.18** *Let $(X, \mathcal{M}, \mu)$ and $(Y, \mathcal{N}, \nu)$ be measure spaces with $\mu(X) = \nu(Y)$, let $f\colon X \to \mathbb{R}$ be $\mathcal{M}$-measurable and $g\colon Y \to \mathbb{R}$ be $\mathcal{N}$-measurable. Suppose that $f$ and $g$ satisfy* (1.1). *Then the following statements are equivalent.*

*(a) $f$ and $g$ are equidistributed, i.e., $\lambda_f(t) = \lambda_g(t)\ \forall t \in \mathbb{R}$.*

*(b) $f$ and $g$ have the same essential infimum, call it $a$, and the induced measures $f_\#\mu$ and $g_\#\nu$ agree on $(a, \infty)$.*

*(c)*

$$\int_X \Phi \circ f\,d\mu = \int_Y \Phi \circ g\,d\nu = \int_{[a,\infty)} \Phi\,d(f_\#\mu)$$

*for all Borel measurable $\Phi\colon [a, \infty) \to \mathbb{R}$ with $\Phi(a) = 0$ for which any of the integrals exist.*

As before, when $a = -\infty$, the interval $[a, \infty)$ is understood to be $\mathbb{R}$. Also when $a = -\infty$, the condition about $\Phi(a)$ is to be omitted.

### A Word about Equidistribution

Let $f\colon (X, \mathcal{M}, \mu) \to \mathbb{R}$ and $g\colon (Y, \mathcal{N}, \nu) \to \mathbb{R}$ be measurable functions with $\mu(X) = \nu(Y)$. The measure $\nu$ is now free-standing, and is no longer induced by $f$. According to Definition 1.6, $f$ and $g$ are equidistributed if $\lambda_f = \lambda_g$ on $\mathbb{R}$, that is, $\mu(f > t) = \nu(g > t)$ for every $t \in \mathbb{R}$. Another way, a priori stronger, in which $f$ and $g$ could be equidistributed would be for $\mu(f \in E) = \nu(g \in E)$ for every $E \in \mathcal{B}(\mathbb{R})$. That is, $f_\#\mu = g_\#\nu$. Consider the following example:

$$X = Y = (0, \infty), \quad \mu = \nu = \mathcal{L}, \quad f(x) = 1/x,$$
$$g(x) = 0 \quad \text{for} \quad 0 < x \le 1, \quad g(x) = 1/(x-1) \quad \text{for} \quad 1 < x < \infty.$$

Then $\lambda_f = \lambda_g$, but $f_{\#}\mu$ and $g_{\#}\nu$ are different, since $f_{\#}\mu(\{0\}) = 0$, $g_{\#}\nu(\{0\}) = \mathcal{L}(g = 0) = 1$. Thus, functions equidistributed in our sense need not be equidistributed in the stronger sense.

Let us say that $f$ and $g$ are *strongly equidistributed* if $f_{\#}\mu = g_{\#}\nu$. When $f$ and $g$ satisfy (1.1), it turns out that the example illustrates the only way in which $f$ and $g$ can be equidistributed but not strongly equidistributed. Proofs of the following assertions are left to the reader.

**Fact 1.19** *Suppose that $f$ and $g$ satisfy* (1.1). *Then*

*(a) If $f$ and $g$ are strongly equidistributed they are equidistributed.*
*(b) Suppose that $f$ and $g$ are equidistributed. Then they have a common essential infimum $a$. If $a = -\infty$ or $\mu(X) < \infty$ then $f$ and $g$ are strongly equidistributed. If $a > -\infty$ and $\mu(X) = \infty$ then $f$ and $g$ are strongly equidistributed if and only if $\mu(f = a) = \nu(g = a)$.*

Proposition 1.18 has a version for strongly equidistributed functions: In (b) change $(a, \infty)$ to $[a, \infty)$, and in (c) drop the condition $\Phi(a) = 0$.

## Joint Distributions

Suppose that we have $k$ real-valued $\mathcal{M}$-measurable functions $f_1, \ldots, f_k$ defined on the measure space $(X, \mathcal{M}, \mu)$. Set $F = (f_1, \ldots, f_k)$. The *joint distribution function* $\lambda_F = \lambda_{(f_1,\ldots,f_k)}$ of the ordered $k$-tuple of functions $(f_1, \ldots, f_k)$, or of the vector function $F$, is the function defined for $(t_1, \ldots, t_k) \in \mathbb{R}^k$ by

$$\lambda_F(t_1, \ldots, t_k) = \mu(\{x \in X \colon f_1(x) > t_1, \ldots, f_k(x) > t_k\}).$$

Two ordered $k$-tuples $(f_1, \ldots, f_k)$ and $(g_1, \ldots, g_k)$ are said to have the same joint distribution if they have the same joint distribution functions. Using the fact that measures on $\mathbb{R}^k$ are determined by their values on sets $(t_1, \infty) \times \cdots \times (t_k, \infty)$, one can mimic the proof of Proposition 1.18 to obtain a version valid for joint distributions. Write

$$a(f_i) = \operatorname{ess\,inf} f_i, \quad a(g_i) = \operatorname{ess\,inf} g_i$$

and

$$R = [a(f_1), \infty) \times \cdots \times [a(f_k), \infty).$$

**Proposition 1.20** *Let $(X, \mathcal{M}, \mu)$ and $(Y, \mathcal{N}, \nu)$ be measure spaces, let $f_i \colon X \to \mathbb{R}$ be $\mathcal{M}$-measurable and $g_i \colon Y \to \mathbb{R}$ be $\mathcal{N}$-measurable, for $i = 1, \ldots, n$. Suppose that each $f_i$ and $g_i$ satisfy* (1.1). *Then the following statements are equivalent.*

*(a)* $f_1, \ldots, f_k$ *and* $g_1, \ldots, g_k$ *have the same joint distribution.*

*(b)* $a(f_i) = a(g_i)$ *for each* $i$*, and the induced measures* $F_{\#}\mu$ *and* $G_{\#}\nu$ *agree on the interior of* $R$*.*

*(c)*

$$\int_X \Phi(f_1(x), \ldots, f_k(x))\, d\mu(x) = \int_Y \Phi(g_1(x), \ldots, g_k(x))\, d\nu(x) = \int_R \Phi\, d(f_{\#}\mu)$$

*for all Borel measurable* $\Phi\colon R \to \mathbb{R}$ *with* $\Phi = 0$ *on* $\partial R$ *for which any of the integrals exists.*

## 1.4 Measure Preserving Transformations

Let $(X, \mathcal{M}, \mu)$ and $(Y, \mathcal{N}, \nu)$ be measure spaces. A function $T\colon X \to Y$ is said to be *measure preserving* from $X$ onto $Y$ if $T$ is $(\mathcal{M}, \mathcal{N})$ measurable, $T$ maps $X$ onto $Y \setminus N$ where $\nu(N) = 0$, and $T_{\#}\mu = \nu$ on $Y$, that is

$$\mu(T^{-1}E) = \nu(E), \quad E \in \mathcal{N}.$$

Taking $E = Y$, we see that if such a $T$ exists, then $\mu(X) = \nu(Y)$. From Folland (1999, pp. 26–27), it is easy to show that a measure preserving transformation from $(X, \mathcal{M}, \mu)$ onto $(Y, \mathcal{N}, \nu)$ is also a measure preserving transformation between the completions of the two measure spaces.

If $T\colon X \to Y$ is measure preserving and $g\colon Y \to \mathbb{R}$ is $\mathcal{N}$-measurable, then Proposition 1.13 shows that

$$\int_Y g\, d\nu = \int_X g \circ T\, d\mu,$$

whenever either integral exists. We also have

**Proposition 1.21** *If* $T\colon (X, \mathcal{M}, \mu) \to (Y, \mathcal{N}, \nu)$ *is measure preserving and* $g\colon Y \to \mathbb{R}$ *is* $\mathcal{N}$*- measurable, then* $g$ *and* $g \circ T$ *are equidistributed, i.e.,* $\lambda_g = \lambda_{g \circ T}$ *on* $\mathbb{R}$*.*

*Proof* Take $t \in \mathbb{R}$. Set $E = g^{-1}(t, \infty)$. Then $T^{-1}(E) = (g \circ T)^{-1}(t, \infty)$, and $\lambda_g(t) = \nu(E) = \mu(T^{-1}(E)) = \lambda_{g \circ T}(t)$. □

Of course, $\lambda_g$ is computed with respect to $\nu$ and $\lambda_{g \circ T}$ with respect to $\mu$. The argument shows that $g$ and $g \circ T$ are in fact strongly equidistributed, as defined at the end of §1.3.

Here are some examples of measure preserving transformations.

**Example 1.22** Let $T$ be a translation of $\mathbb{R}^n$. That is, $T(x) = a + x$ for some fixed $a \in \mathbb{R}^n$ and all $x \in \mathbb{R}^n$. Then $T$ is a measure preserving transformation of $(\mathbb{R}^n, \mathcal{M}^n, \mathcal{L}^n)$ onto itself, where $\mathcal{M}^n$ is the class of Lebesgue measurable subsets of $\mathbb{R}^n$. Likewise, if $T$ is a linear map of $\mathbb{R}^n$ onto itself whose determinant satisfies $|\det T| = 1$, then $T$ is a measure preserving transformation of $(\mathbb{R}^n, \mathcal{M}^n, \mathcal{L}^n)$ onto itself. In particular, if $T \in O(n)$, the group of all rotations and reflections of $\mathbb{R}^n$, then $T$ is $\mathcal{L}^n - \mathcal{L}^n$ measure preserving.

For the results of Example 1.22, see, for example, Folland (1999, §2.6). The maps $T$ in Example 1.22 are 1-1. In general, if $T\colon (X, \mathcal{M}, \mu) \to (Y, \mathcal{N}, \nu)$ is 1-1 and almost onto, then it is easy to see that $T$ is measure preserving if and only if $T^{-1}$ is. Thus, an injective and surjective $T$ is measure preserving if and only if $E \in \mathcal{M} \implies TE \in \mathcal{N}$ and $\nu(TE) = \mu(E)$.

**Example 1.23** Define $T\colon \mathbb{R}^n \to \mathbb{R}^+$ by $T(x) = \alpha_n |x|^n$, where $\alpha_n$ is the volume of the unit ball in $\mathbb{R}^n$:

$$\alpha_n = \mathcal{L}^n(\mathbb{B}^n(1)) = \frac{\pi^{n/2}}{\Gamma(1 + \frac{n}{2})}.$$

The computation of $\alpha_n$ is carried out, for example, in Folland (1999, Cor. 2.55). We claim that $T$ is a measure preserving transformation of $(\mathbb{R}^n, \mathcal{M}^n, \mathcal{L}^n)$ onto $(\mathbb{R}^+, \mathcal{M}^1, \mathcal{L})$. Indeed, for $b \in (0, \infty)$, $T^{-1}[0, b) = \mathbb{B}^n((b/\alpha_n)^{1/n})$. Thus, $\mathcal{L}^n(T^{-1}E) = b = \mathcal{L}(E)$ holds for each interval $[0, b)$. By the usual arguments, it follows that $\mathcal{L}^n(T^{-1}E) = \mathcal{L}(E)$ holds for all $E \in \mathcal{B}[0, \infty)$. Thus, $T$ is measure preserving from $(\mathbb{R}^n, \mathcal{B}(\mathbb{R}^n), \mathcal{L}^n)$ onto $(\mathbb{R}^+, \mathcal{B}[0, \infty), \mathcal{L})$, and also measure preserving between their completions $(\mathbb{R}^n, \mathcal{M}^n, \mathcal{L}^n)$ and $(\mathbb{R}^+, \mathcal{M}^1, \mathcal{L})$.

**Example 1.24** Let $\mathbb{S}^1 = \{z \in \mathbb{C}\colon |z| = 1\}$ denote the unit circle in the complex plane, and $\sigma$ denote arclength measure on $\mathbb{S}^1$. Let $n$ be a nonzero integer, and define $T\colon \mathbb{S}^1 \to \mathbb{S}^1$ by $T(z) = z^n$. Then $T$ is a measure preserving transformation of $(\mathbb{S}^1, \mathcal{M}, \sigma)$ onto itself, where $\mathcal{M}$ denotes the class of Lebesgue measurable subsets of $\mathbb{S}^1$. To prove this, let $E = \{e^{i\theta} : a \le \theta < b\}$, $-\infty < a < b < 2\pi + a$, be an arc on $\mathbb{S}^1$. Then $\sigma(E) = b - a$. One easily checks that $T^{-1}E$ is the disjoint union of $|n|$ arcs, each of which has arclength $\frac{1}{|n|}(b - a)$. Thus, $\sigma(T^{-1}E) = \sigma(E)$ holds for each such arc $E$. Arguing as in Example 1.23, we find that $\sigma(T^{-1}E) = \sigma(E)$ holds in fact for all $E \in \mathcal{B}(\mathbb{S}^1)$, and, more generally, for all Lebesgue measurable $E \subset \mathbb{S}^1$.

## 1.5 Nonatomic Measure Spaces

Our chief goal in this section is to prove that under appropriate hypotheses a function $f\colon X \to \mathbb{R}$ can be factored as $f = f^* \circ T$, where $T$ is a measure preserving transformation of the measure space $X$ onto $[0, \mu(X)]$ and $f^*$ is the decreasing rearrangement of $f$. This result is due essentially to Ryff (1970). An enhanced version will play a key role when we build the theory of the $\star$-function in Chapter 9. In addition to its usefulness, this factorization is very pretty. It is not as well known as one might expect.

The main hypothesis required on $(X, \mathcal{M}, \mu)$ to insure factorization is that $X$ be *nonatomic*. In this section, unless otherwise indicated, all subsets of $X$ that we mention are assumed to be $\mathcal{M}$-measurable. A set $E \subset X$ is called an *atom* for $(X, \mathcal{M}, \mu)$ if $\mu(E) > 0$, and if for each $E_1 \subset E$, either $\mu(E_1) = 0$ or $\mu(E_1) = \mu(E)$. The space $(X, \mathcal{M}, \mu)$ is called *nonatomic* if it contains no atoms.

The space $(\mathbb{R}^n, \mathcal{M}^n, \mathcal{L}^n)$ is nonatomic. To prove this, let $E \subset \mathbb{R}^n$ be a set with $\mathcal{L}^n(E) > 0$. Let $g(r) = \mathcal{L}^n(E \cap \mathbb{B}^n(0, r))$. Then $g$ is continuous on $[0, \infty)$, with $g(0) = 0$ and $\lim_{r\to\infty} g(r) = \mathcal{L}^n(E)$. Thus, there exists $r$ such that $0 < g(r) < \mathcal{L}^n(E)$.

A prototypical space with an atom is $\mathbb{R}^n$ equipped with the measure $\delta_a$, which assigns to each $E \subset \mathbb{R}^n$ the value 1 if $a \in E$ and the value 0 if $a \notin E$. The space $(\mathbb{R}^n, \mathcal{M}^n, \mathcal{L}^n + \delta_a)$ also contains an atom, namely $\{a\}$. Of course, $(X, \mathcal{M}, \mu)$ contains atoms whenever $X$ is at most countable and $\mu$ is not identically zero.

The next two propositions contain some information about nonatomic measure spaces. To keep the statements and proofs uncluttered we will state them under the additional assumption that $\mu(X) < \infty$. At the end of the section, we will present versions for the case when $\mu(X) = \infty$.

**Proposition 1.25** *Let $(X, \mu)$ be a nonatomic measure space for which $\mu(X) < \infty$. Then*

*(a) $X$ contains a family of measurable subsets $\{E_t\colon 0 \le t \le \mu(X)\}$ such that $E_{t_1} \subset E_{t_2}$ whenever $0 \le t_1 \le t_2 \le \mu(X)$, and $\mu(E_t) = t$ for each $t \in [0, \mu(X)]$.*

*(b) There exists a measure preserving transformation $T$ of $(X, \mathcal{M}, \mu)$ onto $([0, \mu(X)], \mathcal{B}, \mathcal{L})$.*

*Proof* If $\mu(X) = 0$ the proposition is trivial, so we assume $0 < \mu(X) < \infty$. Without loss of generality, we may assume that $\mu(X) = 1$. The main step is to prove the following claim:

**Claim** *There exists $E \subset X$ such that $\mu(E) = \frac{1}{2}$.*

*Proof of claim* The proof is in three steps.

*Step 1.* The nonatomicity implies that $X$ can be expressed as the disjoint union of two sets with measure strictly between 0 and 1. Thus, there exists

$E_1 \subset X$ with $0 < \mu(E_1) \leq \frac{1}{2}$. Applying the same argument to $E_1$, then repeatedly, we see that there is a sequence $E_i$ of subsets of $X$ such that $\mu(E_i) > 0$ for each $i$ and $\lim_{i\to\infty} \mu(E_i) = 0$.

*Step 2*. Fix $0 < t < 1$. Let $\mathcal{A} = \{A \subset X : 0 < \mu(A) \leq \frac{1}{2}t\}$. By step 1, $\mathcal{A}$ is nonempty. I claim that there exists $E \in \mathcal{A}$ such that $\mu(E) > \frac{1}{4}t$. To verify this, suppose that no such $E$ exists. Then every $E \in \mathcal{A}$ has $\mu(E) \leq \frac{1}{4}t$. It follows that if two sets are in $\mathcal{A}$ then so is their union. From this, it follows that finite unions of sets in $\mathcal{A}$ are in $\mathcal{A}$ and hence so are countable unions. Let $\alpha = \sup\{\mu(E) : E \in \mathcal{A}\}$. Then $0 < \alpha \leq \frac{1}{4}t$. Take $\{E_i\} \subset \mathcal{A}$ such that $\mu(E_i) > \alpha - \frac{1}{i}$ for $i \geq 1$, and set $E = \bigcup_{i=1}^{\infty} E_i$. Then $\mu(E) \geq \alpha$ since $E$ contains each $E_i$, but also $\mu(E) \leq \alpha$, since $E \in \mathcal{A}$. Hence, $\mu(E) = \alpha \leq \frac{1}{4}t$. By step 1 applied to $X \setminus E$, there exists $F \subset X \setminus E$ such that $0 < \mu(f) < \frac{1}{4}t$. Then $\mu(E \cup F) < \frac{1}{2}t$, so that $E \cup F \in \mathcal{A}$. But $\mu(E \cup F) > \mu(E) = \alpha$, which contradicts the definition of $\alpha$. This contradiction shows that some $E \in \mathcal{A}$ must exist with $\mu(E) > \frac{1}{4}t$.

*Step 3*. Take $E_1 \subset X$ such that $\beta_1 \equiv \mu(E_1) \in (\frac{1}{4}, \frac{1}{2}]$. This is possible by step 2. If $\beta_1 = \frac{1}{2}$ we are done. Suppose that $\beta_1 < \frac{1}{2}$. From step 2 applied to $X \setminus E_1$, with $t = 1 - 2\beta_1$, it follows that there exists $E_2 \subset X \setminus E_1$ such that $\beta_2 \equiv \mu(E_2)$ satisfies $\frac{1}{2}(\frac{1}{2} - \beta_1) < \beta_2 \leq \frac{1}{2} - \beta_1$. Then

$$\mu(E_1 \cup E_2) = \beta_1 + \beta_2 \leq \frac{1}{2}$$

and

$$\mu(E_1 \cup E_2) = \beta_1 + \beta_2 > \frac{1}{4} + \frac{1}{2}\beta_1 > \frac{3}{8}.$$

If $\mu(E_1 \cup E_2) = \frac{1}{2}$ we are done. If $\mu(E_1 \cup E_2) < \frac{1}{2}$, take $E_3 \subset X \setminus (E_1 \cup E_2)$ such that $\beta_3 \equiv \mu(E_3)$ satisfies $\frac{1}{2}(\frac{1}{2} - \beta_1 - \beta_2) < \beta_3 \leq \frac{1}{2} - \beta_1 - \beta_1$. Then, as in the previous estimates, one finds that

$$\frac{7}{16} < \mu(E_1 \cup E_2 \cup E_3) \leq \frac{1}{2}.$$

Continuing, we either encounter an integer $m$ such that $\mu(\bigcup_{i=1}^{m} E_i) = \frac{1}{2}$, or we construct an infinite disjoint sequence of sets $\{E_i\}$ such that $\frac{1}{2} - 2^{-m-1} < \mu(\bigcup_{i=1}^{m} E_i) < \frac{1}{2}$ for each $m \geq 1$. Then $E \equiv \bigcup_{i=1}^{\infty} E_i$ satisfies $\mu(E) = \frac{1}{2}$. The claim is proved. □

*Proof of (a)* For $k = 0, 1, .., 2^m - 1$, $m = 1, 2, \ldots$ we first construct a family of sets $A_{k,m}$. Write $X$ as the union of two disjoint subsets of measure $\frac{1}{2}$. This is possible by the claim. Denote the sets by $A_{0,1}$ and $A_{1,1}$. Write each of $A_{0,1}$ and $A_{1,1}$ as the union of two disjoint subsets of measure $\frac{1}{4}$. The subsets of $A_{0,1}$ are called $A_{0,2}$ and $A_{1,2}$, those of $A_{1,1}$ are called $A_{2,2}$ and $A_{3,2}$. Continue the process. In this fashion, for each $m \geq 1$ $X$ is partitioned into $2^m$ sets $A_{k,m}$ of measure $2^{-m}$, with the property that $A_{k,m} = A_{2k,m+1} \cup A_{2k+1,m+1}$.

Let $D = \{t = k2^{-m} : 0 \leq k < 2^m, m = 1, 2, \dots\}$ denote the subset of $[0, 1)$ consisting of dyadic fractions. For nonzero $t \in D$, write $t = k2^{-m}$, and define

$$B_t = \bigcup_{j=0}^{k-1} A_{j,m}.$$

It is easy to see that $B_t$ depends only on $t$ and not on $k$ and $m$. It is also obvious that $\mu(B_t) = t$ for $t \in D$ and that $B_{t_1} \subset B_{t_2}$ when $t_1, t_2 \in D$ with $0 \leq t_1 \leq t_2 < 1$.

Define $E_0$ to be the empty set, $E_1 = X$, and for $t \in (0, 1) \setminus D$ define

$$E_t = \bigcup_{s \in D, s<t} B_s.$$

Since $D$ is countable, each $E_t$ is measurable. Using density of $D$ in $[0, 1]$, it is easy to show that $\mu(E_t) = t$. The inclusion assertion in (a) is obviously true.

*Proof of (b)* Let $\{E_t\}$ be the family of sets constructed in the proof of (a). Define

$$T(x) = \inf\{t \in [0, 1] : x \in E_t\}.$$

Since $E_t = \bigcup_{0<s<t} E_s$, it follows that $T^{-1}[0, s) = E_s$ for $s \in [0, 1]$, The $(X, \mathcal{M}, \mu) \to (\mathbb{R}^+, \mathcal{B}, \mathcal{L})$- measurability of $T$, and the equation $\mu(T^{-1}E) = \mathcal{L}(E)$ for $\mathcal{B}$-measurable $E \subset [0, 1]$, follow. □

Suppose that $\{a_i\}_{i=1}^k$ is a finite sequence in $\mathbb{R}$, and let $\{a_i^*\}$ be the sequence rearranged in descending order. Then there is a permutation $\pi$ of the integers $\{1, \dots, k\}$ such that $a_i = a^*_{\pi(i)}$ for each $i$. Proposition 1.26 below asserts a continuous analog: Each measurable function $f$ on a nonatomic probability space $X$ with $\mu(X) < \infty$ can be written as $f = f^* \circ T$ a.e., where $T$ is a measure preserving transformation of $X$ onto $[0, \mu(X)]$, which depends on $f$. For use in Chapter 9 we shall in fact provide an algorithm for constructing such a $T$. As in the proof of Proposition 1.25, we shall assume, without loss of generality, that $\mu(X) = 1$.

**Construction** Let $(X, \mathcal{M}, \mu)$ be a nonatomic measure space with $\mu(X) = 1$, and let $f : X \to \mathbb{R}$ be a measurable function. Then $f$ satisfies the finiteness condition (1.1), and hence $f^*$ is well-defined, by (1.2). If $s \in \mathbb{R}$ and $\mu(f = s) > 0$ we shall say that $f$ has a *flat spot* over $s$; see Figure 1.1.

Assume that $f$ has no flat spots. Then $f^*$ is a 1-1 map of $[0, 1]$ onto a set $B$, which equals the real line minus at most countably many half-open intervals. Since $f^*$ is strictly decreasing, its inverse function is a well-defined function from $B$ onto $[0, 1]$, and is Borel measurable since it is decreasing. Moreover,

$f(x) \in B$ for almost every $x \in X$, since $f$ and $f^*$ have the same distribution. Define $T_f\colon X \to \mathbb{R}$ almost everywhere on $X$ by

$$T_f = (f^*)^{-1} \circ f. \tag{1.9}$$

Then obviously,

$$f = f^* \circ T_f, \quad \text{a.e. on} \quad X.$$

The measurability properties of $f$ and $(f^*)^{-1}$ imply that $T = T_f$ is $\mathcal{M} - \mathcal{B}$ measurable. If $E \subset [0,1]$ then $T^{-1}(E) = f^{-1}(f^*(E))$. If in addition $E$ is Borel measurable, then so is $f^*(E)$, and since $\mu(f^{-1}A) = \mathcal{L}((f^*)^{-1}A)$ for all Borel sets $A$, we have $\mu(T^{-1}E) = \mathcal{L}(E)$. This shows that $T_f$ is measure preserving from $(X, \mathcal{M}, \mu)$ onto $([0,1], \mathcal{B}, \mathcal{L})$.

Assume now that $f$ contains flat spots. Let $\psi\colon X \to [0,1]$ be any measurable function on $X$ with no flat spots. For example, $\psi$ could be the function $T$ appearing in the proof of Proposition 1.25. For $i \geq 1$, let $s_i$ run over the set of values of $f$ with $\mu(f = s_i) > 0$, and define

$$X_i = (f = s_i), \quad I_i = (f^* = s_i)\,.$$

Then each $I_i$ is a half-open interval, and $\mu(X_i) = \lambda(I_i)$, since $f$ and $f^*$ are equidistributed. Define

$$X_0 = X \setminus \bigcup_{i=1}^{\infty} X_i, \quad Y = [0,1] \setminus \bigcup_{i=1}^{\infty} I_i.$$

Let $f_0$ be the restriction of $f$ to $X_0$ and $g$ be the restriction of $f^*$ to $Y$. Then $f_0$ and $g$ are equidistributed and have no flat spots. With the notation of (1.9), define

$$T_0 = (T_g)^{-1} \circ T_{f_0}\,.$$

For $i \geq 1$, set

$$\psi_i = \psi|_{X_i}, \quad T_i = T_{\psi_i} + a_i,$$

where $a_i$ is the left endpoint of $I_i$. Define $T_f\colon X \to [0,1]$ a.e. by

$$T_f(x) = T_i(x) \quad \text{for} \quad x \in X_i, \quad i \geq 0. \tag{1.10}$$

**Proposition 1.26** *Let $(X, \mathcal{M}, \mu)$ be a nonatomic measure space with $\mu(X) < \infty$ and $f\colon X \to \mathbb{R}$ be $\mathcal{M}$ measurable. Let $T = T_f$ be given by* (1.9) *if $f$ has no flat spots and by* (1.10) *if $f$ does have flat spots. Then*

*(a) $T$ is a measure preserving map from $(X, \mathcal{M}, \mu)$ onto $([0, \mu(X)], \mathcal{B}, \mathcal{L}^n)$.*
*(b) $f = f^* \circ T$ for almost every $x \in X$.*

*(c) For almost every $x \in X$,*

$$Tx = \mu\big(f > f(x)\big) + \mu\big(f = f(x),\ \psi > \psi(x)\big).$$

*Proof* Assume $\mu(X) = 1$, without loss of generality.

(a) follows from the facts that the $X_i$ partition $X$, the $I_i$ and $Y$ partition $[0, 1]$, and each $T_i$ is measure preserving.

(b) On $X_0$ we have, omitting composition signs,

$$T_0 = T_g^{-1} T_{f_0} = g^{-1} g^* (f_0^*)^{-1} f_0 = g^{-1} f_0. \tag{1.11}$$

The last equality follows from the fact that $f_0$ and $g$ are equidistributed, hence have the same decreasing rearrangement. Since $T_0$ maps $X_0$ into $Y$, it follows that

$$f^* T = g T_0 = f_0 = f, \quad \text{a.e. on } X_0,$$

where the second equality follows from (1.11). This confirms (b) for $x \in X_0$.

For $i \geq 1$ we have $f = s_i$ on $X_i$ and $f^* = s_i$ on $I_i$. Also, $T$ maps $X_i$ onto $I_i$. Thus $f = f^* \circ T = s_i$ on $X_i$, which confirms (b) for almost every $x \in X_i$.

(c) For almost every $x \in X_0$, we have

$$\mu(f > f(x)) = \mu(f^* > f(x)) = \mu(f^* > f^*(Tx)) = Tx.$$

The second equality follows from (b) and the third is true because $Tx$ does not lie in a flat spot.

For $x \in X_i$ with $i \geq 1$, we have $f(x) = s_i$ and $Tx = T_{\psi_i} + a_i$. Since $f^* = s_i$ on $I_i$, we have

$$a_i = \mu(f^* > s_i) = \mu(f > f(x)).$$

Arguing as in the case $x \in X_0$, with $f$ replaced by $\psi_i$ and $X_0$ by $X_i$, we have on $X_i$, $\mu(\psi_i > \psi_i(x)) = T_{\psi_i} x$. Since

$$(\psi_i > \psi_i(x)) = (f = f(x), \psi > \psi(x)),$$

the proof of (c) is finished. □

Proposition 1.26(c) demonstrates in pithy fashion how the map $T_f$ depends only on $f$ and $\psi$. See Figure 1.4.

To get a better understanding of the transformation $T_f$, the reader is advised to return to Examples 1.8 and 1.9. For the simple function in Example 1.8, the reader should write down the $T_f$ obtained from any function $\psi$ on $X$ with no flat spots, while for Example 1.9 he could seek a $T_f$ which is piecewise linear on $(-1, 2)$.

Suppose now that $\mu(X) = \infty$. We will not aim for the most general results along the lines of the two previous propositions, but will confine ourselves to

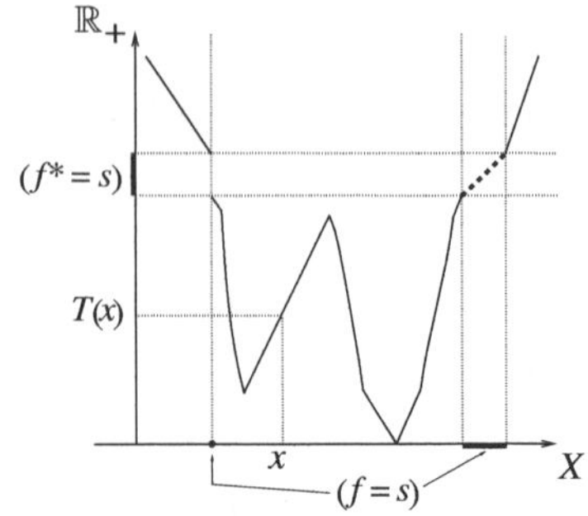


Figure 1.4 A measure preserving transformation $T$ such that $f = f^* \circ T$ for the function $f$ shown in Figure 1.1 (see also Figures 1.2 and 1.3). Since $f^*$ is decreasing, the value of $T$ is large where $f$ is small, and vice versa. Outside of flat spots, $T$ is uniquely determined by $f$ (solid curve). The restriction of $T$ to a flat spot $(f = s)$ is an arbitrary measure preserving map onto the corresponding flat spot of $f^*$ (dashed curve).

stating one proposition of each kind. These propositions cover many situations encountered in practice, and can easily be derived from Propositions 1.25 and 1.26. Details are left to the reader, who is reminded that a measure space $(X, \mathcal{M}, \mu)$ is called $\sigma$-finite if $X$ is the countable union of sets of finite measure. Note also that if $f$ satisfied the finiteness condition (1.1), then $(f > \operatorname{ess\,inf} f)$ is a $\sigma$-finite subset of $X$, even if $X$ is not.

**Proposition 1.27** *Let $(X, \mathcal{M}, \mu)$ be a nonatomic $\sigma$-finite measure space with $\mu(X) = \infty$. Then*

*(a) $X$ contains a family of measurable subsets $\{E_t : t \in [0, \infty)\}$ such that $E_{t_1} \subset E_{t_2}$ whenever $0 \le t_1 \le t_2 < \infty$, $\mu(E_t) = t$ for each $t \in [0, \infty)$, and $\bigcup_{0 \le t < \infty} E_t = X$.*
*(b) There exists a measure preserving transformation $T$ of $(X, \mathcal{M}, \mu)$ onto $([0, \infty), \mathcal{B}, \mathcal{L})$.*

**Proposition 1.28** *Let $(X, \mathcal{M}, \mu)$ be a nonatomic measure space with $\mu(X) = \infty$, and let $f : X \to \mathbb{R}$ be a measurable function satisfying* (1.1). *Set $X_0 = (f > \operatorname{ess\,inf} f)$. Then there exists a measure preserving transformation $T$ of $(X_0, \mathcal{M}, \mu)$ onto $([0, \mu(X_0)), \mathcal{B}, \mathcal{L})$ such that*

$$f = f^* \circ T, \quad \mu - a.e \text{ on } X_0. \tag{1.12}$$

The example $X = \mathbb{R}$, $\mu = \mathcal{L}$, $f(x) = 1/x$ on $(0, \infty)$, $f(x) = 0$ on $(-\infty, 0]$, in which $f^* = f|_{[0,\infty)}$, shows that even if $X$ is $\sigma$-finite we cannot strengthen

Proposition 1.28 to require that (1.12) hold a.e. on $X$. Such a strengthening can be obtained if we permit $T$ to take on the value $\infty$ and define $f^*(\infty) = 0$.

## 1.6 Symmetric Decreasing Rearrangement on $\mathbb{R}^n$

Given $f: \mathbb{R}^n \to \mathbb{R}$ satisfying (1.1), we shall associate to $f$ a rearrangement $f^\#$ with domain $\mathbb{R}^n$ whose level sets $f > t$ are balls centered at the origin. Its definition is in terms of the transformation $T: \mathbb{R}^n \to \mathbb{R}^+$ introduced in §1.4:

$$T(x) = \alpha_n |x|^n, \quad x \in \mathbb{R}^n, \quad \alpha_n = \mathcal{L}^n(\mathbb{B}^n(0,1)). \tag{1.13}$$

**Definition 1.29** Let $f: \mathbb{R}^n \to \mathbb{R}$ be Lebesgue measurable and satisfy (1.1). Then $f^\#: \mathbb{R}^n \to \mathbb{R}$ is the function given by

$$f^\# = f^* \circ T,$$

where $T$ is defined by (1.13).

Note that $f^\#$ is defined at all points of $\mathbb{R}^n$, and not just almost everywhere.

The following proposition lists the principal properties of $f^\#$. Property (b) can be expressed by saying that $f^\#$ is a *radial function*

**Proposition 1.30** *Let $f$ be as in Definition 1.15. Then*

*(a) $f^\#$ and $f$ are equidistributed.*
*(b) $f^\#(x)$ depends only on $|x|$.*
*(c) If $0 \le |x| \le |y|$, then $f^\#(y) \le f^\#(x)$.*
*(d) For each $x \in \mathbb{R}^n$, the function $t \to f^\#(tx)$ is right continuous on $(0,\infty)$.*
*(e) $f^\#$ is the unique function defined everywhere on $\mathbb{R}^n$ which satisfies (a), (b), (c), (d).*

*Proof* In §1.4 we saw that $T$ is $\mathcal{L}^n - \mathcal{L}$ measure preserving. It follows that $f^\#$ and $f^*$ are equidistributed, and hence so are $f^\#$ and $f$. Conclusion (b) follows from the definition of $T$, while (c) and (d) are consequences of the corresponding properties of $f^*$. The proof of (e) is left to the reader. □

In a nutshell, Proposition 1.30 says that $f^\#$ is the unique function on $\mathbb{R}^n$ which is constant on spheres $|x| = r$, is right continuous and decreasing on rays from the origin, and is equidistributed with $f$. When the dimension $n = 1$, we have $f^\#(x) = f^*(2|x|)$.

Functions which are constant on spheres and decreasing on rays from the origin will be called *symmetric decreasing*. Accordingly, we shall call $f^\#$ the

Figure 1.5 A function $f$ on $\mathbb{R}^n$ and its symmetric decreasing rearrangement $f^\#$. Three level surfaces $(f = c)$ are shown. The function in this example is nonnegative, with two local maxima and a saddle point, and vanishes at infinity. By definition, $f^\#$ is the unique radially decreasing, lower semicontinuous function on $\mathbb{R}^n$ that is equidistributed with $f$.

*symmetric decreasing rearrangement of* $f$. Figure 1.5 depicts an example of a nonnegative function $f$ on $\mathbb{R}^2$ and its symmetric decreasing rearrangement $f^\#$.

Next, we consider symmetrization of Lebesgue measurable sets in $\mathbb{R}^n$.

**Definition 1.31** Let $E$ be a Lebesgue measurable subset of $\mathbb{R}^n$. Then $E^\#$ is the open ball centered at the origin of $\mathbb{R}^n$ such that $\mathcal{L}^n(E^\#) = \mathcal{L}^n(E)$.

If $\mathcal{L}^n(E) = 0$ then $E^\#$ is the empty set. If $\mathcal{L}^n(E) = \infty$, then $E^\# = \mathbb{R}^n$. If $0 < \mathcal{L}^n(E) < \infty$, then $E^\# = \mathbb{B}^n(0, R)$, where $R = R(E)$, the *volume radius* of $E$, satisfies

$$\alpha_n R^n = \mathcal{L}^n(E).$$

We call $E^\#$ the *symmetric decreasing rearrangement* of $E$.

One easily verifies the relations

$$(\chi_E)^\# = \chi_{E^\#}$$

and

$$(f^\# > t) = (f > t)^\#, \quad t \in \mathbb{R}.$$

If $E \subset \mathbb{R}^n$ is a Lebesgue measurable set and $f\colon E \to \mathbb{R}$ is a Lebesgue measurable function satisfying (1.1), then one can define $f^\#\colon E^\# \to \mathbb{R}$ by $f^\# = f^* \circ T$, where $T$ is given by (1.13) Later in the book we shall make use of $f^\#$ when $f$ is defined only on $E$. For the present, though, we shall study $f^\#$ only for functions $f$ which are defined a.e. in $\mathbb{R}^n$.

In Chapter 7 we'll take up symmetrization on spheres $\mathbb{S}^n$ and on hyperbolic spaces $\mathbb{H}^n$. The symmetrized objects will again be denoted by $E^\#$ and $f^\#$, and will be referred to as symmetric decreasing rearrangements. When it is necessary to indicate the space with respect to which the process is performed, we shall speak of $\mathbb{R}^n$-symmetric decreasing rearrangement, and so on.

Lamentably, there is no uniformity of terminology or notation in the literature on symmetrization. For example, our functions and sets $f^\#$ and $E^\#$ are called by some authors the *Schwarz symmetrizations* and by others the *point symmetrizations* of $f$ and $E$. They are frequently denoted by $f^*$ and $E^*$. Books and papers employing various notations include Bandle (1980), Kawohl (1985), Lieb and Loss (1997), Pólya and Szegő (1951), Bennett and Sharpley (1988), Alvino et al. (1990), Baernstein (1994), Hayman (1989), and Dubinin (1993).

We derive now a useful representation for symmetric decreasing rearrangements. For simplicity, we will consider only nonnegative $f$, but the procedure can be modified to treat more general $f$ satisfying (1.1).

Consider first nonnegative simple functions $f = \sum_{i=1}^k \alpha_i \chi_{E_i}$, where the $E_i$ are disjoint Lebesgue measurable sets of finite positive measure in $\mathbb{R}^n$, the $\alpha_i$ are distinct, and each $\alpha_i > 0$. Relabel so that the $\alpha_i$ are in descending order: $\alpha_1 > \cdots > \alpha_k > 0$. We are in the situation of Examples 1.2 and 1.8, where the general $X$ there is now $X = \mathbb{R}^n$. Then $f^\# = \alpha_1$ on a ball centered at the origin of measure $\mathcal{L}(E_1)$, $f^\# = \alpha_2$ on a spherical ring with measure $\mu(E_2)$, etc. More precisely, let

$$B_j = \left( \bigcup_{i=1}^{j} E_i \right)^\#, \quad j = 1, \ldots, k.$$

Then $\{B_j\}$ is a strictly increasing sequence of open balls centered at the origin, and we have

$$f^\# = \sum_{j=1}^{k} \alpha_j \chi_{B_j \setminus B_{j-1}} = \sum_{j=1}^{k} (\alpha_j - \alpha_{j+1}) \chi_{B_j}. \tag{1.14}$$

In the first sum $B_0$ denotes the empty set, and in the second sum $\alpha_{k+1} \equiv 0$. The openness of the balls $B_j$ insures right continuity of $f^*$ on rays through the origin. The reader may check that $f^\#$ as given by (1.14) and $f^*$ as given by (1.3) do indeed satisfy the defining relation $f^\# = f^* \circ T$.

Assume now, more generally, that $f\colon \mathbb{R}^n \to \mathbb{R}$ is nonnegative and Lebesgue measurable. For $t > 0$, write $E(t,f) = (f > t)$.

**Proposition 1.32** (Layer cake representation) *Let $f\colon \mathbb{R}^n \to \mathbb{R}$ be nonnegative and Lebesgue measurable. Then, at all $x$ at which $f(x)$ is defined,*

$$f(x) = \int_0^\infty \chi_{E(t,f)}(x)\, dt. \tag{1.15}$$

*Proof* Use the identity

$$\chi_{E(t,f)}(x) = \chi_{[0,f(x))}(t), \quad t \geq 0. \qquad \square$$

Since $E(t,f^\#) = (E(t,f))^\#$, Proposition 1.32 implies, for nonnegative measurable $f$ satisfying (1.1), the representation

$$f^\#(x) = \int_0^\infty \chi_{(E(t,f))^\#}(x)\,dt, \quad x \in \mathbb{R}^n. \tag{1.16}$$

In Lieb and Loss (1997), (1.16) is taken as the definition of $f^\#$. It is not difficult to show directly that the function on the right-hand side of (1.16) satisfies properties (a)–(d) of Proposition 1.30. When $f$ is simple, the integral in (1.16) reduces to the second sum in (1.14).

Sometimes in manipulations involving the layer cake representation one must confront measurability questions. Define $F\colon \mathbb{R}^n \times \mathbb{R}^+ \to \mathbb{R}$ by $F(x,t) = \chi_{E(t,f)}(x) = \chi_{[0,f(x))}(t)$. If one of the variables $t$ or $x$ is held fixed, then $F$ is clearly a Lebesgue measurable function of the other variable. To see how $F$ behaves as a function on the product space, write $F = \chi_G$, where $G = \{(x,t) \in \mathbb{R}^n \times \mathbb{R}^+ : f(x) - t > 0\}$. Since $f(x) - t$ is a $\mathcal{L}^n \times \mathcal{L}$-measurable function, it follows that $G$ is a $\mathcal{L}^n \times \mathcal{L}$-measurable set, and hence $F$ is a $\mathcal{L}^n \times \mathcal{L}$-measurable function on $\mathbb{R}^n \times \mathbb{R}^+$.

We introduced the layer cake representation in the context of functions on $\mathbb{R}^n$. But it works similarly on more general measure spaces $(X, \mathcal{M}, \mu)$. Suppose that $f\colon X \to \mathbb{R}^+$ is measurable. Then (1.15) holds for $x \in X$ at which $f(x)$ is defined. Moreover, if $f$ satisfies (1.1), then its decreasing rearrangement $f^*$ is given by

$$f^*(x) = \int_0^\infty \chi_{(E(t,f))^*}(x)\,dt, \quad x \in \mathbb{R}^+,$$

where $(E(t,f))^*$ denotes the interval $[0, \mu(E(t,f)))$.

## 1.7 Polarization on $\mathbb{R}^n$

Let $H$ be an affine hyperplane in $\mathbb{R}^n$. That is, $H = a + M$, where $a \in \mathbb{R}^n$ and $M$ is an $(n-1)$-dimensional subspace of $\mathbb{R}^n$. The complement of $H$ in $\mathbb{R}^n$ is the union of two open halfspaces; denote one of them by $H^+$, the other by $H^-$. Let

$$\rho_H\colon \mathbb{R}^n \to \mathbb{R}^n$$

denote reflection in $H$. We will write $\rho = \rho_H$ when the context is clear. *Reflections* are isometries of $\mathbb{R}^n$. That is

$$|\rho x - \rho y| = |x - y|, \quad x, y \in \mathbb{R}^n.$$

In general, a map $T\colon \mathbb{R}^n \to \mathbb{R}^n$ is an *isometry* of $\mathbb{R}^n$ onto itself if and only if $T$ has the form

$$T(x) = a + Sx, \quad x \in \mathbb{R}^n,$$

for some $a \in \mathbb{R}^n$ and some $S$ in the orthogonal group $O(n)$. Thus, a map is an isometry of $\mathbb{R}^n$ if and only if it is the composition of translations, reflections and (sense preserving) rotations. As discussed in Example 1.22, isometries are $\mathcal{L}^n$-measure preserving maps of $\mathbb{R}^n$ onto itself. Reflections $\rho$ are special isometries, with the involutory property $\rho \circ \rho = id$.

Let $\mathcal{H}(\mathbb{R}^n)$ denote the set of all affine hyperplanes in $\mathbb{R}^n$. The proposition below records a few additional elementary facts about reflections and hyperplanes.

**Proposition 1.33**

*(a) Let $H_1, H_2 \in \mathcal{H}(\mathbb{R}^n)$. There exists an isometry $T$ of $\mathbb{R}^n$ such that $TH_1 = H_2$.*
*(b) Let $H \in \mathcal{H}(\mathbb{R}^n)$, and suppose that $x \in H$ and $y \in \mathbb{R}^n$. Then $|x - y| = |x - \rho_H(y)|$.*
*(c) Let $H \in \mathcal{H}(\mathbb{R}^n)$, and suppose that $x$ and $y$ are both in $H^+$ or both in $H^-$. Then $|x - y| < |x - \rho_H(y)|$.*
*(d) Let $x$ and $y$ be distinct points of $\mathbb{R}^n$. There exists a unique $H \in \mathcal{H}(\mathbb{R}^n)$ such that $\rho_H(x) = y$.*

*Proof* (a): If $H_1$ and $H_2$ are parallel, then $H_1$ can be mapped onto $H_2$ by a translation. If $H_1$ and $H_2$ are not parallel then they have a point in common and there is a translation which maps $H_1$ and $H_2$ onto subspaces $M_1$ and $M_2$. Then $M_1$ can be mapped onto $M_2$ by a rotation.

(b) and (c): If $H = \{x \in \mathbb{R}^n : x \cdot e_n = 0\}$ the results are easy. Using (a), the case of general $H$ can be reduced to that of the special $H$

(d): Form the line segment with endpoints $x$ and $y$. Let $H$ be the affine hyperplane orthogonal to this segment which passes through its midpoint. Then $\rho_H(x) = y$. Verification of the uniqueness statement is left to the reader. □

We define now one of the central concepts in our approach to symmetrization.

**Definition 1.34** Let $f : \mathbb{R}^n \to \mathbb{R}$ and $H \in \mathcal{H}(\mathbb{R}^n)$. Then $f_H : \mathbb{R}^n \to \mathbb{R}^+$ is the function defined by

$$f_H(x) = \begin{cases} \max(f(x), f(\rho x)), & x \in H^+, \\ \min(f(x), f(\rho x)), & x \in H^-, \\ f(x), & x \in H. \end{cases}$$

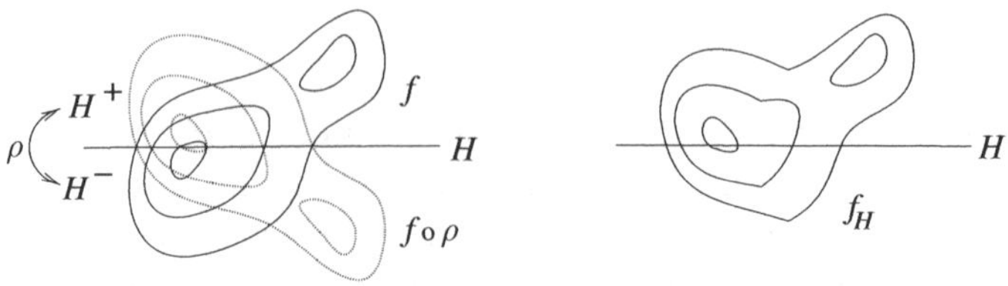


Figure 1.6 A function $f$ on $\mathbb{R}^n$ (left, solid curves), its reflection $f \circ \rho$ at a hyperplane $H$ (left, dotted curves) and its polarization $f_H$ (right). (The function is the same as in Figure 1.5.) By definition, $f_H$ is given by $\max\{f, f \circ \rho\}$ on $H^+$ and by $\min\{f, f \circ \rho\}$ on $H^-$. Thus, $f_H$ is equidistributed with $f$, and $|\nabla f_H|$ is equidistributed with $|\nabla f|$.

We call $f_H$ the *polarization* of $f$ with respect to $H$. In fact, $f_H$ depends also on which complementary halfspace is designated $H^+$, but this ambiguity will usually not cause any difficulties. Figure 1.6 shows an example of a function $f\colon \mathbb{R}^2 \to \mathbb{R}$ and one of its polarizations. When $n = 1$, $H$ is a single point.

For $E \subset \mathbb{R}^n$ the polarization of $\chi_E$ is also the characteristic function of some set. This set, which we denote by $E_H$, is called the polarization of $E$ with respect to $H$. One way to think of $E_H$ is like this: Take $x, y \in \mathbb{R}^n$ with $\rho_H x = y$. If both $x$ and $y$ are in $E$ then both will be in $E_H$. If neither is in $E$ then neither will be in $E_H$. If one is in $E$ but the other is not, then $x$ and $y$ are not in $H$; the member of the pair in $H^+$ will belong to $E_H$ but the member in $H^-$ will not. Note in particular that $E \cap H = E_H \cap H$.

With the situation of Definition 1.34, define a partition $\mathbb{R}^n = A \cup B$ of $\mathbb{R}^n$ as follows:

$$A = H \cup \{x \in H^+ : f(x) \geq f(\rho x)\} \cup \{x \in H^- : f(x) \leq f(\rho(x))\},$$
$$B = \mathbb{R}^n \setminus A.$$

Thus, a point $x \in \mathbb{R}^n$ belongs to $A$ if and only if $f_H(x) = f(x)$, while $x \in B$ if and only if $f_H(x) = f(\rho x)$ and $f(\rho x) \neq f(x)$. Notice that $H \subset A$, and that both $A$ and $B$ are mapped onto themselves by $\rho$.

Define $T_{f,H} = T\colon \mathbb{R}^n \to \mathbb{R}^n$ by $T(x) = x$ if $x \in A$, $T(x) = \rho x$ if $x \in B$. Then

$$f_H = f \circ T. \tag{1.17}$$

**Proposition 1.35** *If $f\colon \mathbb{R}^n \to \mathbb{R}$ is Lebesgue measurable then $T$ is $(\mathbb{R}^n, \mathcal{M}^n, \mathcal{L}^n) \to (\mathbb{R}^n, \mathcal{M}^n, \mathcal{L}^n)$ measure preserving. Consequently, $f_H$ is Lebesgue measurable, and if $f$ satisfies* (1.1), *then $f_H$ is a rearrangement of $f$.*

*Proof* $T|_A$ is the identity, and $T|_B = \rho$. Moreover, $T$ maps each of $A$ and $B$ onto itself. The identity and $\rho$ are each measure preserving maps of

$(\mathbb{R}^n, \mathcal{M}^n, \mathcal{L}^n)$ and the sets $A$ and $B$ are Lebesgue measurable when $f$ is. Thus, $T|_A$ is a measure preserving map of $A$ onto itself, and $T|_B$ is a measure preserving map of $B$ onto itself. Let $E \subset \mathbb{R}^n$ be Lebesgue measurable. Then

$$T^{-1}(E) = T^{-1}(E \cap A) \cup T^{-1}(E \cap B).$$

The sets on the right-hand side are disjoint, each is Lebesgue measurable, and the sum of their Lebesgue measures equals $\mathcal{L}^n(E)$. Thus, $T$ is measure preserving on $(\mathbb{R}^n, \mathcal{M}^n, \mathcal{L}^n)$. The second statement of the proposition follows from (1.17). □

The *diameter* $\operatorname{diam} E$ of a set $E \subset \mathbb{R}^n$ is defined by

$$\operatorname{diam} E = \sup_{x,y \in E} |x - y|.$$

The *modulus of continuity* $\omega(t,f)$ of a function $f : \mathbb{R}^n \to \mathbb{R}$ is defined by

$$\omega(t,f) = \sup\{|f(x) - f(y)| : |x - y| \le t\}.$$

The function is uniformly continuous on $\mathbb{R}^n$ if and only if $\lim_{t \to 0} \omega(t,f) = 0$.

The next two propositions show that polarization has a smoothing effect. The propositions are valid for all real-valued functions on $\mathbb{R}^n$, measurable or not.

**Proposition 1.36** *For $f : \mathbb{R}^n \to \mathbb{R}$ and $H \in \mathcal{H}(\mathbb{R}^n)$ holds*

$$\operatorname{diam}(f_H > t) \le \operatorname{diam}(f > t), \quad t \in \mathbb{R}.$$

**Proposition 1.37** *For $f : \mathbb{R}^n \to \mathbb{R}$ and $H \in \mathcal{H}(\mathbb{R}^n)$ holds*

$$\omega(t, f_H) \le \omega(t,f), \quad t \in \mathbb{R}^+.$$

*Proof of Proposition 1.36* It suffices to show that if $x, y \in \mathbb{R}^n$ satisfy $f_H(x) > t$ and $f_H(y) > t$, then there exist $z, w \in \mathbb{R}^n$ such that

$$f(z) > t, \quad f(w) > t, \quad \text{and} \quad |z - w| \ge |x - y|. \tag{1.18}$$

The proof of this statement proceeds by cases and subcases.

**Case 1.** $x, y \in H^+ \cup H$. There are four possibilities:

(i) $f(x) > t$ and $f(y) > t$.
(ii) $f(x) > t$ and $f(\rho y) > t$.
(iii) $f(\rho x) > t$ and $f(y) > t$.
(iv) $f(\rho x) > t$ and $f(\rho y) > t$.

If (i) holds, take $z = x$ and $w = y$. If (ii), take $z = x$, $w = \rho y$. If (iii) take $z = \rho x$, $w = y$, and if (iv), take $z = \rho x$, $w = \rho y$. Using Proposition 1.33(c), we see that (1.18) holds in each subcase.

**Case 2.** $x, y \in H^- \cup H$. The analysis works as for Case 1.

**Case 3.** $x$ and $y$ lie in different complementary open halfspaces. Say $x \in H^+$, $y \in H^-$. There are two possibilities:

(i) $f(x) = f_H(x)$.
(ii) $f(x) = f_H(\rho x)$.

In subcase (i), take $z = x$ and $w = y$. Since $\rho y \in H^+$, we have $f_H(\rho y) \geq f_H(y) > t$. Thus $\min(f_H(y), f_H(\rho y)) > t$, and so we deduce $\min(f(y), f(\rho y)) > t$. In particular, $f(y) > t$, and (1.18) holds.

In subcase (ii), take $z = \rho x$ and $w = \rho y$. Then $f(z) = f_H(x) > t$. Also, $f(w) > t$, by the argument in (i). Thus, (1.18) holds. □

*Proof of Proposition 1.37* It suffices to show: For each $x, y \in \mathbb{R}^n$, there exists $z, w \in \mathbb{R}^n$ such that

$$|f_H(x) - f_H(y)| \leq |f(z) - f(w)| \quad \text{and} \quad |z - w| \leq |x - y|. \tag{1.19}$$

Again, the proof proceeds by cases and subcases.

**Case 1.** $x, y \in H^+ \cup H$. There are four possibilities.

(i) $f_H(x) = f(x)$ and $f_H(y) = f(y)$.
(ii) $f_H(x) = f(\rho x)$ and $f_H(y) = f(y)$.
(iii) $f_H(x) = f(x)$ and $f_H(y) = f(\rho y)$.
(iv) $f_H(x) = f(\rho x)$ and $f_H(y) = f(\rho y)$.

If (i), take $z = x$, $w = y$. If (iv), take $z = \rho x$, $w = \rho y$.

To handle (ii), we split the argument into two subsubcases. Suppose first that, in addition to the assumptions of Case 1(ii), we have $f_H(x) \geq f_H(y)$. Then

$$|f_H(x) - f_H(y)| = f(\rho x) - f(y) \leq f(\rho x) - f(\rho y).$$

The inequality comes from the assumptions that $y \in H^+ \cup H$ and $f_H(y) = f(y)$. Then (1.19) is satisfied if we take $z = \rho x$ and $w = \rho y$.

If $f_H(x) < f_H(y)$ and the assumptions of Case 1(ii) are satisfied, then

$$|f_H(x) - f_H(y)| = f(y) - f(\rho x) \leq f(y) - f(x).$$

The inequality comes from the assumptions that $x \in H^+ \cup H$ and $f_H(x) = f(\rho x)$. Then (1.19) is satisfied if we take $z = x$ and $w = y$.

**Case 2.** $x, y \in H^- \cup H$. The analysis is like that of Case 1.

**Case 3.** The points $x$ and $y$ lie in different complementary open halfspaces. Say $x \in H^+$, $y \in H^-$. Split into the same subcases (i)–(iv) as for Case 1. Choose $z$ and $w$ as follows: (i) $z = x$, $w = y$. (ii) $z = \rho x$, $w = y$. (iii) $z = x$, $w = \rho y$. (iv) $z = \rho x$, $w = \rho y$.

Since $x$ and $y$ are on different sides of $H$, we have $|z - w| \le |x - y|$ in each subcase. Also, $f(z) = f_H(x)$ and $f(w) = f_H(y)$ in each subcase, so (1.19) is satisfied. □

The next lemma concerns the relation between polarization and symmetric decreasing rearrangement. If a function $f$ agrees with its s.d.r. $f^\#$, then clearly $f = f_H$ for every hyperplane such that $H_+$ contains the origin. The second part of the lemma shows that the converse also holds. The first part will be used in the proof of Theorem 8.11.

**Lemma 1.38** *Let $f$ be a nonnegative measurable function on $\mathbb{R}^n$ satisfying the usual finiteness condition* (1.1). *Suppose that $f$ is not $\mathcal{L}^n$-a.e. equal to its s.d.r. $f^\#$.*

*(a) Let*

$$A_1 = \{(x, y) \in \mathbb{R}^n \times \mathbb{R}^n : f(x) < f(y) \quad \textit{and} \quad |x| < |y|\}.$$

*Then $\mathcal{L}^{2n}(A_1) > 0$.*

*(b) There exists a hyperplane $H \in \mathcal{H}(\mathbb{R}^n)$ with $0 \in H_+$ such that $f_H$ is not $\mathcal{L}^n$-a.e. equal to $f$.*

*Proof* (a) The set $\{x \in \mathbb{R}^n : f(x) \ne f^\#(x)\}$ has positive measure. Since $f$ and $f^\#$ have the same distribution, each set $(f > f^\#)$ and $(f < f^\#)$ also has positive measure (as explained later in the proof of Theorem 2.15b). By Evans and Gariepy (1992, p. 47), almost every point of $\mathbb{R}^n$ is a point of approximate continuity for $f$ and for $f^\#$. For $f$, a point $x_0$ of approximate continuity is a point such that: given $\epsilon > 0$ there exists $\delta > 0$ and a measurable set $E \subset \mathbb{B}^n(x_0, \delta)$ such that $\mathcal{L}^n(E) > (1 - \epsilon)\alpha_n \delta^n$ and

$$|f(x) - f(x_0)| < \epsilon, \quad \forall x \in E.$$

Fix $x_0$ such that $f(x_0) < f^\#(x_0)$ and $x_0$ is a point of approximate continuity of both $f$ and $f^\#$. Set $f^\#(x_0) = b$. Let $B = (f^\# \ge b)$. Then $B$ is a ball centered at the origin, of radius $R$, say, which is closed if $f^\#$ is continuous at $R$ and open if $f^\#$ has a jump at $R$. In any case, $x_0 \in B$, and $\mathcal{L}^n(f \ge b) = \mathcal{L}^n(B)$. The approximate continuity of $f$ at $x_0$ implies that $f(x) < b$ for all $x$ in some set $E_1$ of positive measure contained in $B$. Since $\mathcal{L}^n(f \ge b) = \mathcal{L}^n(f^\# \ge b) = \mathcal{L}^n(B)$, there is a set $E_2 \subset (|x| > R)$ of positive measure such that $f \ge b$ on $E_2$. Then $f(x) < f(y)$ for all $(x, y) \in E_1 \times E_2$. Also, $|x| \le R < |y|$ for $(x, y) \in E_1 \times E_2$. Thus, we have $\mathcal{L}^{2n}(A_1) \ge \mathcal{L}^{2n}(E_1 \times E_2) > 0$.

(b) Let $x_0$ and $y_0$ be as in the proof of (a). By Proposition 1.16.4, there exists a unique $H \in \mathcal{H}(\mathbb{R}^n)$ such that $\rho_H(x_0) = y_0$. Let $H_+$ t be the half space containing $x_0$. Since $|x_0| < |y_0|$, the origin is contained in $H_+$. The approximate continuity of $f$ at $x_0$ and $y_0$ implies that there exists a set $E_2$ of positive measure containing $x_0$ such that $f(x) < f(\rho_H x)$ for all $x \in E_2$. By definition, $f_H(x) = f(\rho_H x) > f(x)$ for all $x \in E_2$. □

## 1.8 Convergence Theorems for Rearrangements

Suppose that $\{f_n\}$ is a sequence of measurable real valued functions defined on a measure space $(X, \mathcal{M}, \mu)$, and that $\{f_n\}$ converges to a function $f$ in some sense. Is it true that $\{f_n^*\}$ converges to $f^*$ on $[0, \mu(X)]$? In this section we will prove two affirmative results and display some counterexamples. Further results will appear later in the book.

The first result concerns monotone convergence.

**Proposition 1.39** *Let $(X, \mathcal{M}, \mu)$ be a measure space and $f_n \colon X \to \mathbb{R}$ be measurable for $n = 1, 2, \ldots$. Suppose that $f_n \nearrow f$ a.e. on $X$, that each $f_n$ satisfies* (1.1), *and that $f$ satisfies* (1.1). *Then $f_n^* \nearrow f^*$ at each point of $[0, \mu(X))$.*

The example $X = [0, \infty)$, $f_n = \chi_{[0,1]\cup[2,3]\cup\cdots\cup[2n,2n+1]}$ shows that $\{f_n\}$ can converge upward to a function $f$ which does not satisfy (1.1), even though each $f_n$ does. Concerning the endpoint $\mu(X)$, the proof will show that

$$\lim_{n\to\infty} f_n^*(\mu(X)) \le f^*(\mu(X)). \tag{1.20}$$

The example $X = [0, 1]$, $f_n = \chi_{[0,1-\frac{1}{n}]}$ shows that strict inequality can hold in (1.20).

*Proof of Proposition 1.39* First, we record a simple inequality for distribution functions and decreasing rearrangements. Let $g$ and $h$ be measurable functions on $X$ satisfying $g \le h$ a.e. Then $\lambda_g(t) \le \lambda_h(t)$. Suppose that $g$ and $h$ satisfy (1.1). Then from formula (1.2) for $g^*$ in terms of $\lambda_g$, we deduce that

$$g^*(x) \le h^*(x), \quad x \in [0, \mu(X)]. \tag{1.21}$$

Returning now to the proof of Proposition 1.39, by changing the $f_n$ and $f$ on a set of measure zero, if necessary, we may assume that $f_n \nearrow f$ at every point of $X$. From (1.21), it follows that the sequence $\{f_n^*\}$ is increasing on $[0, \mu(X)]$. Let $F(x) = \lim_{n\to\infty} f_n^*(x)$. Since $f_n \le f$ on $X$, we have $f_n^* \le f^*$ everywhere on $[0, \mu(X)]$. Hence $F \le f^*$ on $[0, \mu(X)]$.

Take $0 \leq x < \mu(X)$. Suppose that $F(x) < f^*(x)$. Take $s$ with $F(x) < s < f^*(x)$. Since $F$ and $f^*$ are decreasing on $[0, \mu(X)]$, and $f^*$ is right continuous, we have $(f > s) \subset [0, x)$ and $[0, x + \epsilon) \subset (f^* > s)$ for some $\epsilon > 0$. Thus $\lambda_F(s) \leq x$ and $\lambda_{f^*}(s) > x$, so that

$$\lambda_F(s) < \lambda_{f^*}(s). \tag{1.22}$$

On the other hand, since $f_n^* \leq F$ for each $n$, and $\lambda_{f_n^*} = \lambda_{f_n}$, we have

$$\lim_{n\to\infty} \lambda_{f_n}(t) \leq \lambda_F(t), \quad t \in \mathbb{R}. \tag{1.23}$$

But by Proposition 1.5(f), the limit on the left-hand side of (1.23) equals $\lambda_f(t)$. Since $\lambda_f = \lambda_{f^*}$, the choice $t = s$ contradicts (1.22). We conclude that $F(x) \geq f^*(x)$, and hence that $F = f^*$ on $[0, \mu(X))$. This proves Proposition 1.39. □

The second result concerns convergence in measure. The result for decreasing rearrangements will be preceded by a convergence result for distribution functions.

**Proposition 1.40** *Let $(X, \mathcal{M}, \mu)$ be a measure space and $f_n \colon X \to \mathbb{R}$ be measurable for $n = 1, 2, \ldots$. Suppose that $\{f_n\}$ converges in measure on $X$ to $f$. Then*

$$\lambda_f(t) \leq \liminf_{n\to\infty} \lambda_{f_n}(t) \leq \limsup_{n\to\infty} \lambda_{f_n}(t) \leq \lambda_f(t-), \quad t \in \mathbb{R}. \tag{1.24}$$

From (1.24), it follows that $\lim_{n\to\infty} \lambda_{f_n}(t) = \lambda_f(t)$ at each continuity point $t \in \mathbb{R}$ of $\lambda_f$, and hence for almost every $t \in \mathbb{R}$.

*Proof of Proposition 1.40* Take $t \in \mathbb{R}$ and $\epsilon > 0$. Then

$$(f_n > t + \epsilon) \subset (f > t) \cup (|f - f_n| > \epsilon),$$

hence $\lambda_{f_n}(t + \epsilon) \leq \lambda_f(t) + \mu(|f - f_n| > \epsilon)$. Letting $n \to \infty$, we obtain

$$\limsup_{n\to\infty} \lambda_{f_n}(t + \epsilon) \leq \lambda_f(t).$$

Replace $t$ by $t - \epsilon$, then let $\epsilon \to 0$. This gives the third inequality in (1.24). To obtain the first inequality in (1.24), use a similar argument starting with $(f > t + \epsilon) \subset (f_n > t) \cup (|f - f_n| > \epsilon)$, along with the right continuity of $\lambda_f$. □

**Proposition 1.41** *Let $(X, \mathcal{M}, \mu)$ be a measure space and $f_n \colon X \to \mathbb{R}$ be measurable for $n = 1, 2, \ldots$. Suppose that $\{f_n\}$ converges in measure on $X$ to $f$. Suppose also that each $f_n$ satisfies* (1.1). *Then $f$ satisfies* (1.1), *and*

$$f^*(x) \leq \liminf_{n\to\infty} f_n^*(x) \leq \limsup_{n\to\infty} f_n^*(x) \leq f^*(x-), \quad x \in (0, \mu(X)). \tag{1.25}$$

*Moreover, if* $\mu(X) < \infty$ *or if* $\operatorname{ess\,inf} f > -\infty$, *then* $\{f_n^*\}$ *converges in measure on* $[0, \mu(X)]$ *to* $f^*$.

From (1.25), it follows that $f_n^*(x) \to f^*(x)$ at continuity points of $f^*$ in $(0, \mu(X))$, and hence a.e. on $[0, \mu(X)]$. If $\mu(X) = \infty$ and $f^*(\infty) = -\infty$ then $\{f_n^*\}$ need not converge in measure. An example is constructed after the proof.

In (1.25), $x$ is confined to the open interval $(0, \mu(X))$. Let us investigate what happens at endpoints. When $x = 0$ we have $f^*(0) = \operatorname{ess\,sup} f$. It is easy to see that convergence in measure implies that

$$\operatorname{ess\,sup} f \leq \liminf_{n\to\infty} \operatorname{ess\,sup} f_n^*,$$

so that the first inequality in (1.25) holds. Since $f^*(0-)$ is not defined, there is no third inequality. When $x = \mu(X)$ we have $f^*(\mu(X)) = f^*(\mu(X)-) = \operatorname{ess\,inf} f$. The third inequality follows from convergence in measure, but the first inequality can fail. Consider the following example:

$$X = [0, 1], \quad f_n = \chi_{[0,\frac{1}{n}]} + \frac{1}{2}\chi_{[\frac{1}{n},1-\frac{1}{n}]}.$$

Then $f_n = f_n^*$ and $\{f_n\}$ converges in measure to the constant $1/2$. For $x = 0$ strict inequality holds in the first inequality of (1.25) and for $x = \mu(X) = 1$ strict inequality holds in the third inequality.

For more about the case $\mu(X) = \infty$, see (1.27) below.

*Proof of Proposition 1.41* Let $a = \operatorname{ess\,inf} f$. Then $a < \infty$. If $\lambda_f(a) < \infty$ then (1.1) holds for $f$. Suppose that $\lambda_f(a) = \infty$. Take $s$ and $t$ with $a < s < t$. The convergence in measure implies that the set $|f - f_n| > t - s$ has finite measure for all sufficiently large $n$. If $\lambda_f(t)$ were infinite, then $\lambda_{f_n}(s)$ would be infinite for all sufficiently large $n$. But as remarked above, the convergence in measure also implies $\limsup_{n\to\infty} \operatorname{ess\,inf} f_n \leq \operatorname{ess\,inf} f = a < s$. These inequalities contradict the assumption that (1.1) holds for each $f_n$. Thus, $\lambda_f(t) < \infty$ for all $t > a$, and hence (1.1) holds for $f$.

Next, we prove (1.25). Take $x \in (0, \mu(X))$ and $\delta > 0$. Suppose that $f^*(x) > f_n^*(x) + \delta$ for infinitely many $n$. Take $t$ with $f^*(x) - \delta < t < f^*(x)$. Then for these values of $n$,

$$f_n^*(x) < t < f^*(x).$$

Using right continuity of $f^*$, the second inequality implies $\lambda_{f^*}(t) > x$. The first inequality implies $\lambda_{f_n^*}(t) \leq x$ for infinitely many $n$. These inequalities contradict the first inequality in (1.24). Thus, the first inequality in (1.25) is true. If the third inequality in (1.25) were false, there would exist $0 < y < x$, $\delta > 0$ and infinitely many $n$ such that $f^*(y) + \delta < f_n^*(x)$. Take $s, t$ with

$f^*(y) < s < t < f^*(y) + \delta$. Then $\lambda_{f^*}(s) \leq y$. Also, $f_n^*(x) > t$, so that $\lambda_{f_n^*}(t) > x$, for infinitely many $n$. Letting $s \nearrow t$, we obtain a contradiction to the third inequality in (1.24). Thus, the third inequality in (1.25) is also true.

To prove the statement about convergence in measure, take $0 < x_1 < x_2 < \mu(X)$. Since $f^*(x_1-) < \infty$ and $f^*(x_2) > -\infty$, it follows from (1.25) that the sequences $\{f_n^*(x_1)\}$ and $\{f_n^*(x_2)\}$ are bounded. Hence, $\{f_n^*\}$ is uniformly bounded on $[x_1, x_2]$. Since, by (1.25), $\{f_n^*\}$ converges to $f^*$ a.e., the dominated convergence theorem implies that $\{f_n^*\}$ converges to $f^*$ in $L^1[x_1, x_2]$, and hence in measure on $[x_1, x_2]$. From this, it follows that the convergence in measure in fact occurs on every interval $[0, M]$ with $0 < M < \infty$.

If $\mu(X) < \infty$, we just saw that $\{f_n^*\}$ converges in measure to $f^*$ on $[0, \mu(X)]$.

Suppose that $\mu(X) = \infty$. Then, by our hypothesis, ess inf $f = f^*(\infty) > -\infty$. As noted previously, the convergence in measure implies

$$\limsup_{n\to\infty} f_n^*(\infty) \leq f^*(\infty).$$

We claim that the complementary inequality also holds:

$$f^*(\infty) \leq \liminf_{n\to\infty} f_n^*(\infty). \tag{1.26}$$

Suppose that (1.26) is false. Then there exists $\delta > 0$ and infinitely many $n$ with $f_n^*(\infty) + \delta < f^*(\infty)$. Take $s, t$ with $f_n^*(\infty) < s < t < f^*(\infty)$. Then $\lambda_f(t) = \infty$. The argument in the first paragraph of the proof shows that $\lambda_{f_n}(s) = \infty$ for all sufficiently large $n$. For the $n$ with $s > \operatorname{ess\,inf} f_n$, the functions $f_n$ do not satisfy condition (1.1), contrary to hypothesis. Thus (1.26) is true, and we deduce that

$$\lim_{n\to\infty} f_n^*(\infty) = f^*(\infty). \tag{1.27}$$

Given $\epsilon > 0$, take $M \in (0, \infty)$ such that $f^*(M-) < f^*(\infty) + \epsilon$. Using (1.25), then (1.27), we see there exists $n_0$ such that $n \geq n_0$ implies $f^*(M) - \epsilon < f_n^*(M) < f^*(M-) + \epsilon$ and $|f_n^*(\infty) - f^*(\infty)| < \epsilon$. It follows that if $x \geq M$ and $n \geq n_0$, then

$$f^*(\infty) < f_n^*(\infty) + \epsilon \leq f_n^*(x) + \epsilon \leq f_n^*(M) + \epsilon < f^*(M-) + 2\epsilon < f^*(\infty) + 3\epsilon.$$

Hence, $|f_n^*(x) - f^*(x)| < 2\epsilon$ when $x \geq M$ and $n$ is sufficiently large. Since $f_n^* \to f^*$ in measure on $[0, M]$, it follows that $f_n^* \to f^*$ in measure on $[0, \infty)$. □

**Example 1.42** Let $f \in C^1([0, \infty))$ be concave and decreasing, with $f(0) = 0$ and $\lim_{x\to\infty} f'(x) = -\infty$. Then $f^* = f$. Let $\{x_n\}$ be a sequence in $\mathbb{R}^+$ with $\lim_{n\to\infty} x_n = \infty$. Take a sequence $\{\delta_n\}$ in $\mathbb{R}^+$ with $\lim_{n\to\infty} \delta_n = 0$ and $\lim_{n\to\infty} \delta_n f'(\frac{1}{2}x_n) = -\infty$. Set $I_n = (x_n, x_n + \delta_n)$. For $n \geq 1$, define $f_n$ by

$f_n = f$ on $[0,\infty) \setminus I_n$, $f_n = 1$ on $I_n$. Then each $f_n$ satisfies (1.1), and $\{f_n\}$ converges in measure to $f$ on $[0,\infty)$.

We have

$$\begin{aligned} f_n^*(x) &= 1, \quad 0 \le x < \delta_n, \\ &= f(x-\delta_n), \quad \delta_n \le x < x_n + \delta_n, \\ &= f(x), \quad x_n + \delta_n \le x < \infty. \end{aligned}$$

Set $J_n = (\frac{1}{2}x_n + \delta_n, x_n)$. For $x \in J_n$,

$$f^*(x) - f_n^*(x) = f(x) - f(x-\delta_n) \le \delta_n f'(\frac{1}{2}x_n).$$

Thus, $\lim_{n\to\infty} \inf_{J_n} |f^* - f_n^*| = \infty$, and also $\lim_{n\to\infty} \mathcal{L}(J_n) = \infty$. It follows that $\{f_n^*\}$ does not converge to $f^*$ in measure on $[0,\infty)$. Since $\{f_n^*\}$ does converge to $f^*$ in measure on each compact subinterval of $[0,\infty)$, it follows that $\{f_n^*\}$ does not converge in measure to any function on the complete interval $[0,\infty)$.

Propositions 1.39 and 1.41, about decreasing rearrangements of functions on general measure spaces, immediately imply corresponding convergence results for symmetric decreasing rearrangements of functions on $\mathbb{R}^n$. Recall that if $f: \mathbb{R}^n \to \mathbb{R}$ is measurable and satisfies (1.1), then its symmetric decreasing rearrangement $f^\#: \mathbb{R}^n \to \mathbb{R}$ is defined by $f^\#(x) = f^*(\alpha_n |x|^n)$, where $\alpha_n$ is the volume of the unit ball in $\mathbb{R}^n$. For ease of future reference, we state the convergence results in the following proposition.

**Proposition 1.43** *Let $\{f_n\}$ be a sequence of real valued measurable functions on $\mathbb{R}^n$, each of which satisfies* (1.1).

*(a) If $f_n \nearrow f$ a.e. on $\mathbb{R}^n$ and $f$ satisfies* (1.1)*, then $f_n^\# \nearrow f^\#$ at each point of $\mathbb{R}^n$.*
*(b) If $\{f_n\}$ converges in measure to $f$ on $\mathbb{R}^n$, then $f$ satisfies* (1.1)*, and for each $x \in \mathbb{R}^n$,*

$$f^\#(x) \le \liminf_{n\to\infty} f_n^\#(x) \le \limsup_{n\to\infty} f_n^\#(x) \le \lim_{t\to 1-} f^\#(tx). \tag{1.28}$$

*Moreover, if* $\operatorname{ess\,inf} f > -\infty$*, then $\{f_n\}$ converges in measure on $\mathbb{R}^n$ to $f^\#$.*

The last term of (1.28) is to be omitted when $x = 0$.

Finally, we mention a useful compactness property of decreasing functions.

**Lemma 1.44** (cf. Lieb and Loss, 1997, pp. 81, 110) *Let $\{f_k\}$ be a sequence of decreasing nonnegative right-continuous functions on* $(0,\infty)$*, all of which are majorized by a real valued function $g$ on* $(0,\infty)$*. Then there is a subsequence of $\{f_k\}$ which converges $\mathcal{L}$-a.e. on* $(0,\infty)$.

*Proof* Let $Q$ be a countable dense subset of $(0, \infty)$. Via the diagonal argument, we can choose a subsequence of $\{f_k\}$, also denoted $\{f_k\}$, such that $g(x) \equiv \lim_{k\to\infty} f_k(x)$ exists and is finite for each $x \in Q$. Extend $g$ to $(0, \infty)$ by defining $g(x) = \lim_{y\to x+, y\in Q} g(y)$. Then $g$ is decreasing, and it is a simple exercise to show that $g(x) = \lim_{k\to\infty} f_k(x)$ at each point of continuity of $f$. □

## 1.9 Notes and Comments

Rearrangement of sets played a major role in nineteenth-century research of Steiner (1842) and H. A. Schwarz (1890) on isoperimetric problems. The decreasing rearrangement of a function was formally introduced by Hardy and Littlewood (1930). That paper also contains the birth of what we now call the Hardy–Littlewood maximal function. The book of Hardy, Littlewood and Pólya (1952), first published in 1934, contains a chapter on rearrangement inequalities for sequences and for functions of one variable.

The first use I have found of the term "polarization" in our context is in a paper by Dubinin (1987). The technique had been applied earlier to symmetrization problems by Wolontis (1952) and Ahlfors (1973, Lemma 2.2).

Ryff (1970) proves the factorization $f = f^* \circ T$ for $f\colon [0, 1] \to \mathbb{R}$. Chapter 2 of the book by Bennett and Sharpley (1988) contains significant overlap with our Chapter 1. In particular, our Proposition 1.25 appears as Exercise 17, and the result of the exercise is used to prove the Ryff factorization for $f\colon X \to \mathbb{R}$ when $X$ is a general finite non-atomic measure space. Brenier (1991) invented a version of the factorization for $\mathbb{R}^n$-valued functions, which he calls the "polar factorization." Brenier's theory has been applied and further developed by a number of authors, such as R. McCann. See, for example, McCann (1995) and McCann (2001). The latter paper factors functions $f\colon M \to M$, where $M$ is a compact manifold.

The result in Proposition 1.43 that convergence in measure of $\{f_n\}$ implies convergence in measure of the symmetric decreasing rearrangements, when the limit function is essentially bounded below, is due to Almgren and Lieb (1989, p. 694).

# 2

# Main Inequalities on $\mathbb{R}^n$

The main results of this chapter, Theorems 2.12 and 2.15, give general inequalities for symmetric decreasing rearrangements of functions $f\colon \mathbb{R}^n \to \mathbb{R}$. Much of the rest of the book is based on these inequalities and on their spherical and hyperbolic analogues to be proved in Chapter 7. Theorem 2.12 asserts that symmetric decreasing rearrangement decreases the modulus of continuity of $f$, while Theorem 2.15 asserts that certain integral expressions increase when functions are replaced by their symmetric decreasing rearrangements. The proofs of Theorems 2.12 and 2.15 use Theorem 2.9, which is an analogue of Theorem 2.15 for polarization. Theorem 2.9 is, in turn, a consequence of Theorem 2.8, which is an elementary version of Theorem 2.15 involving symmetrization on a space with just two points. In §2.7 we derive some consequences of the general inequalities. Among the simplest is the Hardy–Littlewood type inequality $\int_{\mathbb{R}^n} fg\,dx \le \int_{\mathbb{R}^n} f^\# g^\#\,dx$, where $f$ and $g$ satisfy suitable hypotheses and # denotes symmetric decreasing rearrangement.

The integral expressions in Theorems 2.8, 2.9, and 2.15 involve functions $\Psi(x,y)$ of two real variables whose generalized mixed second partial derivatives satisfy $\Psi_{xy} \ge 0$. We have dubbed these functions AL functions. Section 2.1 contains a few simple results about AL functions, and contains also a resumé of some facts we will use about convex functions.

The proof of parts (a) and (c) of Theorem 2.15 is completed in §2.5 when $\Psi$ is continuous. To extend those results to discontinuous $\Psi$ requires a good deal of additional technical work, which is deferred until the last two sections of the chapter. No essential use of the discontinuous case will appear later in the book, so the reader so inclined may safely skip these last two sections.

## 2.1 Convex and AL Functions

Let $I$ denote an interval in $\mathbb{R}$. A function $f: I \to \mathbb{R}$ is said to be *convex* if for every $x, y \in I$ and $t \in [0, 1]$ holds

$$f(tx + (1-t)y) \leq tf(x) + (1-t)f(y).$$

If strict inequality holds for every $t \in (0, 1)$, then $f$ is said to be *strictly convex*.

Convex functions are continuous on the interior of their domain. For this fact and other information about convex functions, the reader may consult, among other sources, Hardy et al. (1952), Hörmander (1994), Zygmund (1968), or Roberts and Varberg (1973). In this section we will use two additional facts.

**Fact 2.1** *If $I$ is the closed interval $[a, b]$, and $f$ is convex on $I$, then*

$$\max_I f = \max(f(a), f(b)).$$

*Moreover, if $f(x) = \max(f(a), f(b))$ for some $x \in (a, b)$, then $f$ is constant on $I$ (and hence is not strictly convex).*

The proof is left to the reader.

**Fact 2.2** *Suppose that $x_1 < x_2 < x_3 < x_4$ are points of $I$ with $x_1 + x_4 = x_2 + x_3$, and that $f$ is convex on $I$. Then*

$$f(x_2) + f(x_3) \leq f(x_1) + f(x_4). \tag{2.1}$$

*If $f$ is strictly convex on $I$, then strict inequality holds in* (2.1).

*Proof* Let $m = \frac{1}{2}(x_1 + x_4)$. Set $f_1(x) = f(x) + f(2m - x)$. Then $f_1$ is convex on $[x_1, x_4]$, with $f_1(x_1) = f_1(x_4) = f(x_1) + f(x_4)$. Since $x_1 + x_4 = 2m = x_2 + x_3$, we have $f_1(x_2) = f(x_2) + f(x_3)$. Application of Fact 2.1 to $f_1$ yields $f_1(x_2) \leq f_1(x_1)$, which is (2.1).

If equality holds in (2.1), then $f_1$ achieves its maximum on $[x_1, x_4]$ at $x_2$, so that $f_1$ is constant on $[x_1, x_4]$. Hence,

$$f(x_1) + f(x_4) = f_1(x_1) = f_1(m) = 2f(m).$$

Thus, $f$ is not strictly convex on $I$. □

**Definition 2.3** Let $E_1$, $E_2$ be subsets of $\mathbb{R}$, and $f: E_1 \times E_2 \to \mathbb{R}$. We say that $f \in AL(E_1 \times E_2)$ if for every $x_1, x_2 \in E_1$ and $y_1, y_2 \in E_2$ with $x_1 \leq x_2$ and $y_1 \leq y_2$ holds

$$f(x_1, y_2) + f(x_2, y_1) \leq f(x_1, y_1) + f(x_2, y_2). \tag{2.2}$$

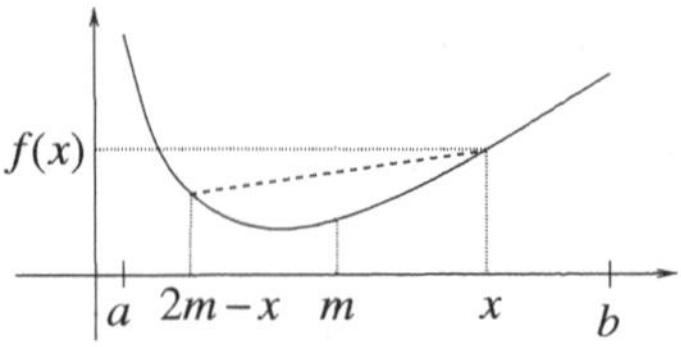


Figure 2.1 Graph of a convex function $f$ on an interval $[a, b]$. By definition, the graph of $f$ lies below any line segment that interpolates between two points on the curve (dashed line). Thus, $f$ assumes its maximum either at $a$ or $b$ (Fact 2.1). Moreover, $f_1(x) = f(x) + f(2m - x)$ is an increasing function of $|x - m|$ (see the proof of Fact 2.2).

We say that $f \in SAL(E_1 \times E_2)$ (the S stands for strict) if strict inequality holds in (2.2) whenever $x_1 < x_2$ and $y_1 < y_2$.

Our first fact about AL functions is an analogue of the fact that a $C^2$ function on an interval $I \subset \mathbb{R}$ is convex if and only if its second derivative is nonnegative on $I$. The subscripts on $f$ below denote partial differentiation.

**Fact 2.4** *If $I_1$, $I_2$ are intervals in $\mathbb{R}$, and $f \in C^2(I_1 \times I_2)$, then $f \in AL(I_1 \times I_2)$ if and only if $f_{xy} \geq 0$ on $I_1 \times I_2$. Moreover, if $f_{xy} > 0$ almost everywhere on $I_1 \times I_2$, then $f \in SAL(I_1 \times I_2)$.*

*Proof* Let $x_1 \leq x_2$ be points of $I_1$ and $y_1 \leq y_2$ be points of $I_2$. Then

$$\int_{y_1}^{y_2} \int_{x_1}^{x_2} f_{xy}(u, v)\, du\, dv = \int_{y_1}^{y_2} (f_y(x_2, v) - f_y(x_1, v))\, dv$$
$$= f(x_2, y_2) - f(x_2, y_1) - f(x_1, y_2) + f(x_1, y_1).$$

The statements in Fact 2.4 follow from the equality of the first and third terms. □

**Example 2.5** The function $f(x, y) = xy$ belongs to $SAL(\mathbb{R} \times \mathbb{R})$.

The next two facts show how $AL$ and convex functions are related.

**Fact 2.6** *Let $g : \mathbb{R} \to \mathbb{R}$ be convex. Then*

$$f(x, y) \equiv g(x) + g(y) - g(x - y) \in AL(\mathbb{R} \times \mathbb{R}).$$

*If $g$ is strictly convex on $\mathbb{R}$, then $f \in SAL(\mathbb{R} \times \mathbb{R})$.*

*Proof* Let $x_1 \leq x_2$ and $y_1 \leq y_2$ be points of $\mathbb{R}$. Then

$$f(x_2, y_2) - f(x_2, y_1) - f(x_1, y_2) + f(x_1, y_1)$$
$$= g(x_1 - y_2) + g(x_2 - y_1) - g(x_1 - y_1) - g(x_2 - y_2).$$

Now

$$x_1 - y_2 \leq x_1 - y_1 \leq x_2 - y_1;$$
$$x_1 - y_2 \leq x_2 - y_2 \leq x_2 - y_1; \text{ and}$$
$$(x_1 - y_2) + (x_2 - y_1) = (x_1 - y_1) + (x_2 - y_2).$$

The conclusion $f \in AL(\mathbb{R} \times \mathbb{R})$ follows from Fact 2.2 applied to $g$. The conclusions involving strictness follow by evident adaptation of the argument. □

**Fact 2.7** *Let* $g\colon \mathbb{R}^+ \to \mathbb{R}$ *be convex and increasing. Then*

$$f(x,y) \equiv g(|x|) + g(|y|) - g(|x-y|) \in AL(\mathbb{R} \times \mathbb{R}).$$

*If* $g$ *is strictly convex on* $\mathbb{R}^+$, *then* $f \in SAL(\mathbb{R} \times \mathbb{R})$.

*Proof* Let $g_1(x) = g(|x|)$. Then $g_1$ is convex on $\mathbb{R}$, and is strictly convex on $\mathbb{R}$ if $g$ is strictly convex on $\mathbb{R}^+$. Fact 2.7 follows from Fact 2.6 applied to $g_1$. □

## 2.2 Main Inequalities for Two-Point Symmetrization

In this section, we let $X$ denote the two-point set $\{1,2\}$, $d$ denote the distance function on $X$ defined by $d(x,y) = 1$ if $x = y$, $d(x,y) = 0$, if $x \neq y$, and $\mu$ denote the measure on $X$ such that $\mu(\{x\}) = 1$ for $x = 1,2$.

For functions $f\colon X \to \mathbb{R}$, define the decreasing rearrangement $f^*\colon X \to \mathbb{R}$ by

$$f^*(1) = \max(f(1), f(2)), \quad f^*(2) = \min(f(1), f(2)).$$

Let $K\colon \{0,1\} \to \mathbb{R}$ and $\Psi\colon f(X) \times g(X) \to \mathbb{R}$ be given functions. For $f, g\colon X \to \mathbb{R}$, define

$$Q(f,g) = Q(f,g,K,\Psi) = \int_{X^2} \Psi(f(x), g(y)) K(d(x,y))\, d\mu(x)\, d\mu(y).$$

**Theorem 2.8** *Suppose that* $K(0) \geq K(1)$ *and that* $\Psi \in AL(f(X) \times g(X))$. *Then, for* $f\colon X \to \mathbb{R}$, $g\colon X \to \mathbb{R}$ *hold*

*(a)* $Q(f,g) \leq Q(f^*, g^*)$.
*(b) If* $K(0) > K(1)$ *and* $\Psi \in SAL((f(X) \times g(X))$ *then strict inequality holds in (a) if and only if*

$$f(1) < f(2) \text{ and } g(1) > g(2), \quad \text{or} \quad f(1) > f(2) \text{ and } g(1) < g(2). \quad (2.3)$$

*(c)* $\int_X \Psi(f,g)\, d\mu \leq \int_X \Psi(f^*, g^*)\, d\mu$.
*(d) If* $\Psi \in SAL((f(X) \times g(X))$, *then strict inequality holds in (c) if and only if (2.3) holds.*

*Proof* First, observe that the set

$$f(X) \times g(X) = \{(f(1), g(1)), (f(2), g(1)), (f(1), g(2)), (f(2), g(2))\}$$

does not change when $f$ and $g$ are replaced by $f^*$ and $g^*$. It follows that

$$\int_{X^2} \Psi(f(x), g(y))\, d\mu(x)\, d\mu(y) = \Psi(f(1), g(1)) + \Psi(f(1), g(2)) + \Psi(f(2), g(1)) + \Psi(f(2), g(2))$$

does not change when $f$ and $g$ are replaced by $f^*$ and $g^*$.

Next, write $\delta = K(0) - K(1)$. Then

$$\begin{aligned} Q(f, g) &= [\Psi(f(1), g(1)) + \Psi(f(2), g(2))]K(0) \\ &\quad + [\Psi(f(1), g(2)) + \Psi(f(2), g(1))]K(1) \\ &= \delta[\Psi(f(1), g(1)) + \Psi(f(2), g(2))] \\ &\quad + K(1) \int_{X^2} \Psi(f(x), g(y))\, d\mu(x)\, d\mu(y). \end{aligned} \tag{2.4}$$

It follows that

$$\begin{aligned} &Q(f^*, g^*) - Q(f, g) \\ &= \delta\, [\Psi(f^*(1), g^*(1)) + \Psi(f^*(2), g^*(2)) - \Psi(f(1), g(1)) - \Psi(f(2), g(2))]. \end{aligned} \tag{2.5}$$

Now $\delta \geq 0$. If either $f$ or $g$ is constant then the expression in the brackets equals zero.

Suppose that neither $f$ nor $g$ is constant, and that (2.3) does not hold. If $f(1) > f(2)$, then $f = f^*$ and $g = g^*$, while if $f(1) < f(2)$, then $f(1) = f^*(2), f(2) = f^*(1)$, $g(1) = g^*(2)$, $g(2) = g^*(1)$. Either way, the expression in the brackets again is zero.

Suppose (2.3) does hold. If $f(1) > f(2)$ and $g(1) < g(2)$, then $f(1) = f^*(1), f(2) = f^*(2)$, $g(1) = g^*(2)$, $g(2) = g^*(1)$. The term in the brackets equals

$$[\Psi(f^*(1), g^*(1)) + \Psi(f^*(2), g^*(2)) - \Psi(f^*(1), g^*(2)) - \Psi(f^*(2), g^*(1))].$$

Since $\Psi \in AL(f(X) \times g(X)) = AL(f^*(X) \times g^*(X))$, the expression above is nonnegative. The same argument shows that it is nonnegative when (2.3) holds with $f(1) < f(2)$ and $g(1) > g(2)$.

We have shown that in all cases the expression on the right-hand side of (2.5) is nonnegative. This proves conclusion (a) of Theorem 2.8. The same argument proves (b).

From (2.4), we see that in the special case when $K(0) = 1$, $K(1) = 0$ we have

$$Q(f,g) = \int_X \Psi(f,g)\, d\mu.$$

Thus, (c) and (d) follow from (a) and (b). □

## 2.3 Main Inequalities for Polarization

Our main goal in this section is to prove Theorem 2.9, an analogue of Theorem 2.8 in which the role of decreasing rearrangement on the two-point space $X$ is taken over by polarization of functions defined on $\mathbb{R}^n$. Once again, we are concerned with an integral functional $Q(f,g)$ defined in terms of fixed functions $K$ and $\Psi$. Because the integrals will be taken over sets of infinite measure, it will be convenient to impose more restrictions on $f, g, K$, and $\psi$ than was the case for Theorem 2.8. Accordingly, we shall assume that $f$, $g$, and $K$ are nonnegative, and shall assume that the function $\Psi$ is an $AL$ function on $\mathbb{R}^+ \times \mathbb{R}^+ = [0,\infty) \times [0,\infty)$ which vanishes on the coordinate axes. The class of all such $\Psi$ will be denoted by $AL_0$. Thus,

$$AL_0 \equiv \{\Psi \in AL(\mathbb{R}^+ \times \mathbb{R}^+) \colon \Psi(x,0) = \Psi(0,y) = 0 \;\; x,\; y \in \mathbb{R}^+\}.$$

In particular, $\Psi(0,0) = 0$ for $\Psi \in AL_0$.

The corresponding class of strong $AL$ functions will be denoted by $SAL_0$. Thus,

$$SAL_0 \equiv AL_0 \cap SAL(\mathbb{R}^+ \times \mathbb{R}^+).$$

For $\Psi \in AL_0$, $x \in \mathbb{R}^+$ and $0 \le y_1 \le y_2$, we have

$$0 \le \Psi(x,y_2) - \Psi(0,y_2) + \Psi(0,y_1) - \Psi(x,y_1) = \Psi(x,y_2) - \Psi(x,y_1).$$

Thus, $\Psi$ is increasing on each vertical line in $\mathbb{R}^+ \times \mathbb{R}^+$. Similarly, $\Psi$ is increasing on each horizontal line. Since it vanishes on the coordinate axes, $\Psi$ is nonnegative on $\mathbb{R}^+ \times \mathbb{R}^+$.

Functions in $AL_0$ can be viewed as two-variable analogues of increasing functions of one variable. An increasing function on $[0,\infty)$ can be decomposed into the sum of a continuous increasing function and a step function. In §2.8 we will prove a decomposition theorem of this type for $AL_0$ functions, Theorem 2.25, which will be needed for analysis involving discontinuous $\Psi$.

Here now are the assumptions and notation for Theorem 2.9.

(i) $f$ and $g$ are nonnegative Lebesgue measurable functions on $\mathbb{R}^n$ with $\lambda_f(t) < \infty$ and $\lambda_g(t) < \infty$ for all $t > 0$.
(ii) $K\colon \mathbb{R}^+ \to \mathbb{R}^+$ is decreasing.
(iii) $\Psi \in AL_0$.
(iv) $Q(f,g) = Q(f,g,K,\Psi) \equiv \int_{\mathbb{R}^{2n}} \Psi(f(x),g(y))K(|x-y|)\,dx\,dy$.
(v) $H \in \mathcal{H}(\mathbb{R}^n)$. $H^+$ and $H^-$ are the complementary open halfspaces cut out by $H$.
(vi) $\rho = \rho_H$ (reflection in $H$) and

$$A \equiv \{(x,y) \in (H^+)^2\colon \text{ either } f(x) < f(\rho x) \text{ and } g(y) > g(\rho y), \\ \text{or } f(x) > f(\rho x) \text{ and } g(y) < g(\rho y)\}.$$

From the decomposition Theorem 2.25 for $AL_0$ functions, it follows that functions in $AL_0$ are Borel measurable. Thus, the integrand in (iv) is nonnegative and Borel measurable.

Concerning (v), recall from §1.7 that $\mathcal{H}(\mathbb{R}^n)$ is the set of all affine hyperplanes in $\mathbb{R}^n$. The polarization $f_H\colon \mathbb{R}^n \to \mathbb{R}$ of $f\colon \mathbb{R}^n \to \mathbb{R}$ was defined in §1.7 by the formula

$$\begin{aligned} f_H(x) &= \max(f(x), f(\rho x)), \quad \text{if } x \in H^+, \\ &= \min(f(x), f(\rho x)), \quad \text{if } x \in H^-, \\ &= f(x), \quad \text{if } x \in H, \end{aligned}$$

where $\rho = \rho_H\colon \mathbb{R}^n \to \mathbb{R}^n$ denotes reflection in $H$.

The set $A$ in (vi) depends on $H$, $f$, and $g$. A pair $(x,y)$ is in $A$ if and only if when we pass to the reflected pair, one of $f$ or $g$ strictly increases and the other strictly decreases.

Define also, for fixed $H$,

$$A_0 = \{x \in H^+\colon (x,x) \in A\}. \tag{2.6}$$

**Theorem 2.9** *Let assumptions* (i)–(vi) *be satisfied. Then*

*(a)* $Q(f,g) \le Q(f_H, g_H)$.
*(b) If $K$ is strictly decreasing on $\mathbb{R}^+$, $\Psi \in SAL_0(\mathbb{R}^+ \times \mathbb{R}^+)$, and $Q(f,g) < \infty$, then strict inequality holds in (a) if and only if $\mathcal{L}^{2n}(A) > 0$.*
*(c)* $\int_{\mathbb{R}^n} \Psi(f(x),g(x))\,dx \le \int_{\mathbb{R}^n} \Psi(f_H(x),g_H(x))\,dx$.
*(d) If $\Psi \in SAL_0(\mathbb{R}^+ \times \mathbb{R}^+)$ and $\int_{\mathbb{R}^n} \Psi(f(x),g(x))\,dx < \infty$, then strict inequality holds in (c) if and only if $\mathcal{L}^n(A_0) > 0$.*

*Proof* In (iv), split the integral defining $Q$ into integrals over $H^+ \times H^+$, $H^+ \times H^-$, $H^- \times H^+$, and $H^- \times H^-$. As noted in §1.7, $\rho$ is a measure preserving map of $(H^+, \mathcal{L}^n)$ onto $(H^-, \mathcal{L}^n)$ with $|x-y| = |\rho x - \rho y|$. Thus, we can write

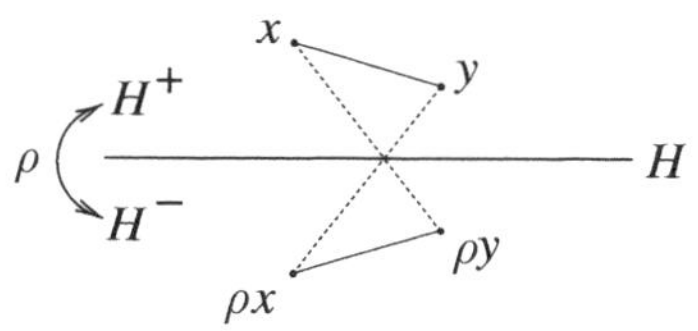


Figure 2.2 A pair of points $x, y \in H^+$ and their mirror images $\rho x, \rho y$ in the complementary halfspace $H^-$. Since the reflection $\rho$ is an isometry, the distances from $x$ to $y$ and from $\rho x$ to $\rho y$ are equal (solid lines) and are strictly shorter than the diagonal distances from $x$ to $\rho y$ and from $\rho x$ to $y$ (dashed lines).

$$Q(f,g) = \int_{H^+\times H^+} [K(|x-y|)\,(\Psi(f(x),g(y)) + \Psi(f(\rho x),g(\rho y))) \\ + K(|x-\rho y|)\,(\Psi(f(x),g(\rho y)) + \Psi(f(\rho x),g(y)))]\,dx\,dy. \tag{2.7}$$

Fix $x,\ y \in H^+$, as in Figure 2.2. Define $f_1$ and $g_1$ on $X = \{1,2\}$ by

$$f_1(1) = f(x),\ f_1(2) = f(\rho x),\ g_1(1) = g(y),\ g_1(2) = g(\rho y), \tag{2.8}$$

and define $K_1$ on $\{0,1\}$ by $K_1(0) = K(|x-y|)$, $K_1(1) = K(|x-\rho y|)$. With the notation of §2.2, the integrand in (2.7) equals $Q(f_1, g_1, K_1, \Psi)$. When $f$ and $g$ in (2.7) are replaced by $f_H$ and $g_H$, then $f_1$ and $g_1$ are replaced in the integrand by $(f_1)^*$ and $(g_1)^*$. Using the fact that $|x-y| \le |x-\rho y|$, it follows that the hypotheses of Theorem 2.8 are satsified for each $x,\ y \in H^+$. Conclusions (a) and (b) of Theorem 2.9 follow from (a) and (b) of Theorem 2.8.

Write

$$\int_{\mathbb{R}^n} \Psi(f(x),g(x))\,dx = \int_{H^+} \Psi(f(x),g(x))\,dx + \int_{H^-} \Psi(f(x),g(x))\,dx \\ = \int_{H^+} [\Psi(f(x),g(x)) + \Psi(f(\rho x),g(\rho x))]\,dx.$$

Define $f_1$ as in (2.8), and define $g_1$ as in (2.8) but with $y$ replaced by $x$. Then conclusions (c) and (d) of Theorem 2.9 follow from (c) and (d) of Theorem 2.8. □

Theorem 2.9 remains true under some less restrictive hypotheses on $f, g, K$ and $\Psi$. A discussion appears in §7.3. Here we just give a sample. Suppose that $f, g$ and $K$ satisfy (i) and (ii), but for $\Psi$ we require only that $\Psi \in AL(\mathbb{R}^+ \times \mathbb{R}^+)$, that $\Psi \ge 0$ on $\mathbb{R}^+ \times \mathbb{R}^+$, and that $\Psi(0,0) = 0$. We will verify that Theorem 2.9(a) remains valid.

Define $\Psi_1(x,y) = \Psi(x,y) - \Psi(x,0) - \Psi(0,y)$, and observe $\Psi_1(x,0) = \Psi_1(0,y) = 0$. From the definition of $AL$ function, one easily checks that

$\Psi_1 \in AL(\mathbb{R}^+ \times \mathbb{R}^+)$, so that $\Psi_1 \in AL_0$. Let $Q_1$ denote the integral in (iv) when $\Psi$ is replaced by $\Psi_1$, and write $c = c(K) = \int_{\mathbb{R}^n} K(|x|)\,dx$. Assume that $c < \infty$. Then

$$Q(f,g) = Q_1(f,g) + c\int_{\mathbb{R}^n} \Psi(f(x),0)\,dx + c\int_{\mathbb{R}^n} \Psi(0,g(y))\,dy.$$

The last two terms on the right do not change when $f$ and $g$ are replaced by their rearrangements $f_H, g_H$. The first term increases, by Theorem 2.9(a). Thus, $Q(f,g) \le Q(f_H, g_H)$ holds when $c(K) < \infty$.

For general $K$ satisfying (ii) and positive integers $m$, define $K_m(t) = \min(K(t), m)$ if $0 \le t \le m$, $K_m(t) = 0$ if $t \ge m$. Then $c(K_m) < \infty$, so $Q(f,g,K_m) \le Q(f_H, g_H, K_m)$. Moreover, $K_m$ increases pointwise to $K$ on $\mathbb{R}^+$. Since $\Psi \ge 0$, the monotone convergence theorem is applicable, and gives

$$Q(f,g,K) = \lim_{m\to\infty} Q(f,g,K_m) \le \lim_{m\to\infty} Q(f_H,g_H,K_m) = Q(f_H,g_H,K).$$

Thus, conclusion (a) of Theorem 2.9 holds for $f, g, K$, and $\Psi$.

We shall prove now some results about polarization and symmetric decreasing rearrangement on $\mathbb{R}^n$ which will be needed in the two following sections. Let $f\colon \mathbb{R}^n \to \mathbb{R}$ be measurable and satisfy $\lambda_f(t) < \infty$ for all $t > \operatorname{ess\,inf} f$. The symmetric decreasing rearrangement $f^\#\colon \mathbb{R}^n \to \mathbb{R}$ of $f$ was defined in §1.6. Suppose that the hyperplane $H \in \mathcal{H}(\mathbb{R}^n)$ does not contain the origin. Let $H^+$ denote the halfspace which contains the origin. Then $|\rho_H x| > |x|$ for each $x \in H^+$, so that $f^\#(x) \ge f^\#(\rho x)$. Thus, by the definition of polarization, $(f^\#)_H = f^\#$. For later reference we give a formal statement of this fact.

**Proposition 2.10** *Suppose that $H \in \mathcal{H}(\mathbb{R}^n)$ with $0 \notin H$ and $0 \in H^+$. Then*

$$(f^\#)_H = f^\#$$

*for each measurable $f\colon \mathbb{R}^n \to \mathbb{R}$ with $\lambda_f(t) < \infty$ for all $t > \operatorname{ess\,inf} f$.*

We note also that for $f$ as in Proposition 2.10, Proposition 1.35 of §1.7 states that $f$ and $f_H$ are rearrangements of each other, so that $(f_H)^\# = f^\#$.

The other results we need are about continuous nonnegative functions on $\mathbb{R}^n$ with compact support. Symmetric decreasing functions in this class can be conveniently characterized in terms of polarization inequalities.

**Proposition 2.11** *Let $f \in C_c(\mathbb{R}^n, \mathbb{R}^+)$. Then*

*(a) $f^\# \in C_c(\mathbb{R}^n, \mathbb{R}^+)$.*
*(b) If $f \ne f^\#$, there exists $H \in \mathcal{H}(\mathbb{R}^n)$ with $0 \notin H$ and $0 \in H^+$ such that*

$$\int_{\mathbb{R}^n} f f^\#\,dx < \int_{\mathbb{R}^n} f_H f^\#\,dx.$$

*Proof of (a)* It is clear that $f^\#$ is nonnegative. We assume in the rest of the proof that $f$ is not identically zero. Then, since $\mathcal{L}^n(f^\# > 0) = \mathcal{L}^n(f > 0) < \infty$, it follows that $(f^\# > 0)$ is a ball of finite measure. Hence, $f^\#$ has compact support. Let $M = \sup_{\mathbb{R}^n} f = \sup_{\mathbb{R}^n} f^\#$ and $(a, b)$ be a nonempty open subinterval of $(0, M)$. Since $f$ is continuous and $\min_{\mathbb{R}^n} f = 0$, the set $(f \in (a, b))$ is open and nonempty, so that $\mathcal{L}^n(f \in (a, b)) > 0$. Thus, $\mathcal{L}^n(f^\# \in (a, b)) > 0$ for all such $(a, b)$. Since $f^\#$ is symmetric decreasing, it must therefore be continuous. □

*Proof of (b)* For $H \in \mathcal{H}(\mathbb{R}^n)$ with $0 \notin H$ and $0 \in H^+$, Theorem 2.9(c) applied with $\Psi(x, y) = xy$ and Proposition 2.10 give

$$\int_{\mathbb{R}^n} f f^\# \, dx \leq \int_{\mathbb{R}^n} f_H \, (f^\#)_H \, dx = \int_{\mathbb{R}^n} f_H f^\# \, dx. \tag{2.9}$$

By Theorem 2.9(d), to find an $H$ for which strict inequality holds in (2.9) is equivalent to finding an $H$ whose associated set $A_0$ has positive $\mathcal{L}^n$ measure, when $g = f^\#$.

Since $f \neq f^\#$, from uniqueness of the symmetric decreasing rearrangement (Proposition 1.30(e)), it follows that $f$ is not symmetric decreasing on $\mathbb{R}^n$. Thus, there exist $x, y \in \mathbb{R}^n$ with $|x| \leq |y|$ and $f(x) < f(y)$. Take $t$ such that $f(x) < t < f(y)$. Let

$$E_1 = (f > t), \quad E_2 = (f^\# > t).$$

Then $E_1$ and $E_2$ are open sets with the same positive finite $\mathcal{L}^n$ measure, $E_2$ is a ball centered at the origin, $y \in E_1$, and $x \notin \overline{E}_1$. Since $|x| \leq |y|$, it follows that $E_1$ is not a ball centered at the origin, and hence $E_1 \neq E_2$. Next, we shall show the stronger conclusion that neither of the sets $E_1$ nor $E_2$ contains the other.

Suppose we had $E_2 \subset E_1$. Then $E_1$ would be an open set containing the open ball $E_2$ which has the same measure as $E_2$. This is possible only if $E_1$ coincides with $E_2$. But we saw above that $E_1 \neq E_2$. Thus, $E_2$ is not a subset of $E_1$.

Suppose we had $E_1 \subset E_2$. Then $\mathcal{L}^n(E_2 \setminus E_1) = 0$, since the sets have the same measure. Thus, $E_1$ is dense in $E_2$. Since $f > t$ on $E_1$, we have $f \geq t$ on $\overline{E}_2$. Hence, $x \notin \overline{E}_2$. On the other hand, $y \in E_1 \subset E_2$. Since $E_2$ is a ball centered at the origin, we would have $|y| < |x|$. But this contradicts our earlier specification that $|x| \leq |y|$. Thus, $E_1$ is not a subset of $E_2$.

We have shown now that there exist points $x_1$ and $y_1$ with $x_1 \in E_2 \setminus E_1$, $y_1 \in E_1 \setminus E_2$. Let $H$ be the affine hyperplane whose associated reflection $\rho$ satisfies $\rho x_1 = y_1$. Since $E_2$ is a ball centered at the origin, we have $|x_1| < |y_1|$, from which it follows that $x_1$ and the origin lie in the same open halfspace defined by $H$, call it $H^+$. Now

$$f(x_1) \leq t < f^\#(x_1) \quad \text{and} \quad f^\#(y_1) \leq t < f(y_1). \tag{2.10}$$

From (2.10), it follows that $f(x_1) < f(y_1)$ and $f^{\#}(y_1) < f^{\#}(x_1)$. Since $y_1 = \rho x_1$ and $f, f^{\#}$ are continuous, there is an open neighborhood $U$ of $x_1$ with $U \subset H^+$ such that, for every $x \in U$,

$$f(x) < f(\rho x) \quad \text{and} \quad f^{\#}(\rho x) < f^{\#}(x).$$

Thus, the set $A_0$ in (2.6) for $f$ and $f^{\#}$ contains $U$, and hence $\mathcal{L}^n(A_0) > 0$. As noted at the beginning of the proof, Proposition 2.11(b) now follows from Theorem 2.9(d). □

## 2.4 Symmetrization Decreases the Modulus of Continuity

Recall that the modulus of continuity $\omega(t,f)$ of a function $f\colon \mathbb{R}^n \to \mathbb{R}$ is defined for $t > 0$ by

$$\omega(t,f) = \sup\{|f(x) - f(y)|\colon\ x, y \in \mathbb{R}^n,\ |x - y| \le t\}.$$

In Proposition 1.37 we proved that polarization decreases the modulus of continuity. That is, for all real functions $f$ on $\mathbb{R}^n$, $t > 0$ and affine hyperplanes $H$ holds

$$\omega(t, f_H) \le \omega(t,f).$$

In this section we will prove that, for appropriate functions $f$ on $\mathbb{R}^n$, the modulus of continuity decreases when $f$ is changed to its symmetric decreasing rearrangement $f^{\#}$. This is one of the fundamental results of symmetrization theory. It will, for example, play a prominent role in our proof of the isoperimetric inequality in Chapter 4.

**Theorem 2.12** *Let $f \in C(\mathbb{R}^n, \mathbb{R})$, with $\lambda_f(t) < \infty$ for all $t >$ ess inf $f$. Then, for every $t > 0$ holds*

$$\omega(t, f^{\#}) \le \omega(t,f). \tag{2.11}$$

*Proof* Assume first that $f$ is nonnegative and has compact support, so that $f \in C_c(\mathbb{R}^n, \mathbb{R}^+)$. If $f \equiv 0$ then (2.11) holds. In the sequel, we assume that $f$ is not identically zero. By Proposition 2.11(a), we already know that $f^{\#}$ is continuous; (2.11) will provide a sharp version of this Proposition.

Let $R = \operatorname{diam} \operatorname{supp} f$, the diameter of the support of $f$. Define a set $\mathcal{S} = \mathcal{S}(f) \subset C_c(\mathbb{R}^n, \mathbb{R}^+)$ as follows:

$$\begin{aligned} \mathcal{S} = \{F \in C_c(\mathbb{R}^n, \mathbb{R}^+)\colon\ & \omega(\cdot, F) \le \omega(\cdot, f) \text{ on } (0,\infty), \\ & \lambda_F = \lambda_f \text{ on } (0,\infty), \text{ and } \operatorname{diam} \operatorname{supp} F \le R\}. \end{aligned} \tag{2.12}$$

Define also a number $d = d(f)$ by

$$d = \inf_{F \in \mathcal{S}} \|F - f^{\#}\|_2,$$

where $\|g\|_2$ denotes the $L^2(\mathbb{R}^n, \mathcal{L}^n)$ norm of a function $g$.

**Claim 1.** There exists $F_0 \in \mathcal{S}$ such that $d = \|F_0 - f^{\#}\|_2$.

**Claim 2.** $F_0 = f^{\#}$.

More precisely, Claim 2 asserts that any $F_0 \in \mathcal{S}$ which has minimal $L^2$ distance to $f^{\#}$ must equal $f^{\#}$.

If the claims are true, then $f^{\#} \in \mathcal{S}$, so that $\omega(\cdot, f^{\#}) \leq \omega(\cdot, f)$ on $(0, \infty)$, which is (2.11). Thus, Theorem 2.12 will be proved for compactly supported $f$ once we have proved Claims 1 and 2. □

*Proof of Claim 1* Let $\{f_k\}_{k=1}^{\infty}$ be a sequence in $\mathcal{S}$ such that $\lim_{k\to\infty}\|f_k - f^{\#}\|_2 = d$. We would like to obtain a subsequence of $\{f_k\}$ which converges uniformly on $\mathbb{R}^n$. An obstacle to doing this is that the supports of the $f_k$ might drift off to infinity. Accordingly, we define a possibly new minimizing sequence $\{F_k\}$ as follows: For a given $k \geq 1$, if $\operatorname{supp} f_k$ contains a point $x$ with $|x| \leq R$, let $F_k = f_k$. If not, choose a point $x_k$ such that $f_k(x_k) > 0$. Such a point exists because $f_k$ has the same distribution as $f$, and hence is not identically zero. We have $|x_k| > R$. Let $H = H_k \in \mathcal{H}(\mathbb{R}^n)$ be the affine hyperplane such that $\rho_H x_k = 0$, and let $H^+$ be the corresponding halfspace which contains the origin. Let $F_k = (f_k)_H$.

We saw in §1.7 that polarization decreases the diameter of the support, decreases the modulus of continuity, and preserves the distribution function. Thus, each $F_k \in \mathcal{S}$. Moreover, if $k$ is an index for which $F_k = (f_k)_H$, then by Theorem 2.9(c), with $\Psi(x, y) = xy$, Proposition 2.10, and the fact that $f_k$ and $F_k$ are rearrangements of each other, we have

$$\begin{aligned} \int_{\mathbb{R}^n} (F_k - f^{\#})^2 &= \int_{\mathbb{R}^n} F_k^2 + \int_{\mathbb{R}^n} (f^{\#})^2 - 2\int_{\mathbb{R}^n} F_k f^{\#} \\ &\leq \int_{\mathbb{R}^n} f_k^2 + \int_{\mathbb{R}^n} (f^{\#})^2 - 2\int_{\mathbb{R}^n} f_k f^{\#} = \int_{\mathbb{R}^n} (f_k - f^{\#})^2. \end{aligned} \tag{2.13}$$

Thus, $d \leq \|F_k - f^{\#}\|_2 \leq \|f_k - f^{\#}\|_2$, and hence $\lim_{k\to\infty}\|F_k - f^{\#}\|_2 = d$, so that $\{F_k\}$ is also a minimizing sequence.

Let $k$ be an index such that $F_k = (f_k)_H$. Since $f_k(x_k) > 0$ and $\operatorname{diam} \operatorname{supp} f_k \leq R$, we have $f_k(0) = 0$. It follows that $F_k(0) = f_k(x_k) > 0$. Since $\operatorname{diam} \operatorname{supp} F_k \leq R$, it follows that $\operatorname{supp} F_k \subset \overline{B}(R) = \{x \in \mathbb{R}^n : |x| \leq R\}$. If $F_k = f_k$, then $\operatorname{supp} f_k \subset \overline{B}(2R)$. Thus,

$$\operatorname{supp} F_k \subset \overline{B}(2R), \quad \forall\, k \geq 1. \tag{2.14}$$

Since $\omega(\cdot, F_k) \leq \omega(\cdot, f)$ and $\sup_{\mathbb{R}^n} F_k = \sup_{\mathbb{R}^n} f$, the functions $F_k$ form a uniformly bounded equicontinuous sequence in $\mathbb{R}^n$. By the Arzelà–Ascoli Theorem, see for example Rudin (1964, p. 144), there exists a subsequence of $\{F_k\}$, which we will also denote by $\{F_k\}$, which converges uniformly on $\overline{B}(2R)$. By (2.14), the convergence is in fact uniform on $\mathbb{R}^n$.

Let $F_0 = \lim_{k\to\infty} F_k$. One easily sees that

$$\operatorname{diam} \operatorname{supp} F_0 \leq \liminf_{k\to\infty} \operatorname{diam} \operatorname{supp} F_k$$

and that $\omega(\cdot, F_0) \leq \liminf_{k\to\infty} \omega(\cdot, F_k)$. Thus, $\operatorname{diam} \operatorname{supp} F_0 \leq R$ and $\omega(\cdot, F_0) \leq \omega(\cdot, f)$. Moreover, uniform convergence implies convergence in measure. From Proposition 1.41 and the fact that the decreasing rearrangements $F_k^*$ are all equal to $f^*$, we deduce that $F_0^*(x) \leq f^*(x) \leq F_0^*(x-)$ for each $x \in [0,\infty)$. From Proposition 2.11(a) and Definition 1.29 of symmetric decreasing rearrangement, we see that $f^*$ and $F_0^*$ are continuous on $[0,\infty)$. Thus, ${F_0}^* = f^*$, so that $\lambda_{F_0} = \lambda_f$ and thus $F_0 \in \mathcal{S}$. Since the $F_k$ are uniformly bounded in $\overline{B}(2R)$ and are all zero outside $\overline{B}(2R)$, and $f^\#$ is bounded and has compact support, the dominated convergence theorem may be invoked to give $\lim_{k\to\infty} \|F_k - f^\#\|_2 = \|F_0 - f^\#\|_2$. Since the limit on the left equals $d$, Claim 1 is proved. □

*Proof of Claim 2* Suppose that $F_0 \neq f^\#$. By Proposition 2.11(b), there exists $H \in \mathcal{H}(\mathbb{R}^n)$ with $0 \in H^+$ such that $\int_{\mathbb{R}^n} F_0 f^\# < \int_{\mathbb{R}^n} (F_0)_H f^\#$. By the argument in (2.13), it follows that

$$\|(F_0)_H - f^\#\|_2 < \|F_0 - f^\#\|_2. \tag{2.15}$$

But, as noted in the second paragraph of the proof of Claim 1, the polarization of a function in $\mathcal{S}$ is also in $\mathcal{S}$. Thus $(F_0)_H \in \mathcal{S}$. The right-hand side of (2.15) equals $d$, so we have a contradiction to the definition of $d$. We conclude, therefore, that $F_0 = f^\#$. Claim 2 is proved. □

*Conclusion of the proof of Theorem 2.12* We have now verified Theorem 2.12 for compactly supported $f \in C(\mathbb{R}^n, \mathbb{R}^+)$. Suppose that $f$ is an arbitrary function satisfying the hypotheses of Theorem 2.12. One easily sees that $\omega(2t,f) \leq 2\omega(t,f)$. It follows that if $\omega(t,f) = \infty$ for some $t > 0$, then $\omega(t,f) = \infty$ for all $t > 0$, so that (2.11) is automatically satisfied. We assume henceforth that $\omega(t,f) < \infty$, for all $t > 0$.

Let $\{\alpha_m\}$ be a strictly decreasing sequence with $\lim_{m\to\infty} \alpha_m = \operatorname{ess\,inf} f$. The hypotheses $\lambda_f(t) < \infty$ for $t > \operatorname{ess\,inf} f$ and $\omega(t,f) < \infty$ for $t > 0$ imply, via a simple argument by contradiction, that $\lim_{|x|\to\infty} f(x)$ exists and equals $\operatorname{ess\,inf} f$.

Let $f_m = \max(f, \alpha_m)$. Then $f_m - \alpha_m$ is continuous, nonnegative, and compactly supported. Thus, writing $f_m^\# = (f_m)^\#$,

$$\omega(\cdot, f_m^\#) = \omega(\cdot, f_m^\# - \alpha_m) = \omega(\cdot, (f_m - \alpha_m)^\#) \leq \omega(\cdot, f_m - \alpha_m) = \omega(\cdot, f_m).$$

One easily sees that $f_m^\# = \max(f^\#, \alpha_m)$ and that $\omega(\cdot, f_m) \leq \omega(\cdot, f)$.

Fix $t > 0$, and let $x, y \in \mathbb{R}^n$ satisfy $|x - y| \leq t$. Then

$$\begin{aligned} |f^\#(x) - f^\#(y)| &= \lim_{m\to\infty} |f_m^\#(x) - f_m^\#(y)| \leq \liminf_{m\to\infty} \omega(t, f_m^\#) \leq \liminf_{m\to\infty} \omega(t, f_m) \\ &\leq \omega(t, f). \end{aligned}$$

Thus, $\omega(t, f^\#) \leq \omega(t, f)$. The proof of Theorem 2.12 is complete. □

As a by-product of the proof, we have a corollary.

**Corollary 2.13** *Let $f \in C(\mathbb{R}^n, \mathbb{R}^+)$, with $\lambda_f(t) < \infty$ for all $t > 0$. Then*

$$\operatorname{diam} \operatorname{supp} f^\# \leq \operatorname{diam} \operatorname{supp} f. \tag{2.16}$$

*Proof* If $f$ has compact support, then in the proof above we showed that $f^\# = F_0 \in \mathcal{S}$, so that in particular (2.16) holds. For general $f \in C(\mathbb{R}^n, \mathbb{R}^+)$, (2.16) holds for the functions $f_m$ at the end of the proof of Theorem 2.12. Since $f_m$ and $f_m^\#$ converge pointwise to $f$ and $f^\#$, respectively, and diam supp $f_m \leq$ diam supp$f$, (2.16) easily follows for $f$. □

Using Corollary 2.13, we can prove the *isodiametric inequality*:

**Corollary 2.14** *For each measurable set $E \subset \mathbb{R}^n$ holds*

$$\operatorname{diam} E^\# \leq \operatorname{diam} E. \tag{2.17}$$

*An equivalent form of* (2.17) *is*

$$\mathcal{L}^n(E) \leq \alpha_n \left( \frac{\operatorname{diam} E}{2} \right)^n$$

*where $\alpha_n$ is the volume of the unit ball.*

*Proof* We may assume that $E$ is bounded, for otherwise (2.17) is trivial. For $n = 1, 2, \ldots$ define $f_n(x) = (1 - n\,d(x, E))^+$ where $d(x, E) = \inf\{|x-y| : y \in E\}$. Since $f_n = 1$ on $E$, it follows that $E^\# \subset \operatorname{supp} f_n^\#$. By Corollary 2.13,

$$\operatorname{diam} E^\# \leq \operatorname{diam} \operatorname{supp} f^\# \leq \operatorname{diam} \operatorname{supp} f_n^\#.$$

Letting $n \to \infty$ yields (2.17). □

## 2.5 Symmetrization Increases Certain Integrals in $\mathbb{R}^n$

Theorem 2.15 in this section contains the main integral inequalities in our approach to symmetrization on $\mathbb{R}^n$. The theorem has the same structure as Theorem 2.9, but now integrals involving two functions $f$ and $g$ on $\mathbb{R}^n$ will be compared with integrals involving their symmetric decreasing rearrangements $f^\#$ and $g^\#$ instead of with their polarizations $f_H$ and $g_H$. Theorem 2.9 will be a principal tool in the proof of Theorem 2.15.

Here are the hypotheses and definitions for Theorem 2.15. The first four hypotheses are the same as the first four hypotheses for Theorem 2.9. But the set $A$ defined in (V) is different from the set $A$ in Theorem 2.9.

(I) $f$ and $g$ are nonnegative Lebesgue measurable functions on $\mathbb{R}^n$ with $\lambda_f(t) < \infty$ and $\lambda_g(t) < \infty$ for all $t > 0$.
(II) $K\colon \mathbb{R}^+ \to \mathbb{R}^+$ is decreasing.
(III) $\Psi\colon \mathbb{R}^+ \times \mathbb{R}^+ \to \mathbb{R}^+$ is in $AL_0$.
(IV) $Q(f,g) = Q(f,g,K,\Psi) \equiv \int_{\mathbb{R}^{2n}} \Psi(f(x),g(y))K(|x-y|)\,dx\,dy$.
(V) $A \equiv \{(x,y) \in \mathbb{R}^{2n} : f(x) < f(y) \text{ and } g(x) > g(y)\}$.

Concerning (III), recall that $AL_0$, defined in §2.3, is the subclass of $AL$ functions on $\mathbb{R}^+ \times \mathbb{R}^+$ which vanish on the coordinate axes.

**Theorem 2.15** *Let assumptions* (I)–(IV) *be satisfied, and the set A be defined by* (V). *Then*

*(a)* $Q(f,g) \le Q(f^\#,g^\#)$.
*(b) Suppose that neither $f$ nor $g$ is identically zero, that $K$ is strictly decreasing on $\mathbb{R}^+$, that $\Psi \in SAL_0$, and that $Q(f,g) < \infty$. Then equality holds in (a) if and only if there exists a translation $T$ of $\mathbb{R}^n$ such that*

$$f = f^\# \circ T \quad \text{and} \quad g = g^\# \circ T, \quad \text{a.e. on } \mathbb{R}^n. \tag{2.18}$$

*(c)* $\int_{\mathbb{R}^n} \Psi(f(x),g(x))\,dx \le \int_{\mathbb{R}^n} \Psi(f^\#(x),g^\#(x))\,dx$.
*(d) If $\Psi \in SAL_0$ and $\int_{\mathbb{R}^n} \Psi(f(x),g(x))\,dx < \infty$, then equality holds in (c) if and only if $\mathcal{L}^{2n}(A) = 0$.*

Observe that inequalities (a) and (c) in Theorems 2.9 and 2.15 are parallel, but the equality conditions (b) and (d) differ a bit from theorem to theorem.

Concerning (b), note that translations are $\mathcal{L}^n$-measure preserving and satisfy $K(|Tx - Ty|) = K(|x-y|)$. Thus, the "if" part of (b) is trivial.

To get some idea of what (d) is about, we confirm half of it in a special case. If $g = \phi \circ f$ for some increasing function $\phi$, then $A$ is empty. Also in this case, $g^\# = \phi \circ f^\#$. Now, if functions $f$ and $h$ have the same distribution, and if $\phi$ is increasing, then it is not difficult to show that the pairs $(f, \phi \circ f)$ and $(h, \phi \circ h)$ have the same joint distribution. Thus, $(f, g)$ and $(f^\#, g^\#)$ have the same joint distribution. By Proposition 1.18(a), $\int_{\mathbb{R}^n} \Psi(f(x), g(x))\, dx = \int_{\mathbb{R}^n} \Psi(f^\#(x), g^\#(x))\, dx$.

As with Theorem 2.9, the hypotheses on $f, g, K$, and $\Psi$ can be relaxed. For example, Theorem 2.15 can be extended to general nonnegative $\Psi \in AL(\mathbb{R}^+ \times \mathbb{R}^+)$ with $\Psi(0,0) = 0$ in the same way that Theorem 2.9 was extended in the paragraphs after its proof. A discussion of extensions appears in §7.3.

Our scheme for proving Theorem 2.15 is as follows:

1. In the remainder of this section, we prove (a) and (c) under the additional assumption that $\Psi$ is continuous. The proof will be similar to the proof in §2.4 that symmetric decreasing rearrangement decreases the modulus of continuity.
2. The proofs of (b) and (d) will appear in the next section.
3. The extension to discontinuous $\Psi$ requires a decomposition theorem for $AL_0$ functions which is deferred to §2.8. The proof of Theorem 2.15 is completed in §2.9.

For most applications of Theorem 2.15 one can assume that $\Psi$ is in fact continuous. The reader who chooses to skip the rather technical Sections 2.8 and 2.9 can do so without missing anything essential for the remainder of this book.

**Proof of (a) and (c) when $\Psi$ is Continuous**

Assume, in addition to (I), (II), (IV), that $\Psi \in AL_0 \cap C(\mathbb{R}^+ \times \mathbb{R}^+)$. It will suffice to prove (a) and (c) under the additional assumptions that

$$K(0) < \infty \quad \text{and} \quad K(t) = 0, \quad \text{for all sufficiently large } t. \tag{2.19}$$

For suppose we have proved 2.15(a) under the stated restrictions on $K$. Then, for general decreasing $K\colon \mathbb{R}^+ \to \mathbb{R}^+$, for positive integers $m$ set $K_m(t) = \min(K(t), m)$ for $0 \le t \le m$, $K_m(t) = 0$ for $t \ge m$. Then 2.15(a) holds when $K$ is replaced by $K_m$. Moreover, $K_m(t) \nearrow K(t)$ for each $t \in [0, \infty)$. In §2.3 we noted that $AL_0$ functions are nonnegative. Thus, for almost every $x, y \in \mathbb{R}^n$, $0 \le \Psi(f(x), g(y))K_m(|x-y|) \nearrow \Psi(f(x), g(y))K(|x-y|)$ as $m \to \infty$. and likewise when $f$ and $g$ are replaced by $f^\#$ and $g^\#$. Using the monotone convergence theorem, we obtain

$$\begin{aligned} Q(f, g, K, \Psi) &= \lim_{m\to\infty} Q(f, g, K_m, \Psi) \\ &\le \lim_{m\to\infty} Q(f^\#, g^\#, K_m, \Psi) = Q(f^\#, g^\#, K, \Psi). \end{aligned}$$

Similar considerations show that it suffices to prove (c) when $K$ satisfies (2.19).

**Proof of Theorem 2.15(a), Part 1**

Assume first that $f$ and $g$ are in $C_c(\mathbb{R}^n, \mathbb{R}^+)$. Let $\mathcal{S}(f)$ and $\mathcal{S}(g)$ denote the subsets of $C_c(\mathbb{R}^n, \mathbb{R}^+)$ associated with $f$ and $g$ respectively via (2.12). Define $\mathcal{S} \subset \mathcal{S}(f) \times \mathcal{S}(g)$ and $d \geq 0$ by

$$\mathcal{S} = \mathcal{S}(f,g) = \{(F,G) \in \mathcal{S}(f) \times \mathcal{S}(g) \colon Q(f,g) \leq Q(F,G)\},$$

$$d^2 = \inf_{(F,G)\in\mathcal{S}} \|f^\# - F\|_2^2 + \|g^\# - G\|_2^2,$$

where $\|\cdot\|_2$ is the norm in $L^2(\mathbb{R}^n, \mathcal{L}^n)$.

Let $\{f_k\}_{k=1}^\infty$ and $\{g_k\}_{k=1}^\infty$ be sequences in $\mathcal{S}(f)$ and $\mathcal{S}(g)$ respectively such that

$$d^2 = \lim_{k\to\infty} \|f^\# - f_k\|_2^2 + \|g^\# - g_k\|_2^2. \tag{2.20}$$

We shall choose possibly new minimizing sequences $\{F_k\}$ and $\{G_k\}$ consisting of functions with uniformly bounded supports. Fix a number $R$ so large that $2R \geq \max(\operatorname{diam} \operatorname{supp} f, \operatorname{diam} \operatorname{supp} g)$. For $k \geq 1$, let $\overline{B}_k$ and $\overline{B'_k}$ be closed balls of radius $R$ such that $\overline{B}_k$ contains the support of $f_k$ and $\overline{B'_k}$ contains the support of $g_k$. We consider three cases.

**Case 1.** If $\overline{B}_k \cap \overline{B'_k}$ is nonempty and $\overline{B}_k \cup \overline{B'_k}$ intersects $\overline{B}(2R)$, the closed ball with radius $2R$ centered at the origin, take $F_k = f_k$ and $G_k = g_k$.

**Case 2.** Suppose that $\overline{B}_k \cap \overline{B'_k}$ is nonempty and $\overline{B}_k \cup \overline{B'_k}$ does not intersect $\overline{B}(2R)$. Let $x_k$ be a point of $\overline{B}_k \cup \overline{B'_k}$ closest to the origin, and let $H$ be the affine hyperplane passing through the midpoint of the line segment from 0 to $x_k$ and orthogonal to it. One easily verifies that the origin and $\overline{B}_k \cup \overline{B'_k}$ lie on different sides of $H$. Let $H^+$ be the halfspace containing the origin. Define $F_k = (f_k)_H$ and $G_k = (g_k)_k$.

**Case 3.** Suppose that $\overline{B}_k \cap \overline{B'_k}$ is empty. Let $H_1$ be the affine hyperplane passing through the midpoint of the line segment connecting the centers of $\overline{B}_k$ and $\overline{B'_k}$ and orthogonal to it. If $0 \notin H_1$, let $H_1^+$ be the halfspace which contains the origin. If $0 \in H_1$ we can take $H_1^+$ to be either of the complementary halfspaces. Suppose that $\overline{B}_k \subset H_1^+$; the argument when $\overline{B'_k} \subset H_1^+$ is analogous. The supports of $(f_k)_{H_1}$ and $(g_k)_{H_1}$ are each contained in $\overline{B}_k$. If $\overline{B}_k \cap \overline{B}(2R)$ is nonempty, define $F_k = (f_k)_{H_1}$ and $G_k = (g_k)_{H_1}$. If $\overline{B}_k \cap \overline{B}(2R)$ is empty, let $x_k$ be the point of $\overline{B}_k$ closest to the origin, and define $F_k$ and $G_k$ to be the respective polarizations of $(f_k)_{H_1}$ and $(g_k)_{H_1}$ in the affine hyperplane $H$ passing through the midpoint of the line segment joining the origin to $x_k$ and orthogonal to that segment. Take $H^+$ to be the halfspace containing the origin.

In Case 1, the supports of $F_k$ and $G_k$ are contained in $\overline{B}(6R)$; in Cases 2 and 3 they are contained in $\overline{B}(4R)$. Thus the supports of all the $F_k$ and $G_k$ are contained in $\overline{B}(6R)$.

As in the proof of Theorem 2.12, each set $\mathcal{S}(f)$ and $\mathcal{S}(g)$ is mapped into itself by polarization. By Theorem 2.9, polarization increases the $Q$ integral. Thus, $(F_k, G_k) \in \mathcal{S}$ for $k \geq 1$. The computation in (2.13) shows that $\|F_k - f^\#\|_2 \leq \|f_k - f^\#\|_2$ and $\|G_k - g^\#\|_2 \leq \|g_k - g^\#\|_2$. Thus, (2.20) holds when $f_k$ and $g_k$ are replaced by $F_k$ and $G_k$.

By the Arzelà–Ascoli Theorem, there is a subsequence of $\{(F_k, G_k)\}$, denoted also by $\{(F_k, G_k)\}$, such that $\{F_k\}$ converges uniformly in $\mathbb{R}^n$ to a function $F_0 \in C_c(\mathbb{R}^n, \mathbb{R}^+)$ and $\{G_k\}$ converges uniformly in $\mathbb{R}^n$ to a function $G_0 \in C_c(\mathbb{R}^n, \mathbb{R}^+)$. Since the sequences are uniformly bounded and have uniformly bounded supports, the dominated convergence theorem can be applied to prove that

$$d^2 = \|F_0 - f^\#\|_2^2 + \|F_0 - f^\#\|_2^2. \tag{2.21}$$

As in the proof of Theorem 2.12, $F_0$ and $G_0$ belong respectively to $\mathcal{S}(f)$ and $\mathcal{S}(g)$. To obtain $(F_0, G_0) \in \mathcal{S}$, we need to know that $Q(F_0, G_0) \geq Q(f, g)$. Let $M$ be a number with $M \geq \max(\|f\|_\infty, \|g\|_\infty)$. Then $M$ is an upper bound for each $F_k$ and $G_k$. As noted at the beginning of §2.3, $\Psi$ is increasing on each horizontal and each vertical line, from which we deduce that $0 \leq \Psi(F_k(x), G_k(y)) \leq \Psi(M, M)$. Moreover, $\Psi(F_k(x), G_k(y)) = 0$ if either $x$ or $y$ lies outside $\overline{B}(6R)$, $K$ is nondecreasing, and $K(0) < \infty$, by assumption (2.19). Thus,

$$0 \leq \Psi(F_k(x), G_k(y))K(|x - y|) \leq \Psi(M, M)K(0)\chi_{\overline{B}(6R)\times\overline{B}(6R)}(x, y). \tag{2.22}$$

The dominated convergence theorem is again applicable, and yields $\lim_{k\to\infty} Q(F_k, G_k) = Q(F_0, G_0)$. Thus, $Q(F_0, G_0) \geq Q(f, g)$, so that $(F_0, G_0) \in \mathcal{S}$.

Suppose that $F_0 \neq f^\#$ or that $G_0 \neq g^\#$, say the former. Proceed as in the proof of Claim 2 in §2.4. There exists $H \in \mathcal{H}(\mathbb{R}^n)$ such that $\|(F_0)_H - f^\#\|_2 < \|F_0 - f^\#\|_2$. Also, $\|(G_0)_H - g^\#\|_2 \leq \|G_0 - g^\#\|_2$, as in (2.13). Thus, $((F_0)_H, (G_0)_H) \in \mathcal{S}$, so that from (2.21) it follows that $\|(F_0)_H - f^\#\|_2^2 + \|(G_0)_H - g^\#\|_2^2 < d^2$.

This contradiction to the definition of $d$ shows that $(F_0, G_0) = (f^\#, g^\#)$, so that $(f^\#, g^\#) \in \mathcal{S}$. In particular, $Q(f, g) \leq Q(f^\#, g^\#)$. Theorem 2.15 is proved when $f$ and $g$ are in $C_c(\mathbb{R}^n, \mathbb{R}^+)$. □

**Proof of Theorem 2.15(a), Part 2**

Assume that $f, g \in L^\infty(\mathbb{R}^n)$, that $K$ satisfies (II) and (2.19), and that $f$ and $g$ have compact support, say $\operatorname{supp} f \subset \overline{B}(R)$ and $\operatorname{supp} g \subset \overline{B}(R)$ for some $R \in (0, \infty)$. Take sequences $\{f_k\}_{k=1}^\infty$ and $\{g_k\}_{k=1}^\infty$ in $C_c(\mathbb{R}^n, \mathbb{R}^+)$ such that $f_k \to f$ in $L^2(\mathbb{R}^n, \mathcal{L}^n)$ and $g_k \to g$ in $L^2(\mathbb{R}^n, \mathcal{L}^n)$. We assume also that the support of each $f_k$ and $g_k$ is contained in $\overline{B}(R)$, and that $\|f_k\|_\infty \leq \|f\|_\infty$, $\|g_k\|_\infty \leq \|g\|_\infty$. By passing to a subsequence, if necessary, we may assume further that $f_k \to f$ and $g_k \to g$ a.e. on $\mathbb{R}^n$. Then estimate (2.22) holds with $F_k$ and $G_k$ replaced by $f_k$ and $g_k$ and $M \geq \max(\|f\|_\infty, \|g\|_\infty)$. The dominated convergence theorem gives

$$\lim_{k\to\infty} Q(f_k, g_k) = Q(f, g). \tag{2.23}$$

Now $f_k$ and $g_k$ converge in measure to $f$ and $g$, respectively. From Proposition 1.41 and Definition 1.29, it follows that $f_k^\#$ and $g_k^\#$ converge in measure to $f^\#$ and $g^\#$, respectively. Choose subsequences which converge a.e. The functions $f_k^\#$ and $g_k^\#$ are all supported in $\overline{B}(R)$ and have sup norms bounded above by $M$. Thus, (2.22) holds for $f_k^\#$ and $g_k^\#$, and so does (2.23), with $Q(f, g)$ replaced by $Q(f^\#, g^\#)$. Since $Q(f_k, g_k) \leq (f_k^\#, g_k^\#)$ for each $k$, we have $Q(f, g) \leq Q(f^\#, g^\#)$, as desired.

**Proof of Theorem 2.15(a), Part 3**

Assume that $f$ and $g$ are arbitrary functions satisfying (I) and that $K$ satisfies (II) and (2.19). For $k \geq 1$, Let

$$f_k = \chi_{B(k)} \min(f, k), \quad g_k = \chi_{B(k)} \min(g, k).$$

Then $f_k \nearrow f$ and $g_k \nearrow g$ a.e. As remarked at the beginning of the proof, we know that $\Psi$ is increasing on each horizontal and vertical line. Thus, $\Psi(f_k(x), g_k(y)) \leq \Psi(f_k(x), g_{k+1}(y)) \leq \Psi(f_{k+1}(x), g_{k+1}(y))$, so that $0 \leq \Psi(f_k(x), g_k(y))K(|x-y|) \nearrow \Psi(f(x), g(y))K(|x-y|)$ a.e. on $\mathbb{R}^{2n}$. The monotone convergence theorem gives $Q(f_k, g_k) \nearrow Q(f, g)$. By Proposition 1.39 and Definition 1.29, monotone increasing convergence of a nonnegative sequence of functions implies monotone increasing convergence of the sequence of symmetric decreasing rearrangements. Thus $Q(f_k^\#, g_k^\#) \nearrow Q(f^\#, g^\#)$. Since $Q(f_k, g_k) \leq Q(f_k^\#, g_k^\#)$ for each $k$, we have $Q(f, g) \leq Q(f^\#, g^\#)$. Theorem 2.15(a) is completely proved for continuous $\Psi$. $\square$

**Proof of Theorem 2.15(c)**

For continuous, compactly supported $f$ and $g$, the proof that the integral $\int_{\mathbb{R}^n} \Psi(f(x), g(x))\, dx$ increases under symmetric decreasing rearrangement can be accomplished in the same way that $Q(f, g)$ is shown to increase. The only noteworthy difference is that in the definition of $\mathcal{S}(f, g)$ the condition involving $Q$ is replaced by the condition

$$\int_{\mathbb{R}^n} \Psi(f(x), g(x))\, dx \le \int_{\mathbb{R}^n} \Psi(F(x), G(x))\, dx.$$

The approximation steps Parts 2 and 3 are carried out just as in the proof of Theorem 2.15(a). □

Another way to establish Theorem 2.15(c) is to observe that it is a limiting case of Theorem 2.15(a). For $\epsilon > 0$, take $K = \chi_{[-\epsilon,\epsilon]}$. Then for $f, g$ satisfying (I) and $\Psi$ satisfying (III), Theorem 2.15(a) gives

$$\int_{|x-y|\le\epsilon} \Psi(f(x), g(y))\, dx\, dy \le \int_{|x-y|\le\epsilon} \Psi(f^{\#}(x), g^{\#}(y))\, dx\, dy. \tag{2.24}$$

If, for example, $f,\ g \in C_c(\mathbb{R}^n, \mathbb{R}^+)$, then, as the reader is invited to show,

$$\lim_{\epsilon\to 0} \frac{1}{\alpha_n \epsilon^n} \int_{|x-y|\le\epsilon} \Psi(f(x), g(y))\, dx\, dy = \int_{\mathbb{R}^n} \Psi(f(x), g(x))\, dx, \tag{2.25}$$

where $\alpha_n$ is the volume of the unit ball in $\mathbb{R}^n$. Identity (2.25) also holds when $f$ and $g$ are replaced by $f^{\#}$ and $g^{\#}$. Thus, Theorem 2.15(c) for $f, g \in C_c(\mathbb{R}^n, \mathbb{R}^+)$, follows from (2.24) and (2.25).

## 2.6 Proofs of the Uniqueness Statements

**Proof of Theorem 2.15(b)**

Let $f, g, K$, and $\Psi$ satisfy the assumptions of Theorem 2.15(b). As noted after the statement of Theorem 2.15, if a translation $T$ exists for which (2.18) holds for $f$ and $g$ then $Q(f,g) = Q(f^{\#}, g^{\#})$. Let us assume then that no such $T$ exists and that $Q(f,g) < \infty$. We shall prove that

$$Q(f,g) < Q(f^{\#}, g^{\#}). \tag{2.26}$$

Write $E(t,f) = (f > t)$, $E(t,g) = (g > t)$. Assume for now that the sets $E(t,f)$ and $E(t,g)$ are bounded for every $t > 0$. For bounded measurable sets $E \subset \mathbb{R}^n$ we shall let $c(E)$ denote the center of mass of $E$. Thus,

$$c(E) = \frac{1}{\mathcal{L}^n(E)} \int_E x\, d\mathcal{L}^n.$$

The proof of (2.26) under the added boundedness hypothesis is split into two cases.

**Case 1.** Assume there exist $s \in (0, \operatorname{ess\,sup} f)$ and $t \in (0, \operatorname{ess\,sup}\, g)$ such that $c(E(s,f)) \neq c(E(t,g))$. Performing a translation and rotation, if necessary, we may assume that $c(E(t,g)) = -Re_1$ and $c(E(s,f)) = Re_1$ for some $0 < R < \infty$.

Write $x_1 = x\cdot e_1$. Let $H$ be the hyperplane $x_1 = 0$, and $H^+ = \{x \in \mathbb{R}^n : x_1 > 0\}$. Then, with $\rho$ denoting reflection in $H$,

$$R\mathcal{L}^n(E(s,f)) = \int_{E(s,f)} x_1\,dx = \int_{E(s,f)\cap H^+} x_1\,dx - \int_{\rho E(s,f)\cap H^+} x_1\,dx. \quad (2.27)$$

Let $A_1 = (E(s,f)\setminus\rho E(s,f))\cap H^+$. Since $R > 0$ and $x_1 > 0$ on $H^+$, it follows from (2.27) that $\mathcal{L}^n(A_1) > 0$. Similarly, letting $A_2 = (E(t,g)\setminus\rho E(t,g))\cap H^-$, we have $\mathcal{L}^n(A_2) > 0$. If $x \in A_1$ and $z \in A_2$, then $f(\rho x) \le s < f(x)$ and $g(\rho z) \le t < g(z)$. Thus, if $(x,y) \in A_1 \times \rho A_2$, then $x, y \in H^+$, $f(\rho x) < f(x)$, and $g(y) < g(\rho y)$. Hence, $A_1 \times \rho A_2 \subset A$, where $A$ is the set in (v), so that $\mathcal{L}^{2n}(A) \ge \mathcal{L}^{2n}(A_1 \times \rho A_2) > 0$. From Theorems 2.9(b) and 2.15(a), we deduce $Q(f,g) < Q(f_H, g_H) \le Q(f^\#, g^\#)$, which confirms (2.26).

**Case 2.** Assume that $c(E(s,f)) = c(E(t,g))$ for every $s \in (0, \operatorname{ess\,sup} f)$ and $t \in (0, \operatorname{ess\,sup} g)$. Then, by fixing $s$ and letting $t$ vary, and vice versa, we see that all the sets $E(s,f)$ and $E(t,g)$ have a common center of mass, which we may take to be the origin. If $f$ and $g$ were both symmetric decreasing then (2.18) would hold with $T$ the identity map, contrary to our assumption. Thus, at least one of $f$ or $g$ is not a.e. equal to its symmetric decreasing rearrangement, say $f \ne f^\#$. Now

$$(f \ne f^\#) = (f < f^\#) \cup (f^\# < f) = \bigcup_s (f \le s < f^\#) \cup \bigcup_s (f^\# \le s < f),$$

where $s$ runs through the set of positive rational numbers. Thus, there exists $s > 0$ such that at least one of the sets $E_1 \equiv (f^\# \le s < f)$ or $E_2 \equiv (f \le s < f^\#)$ has positive $\mathcal{L}^n$-measure. The sets $E(s,f)$ and $E(s,f^\#)$ have the same measure, and $E_1 = E(s,f)\setminus E(s,f^\#)$, $E_2 = E(s,f^\#)\setminus E(s,f)$. We conclude that both $E_1$ and $E_2$ have positive measure. Now $E(s,f^\#)$ is an open ball $B(R)$ for some $R > 0$. If $x \in E_1$ and $z \in E_2$, then $|z| < R \le |x|$, and $f(z) \le s < f(x)$. Hence,

$$(x,z) \in E_1 \times E_2 \implies |z| < |x| \quad \text{and} \quad f(z) < f(x). \quad (2.28)$$

Define $F\colon \mathbb{R}^n \times (\mathbb{R}^n \setminus \{0\}) \to \mathbb{R}^n \times \mathbb{R}^n$ by

$$F(x,w) = (x, \rho(x,w)), \quad (2.29)$$

where $\rho(\cdot, w)$ denotes reflection in the hyperplane $H(w)$ orthogonal to the line through 0 and $w$ which passes through $\frac{1}{2}w$. It is easy to verify that $F$ is Lipschitz on each compact subset of $\mathbb{R}^n \times (\mathbb{R}^n \setminus \{0\})$, and that $F$ maps $\mathbb{R}^n \times (\mathbb{R}^n \setminus \{0\})$ onto $(\mathbb{R}^n \times \mathbb{R}^n)\setminus\{(x,y) : |x| = |y|\}$.

In particular, $F(\mathbb{R}^n \times (\mathbb{R}^n \setminus \{0\}))$ contains $E_1 \times E_2$. Let $E_3 = F^{-1}(E_1 \times E_2)$. Since locally Lipschitz maps take nullsets to nullsets, it follows that $\mathcal{L}^{2n}(E_3) > 0$. Hence, $\int_{\mathbb{R}^n} dw \int_{\mathbb{R}^n} \chi_{E_3}(x,w)\,dx > 0$. Take $w$ such that

$E_4 \equiv \{x \in \mathbb{R}^n : (x, w) \in E_3\}$ has $\mathcal{L}^n(E_4) > 0$, and write $\rho x = \rho(x, w)$. If $x \in E_4$, then $(x, \rho x) \in E_1 \times E_2$. From (2.28), we see that

$$x \in E_4 \implies |\rho x| < |x| \quad \text{and} \quad f(\rho x) < f(x). \tag{2.30}$$

Write $H = H(w)$, and let $H^+$ denote the halfspace which contains the origin. Note that if $x \in E_4$, then the inequality $|\rho x| < |x|$ implies that $x \in H^-$ and $\rho x \in H^+$. Take $t \in (0, \operatorname{ess\,sup} g)$. Let $E_5 = (E(t, g) \setminus \rho E(t, g)) \cap H^+$. If $\mathcal{L}^n(E_5)$ were zero, a "balancing" argument like the one in Case 1 would imply that the center of mass of $E(t, g)$ is in $H \cup H^-$. This contradicts our assumption that the center of mass of $E(t, g)$ is the origin, Thus, $\mathcal{L}^n(E_5) > 0$. Using (2.30), we see that $\rho E_4 \times E_5 \subset A$, where $A$ is the set in (v). As in Case 1, we deduce that $Q(f, g) < Q(f^\#, g^\#)$.

We have now proved Theorem 2.15(b) under the additional assumption that all of the sets $E(t,f)$ or $E(t, g)$ are essentially bounded. Let us address the remaining case, when at least one of these sets is essentially unbounded. Suppose, say, that $E(s,f)$ is essentially unbounded for a certain $s \in (0, \operatorname{ess\,sup} f)$. Write $E_1 = E(s,f)$. Take any $t \in (0, \operatorname{ess\,sup} g)$. Write $E_2 = E(t, g)$. Since $E_1$ and $E_2$ have finite measure, there exists $R_0$ such that for each $R \geq R_0$ the ball $B(R) = (|x| < R)$ satisfies

$$\mathcal{L}^n(B(R) \setminus E_1) > 0 \quad \text{and} \quad \mathcal{L}^n(B(R) \cap E_2) > \frac{1}{2}\mathcal{L}^n(E_2). \tag{2.31}$$

There exist points of density $x_0$ of $E_1$ such that $|x_0|$ is arbitrarily large. Take such an $x_0$ with $|x_0| > 4R_0$. After rotation, we may assume that $x_0$ lies on the positive $x_1$-axis. Write $x_0 = 4Re_1$. Let $w_0$ be a point of density of $B(R) \setminus E_1$, and let $H$ be the affine hyperplane orthogonal to the line through $x_0$ and $w_0$ which passes through their midpoint. Let $\rho$ denote reflection in $H$. Then $\rho$ maps balls centered at $x_0$ to balls centered at $w_0$ with the same radius. Choose $\delta$ so small that $B \equiv B(x_0, \delta)$ satisfies $\mathcal{L}^n(B \cap E_1) > \frac{1}{2}\mathcal{L}^n(B)$, $\mathcal{L}^n((\rho B) \setminus E_1) > \frac{1}{2}\mathcal{L}^n(B)$, $B \subset \mathbb{R}^n \setminus B(3R)$, and $\rho B \subset B(R)$. Let $E_3 = \{x \in B \cap E_1 : \rho x \notin E_1\}$. Then $\mathcal{L}^n(E_3) > 0$, since $E_3$ is the intersection of two subsets of $B$ each of which has measure larger than $\frac{1}{2}\mathcal{L}^n(B)$. Moreover,

$$x \in E_3 \implies f(x) > s \geq f(\rho x). \tag{2.32}$$

Let $z \in B(R)$. Then $|z - w_0| \leq 2R$ and $|z - x_0| \geq 3R$. Hence, $|z - w_0| < |z - x_0|$, so that $z$ and $x_0$, hence $B(R)$ and $x_0$, lie on different sides of $H$. Let $H^+$ be the halfspace which contains $B(R)$, and let $E_4 = \{y \in B(R) \cap E_2 : \ \rho y \notin E_2\}$. From the second inequality in (2.31), and the fact that $B(R)$ and $\rho B(R)$ are disjoint, it follows that $\mathcal{L}^n(E_4) > 0$. Moreover,

$$y \in E_4 \implies g(y) > t \geq g(\rho y). \tag{2.33}$$

From (2.32) and (2.33) it follows that the set $A$ of (v) contains $\rho E_3 \times E_4$. Hence $\mathcal{L}^{2n}(A) > 0$. As in Cases 1 and 2, we deduce that $Q(f,g) < Q(f^\#,g^\#)$. The proof of part Theorem 2.15(b) is complete. □

**Proof of Theorem 2.15(d)**

Let $f, g$, and $\Psi$ satisfy the assumptions of Theorem 2.15(d). Recall that the set $A$ of (V) is

$$A = \{(x,y) \in \mathbb{R}^{2n} : f(x) < f(y) \text{ and } g(x) > g(y)\}.$$

Suppose that $\mathcal{L}^{2n}(A) = 0$. Take $s \in (0, \operatorname{ess\,sup} f)$, $t \in (0, \operatorname{ess\,sup} g)$. Suppose that $\lambda_f(s) \le \lambda_g(t)$; if the contrary holds, we reverse the roles of $f$ and $g$ in the argument to follow. Then $E(s,f^\#) \subset E(t,g^\#)$, so that

$$\mathcal{L}^n(f^\# > s,\ g^\# > t) = \lambda_f(s). \tag{2.34}$$

Also, $(f > s) = (f > s,\ g > t) \cup (f > s,\ g \le t)$. Suppose that the second set on the right has positive measure. Then $\lambda_g(t) \ge \lambda_f(s) > \mathcal{L}^n(f > s,\ g > t)$. Hence, $\mathcal{L}^n(g > t, f \le s) > 0$. Now $(g > t, f \le s) \times (f > s,\ g \le t) \subset A$. So if $(f > s,\ g \le t)$ had positive $\mathcal{L}^n$-measure then $A$ would have positive $\mathcal{L}^{2n}$-measure, contrary to our assumption. Thus, the sets $(f > s)$ and $(f > s,\ g > t)$ differ by a nullset, and hence (2.34) holds when $f^\#$ and $g^\#$ are replaced with $f$ and $g$.

We have shown that $\mathcal{L}^n(f > s, g > t) = \mathcal{L}^n(f^\# > s, g^\# > t)$ for every $s$ and $t$. Thus, $(f,g)$ and $(f^\#,g^\#)$ have the same joint distribution. By Proposition 1.18,

$$\int_{\mathbb{R}^n} \Psi(f(x),g(x))\,dx = \int_{\mathbb{R}^n} \Psi(f^\#(x),g^\#(x))\,dx.$$

Conversely, suppose that $\mathcal{L}^{2n}(A) > 0$. Let $F$ be the map (2.29). Then, as in the proof of Case 2 of Theorem 2.15(b), $\mathcal{L}^{2n}(F^{-1}(A)) > 0$, and there exists $w \in \mathbb{R}^n \setminus \{0\}$ such that $E \equiv \{x \in \mathbb{R}^n : (x,w) \in F^{-1}(A)\}$ has $\mathcal{L}^n(E) > 0$. Let $H = H(w)$, and let $\rho$ denote reflection in $H$, as in the proof of Theorem 2.15(b). If $x \in E$, then $(x, \rho x) \in A$, and hence $f(x) < f(\rho x)$, $g(x) > g(\rho x)$. Take $H^+$ to be a halfspace defined by $H$ which intersects $E$ in a set of positive measure. Then $E \cap H^+$ is a subset of the set $A_0$ of (2.6). Thus $\mathcal{L}^n(A_0) > 0$, and from Theorems 2.9(d) and 2.15(c) follow

$$\begin{aligned}\int_{\mathbb{R}^n} \Psi(f(x),g(x))\,dx &< \int_{\mathbb{R}^n} \Psi(f_H(x),g_H(x))\,dx \\ &\le \int_{\mathbb{R}^n} \Psi(f^\#(x),g^\#(x))\,dx.\end{aligned}$$

The proof of Theorem 2.15(d) is complete. □

## 2.7 Direct Consequences of the Main Inequalities

This section contains some inequalities that follow immediately or with modest additional effort from Theorem 2.15. As always for us, $f^\#$ denotes the symmetric decreasing rearrangement of a function $f$ defined on $\mathbb{R}^n$, and $f^*$ the decreasing rearrangement of a function defined on an arbitrary measure space $X$.

The inequalities have associated equality statements. We will provide these statements for Corollaries 2.16 and 2.19; the reader is invited to supply others with the aid of parts (b) and (d) of Theorem 2.15.

**Corollary 2.16** *Let $f$ and $g$ be nonnegative Lebesgue measurable functions on $\mathbb{R}^n$ with $\lambda_f(t) < \infty$ and $\lambda_g(t) < \infty$ for all $t > 0$. Then*

$$\int_{\mathbb{R}^n} fg\,dx \le \int_{\mathbb{R}^n} f^\# g^\#\,dx = \int_{\mathbb{R}^+} f^* g^*\,dx. \tag{2.35}$$

*If $\int_{\mathbb{R}^n} fg\,dx < \infty$, then equality holds in (2.35) if and only if*

$$\mathcal{L}^{2n}(\{(x,y) \in \mathbb{R}^{2n} : f(x) < f(y) \text{ and } g(x) > g(y)\}) = 0.$$

*Proof* The inequality in (2.35) is from Theorem 2.15(c) with $\Psi(x,y) = xy$. Equality of the second and third integrals comes from the representations given by Definition 1.29: $f^\# = f^* \circ T$, $g^\# = g^* \circ T$, where $T(x) = \alpha_n |x|^n$ is measure preserving from $(\mathbb{R}^n, \mathcal{L}^n)$ to $(\mathbb{R}^+, \mathcal{L})$. The uniqueness statement comes from Theorem 2.15(d). □

The next corollary may be regarded as a generalization of (2.35).

**Corollary 2.17** *Let $(X, \mu)$ be a measure space, and $f$ and $g$ be nonnegative measurable functions on $X$ with $\lambda_f(t) < \infty$ and $\lambda_g(t) < \infty$ for all $t > 0$. Then*

$$\int_X fg\,d\mu \le \int_{\mathbb{R}^+} f^* g^*\,dx. \tag{2.36}$$

When $\mu(X) < \infty$, the integral on the right need be taken only over $[0, \mu(X)]$.

*Proof of Corollary 2.17* Assume first that $f$ and $g$ are simple. Write

$$f = \sum_{i=1}^{k} \alpha_i \chi_{E_i}, \quad g = \sum_{i=1}^{m} \beta_i \chi_{F_i},$$

where the $\alpha_i$ and $\beta_i$ are positive, the $E_i$ are disjoint, the $F_i$ are disjoint, and each $E_i$ and $F_i$ have positive $\mu$-measure. For $i = 1, \ldots, k$ take disjoint closed intervals $I_i \subset \mathbb{R}$ such that $\mathcal{L}(I_i) = E_i$. Let

$$f_1 = \sum_{i=1}^{k} \alpha_i \chi_{I_i}.$$

For each $i \in \{1, \ldots, k\}$, let $J_{ij}$, $j = 1, \ldots, m$ be disjoint closed subintervals of $I_i$ such that $\mathcal{L}(J_{ij}) = \mu(E_i \cap F_j)$. Let

$$g_1 = \sum_{i=1}^{k} \sum_{j=1}^{m} \beta_j \chi_{J_{ij}}.$$

Let $X_1 = \bigcup_{i=1}^{k} E_i \cup \bigcup_{i=1}^{m} F_i$ and $S = \bigcup_{i=1}^{k} I_i$. The joint distribution of $(f, g)$ on $(X_1, \mu)$ is the same as that of $(f_1, g_1)$ on $(S, \mathcal{L})$. Thus, using (2.35),

$$\begin{aligned}\int_X fg\, d\mu = \int_{X_1} fg\, d\mu &= \int_S f_1 g_1\, dx \\ &= \int_{\mathbb{R}} f_1 g_1\, dx \leq \int_{\mathbb{R}^+} f_1^* g_1^*\, dx = \int_{\mathbb{R}^+} f^* g^*\, dx.\end{aligned}$$

This proves (2.36) for simple $f$ and $g$. For general $f$ and $g$ satisfying the assumptions of Corollary 2.17, we use a standard method for approximating nonnegative measurable functions by simple functions. For $k \geq 1$ and $x \in X$, if $f(x) \geq k$, define $f_k(x) = k$. If $f(x) \leq k$ and $i + j2^{-k} \leq f(x) < i + (j+1)2^{-k}$, for integers $i \in \{0, \ldots, k-1\}$ and $j \in \{0, \ldots, 2^k - 1\}$, define $f_k(x) = i + j2^{-k}$. Then $f_k \nearrow f$ at every point of $X$ where $f$ is defined, and hence a.e. By Proposition 1.39, $f_k^* \nearrow f^*$ on $\mathbb{R}^+$. Let $g_k$ denote the corresponding approximants to $g$. Then, using the monotone convergence theorem,

$$\int_X fg\, d\mu = \lim_{k\to\infty} \int_X f_k g_k\, d\mu \leq \liminf_{k\to\infty} \int_{\mathbb{R}^+} f_k^* g_k^*\, dx = \int_{\mathbb{R}^+} f^* g^*\, dx. \qquad \square$$

Next, we prove an inequality that goes in the opposite direction from Corollary 2.17. Assume that $\mu(X) < \infty$. Set $A = \mu(X)$. For nonnegative measurable $f$ on $X$, define a function $f_*$ on $[0, A]$ by

$$f_*(x) = f^*(A - x).$$

Then $f_*$ is increasing, and, since $x \to A - x$ is $\mathcal{L}$-measure preserving, $f_*$ is a rearrangement of $f^*$ and hence of $f$. We call $f_*$ the *increasing rearrangement* of $f$.

**Corollary 2.18** *Let $(X, \mu)$ be a measure space with $\mu(X) < \infty$, and $f$ and $g$ be nonnegative measurable functions on $X$. Then, with $A = \mu(X)$,*

$$\int_0^A f_* g^*\, dx \leq \int_X fg\, d\mu. \tag{2.37}$$

*Proof* Assume first that $f \in L^\infty(X)$ and $g \in L^1(X)$ Take a number $M \geq \|f\|_\infty$, and define $h = M - f$. Then $h \geq 0$ and $h^* = M - f_*$. By Corollary 2.17 and the fact that $g^*(x) = 0$ for $x \geq A$,

$$\int_X gh\, d\mu \leq \int_0^A g^* h^*\, dx = M \int_0^A g^*\, dx - \int_0^A g^* f_*\, dx.$$

Since $\int_X gh\,d\mu = M\int_X g\,d\mu - \int_X gf\,d\mu$ and $\int_X g\,d\mu = \int_0^A g^*\,dx$, (2.37) follows. Corollary 2.18 is proved when $f \in L^\infty$, $g \in L^1$.

For general nonnegative $f$ and $g$, set $f_k = \min(f,k)$, $g_k = \min(g,k)$ for $k \geq 1$. Then $f_k$ and $g_k$ increase a.e. to $f$, respectively $g$. Also, $(f_k)_* = \min(f_*,k)$, $g_k^* = \min(g^*,k)$, so $(f_k)_*$ and $g_k^*$ increase a.e. to $f_*$ and $g^*$, respectively. The validity of (2.37) for $f$ and $g$ follows from its validity for $f_k$, $g_k$ and the monotone convergence theorem. □

**Corollary 2.19** *Let $f$ and $g$ be nonnegative Lebesgue measurable functions on $\mathbb{R}^n$ with $\lambda_f(t) < \infty$ and $\lambda_g(t) < \infty$ for all $t > 0$, and let $K\colon \mathbb{R}^+ \to \mathbb{R}^+$ be decreasing. Then*

$$\int_{\mathbb{R}^{2n}} f(x)g(y)K(|x-y|)\,dx\,dy \leq \int_{\mathbb{R}^{2n}} f^\#(x)g^\#(y)K(|x-y|)\,dx\,dy.$$

*Suppose that neither $f$ nor $g$ is identically zero, that $K$ is strictly decreasing on $\mathbb{R}^+$ and that $Q(f,g) < \infty$. Then equality holds if and only if there exists a translation $T$ of $\mathbb{R}^n$ such that $f = f^\# \circ T$ and $g = g^\# \circ T$ a.e. on $\mathbb{R}^n$.*

*Proof* Apply Theorem 2.15(a) and (b) with $\Psi(x,y) = xy$. □

The inequality in Corollary 2.19 is a special case of the Riesz convolution inequality, which will be proved in Chapter 8.

The next three corollaries contain inequalities for convex integrands. Let $\Phi\colon \mathbb{R}^+ \to \mathbb{R}^+$ be convex and increasing. Then, by Fact 2.7,

$$\Psi(x,y) \equiv \Phi(x) + \Phi(y) - \Phi(|x-y|) \in AL(\mathbb{R}^+ \times \mathbb{R}^+). \tag{2.38}$$

**Corollary 2.20** *Let $f$ and $g$ be nonnegative Lebesgue measurable functions on $\mathbb{R}^n$ with $\lambda_f(t) < \infty$ and $\lambda_g(t) < \infty$ for all $t > 0$, and let $\Phi\colon \mathbb{R}^+ \to \mathbb{R}^+$ be convex and increasing with $\Phi(0) = 0$. Then*

$$\int_{\mathbb{R}^n} \Phi(|f^\#(x) - g^\#(x)|)\,dx \leq \int_{\mathbb{R}^n} \Phi(|f(x) - g(x)|)\,dx. \tag{2.39}$$

The choice $\Phi(t) = t^p$ shows that, for $1 \leq p < \infty$, the map $f \to f^\#$ is a contraction on $L^p(\mathbb{R}^n, \mathcal{L}^n)$:

$$\int_{\mathbb{R}^n} |f^\#(x) - g^\#(x)|^p\,dx \leq \int_{\mathbb{R}^n} |f(x) - g(x)|^p\,dx, \quad 1 \leq p < \infty.$$

In Corollary 2.23 below it is shown that contractivity also holds for $p = \infty$. But for $0 < p < 1$ it fails. Consider, for example, on $\mathbb{R}$, $f = \chi_{[-\frac{1}{2},\frac{1}{2}]}$, $g = 2\chi_{[-\frac{1}{2},\frac{1}{2}]} + 3\chi_{[-1,-\frac{1}{2}]\cup[\frac{1}{2},1]}$. Then, we have

$$\|f-g\|_p^p = 1^p + 3^p, \quad \|f^\# - g^\#\|_p^p = 2\cdot 2^p.$$

Since $x^p$ is concave for $0 < p < 1$, it follows that $\|f - g\|_p < \|f^\# - g^\#\|_p$. Using the same kind of argument, one can show that for every $\Phi$ which fails to be convex, (2.39) will fail for some $f$ and $g$.

*Proof of Corollary 2.20* Suppose first that $f$ and $g$ are bounded and have compact support. Let $\Psi$ be the function in (2.38). Then

$$\int_{\mathbb{R}^n} \Psi\big(f(x), g(x)\big)\, dx$$
$$= \int_{\mathbb{R}^n} \Phi\big(f(x)\big)\, dx + \int_{\mathbb{R}^n} \Phi\big(g(x)\big)\, dx - \int_{\mathbb{R}^n} \Phi\big(|f(x) - g(x)|\big)\, dx.$$

When $f$ and $g$ are changed to $f^\#$ and $g^\#$, Theorem 2.15(c) implies that the left-hand side increases, while the first two integrals on the right-hand side do not change. Since all the integrals are finite, (2.39) follows.

For general $f$ and $g$ satisfying the hypotheses of Corollary 2.20, define, for $k \geq 1$, $f_k = \chi_{B(k)} \min(f, k)$, $g_k = \chi_{B(k)} \min(g, k)$. One easily checks that $|f_k - g_k| \leq |f_{k+1} - g_{k+1}|$, so that $\Phi(|f_k - g_k|) \nearrow \Phi(|f - g|)$ a.e. By Proposition 1.39 and Definition 1.29, $f_k^\# \nearrow f^\#$ and $g_k^\# \nearrow g^\#$. By Fatou's lemma and the monotone convergence theorem,

$$\int_{\mathbb{R}^n} \Phi(|f^\# - g^\#|)\, dx \leq \liminf_{k\to\infty} \int_{\mathbb{R}^n} \Phi(|f_k^\# - g_k^\#|)\, dx$$
$$\leq \liminf_{k\to\infty} \int_{\mathbb{R}^n} \Phi(|f_k - g_k|)\, dx = \int_{\mathbb{R}^n} \Phi(|f - g|)\, dx. \qquad \square$$

**Corollary 2.21** *Let $(X, \mu)$ be a measure space, $f$ and $g$ be nonnegative measurable functions on $X$ with $\lambda_f(t) < \infty$ and $\lambda_g(t) < \infty$ for all $t > 0$, and $\Phi\colon \mathbb{R}^+ \to \mathbb{R}^+$ be convex and increasing with $\Phi(0) = 0$. Then*

$$\int_{\mathbb{R}^+} \Phi(|f^* - g^*|)\, dx \leq \int_X \Phi(|f - g|)\, d\mu. \tag{2.40}$$

*Proof* Suppose first that $f$ and $g$ are simple. Let $f_1 : \mathbb{R} \to \mathbb{R}^+$ and $g_1 : \mathbb{R} \to \mathbb{R}^+$ be the functions defined in the proof of Corollary 2.17. Then $(f, g)$ and $(f_1, g_1)$ have the same joint distribution, and so do $(f^\#, g^\#)$ and $(f_1^*, g_1^*)$. Using Proposition 1.18(a) and Corollary 2.20, we have

$$\int_{\mathbb{R}^+} \Phi(|f^* - g^*|)\, dx = \int_{\mathbb{R}^+} \Phi(|f_1^* - g_1^*|)\, dx = \int_{\mathbb{R}} \Phi(|f_1^\# - g_1^\#|)\, dx$$
$$\leq \int_{\mathbb{R}} \Phi(|f_1 - g_1|)\, dx = \int_X \Phi(|f - g|)\, d\mu.$$

Suppose next that $f$ and $g$ are bounded, and that $\mu(f > 0) < \infty$, $\mu(g > 0) < \infty$. Let $f_k$ and $g_k$ be the simple approximants to $f$ and $g$ defined in the proof of Corollary 2.17. By Proposition 1.39, $f_k^* \nearrow f^*$ and $g_k^* \nearrow g^*$.

Moreover, all the $f_k^*$ and $g_k^*$ are zero on $[A, \infty)$, where $A = \max(\lambda_f(0), \lambda_g(0))$. In (2.40), replace $f$ and $g$ by $f_k$ and $g_k$. The dominated convergence theorem can be applied to pass to the limit on each side, and we obtain (2.40) for $f$ and $g$.

Next, assume that $f$ and $g$ are arbitrary functions satisfying the hypotheses of Corollary 2.21, and that $(X, \mu)$ is $\sigma$-finite. Let $X_k$ be an increasing sequence of subsets of $X$ with finite measure whose union is $X$. Let $f_k = \chi_{X_k} \min(f, k)$, $g_k = \chi_{X_k} \min(g, k)$. Then Proposition 1.39 and the argument at the end of the proof of Corollary 2.20 establish (2.40).

Finally, if $(X, \mu)$ is an arbitrary measure space, let $Y = (f > 0) \cup (g > 0)$. The hypotheses $\lambda_f(t) < \infty$ and $\lambda_g(t) < \infty$ for all $t > 0$ imply that $Y$ is $\sigma$-finite. Moreover, $(f|_Y)^* = f^*$, and similarly for $g$. Thus,

$$\int_{\mathbb{R}^+} \Phi(|f^* - g^*|)\, dx \le \int_Y \Phi(|f - g|)\, d\mu = \int_X \Phi(|f - g|)\, d\mu. \qquad \square$$

**Corollary 2.22** *Let $f$ and $g$ be nonnegative Lebesgue measurable functions on $\mathbb{R}^n$ with $\lambda_f(t) < \infty$ and $\lambda_g(t) < \infty$ for all $t > 0$, let $\Phi\colon \mathbb{R}^+ \to \mathbb{R}^+$ be convex and increasing with $\Phi(0) = 0$, and let $K\colon \mathbb{R}^+ \to \mathbb{R}^+$ be decreasing. Then*

$$\int_{\mathbb{R}^{2n}} \Phi(|f^\#(x) - g^\#(y)|)K(|x - y|)\, dx\, dy \le \int_{\mathbb{R}^{2n}} \Phi(|f(x) - g(y)|)K(|x - y|)\, dx\, dy.$$

*Proof* Like that of Corollary 2.20, except that Theorem 2.15(a) is used instead of Theorem 2.15(c). $\square$

We turn now to the contractivity of the decreasing and symmetric decreasing rearrangements on $L^\infty$.

**Corollary 2.23**

(a) *Let $(X, \mu)$ be a measure space, and let $f, g\colon X \to \mathbb{R}^+$ with $\lambda_f(t) < \infty$ and $\lambda_g(t) < \infty$ for all $t > 0$. Then*

$$\|f^* - g^*\|_{L^\infty(\mathbb{R}^+)} \le \|f - g\|_{L^\infty(X,\mu)}. \tag{2.41}$$

(b) *Let $f, g\colon \mathbb{R}^n \to \mathbb{R}^+$ with $\lambda_f(t) < \infty$ and $\lambda_g(t) < \infty$ for all $t > 0$. Then*

$$\|f^\# - g^\#\|_{L^\infty(\mathbb{R}^n)} \le \|f - g\|_{L^\infty(\mathbb{R}^n)}.$$

*Proof* From Definition 1.29, (b) is a consequence of (a). So we just need to prove (a). Suppose first that $Y \equiv (f > 0) \cup (g > 0)$ has finite measure. Then $\|f - g\|_{L^\infty(Y,\mu)} = \lim_{p\to\infty} \|f - g\|_{L^p(Y,\mu)}$ and $\|f^* - g^*\|_{L^\infty([0,\mu(Y)])} = \lim_{p\to\infty} \|f^* - g^*\|_{L^p([0,\mu(Y)])}$.

Since $\|f - g\|_{L^\infty(Y,\mu)} = \|f - g\|_{L^\infty(X,\mu)}$ and $\|f^* - g^*\|_{L^\infty([0,\mu(Y)])} = \|f^* - g^*\|_{L^\infty(\mathbb{R}^+)}$, (2.41) follows.

For general $f$ and $g$ satisfying the hypotheses of Corollary 2.23(b), define, for $k \geq 1$, $f_k = (f - \frac{1}{k})^+$, $g_k = (g - \frac{1}{k})^+$. Then the set where $f_k$ or $g_k$ is positive has finite measure. Also, $f_k^* = (f^* - \frac{1}{k})^+$ and $g_k^* = (g^* - \frac{1}{k})^+$,

$$\|f - g\|_{L^\infty(X,\mu)} = \lim_{k\to\infty} \|f_k - g_k\|_{L^\infty(X,\mu)}$$

and

$$\|f^* - g^*\|_{L^\infty(\mathbb{R}^+)} = \lim_{k\to\infty} \|f_k^* - g_k^*\|_{L^\infty(\mathbb{R}^+)}.$$

Thus, the validity of (2.41) for $f$ and $g$ follows from its validity for each $f_k$ and $g_k$. □

## 2.8 Decomposition of Monotone and $AL_0$ Functions

Let $\mathcal{M}$ denote the set of all increasing functions $f\colon \mathbb{R}^+ \to \mathbb{R}^+$ with $f(0) = 0$. Recall that $\mathbb{R}^+ = [0,\infty)$, and that "increasing" means nondecreasing. Let $P$ be the set of points in $\mathbb{R}^+$ at which $f$ has a jump. Then $P$ is at most countable, and $f$ is continuous on $\mathbb{R}^+ \setminus P$.

For $p \in P$, define a function $\phi_p \in \mathcal{M}$ by

$$\phi_p(x) = \begin{cases} 0 & \text{if } 0 \leq x < p, \\ f(p) - f(p-) & \text{if } x = p, \\ f(p+) - f(p-) & \text{if } x > p. \end{cases}$$

For $p = 0$, we interpret $p- = 0$.

It is easy to see that $\phi_p \in \mathcal{M}$, that $f - \phi_p \in \mathcal{M}$ and that $f - \phi_p$ is continuous at $p$ as well as on $\mathbb{R}^+ \setminus P$. In similar fashion, defining $\tau = f - \sum_{p\in P} \phi_p$, one can verify that $\tau$ is continuous on $\mathbb{R}^+$ and that $\tau \in \mathcal{M}$. We state this result as a decomposition theorem:

**Theorem 2.24** *Let $f \in \mathcal{M}$ and $P$ be its set of jump points. Then*

$$f(x) = \tau(x) + \sum_{p\in P} \phi_p(x), \quad \forall\, x \in \mathbb{R}^+,$$

*where $\tau \in \mathcal{M} \cap C(\mathbb{R}^+)$.*

Associate to $f$ a nonnegative Borel measure $\mu$ on $\mathbb{R}^+$ by setting $\mu([0,x)) = f(x-)$. See, for example, Folland (1999, §1.5). Then $p \in P$ if and only if $\mu(\{p\}) > 0$. In fact, $\mu(\{p\}) = f(p+) - f(p-)$. Under suitable additional hypotheses on $f$, such as right continuity, there is a 1–1 correspondence between $f$ and $\mu$. But in general, since $\mu$ can see only $f(p+) - f(p-)$ and cannot detect $f(p)$, different $f$ can have the same $\mu$. Let $\kappa$ be the measure obtained from $\mu$ by subtracting all its point masses. Then for $\mu$, we have the decomposition

$$\mu = \kappa + \sum_{p \in P} \mu(\{p\})\delta_p,$$

where $\kappa$ is a continuous nonnegative Borel measure on $\mathbb{R}^+$ (continuous means that the measure of each single point is zero) and $\delta_p$ denotes the unit point mass at $p$.

Next, recall that $AL_0$ is the set of all functions in $AL(\mathbb{R}^+ \times \mathbb{R}^+)$ which vanish on the coordinate axes. Our main goal in this section is to prove decomposition results for $AL_0$ akin to the ones proved above for $\mathcal{M}$. For $z = x+iy \in \mathbb{R}^+ \times \mathbb{R}^+$, define rectangles $R(z)$ and $\overline{R}(z)$ by

$$R(z) = [0, x) \times [0, y), \quad \overline{R}(z) = [0, x] \times [0, y].$$

Let $f \in AL_0$. In this section, we will usually write $f(x, y)$ instead of $f(z)$. Associate to $f$ a nonnegative Borel measure $\mu$ on $\mathbb{R}^+ \times \mathbb{R}^+$ by setting

$$\mu(R(z)) = \lim_{t \to 0+} f(x - t, y - t). \tag{2.42}$$

The limit exists since $f$ is increasing on horizontal and vertical lines

For $a \in \mathbb{R}^+$, $b \in \mathbb{R}^+$, let $V(a)$ be the vertical half-line $a + i\mathbb{R}^+$ through $x = a$ and let $H(b) = \mathbb{R}^+ + ib$ be the horizontal half-line through $y = b$. Define sets $P, A$, and $B$, associated with $f$, by

$$P = \{p \in \mathbb{R}^+ \times \mathbb{R}^+ : \mu(\{p\}) > 0\}, \quad A = \{a \in \mathbb{R}^+ : \mu(V(a) \setminus P) > 0\},$$
$$B = \{b \in \mathbb{R}^+ : \mu(H(b) \setminus P) > 0\}.$$

The sets $P, A$ and $B$ are at most countable. For $p \in P$, $a \in A$, $b \in B$, define nonnegative Borel measures $\nu_p$, $\lambda_a$, $\lambda^b$ on $\mathbb{R}^+ \times \mathbb{R}^+$ as follows:

- $\nu_p = \mu(\{p\})\delta_p$.
- $\lambda_a$ is the restriction of $\mu$ to $V(a) \setminus P$.
- $\lambda^b$ is the restriction of $\mu$ to $H(b) \setminus P$.

Thus, for example, for each Borel set $E \subset \mathbb{R}^+ \times \mathbb{R}^+$,

$$\lambda_a(E) = \mu((V(a) \setminus P) \cap E).$$

Then $\nu_p$ is a point mass at $p$, while $\lambda_a$, $\lambda^b$ are line masses that assign measure zero to each point. Let $\kappa$ be the measure obtained by subtracting all the $\nu_p$, $\lambda_a$

and $\lambda^b$ from $\mu$. Then $\kappa$ is a nonnegative measure that assigns measure zero to each point and to each horizontal and each vertical line in $\mathbb{R}^+ \times \mathbb{R}^+$. We have

$$\mu = \kappa + \sum_{a \in A} \lambda_a + \sum_{b \in B} \lambda^b + \sum_{p \in P} \nu_p. \tag{2.43}$$

To construct a corresponding decomposition of $f$, one must pay close attention to detail. First, we need an analogue of the fact that increasing functions on $\mathbb{R}^+$ have left- and right-hand limits at each point. Take $w = u + iv \in \mathbb{R}^+ \times \mathbb{R}^+$. Since $f$ is increasing on coordinate lines, it follows that the limit

$$f(u+, v) \equiv \lim_{t \to 0+} f(u + t, v)$$

exists, as do the corresponding limits $f(u, v+), f(u-, v)$ and $f(u, v-)$. Limits also exist for $f(z)$ when $z$ approaches $w$ within an open coordinate quadrant with origin $w$. For example,

$$\lim_{(s,t) \to (0,0),\, s>0,\, t>0} f(u - s, v - t) = \mu(R(w)). \tag{2.44}$$

To prove (2.44), let $s$ and $t$ be small positive numbers. Say $0 < t < s$. Since $f$ is increasing on coordinate lines, we have

$$f(u - s, v - s) \leq f(u - s, v - t) \leq f(u - t, v - t).$$

By (2.42), as $s$ and $t$ approach zero, the terms on the left and the right approach the common limit $\mu(R(w))$.

Denote the left-hand side of (2.44) by $f(u-, v-)$. The limits associated with the other three open coordinate quadrants will be denoted by $f(u+, v-)$, etc. The reader may verify that they exist, and that we have

$$\begin{aligned} f(u+, v+) &= \mu(\overline{R}(w)), \\ f(u+, v-) &= \mu(R(w) \cup V_1(w)), \\ f(u-, v+) &= \mu(R(w) \cup H_1(w)), \\ f(u-, v-) &= \mu(R(w)), \end{aligned}$$

where $H_1(w)$ is the horizontal segment from $(0, v)$ to $w$, closed on the left, open on the right, and $V_1(w)$ the vertical segment from $(u, 0)$ to $w$, closed on the bottom, open on the top. We also obtain the easily checked formulas

$$\begin{aligned} \mu(H_1(w)) &= f(u-, v+) - f(u-, v-), \\ \mu(V_1(w)) &= f(u+, v-) - f(u-, v-), \\ \mu(\{w\}) &= f(u+, v+) - f(u-, v+) + f(u-, v-) - f(u+, v-). \end{aligned} \tag{2.45}$$

We are now ready to construct our decomposition of $f$. Take $p \in P$. Write $p = u + iv$, where $u, v \in \mathbb{R}^+$. Define $\phi_p : \mathbb{R}^+ \times \mathbb{R}^+ \to \mathbb{R}$ by

$$\phi_p(x,y) = \begin{cases} 0, & \text{if } 0 < x < u \text{ or } 0 < y < v, \\ c_1, & \text{if } x = u \text{ and } y > v, \\ c_2, & \text{if } x = u \text{ and } y = v, \\ c_3, & \text{if } x > u \text{ and } y = v, \\ c_4, & \text{if } x > u \text{ and } y > v, \end{cases} \tag{2.46}$$

where

$$\begin{aligned} c_1 &= f(u, v+) - f(u-, v+) + f(u-, v-) - f(u, v-), \\ c_2 &= f(u, v) - f(u-, v) + f(u-, v-) - f(u, v-), \\ c_3 &= f(u+, v) - f(u-, v) + f(u-, v-) - f(u+, v-), \\ c_4 &= f(u+, v+) - f(u-, v+) + f(u-, v-) - f(u+, v-). \end{aligned} \tag{2.47}$$

**Claim** *$\phi_p \in AL_0$ and $f - \phi_p \in AL_0$.*

*Proof* Let $R$ denote a closed rectangle in $\mathbb{R}^+ \times \mathbb{R}^+$ with sides parallel to the coordinate axes and vertices $z_1, z_2, z_3, z_4$, where $z_1$ is the northeast vertex and the remaining vertices are labelled in counterclockwise order. Let $\mathcal{R}$ denote the set of all such $R$. Then $\phi_p$ is in $AL_0$ if and only if

$$\phi_p(R) \equiv \phi_p(z_1) - \phi_p(z_2) + \phi_p(z_3) - \phi_p(z_4) \geq 0 \tag{2.48}$$

for all $R \in \mathcal{R}$, and $f - \phi_p \in AL_0$ if and only if

$$\phi_p(R) \leq f(R), \quad \forall\, R \in \mathcal{R}. \tag{2.49}$$

If $R$ does not contain $p$ then $\phi_p(R) = 0$, so (2.48) and (2.49) hold. If $R$ does contain $p$, one considers subcases according to whether $p$ is a vertex, belongs to the interior of a boundary side, or belongs to the interior of $R$. If $R$ is subdivided into two rectangles $R_1$, $R_2$ by insertion of a new horizontal or vertical boundary side, then $g(R) = g(R_1) + g(R_2)$ for all functions $g$. From this, one sees that it suffices to verify (2.48) and (2.49) when $p$ is a vertex of $R$. If $p$ is the southwest vertex, then, using (2.46)–(2.47),

$$\begin{aligned} \phi_p(R) &= c_4 - c_1 + c_2 - c_3 = f(u+, v+) - f(u, v+) + f(u, v) - f(u+, v) \\ &= \lim_{t \to 0+} f(R(t)), \end{aligned} \tag{2.50}$$

where $R(t)$ is the rectangle with southwest vertex at $(u, v)$, northeast vertex at $(u + t, v + t)$. Since $f \in AL_0$, the last expression is $\geq 0$. Thus (2.48) holds.

In general, if $R, R' \in \mathcal{R}$ with $R \subset R'$, then the definition of $AL$ and a subdivision argument show that $f(R) \leq f(R')$. Thus, in (2.50), $f(R(t)) \leq f(R)$ for small $t$, so that (2.49) also holds.

In similar fashion, one shows that (2.48) and (2.49) hold when $p$ is the southeast, northeast, or northwest vertex of $R$. The claim is proved. □

Let $\nu_p$ be the measure associated to $\phi_p$. Then $\nu_p$ assigns mass $c_4$ to $R(w)$ if $p \in R(w)$, and assigns mass zero to $R(w)$ for other $w \in \mathbb{R}^+ \times \mathbb{R}^+$. From (2.45) and (2.47) it follows that $c_4 = \mu(\{p\})$. Thus, $\nu_p = \mu(\{p\})\delta_p$.

Supposing that the set $P$ of point masses for $\mu$ is non-empty, enumerate it as $P = \{p_1, p_2, \dots\}$. Let $\phi_1$ be the function constructed from $f$ and $p_1$ as above. Set $f_1 = f - \phi_1$. Then $f_1$ and $\phi_1$ are in $AL_0$, by the claim. Suppose that $f_n$ has been constructed for $n \geq 1$. Let $\phi_{n+1}$ be the function constructed from $f_n$ and $p_{n+1}$, and let $f_{n+1} = f_n - \phi_{n+1}$. Define a function $F$ on $\mathbb{R}^+ \times \mathbb{R}^+$ by

$$F = f - \sum_{p \in P} \phi_p. \tag{2.51}$$

The series $\sum_{p \in P} \phi_p$ has nonnegative terms. Each $f_n$ is in $AL_0$, hence is nonnegative. The $n$th partial sum of the series equals $f - f_n$, so the partial sums are bounded above by $f$. Thus, the series converges pointwise on $\mathbb{R}^+ \times \mathbb{R}^+$. $F$ is the pointwise limit of the functions $f_n \in AL_0$, hence $F \in AL_0$. The measure associated to two functions in $AL_0$ is the sum of their associated measures. It follows that the measure associated to $f_n$ is the measure $\mu$ associated to $f$ with the point masses at $p_1, \dots, p_n$ removed, and the measure associated to $F$ is $\mu$ with all of its point masses removed. From (2.43), we deduce that the measure associated to $F$ is

$$\mu_1 \equiv \kappa + \sum_{a \in A} \lambda_a + \sum_{b \in B} \lambda^b. \tag{2.52}$$

We now turn attention to the line masses in $\mu$. Recall that $a \in A$ means that $\mu(V(a) \setminus P) > 0$, where $V(a)$ is the vertical half-line through $x = a$. Define, for $a \in A$,

$$\sigma_a(x, y) = \begin{cases} 0, & \text{if } 0 \leq x < a, \\ F(a, y) - F(a-, y), & \text{if } x = a, \\ F(a+, y) - F(a-, y), & \text{if } x > a. \end{cases} \tag{2.53}$$

For $b \in B$, define $\sigma^b$ by

$$\sigma^b(x, y) = \begin{cases} 0, & 0 \leq y < b, \\ F(x, b) - F(x, b-), & y = b, \\ F(x, b+) - F(x, b-), & y > b. \end{cases} \tag{2.54}$$

For fixed $y$, $\sigma_a(x, y)$ takes on at most three different values: It is zero to the left of the line $x = a$. At $(a, y)$ and at $(x, y)$ for $x > a$ its values are chosen so that the jumps $\sigma_a(a, y) - \sigma_a(a-, y)$ and $\sigma_a(a+, y) - \sigma_a(a, y)$ are equal to the corresponding jumps of $F$ to the left and right of $(a, y)$. Corresponding remarks apply to the behavior of $\sigma^b$ along vertical lines.

**Claim** *$\sigma_a$, $\sigma^b$, $F - \sigma_a$, and $F - \sigma^b$ are in $AL_0$.*

*Proof* We will give the proofs for $\sigma_a$ and $F - \sigma_a$. As with $\phi_p$, the claims are equivalent to the inequalities $0 \le \sigma_a(R) \le F(R)$ for $R \in \mathcal{R}$. If $R$ does not intersect the line $V(a)$, then $\sigma_a(R) = 0$, and we are done. To prove the inequalities when $R$ intersects $V(a)$, it suffices to consider the two cases when the eastern boundary or the western boundary of $R$ lies on $V(a)$. Consider the first case. For $t > 0$, let $R_1(t)$ be the rectangle whose eastern boundary coincides with that of $R$ and whose western boundary is on the line $V(a - t)$. Then, using (2.53),

$$\sigma_a(R) = \lim_{t \to 0+} F(R(t)).$$

Since $0 \le F(R(t)) \le F(R)$ for small $t$, the desired inequalities follow. The proof of the second case is analogous. □

Next, take $a \in \mathbb{R}^+$ and consider a rectangle $R(x, y) = [0, x) \times [0, y)$ with $(x, y) \in \mathbb{R}^+ \times \mathbb{R}^+$. If $x \le a$, then both $\sigma_a(x-, y-)$ and $\lambda_a(R(x, y))$ equal zero. If $x > a$, then

$$\begin{aligned}\sigma_a(x-, y-) = \sigma_a(a+, y-) &= F(a+, y-) - F(a-, y-)\\ &= \mu_1(V_1(a, y)) = \lambda_a(R(x, y)).\end{aligned}$$

Thus, the measure associated with $\sigma_a$ is $\lambda_a$. Similarly, the measure associated with $\sigma^b$ is $\lambda^b$.

Take again $a \in \mathbb{R}^+$. For $y \in \mathbb{R}^+$,

$$\begin{aligned}\sigma_a(a, y+) - \sigma_a(a, y-) &= F(a, y+) - F(a-, y+) - F(a, y-) + F(a-, y+)\\ &\le F(a+, y+) - F(a-, y-) - F(a-, y-) + F(a-, y+),\end{aligned} \tag{2.55}$$

where the inequality follows from the fact that for the $AL_0$ function $F$, $F(x, t+) - F(x, t-)$ is an increasing function of $x$. Since $F$ has no point masses in its associated measure, from the third equation in (2.45) we deduce that the last expression in (2.55) is zero. Since $\sigma_a \in AL_0$, $\sigma_a(a, y)$ is an increasing function of $y$. It follows that $\sigma_a(a, y)$ is a continuous function of $y$. If $x > a$, a similar argument shows that $\sigma_a(x, y) = \sigma_a(a+, y)$ is an increasing continuous function of $y$. If $x < a$ then $\sigma_a(x, y) = 0$. Thus, we have shown that:

*For each fixed* $x \in \mathbb{R}^+$, $\sigma_a(x,y)$ *is a continuous increasing function of* $y \in \mathbb{R}^+$.

Similarly, for each fixed $y \in \mathbb{R}^+$, $\sigma^b(x,y)$ is a continuous increasing function of $x \in \mathbb{R}^+$.

Finally, define $\tau = F - \sum_{a \in A} \sigma_a - \sum_{b \in B} \sigma^b$. Then, as in the analysis of the point masses, $\tau \in AL_0$. Using (2.52), we find that the measure associated to $\tau$ is $\kappa$. From $\tau(x-,y-) = \kappa(R(z))$ and the fact observed before (2.43) that $\kappa$ assigns zero measure to points and coordinate lines, it follows that $\tau \in C(\mathbb{R}^+ \times \mathbb{R}^+)$. Recalling (2.51), we have proved the result we sought about the decomposition of $f$:

**Theorem 2.25** *For* $f \in AL_0$, *we can write*

$$f = \tau + \sum_{a \in A} \sigma_a + \sum_{b \in B} \sigma^b + \sum_{p \in P} \phi_p, \tag{2.56}$$

*where* $\tau \in AL_0 \cap C(\mathbb{R}^+ \times \mathbb{R}^+)$, *the equation holds at each point of* $\mathbb{R}^+ \times \mathbb{R}^+$, *and the functions* $\sigma_a, \sigma^b, \phi_p$ *are defined in* (2.53), (2.54) *and* (2.46) *respectively.*

For the benefit of readers who begin to read this section here at its end, we note that the discussion of $AL_0$ functions $f$ and their associated objects $\mu, \kappa, \tau, \phi_p$ begins about one page after the beginning of the section. The beginning of the section is devoted to analogous objects for increasing functions of one variable.

## 2.9 Proof of Theorem 2.15 for Discontinuous $\Psi$

Let $f, g, K$ be as in the statement of Theorem 2.15 and $\Psi \in AL_0$. We shall prove inequality (a) of Theorem 2.15:

$$Q(f,g) \equiv \int_{\mathbb{R}^{2n}} \Psi(f(x), g(y)) K(|x-y|)\, dx\, dy \le Q(f^\#, g^\#). \tag{2.57}$$

The proof of (c) is similar. Once (a) and (c) are established, the proofs of (b) and (d) given in §2.6 remain valid for discontinuous $\Psi$.

In our proof of (2.57) we may assume, as in §2.5, that $K(0) < \infty$ and that $K(t) = 0$ for all sufficiently large $t$. For positive integers $m$, set $\Psi_m(x,y) = \Psi(x,y)$ if $x \ge 1/m$ and $y \ge 1/m$, $\Psi_m(x,y) = 0$ for other $(x,y) \in \mathbb{R}^+ \times \mathbb{R}^+$. Then each $\Psi_m \in AL_0$, and $\Psi$ is the increasing pointwise limit of $\{\Psi_m\}$. By the monotone convergence theorem, if (2.57) is true for each $\Psi_m$ it is true for $\Psi$. Thus, we may assume that there exists $a_0 > 0$ such that $\Psi(x,y) = 0$ if either $x < a_0$ or $y < a_0$.

Decompose $\Psi$ as in (2.56). Each $a \in A$, each $b \in B$, and the real and imaginary parts of each $p \in P$ are $\geq a_0$. Each function in the decomposition is nonnegative. Equation (2.57) has already been proved when $\Psi$ is continuous. So to prove (2.57) for general $\Psi$, we just need to prove it when $\Psi$ is replaced by $\sigma_a$, $\sigma^b$ or $\phi_p$.

First, we examine $\phi_p$, which we shall call simply $\phi$. By its construction, $\phi$ belongs to $AL_0$, takes on at most four different positive values, and $\phi(s,t) = 0$ if $0 \leq s < a_0$ or $0 \leq t < a_0$. Fix $\delta$ with $0 < \delta < a_0$. Then, for $(x, y) \in \mathbb{R}^{2n}$,

$$\phi(f(x), g(y))K(|x - y|) \leq (\max \phi)K(0)\chi_{f>\delta}(x)\chi_{g>\delta}(y).$$

By assumption (I), each of the sets $f > \delta$ and $g > \delta$ has finite Lebesgue measure in $\mathbb{R}^n$. With the aid of the dominated convergence theorem, to prove (2.57) for $\phi$ it will suffice to express $\phi$ as the pointwise limit on $\mathbb{R}^+ \times \mathbb{R}^+$ of a uniformly bounded sequence of continuous functions which vanish on $\{(s,t) \in \mathbb{R}^+ \times \mathbb{R}^+ : s \leq \delta \text{ or } t \leq \delta\}$.

For our proof to work, the approximating sequence must converge to $\phi$ at *every* point of $\mathbb{R}^+ \times \mathbb{R}^+$. Almost everywhere convergence will not suffice.

Write $p = p_1 + ip_2$. From the definition of $\phi$ in §2.8, one can check that

$$\phi = \alpha_1\eta_1 + \alpha_2\eta_2 + \alpha_3\eta_3 + \alpha_4\eta_4,$$

where the constants $\alpha_i$ are defined by

$$\alpha_1 = c_4 - c_1 + c_2 - c_3, \quad \alpha_2 = c_1 - c_2, \quad \alpha_3 = c_3 - c_2, \quad \alpha_4 = c_2,$$

and the functions $\eta_1, \ldots, \eta_4$ are the characteristic functions of the respective sets

$$(p_1, \infty) \times (p_2, \infty), \quad [p_1, \infty) \times (p_2, \infty),$$
$$(p_1, \infty) \times [p_2, \infty), \quad [p_1, \infty) \times [p_2, \infty).$$

For $c > 0$, each of the one-variable functions $\chi_{(c,\infty)}$ and $\chi_{[c,\infty)}$ is the pointwise limit on $\mathbb{R}^+$ of a sequence of continuous functions uniformly bounded by 1. If $c > \delta$, we can also require the approximating functions to vanish on $[0, \delta]$. Suitable products of the one-variable functions furnish appropriate two-variable approximations to the $\eta_j$. Thus, (2.57) holds for each $\eta_j$, and hence for $\phi$.

Finally, we shall verify (2.57) for $\sigma_a$. The argument for $\sigma^b$ requires obvious changes. In (2.51), replace $f$ by $\Phi$. We shall continue to denote the function on the left of (2.51) by $F$. Define functions $h_1$, $h_2$ on $\mathbb{R}^+$ by

$$h_1(t) = F(a, t) - F(a-, t), \; h_2(t) = F(a+, t) - F(a, t).$$

By (2.53), for $(s,t) \in \mathbb{R}^+ \times \mathbb{R}^+$,

$$\sigma_a(s,t) = \chi_{[a,\infty)}(s)h_1(t) + \chi_{(a,\infty)}(s)h_2(t). \tag{2.58}$$

Near the end of §2.8, we showed that $\sigma_a$ is continuous and increasing on vertical lines. It follows that $h_1$ and $h_2$ are continuous and increasing on $\mathbb{R}^+$. Also, $h_1$ and $h_2$ vanish on $[0,a)$, and hence on $[0,a_0)$.

Replacing $h_1$ and $h_2$ by $\max(h_1,m)$ and $\max(h_2,m)$ where $m$ is a positive integer, and bearing in mind the monotone convergence theorem, we may assume that the $h_j$ in (2.58) are bounded on $\mathbb{R}^+$, as well as continuous. Approximate $\chi_{[a,\infty)}$ and $\chi_{(a,\infty)}$ by suitable one-variable uniformly bounded continuous functions, as in the proof for $\phi$. Replace the characteristic functions in (2.58) by their approximants. The resulting two-variable functions are continuous, uniformly bounded, belong to $AL_0$, and converge to $\sigma_a$ pointwise on $\mathbb{R}^+ \times \mathbb{R}^+$. Equation (2.57) holds when these two-variable functions are substituted for $\Psi$. Applying the dominated convergence theorem, as in the proof for $\phi$, we obtain (2.57) when $\sigma_a$ is substituted for $\Psi$. □

## 2.10 Notes and Comments

For $\Psi(x,y) = xy$, Theorem 2.15(a) is a special case of an inequality of F. Riesz (1930) when $n = 1$ and of Sobolev (1938) when $n \geq 1$. We shall take up the general Riesz–Sobolev inequality and some of its generalizations in Chapter 8.

The special case of Theorem 2.15(c) when $f = g = 1$ and $\Psi$ is continuous is essentially due to Crowe, Rosenblum and Zweibel (1986, Thm. 3). Again with continuous $\Psi$, Almgren and Lieb (1989, p. 691) state Theorem 2.15(c) and a more general version of Theorem 2.15(a) in which $K(|x-y|)$ is replaced on the left side of the inequality by $W(ax+by)$ and on the right side by $W^{\#}(ax+by)$, with $a, b \in \mathbb{R}$ and $W\colon \mathbb{R}^n \to \mathbb{R}^+$. They prove their inequality for general $\Psi$ by reducing it to the case $\Psi(x,y) = xy$, then invoking the Riesz–Sobolev Theorem. By contrast, the proof given here for continuous $\Psi$ makes use only of elementary results for polarization and two-point symmetrization, and of the Arzelà–Ascoli Theorem to supply a compactness ingredient. In Chapter 7 we shall see that Theorem 2.15 carries over to spheres and hyperbolic spaces, with essentially the same proof. The Almgren–Lieb version of Theorem 2.15(a), with $W(ax+by)$ instead of $K(|x-y|)$, has an analogue for circles, to be presented in Chapter 8, but I know of no comparable theorem for higher dimensional spheres or hyperbolic spaces.

The use of polarization to obtain Riesz-type integral inequalities appears in Baernstein and Taylor (1976) in the context of spheres. The proof in that paper was inspired by the proof of Ahlfors (1973, Lemma 2.2), although his proof appears to be incomplete. Other papers using polarization to prove integral inequalities, by Beckner, Brock, Solynin, and others, will be cited in Chapters 4, 7, and 8.

The inequality of Corollary 2.16, $\int fg \leq \int f^{\#}g^{\#}$, is in Hardy, Littlewood, and Pólya (1952, p. 278) for functions of one variable. The contraction inequalities, Corollaries 2.20 and 2.21, were apparently first proved, independently and respectively, by Chiti (1979) and Crandall and Tartar (1980). A result like Corollary 2.22 is in Almgren and Lieb (1989, p. 693).

A number of authors have used isoperimetric or Brunn–Minkowski-type inequalities to prove that symmetrization decreases the Lipschitz norm of functions. The proofs can usually be easily modified to obtain the more general modulus of continuity decrease asserted by Theorem 2.12. As in our proof of Theorem 2.15, the proof of Theorem 2.12 presented here uses only elementary inequalities, together with the Arzelà–Ascoli Theorem. In Chapter 8 we shall use the general Riesz inequality in $\mathbb{R}^n$ to prove the general Brunn–Minkowski inequality $|A+B|^{1/n} \geq |A|^{1/n} + |B|^{1/n}$ in $\mathbb{R}^n$.

# 3

# Dirichlet Integral Inequalities

For functions $f : \mathbb{R}^n \to \mathbb{R}$ and $\Phi : \mathbb{R}^+ \to \mathbb{R}^+$, the $\Phi$-Dirichlet integral of $f$ is the integral $\int_{\mathbb{R}^n} \Phi(|\nabla f|)\, dx$. When $\Phi(t) = t^p$, $\int_{\mathbb{R}^n} |\nabla f|^p\, dx$ is called simply the $p$-Dirichlet integral of $f$. The principal results of this chapter are Theorems 3.7, 3.11, and 3.20. Theorem 3.7 asserts that if $f$ is Lipschitz and its distribution function satisfies mild assumptions, then for $p \geq 1$ the $p$-Dirichlet integral decreases when $f$ is replaced by its symmetric decreasing rearrangement $f^{\#}$. The proof is based on the double integral inequality Corollary 2.22. Theorem 3.11 is an extension of Theorem 3.7: it says that for Lipschitz functions, symmetrization decreases $\Phi$-Dirichlet integrals for all convex $\Phi$. In Theorem 3.20 we return to $p$-Dirichlet integrals, and prove that, when $p \geq 1$, symmetrization decreases the $p$-Dirichlet integral of nonnegative functions in the first order Sobolev space $W^{1,p}(\mathbb{R}^n)$.

Sections 3.1 and 3.4 provide introductory material on Lipschitz and first order Sobolev functions, respectively. Section 3.6 contains a discussion of the phenomenon that while $\|f_n - f\|_{L^p(\mathbb{R}^n)} \to 0$ implies $\|f_n^{\#} - f^{\#}\|_{L^p(\mathbb{R}^n)} \to 0$, (under reasonable conditions), it is not true in general that $\|f_n - f\|_{L^p(\mathbb{R}^n)} \to 0$ and $\|\nabla f_n - \nabla f\|_{L^p(\mathbb{R}^n)} \to 0$ imply $\|\nabla f_n^{\#} - \nabla f^{\#}\|_{L^p(\mathbb{R}^n)} \to 0$.

## 3.1 Lipschitz Functions

Let $(X, d_X)$ and $(Y, d_Y)$ be metric spaces. A function $f : X \to Y$ is said to be *Lipschitz* if there is a constant $C$ such that

$$d_Y(f(x), f(y)) \leq C d_X(x, y), \quad \forall\, x, y \in X.$$

Of course, Lipschitz functions are uniformly continuous.

It is easy to see that if one such constant exists, then there is a least such constant, called the *Lipschitz constant* of $f$. Denote this constant by $\|f\|_{\mathrm{Lip}(X,Y)}$, and let Lip($X$, $Y$) denote the set of all Lipschitz functions from $X$ to $Y$. General references for Lipschitz functions include Evans and Gariepy (1992) and Federer (1969).

In this chapter, we shall have occasion to use two basic results about Lipschitz functions, the *Extension Theorem* and *Rademacher's Theorem.*

**Theorem 3.1** (Extension Theorem) *Let* $(X, d_X)$ *be a metric space,* $A \subset X$, *and* $f \in \mathrm{Lip}(A, \mathbb{R})$. *There exists* $\bar{f} \in \mathrm{Lip}(X, \mathbb{R})$ *such that* $f = \bar{f}$ *on* $A$ *and* $\|f\|_{\mathrm{Lip}(A,\mathbb{R})} = \|\bar{f}\|_{\mathrm{Lip}(X,\mathbb{R})}$.

The proof is simple: Take

$$\bar{f}(x) = \inf_{a \in A}\{f(a) + \|f\|_{\mathrm{Lip}(A,\mathbb{R})} d_X(x, a)\}.$$

A proof that $\bar{f}$ has the required properties when $X = \mathbb{R}^n$ is in Federer (1969, p. 202); the same proof works for any metric space $X$.

When the target space $\mathbb{R}$ is replaced by $\mathbb{R}^n$, $n \geq 2$, it is still possible to extend $f \in \mathrm{Lip}(A, \mathbb{R}^n)$ to some $\bar{f} \in \mathrm{Lip}(X, \mathbb{R}^n)$, but the Lipschitz constant of any such extension must sometimes be strictly larger than the Lipschitz constant of $f$. An example is in Federer (1969, p. 202). When $X = \mathbb{R}^m$, $Y = \mathbb{R}^n$, norm preserving extensions do exist. This is known as *Kirszbraun's Theorem.* It is proved in Federer (1969, p. 202).

In general, Lipschitz extensions of $f \in \mathrm{Lip}(A, Y)$ to supersets of $A$ need not be unique. If, however, $Y$ is complete, then it is easy to see that $f$ has a unique continuous extension to the closure $\bar{A}$ of $A$, and that this extension still has Lipschitz constant $\|f\|_{\mathrm{Lip}(A,Y)}$.

Rademacher's theorem asserts that real-valued Lipschitz functions on $\mathbb{R}^n$ are differentiable almost everywhere. Proofs may be found, for example, in Evans and Gariepy (1992) or Federer (1969).

**Theorem 3.2** (Rademacher) *Let* $f \in \mathrm{Lip}(\mathbb{R}^n, \mathbb{R})$. *Then* $f$ *is differentiable at* $\mathcal{L}^n$*-almost every* $x \in \mathbb{R}^n$.

Recall that a function $f\colon \mathbb{B}^n(x, R) \to \mathbb{R}$ is said to be *differentiable* at $x$ if there exists a linear map $L\colon \mathbb{R}^n \to \mathbb{R}$ such that

$$f(y) = f(x) + L(y - x) + o(|x - y|), \quad y \to x. \tag{3.1}$$

One sometimes writes $L = Df(x)$, and calls $Df(x)$ the derivative of $f$ at $x$. There exists $v \in \mathbb{R}^n$ such that $L(y) = v \cdot y$ for all $y \in \mathbb{R}^n$. We shall denote $v$ by $\nabla f(x)$, and call $\nabla f(x)$ the gradient of $f$ at $x$. In our context, $Df(x)$ and $\nabla f(x)$ are essentially the same object, though technically $Df(x) \in (\mathbb{R}^n)^*$, the vector space dual to $\mathbb{R}^n$, whereas $\nabla f(x) \in \mathbb{R}^n$.

Equation (3.1) can be written as

$$f(y) = f(x) + \nabla f(x) \cdot (y - x) + o(|y - x|), \quad y \to x. \tag{3.2}$$

Taking $y = x + te_i$, $t \in \mathbb{R}$, in (3.2), one sees that if $f$ is differentiable at $x$, then for $i = 1, \ldots, n$, the $i$th partial derivative $\frac{\partial f}{\partial x_i} \equiv \partial_i f$ evaluated at $x$ exists, and is equal to the $i$th component of $\nabla f(x)$. Thus,

$$\nabla f(x) = (\partial_1 f(x), \ldots, \partial_n f(x)), \tag{3.3}$$

and $|\nabla f(x)|$ is defined by

$$|\nabla f(x)|^2 = \sum_{i=1}^{n} (\partial_i f(x))^2.$$

There are occasions when all the partial derivatives of $f$ exist at $x$, but $f$ fails to be differentiable at $x$. In that case we still define $\nabla f(x)$ by (3.3). Later on, when we study weak derivatives, the functions $\partial_i f$ will exist in some generalized sense, and $\nabla f$ will be defined to be the object $(\partial_1 f, \ldots, \partial_n f)$.

For $v \in \mathbb{R}^n$ let

$$\partial_v f(x) = \lim_{t \to 0} \frac{f(x + tv) - f(x)}{t}$$

denote the derivative of $f$ at $x$ in the direction $v$. If $f$ is differentiable at $x$ then $\partial_v f(x)$ exists, and $\partial_v f(x) = \nabla f(x) \cdot v$. Moreover, if $f$ is Lipschitz in a neighborhood of $x$ with constant $C$, then $|\partial_v f(x)| \leq C|v|$. From Rademacher's Theorem 3.2, we deduce

**Corollary 3.3** *Let $f \in \mathrm{Lip}(\mathbb{R}^n, \mathbb{R})$. Then $\partial_v f$ exists a.e. for each $v \in \mathbb{R}^n$, and*

$$\|\partial_v f\|_\infty \leq \|f\|_{\mathrm{Lip}(\mathbb{R}^n, \mathbb{R})} |v|.$$

In particular, if $f \in \mathrm{Lip}(\mathbb{R}^n, \mathbb{R})$, then all the partial derivatives of $f$ exist at almost all points of $\mathbb{R}^n$, and we have $\|\partial_i f\|_\infty \leq \|f\|_{\mathrm{Lip}(\mathbb{R}^n, \mathbb{R})}$ for $i = 1, \ldots, n$.

If $f \in \mathrm{Lip}(\mathbb{R}^n, \mathbb{R})$, then the $\mathbb{R}^n$-valued function $\nabla f$ is defined a.e. on $\mathbb{R}^n$, and is Lebesgue measurable (each $\partial_i f$ is the a.e. limit of the measurable sequence $g_k(x) \equiv k(f(x + k^{-1} e_i) - f(x))$). Write $\|\nabla f\|_\infty$ for the essential supremum of $|\nabla f|$ over $\mathbb{R}^n$.

**Corollary 3.4** *Let $f \in \mathrm{Lip}(\mathbb{R}^n, \mathbb{R})$. Then $\nabla f \in L^\infty(\mathbb{R}^n, \mathbb{R}^n)$, and*

$$\|\nabla f\|_\infty = \|f\|_{\mathrm{Lip}(\mathbb{R}^n, \mathbb{R})}.$$

*Proof* At each point of differentiability of $f$ it follows from (3.2) that

$$|\nabla f(x)| = \sup_{|v|=1} |\partial_v f(x)|.$$

Thus, from Corollary 3.3 we have $|\nabla f(x)| \le \|f\|_{\mathrm{Lip}(\mathbb{R}^n,\mathbb{R})}$ at points of differentiability, so that $\|\nabla f\|_\infty \le \|f\|_{\mathrm{Lip}(\mathbb{R}^n,\mathbb{R})}$.

If $\|f\|_{\mathrm{Lip}(\mathbb{R}^n,\mathbb{R})} = 0$ the opposite inequality is also true. If $\|f\|_{\mathrm{Lip}(\mathbb{R}^n,\mathbb{R})} > 0$, take $0 < \alpha < \|f\|_{\mathrm{Lip}(\mathbb{R}^n,\mathbb{R})}$. There exist $x, y \in \mathbb{R}^n$ such that $|f(x) - f(y)| > \alpha|x - y|$. After translation and rotation, we may assume that $x = 0$, $y = re_1$, for some $r > 0$, and that $f(re_1) > f(0)$. By continuity, there exists $\epsilon > 0$ such that for each $x$ in the cube $Q = [-\epsilon, \epsilon]^n$ we have $f(x + re_1) - f(x) > r\alpha$. Let $Q' = \{x \in Q \colon x_1 = 0\}$, and write $x = (x_1, y)$, where $y \in \mathbb{R}^{n-1}$. Then

$$\int_{[0,r]\times Q'} \partial_1 f(x)\,dx = \int_{Q'} dy \int_0^r \partial_1 f(x_1, y)\,dx_1 = \int_{Q'} (f(r, y) - f(0, y))\,dy.$$

The first equality is from Fubini's theorem. The second equality comes from the fact that the restriction of a Lipschitz function to a line is a Lipschitz function of one variable, hence is an absolutely continuous function of one variable, and thus is the indefinite integral of its derivative. The third term is $> r\alpha\mathcal{L}^{n-1}(Q')$, and the first term is $\le r\mathcal{L}^{n-1}(Q')\|\nabla f\|_\infty$. We conclude that $\|\nabla f\|_\infty > \alpha$, and hence that $\|\nabla f\|_\infty \ge \|f\|_{\mathrm{Lip}(\mathbb{R}^n,\mathbb{R})}$. □

**Corollary 3.5** *Let $f \in \mathrm{Lip}(\mathbb{R}^n, \mathbb{R})$. For fixed $c \in \mathbb{R}$, let $E = (f = c)$. Then $\nabla f = 0$ a.e. on $E$.*

*Proof* Let

$$E_1 = \{x \in E \colon f \text{ is differentiable at } x\} \quad \text{and}$$
$$E_2 = \{x \in E_1 : x \text{ is a point of density of } E_1\}.$$

Then $\mathcal{L}^n(E \setminus E_1) = 0$ and $\mathcal{L}^n(E_1 \setminus E_2) = 0$. To prove the corollary, it suffices to prove that $\nabla f(x_0) = 0$ for each $x_0 \in E_2$.

Take $x_0 \in E_2$. We may assume that $x_0 = 0$. Suppose that $\nabla f(0) \ne 0$. Let

$$K = \left\{ y \in \mathbb{S}^{n-1} : y \cdot \nabla f(0) \ge \frac{1}{2}|\nabla f(0)| \right\}.$$

Choose $R$ so small that if $0 < |x| \le R$ then

$$|f(x) - f(0) - x \cdot \nabla f(0)| < |x|\frac{|\nabla f(0)|}{4}.$$

If $x \in E_1$ then $f(x) = f(0) = c$, so if $|x| \le R$ and $x \in E_1$ then $|x \cdot \nabla f(0)| < \frac{1}{4}|x|\,|\nabla f(0)|$. Let $K(R) = \{x \in \mathbb{R}^n \colon 0 < |x| < R,\ x/|x| \in K\}$. Then $K(R) \cap E_1$ is empty. This is incompatible with our assumption that 0 is a point of density of $E_1$. We conclude that $\nabla f(0) = 0$. □

Here now is our first result on the theme that symmetrization decreases the size of the gradient. For $f \in \mathrm{Lip}(\mathbb{R}^n, \mathbb{R})$, note that $\inf f = \operatorname{ess\,inf} f$.

**Theorem 3.6** *Let $f \in \mathrm{Lip}(\mathbb{R}^n, \mathbb{R})$, and suppose that $\lambda_f(t) < \infty$ for all $t > \inf f$. Then $f^\# \in \mathrm{Lip}(\mathbb{R}^n, \mathbb{R})$, and*

$$\|\nabla f^\#\|_\infty \leq \|\nabla f\|_\infty.$$

*Proof* By Theorem 2.12, the moduli of continuity of $f$ and $f^\#$ satisfy

$$\omega(t, f^\#) \leq \omega(t, f), \quad t > 0.$$

Also, it follows from the definitions that

$$\|f\|_{\mathrm{Lip}(\mathbb{R}^n,\mathbb{R})} = \sup_{t>0} \frac{\omega(t,f)}{t}.$$

Thus, $\|f^\#\|_{\mathrm{Lip}(\mathbb{R}^n,\mathbb{R})} \leq \|f\|_{\mathrm{Lip}(\mathbb{R}^n,\mathbb{R})}$. By Corollary 3.4, this inequality is equivalent to the conclusion of Theorem 3.6. □

## 3.2 Symmetrization Decreases the $p$-Dirichlet Integral of Lipschitz Functions

**Theorem 3.7** *Let $f \in \mathrm{Lip}(\mathbb{R}^n, \mathbb{R})$ and suppose that $\lambda_f(t) < \infty$ for all $t > \inf f$. Then*

$$\int_{\mathbb{R}^n} |\nabla f^\#|^p \, dx \leq \int_{\mathbb{R}^n} |\nabla f|^p \, dx, \quad 1 \leq p < \infty. \tag{3.4}$$

For $0 < p < 1$, Theorem 3.7 is false. To get counterexamples for $n = 1$, take $a \in [0, 1)$. Let $f_a \colon \mathbb{R} \to \mathbb{R}^+$ be the piecewise linear function which is zero for $|x| \geq 1$, has $f_a(a) = 1$, and is linear on each of $[-1, a]$ and $[a, 1]$. Then $f_a' = 1/(1 + a)$ on $(-1, a)$, $f_a' = -1/(1 - a)$ on $(a, 1)$, and $f_a' = 0$ on $|x| > 1$. Thus,

$$\int_{\mathbb{R}} |f_a'|^p \, dx = (1 + a)^{1-p} + (1 - a)^{1-p}.$$

One easily checks that all the $f_a$ for $0 \leq a < 1$ have the distribution function $\lambda(t) = 2(1 - t)^+$ for $t > 0$, $\lambda(t) = 2$ for $t \leq 0$. Thus, $f_a^\# = f_0$. Since $x \to x^{1-p}$ is strictly concave on $\mathbb{R}^+$, we see that

$$\int_{\mathbb{R}} |f_a'|^p \, dx < \int_{\mathbb{R}} |f_0'|^p \, dx = \int_{\mathbb{R}} |(f_a^\#)'|^p \, dx$$

when $0 < a < 1$.

Counterexamples of the same type can be constructed in $\mathbb{R}^n$ as follows. Take again $a \in [0, 1)$. Define $S = S_a \colon \mathbb{B}^n(1) \to \mathbb{B}^n(1)$ by

$$S(x) = (1 - |x|)ae_1 + x, \quad x \in \mathbb{B}^n(1).$$

Then for $0 < r \le 1$, $S$ maps the sphere $|x| = r$ 1-1 onto the sphere with radius $r$, center $(1-r)ae_1$. If $0 < r_1 < r_2 \le 1$ then, as is easily shown, the sphere $S(|x| = r_1)$ lies strictly inside the sphere $S(|x| = r_2)$. Moreover, $S(|x| = 1) = (|x| = 1)$. We conclude that $S$ maps $\mathbb{B}^n(1)$ bijectively onto itself.

Define $f_0 \colon \mathbb{R}^n \to \mathbb{R}^+$ by $f_0(x) = (1-|x|)^+$, and set $f_a = f_0 \circ S_a^{-1}$. For $t \in (0,1)$, $(f_a > t)$ is a ball of radius $1-t$ contained in $\mathbb{B}^n(1)$. So we have again $f_a^\# = f_0$. Moreover, $(f_a = t)$ and $(f_0 = t)$ have the same $(n-1)$-dimensional Hausdorff measure for each $t$, $|\nabla f_a|$ is nonconstant on $(f_a = t)$ for $0 < t < 1$, and $t \to t^p$ is strictly convex. An argument with the coarea formula, to be given in §4.5, shows that

$$\int_{\mathbb{R}^n} |\nabla f_a|^p\,dx < \int_{\mathbb{R}^n} |\nabla f_0|^p\,dx = \int_{\mathbb{R}^n} |\nabla f_a^\#|^p\,dx, \quad 0 < p < 1,\ 0 < a < 1.$$

For *decreasing* rearrangements $f^*$ of functions of one variable the analogue of (3.4) holds also for $0 < p < 1$. See Theorem 3.13 in §3.3.

The uniqueness problem associated with Theorem 3.7 is surprisingly complicated, as discovered by Brothers and Ziemer (1988). They show among other things that if $f \ge 0$ and $\lambda_f$ is absolutely continuous on $\mathbb{R}^+$ then finite equality in (3.4) for some $p \in (1,\infty)$ implies that $f = f^\# \circ T$ for some isometry $T$ of $\mathbb{R}^n$, but that such a uniqueness result need not hold when $\lambda_f$ has a singular component. For more recent results on this problem, see Cianchi and Fusco (2006) and Burchard and Ferone (2015).

*Proof of Theorem 3.7* For $\epsilon > 0$, let

$$E(\epsilon) = \{(x,y) \in \mathbb{R}^n \times \mathbb{R}^n \colon |x-y| < \epsilon\}.$$

Assume for now that $\inf f = 0$. Apply Corollary 2.22 with $f = g$, $\Phi(t) = t^p$, and $K(x) = \chi_{[0,\epsilon]}$. The result is

$$\int_{E(\epsilon)} |f^\#(x) - f^\#(y)|^p dx\,dy \le \int_{E(\epsilon)} |f(x) - f(y)|^p dx\,dy. \tag{3.5}$$

For fixed $x$, let $y = x + \epsilon z$. Then $dy = \epsilon^n\,dz$. Divide (3.5) by $\epsilon^{-n-p}$. Fubini's theorem gives

$$\begin{aligned}\int_{\mathbb{R}^n} dx \int_{\mathbb{B}^n(1)} &\left|\frac{f^\#(x+\epsilon z) - f^\#(x)|}{\epsilon}\right|^p dz \\ &\le \int_{\mathbb{R}^n} dx \int_{\mathbb{B}^n(1)} \left|\frac{f(x+\epsilon z) - f(x)|}{\epsilon}\right|^p dz.\end{aligned} \tag{3.6}$$

We want to pass to the limit $\epsilon \to 0$. Assume, in addition to $f \ge 0$, that $f$ has compact support, say $f(x) = 0$ for $|x| \ge R$. Suppose also that $\epsilon < 1$. Then for $x \in \mathbb{R}^n$ and $z \in \mathbb{B}^n(1)$, we have

$$\left|\frac{f(x+\epsilon z)-f(x)}{\epsilon}\right|^p \leq \|f\|^p_{\mathrm{Lip}(\mathbb{R}^n,\mathbb{R})}|z|^p \chi_{\mathbb{B}^n(R+1)}(x).$$

Since $\int_{\mathbb{R}^n} \chi_{\mathbb{B}^n(R+1)}(x)\,dx \int_{\mathbb{B}^n(1)} |z|^p\,dz < \infty$, the dominated convergence theorem permits passage to the limit inside the integral on the right-hand side of (3.6). At points of differentiability of $f$, the integrand approaches $|\nabla f(x)\cdot z|$. Since, by Rademacher's Theorem 3.2, almost all points are points of differentiability, the right-hand side of (3.6) tends to the limit

$$\int_{\mathbb{R}^n} dx \int_{\mathbb{B}^n(1)} |\nabla f(x)\cdot z|^p\,dz.$$

Write $z = (z_1, \ldots, z_n)$, and set

$$C_p = \int_{\mathbb{B}^n(1)} |z_1|^p\,dz.$$

Then, for $a \in \mathbb{R}^n$,

$$\int_{\mathbb{B}^n(1)} |a\cdot z|^p\,dz = C_p|a|^p.$$

We conclude that

$$\lim_{\epsilon\to 0}\int_{\mathbb{R}^n} dx \int_{\mathbb{B}^n(1)} \left|\frac{f(x+\epsilon z)-f(x)}{\epsilon}\right|^p dz = C_p \int_{\mathbb{R}^n} |\nabla f|^p\,dx. \tag{3.7}$$

Since $f$ and $f^\#$ have the same distribution, $f^\#$ is also nonnegative with compact support. By Theorem 3.6, $f^\# \in \mathrm{Lip}(\mathbb{R}^n,\mathbb{R})$. Thus, the same argument yields (3.7) when $f$ is replaced by $f^\#$. From (3.6), we see that (3.4) holds for nonnegative compactly supported $f$.

Let now $f$ be a general function satisfying the assumptions of Theorem 3.7. Then, from the assumptions $f \in \mathrm{Lip}(\mathbb{R}^n,\mathbb{R})$ and $\lambda_f(t) < \infty$ for $t > \inf f$, it follows that

$$\lim_{x\to\infty} f(x) = \inf f. \tag{3.8}$$

Take a strictly decreasing sequence $\{\alpha_m\}$, $m \geq 1$, with $\alpha_m \searrow \inf f$. Let $f_m = (f-\alpha_m)^+$, $\chi_m = \chi_{(f>\alpha_m)}$, and $\overline{\chi}_m = \chi_{(f^\#>\alpha_m)}$. Then $f_m \in \mathrm{Lip}(\mathbb{R}^n,\mathbb{R})$, and Corollary 3.5 implies that $\nabla f_m = 0$ a.e. on $(f = \alpha_m)$. It follows that $\nabla f_m = \chi_m \nabla f$ a.e. in $\mathbb{R}^n$. Also, from the uniqueness of symmetric decreasing rearrangements (see Proposition 1.30), we have $f_m^\# = (f^\# - \alpha_m)^+$. Thus $\nabla f_m^\# = \overline{\chi}_m \nabla f^\#$. By (3.8), $f_m$ has compact support, so (3.4) holds for each $f_m$. Since $\chi_m \nearrow 1$ and $\overline{\chi}_m \nearrow 1$ on the sets $(f > \inf f)$ and $(f^\# > \inf f^\#)$ respectively, and since $\nabla f = 0$ a.e. on $(f = \inf f)$, $\nabla f^\# = 0$ a.e. on $(f^\# = \inf f^\#)$, the monotone convergence theorem shows that (3.4) holds for $f$. □

For $f \in \mathrm{Lip}(\mathbb{R}^n, \mathbb{R})$ and $-\infty \le a < b \le \infty$, define $g\colon \mathbb{R}^n \to \mathbb{R}$ by

$$g(x) = \begin{cases} b, & \text{if } f(x) \ge b, \\ f(x), & \text{if } a \le f(x) \le b, \\ a, & \text{if } f(x) \le a. \end{cases}$$

Thus, $g$ is a truncation of $f$. Then $\|g\|_{\mathrm{Lip}(\mathbb{R}^n,\mathbb{R})} \le \|f\|_{\mathrm{Lip}(\mathbb{R}^n,\mathbb{R})}$, and from Corollary 3.5 it follows that $\nabla g = \chi_{f^{-1}(a,b)} \nabla f$ a.e. Also, from uniqueness of symmetric decreasing rearrangements it follows that $g^\#$ is obtained by truncating $f^\#$ in the same way that $g$ is obtained by truncating $f$. That is, the truncation and symmetric decreasing rearrangement operators commute. Thus, $\nabla g^\# = \chi_{(f^\#)^{-1}(a,b)} \nabla f^\#$. Applying (3.4) to $g$, we see that Theorem 3.7 implies a generalization of itself:

**Corollary 3.8** *Let $f \in \mathrm{Lip}(\mathbb{R}^n, \mathbb{R})$ and suppose that $\lambda_f(t) < \infty$ for all $t > \inf f$. Then, for $-\infty \le a < b \le \infty$,*

$$\int_{(f^\#)^{-1}(a,b)} |\nabla f^\#|^p \, dx \le \int_{f^{-1}(a,b)} |\nabla f|^p \, dx, \quad 1 \le p < \infty.$$

Next, we shall establish a symmetrization inequality for Dirichlet integrals of functions defined on *subsets* of $\mathbb{R}^n$. For $E \subset \mathbb{R}^n$ a Lebesgue measurable set and $f\colon E \to \mathbb{R}$ a Lebesgue measurable function satisfying

$$\lambda_f(t) < \infty, \quad \forall t > \operatorname{ess\,inf}_E f,$$

we have defined, after Definition 1.31, the symmetric decreasing rearrangement $f^\#\colon E^\# \to \mathbb{R}$ by

$$f^\#(x) = f^*(\alpha_n |x|^n), \quad x \in E^\#.$$

Here $\alpha_n = \mathcal{L}^n(\mathbb{B}^n(1))$ and $E^\#$ is the ball in $\mathbb{R}^n$ centered at the origin with $\mathcal{L}^n(E^\#) = \mathcal{L}^n(E)$. The distribution function $\lambda_f$ and the decreasing rearrangement $f^*$ are computed with respect to $\mathcal{L}^n$ on $E$.

**Corollary 3.9** *Let $\Omega$ be an open subset of $\mathbb{R}^n$ with $\mathcal{L}^n(\Omega) < \infty$, and let $f$ be a nonnegative Lipschitz function on $\overline{\Omega}$ such that $f = 0$ on $\partial\Omega$. Then*

$$\int_{\Omega^\#} |\nabla f^\#|^p \, dx \le \int_\Omega |\nabla f|^p \, dx, \quad 1 \le p < \infty. \tag{3.9}$$

*Proof* Define $g\colon \mathbb{R}^n \to \mathbb{R}$ by $g = f$ on $\Omega$, $g = 0$ on $\mathbb{R}^n \setminus \Omega$. Then $g \in \mathrm{Lip}(\mathbb{R}^n, \mathbb{R})$, $\nabla g = \nabla f$ a.e. on $\Omega$, and $\nabla g = 0$ a.e. on $\mathbb{R}^n \setminus \Omega$. Here, one uses Corollary 3.5 to insure that $\nabla g = 0$ a.e. on $\partial\Omega$. Moreover, we have $g^\# = f^\#$ on

$\Omega^\#$, $g^\# = 0$ on $\mathbb{R}^n \setminus \Omega^\#$, where $g^\#$ is the symmetric decreasing rearrangement computed with respect to $\mathbb{R}^n$. Applying Theorem 3.7 to $g$, we obtain

$$\int_{\Omega^\#} |\nabla f^\#|^p \, dx = \int_{\mathbb{R}^n} |\nabla g^\#|^p \, dx \leq \int_{\mathbb{R}^n} |\nabla g|^p \, dx = \int_{\Omega} |\nabla f|^p \, dx. \qquad \square$$

In Corollary 3.9 we assumed that $f \geq 0$. If $f$ is strictly negative somewhere in $\Omega$ and $f = 0$ on $\partial\Omega$, then the extension $g$ in the proof is still in $\mathrm{Lip}(\mathbb{R}^n, \mathbb{R})$, but does not satisfy all the hypotheses of Theorem 3.7, since $\inf g < 0$, but $\lambda_g(t) = \infty$ for $t < 0$. Thus, the proof of Corollary 3.9 just given breaks down. For $n = 1$, however, Corollary 3.9 remains true even without the nonnegativity assumption on $f$ (assuming $f = 0$ on $\partial\Omega$). This is a consequence of the result to be proved in Chapter 7 that circular symmetrization decreases Dirichlet integrals for Lipschitz functions defined on a circle (Theorem 7.4). By contrast, for $n \geq 2$, vanishing of $f$ on $\partial\Omega$ is not sufficient for (3.9) to hold.

**Example 3.10** Let $\Omega = \mathbb{B}^n(1)$, $f(x) = |x| - 1, |\nabla f(x)| = 1$. Then (3.9) is false for every $n \geq 2$ and every $1 \leq p < \infty$. We carry out the computations for $n = 2$. Write $r = |x|$.

$$\lambda_f(t) = \pi(1 - (1+t)^2), \quad -1 \leq t \leq 0,$$

$$f^\#(x) = (1 - r^2)^{1/2} - 1, \quad |\nabla f^\#(x)| = \frac{r}{(1 - r^2)^{1/2}}, \quad x \in \mathbb{B}^2(1).$$

$$\int_{\mathbb{B}^2(1)} |\nabla f|^p = \pi,$$

$$\int_{\mathbb{B}^2(1)} |\nabla f^\#|^p = 2\pi \int_0^1 \left( \frac{r}{(1 - r^2)^{1/2}} \right)^p r \, dr = \pi \int_0^1 \left( \frac{s}{1 - s} \right)^{p/2} ds.$$

For $p = 1$, $\int_0^1 (\frac{s}{1-s})^{1/2} \, ds = \frac{\pi}{2} > 1$. For $p \geq 1$, by convexity,

$$\int_0^1 \left( \frac{s}{1-s} \right)^{p/2} ds \geq \left( \int_0^1 \left( \frac{s}{1-s} \right)^{1/2} ds \right)^p > 1.$$

Thus,

$$\int_{\mathbb{B}^2(1)} |\nabla f|^p < \int_{\mathbb{B}^2(1)} |\nabla f^\#|^p, \quad \forall p \in [1, \infty).$$

Corollary 3.9 is of course still true for Lipschitz functions in $\Omega$ satisfying $f \geq c$ in $\Omega, f = c$ on $\partial\Omega$. The corollary is also still true for nonnegative Lipschitz functions on domains of infinite volume, vanishing at the boundary, if we add the assumption $\mathcal{L}^n(f > t) < \infty$ for all $t > 0$.

## 3.3 Symmetrization Decreases the Φ-Dirichlet Integral of Lipschitz Functions

**Theorem 3.11** *Let* $f \in \mathrm{Lip}(\mathbb{R}^n, \mathbb{R})$, *and suppose that* $\lambda_f(t) < \infty$ *for all* $t > \inf f$. *Let* $\Phi\colon \mathbb{R}^+ \to \mathbb{R}^+$ *be convex and increasing, with* $\Phi(0) = 0$. *Then*

$$\int_{\mathbb{R}^n} \Phi(|\nabla f^\#|)\, dx \le \int_{\mathbb{R}^n} \Phi(|\nabla f|)\, dx. \tag{3.10}$$

More generally, it is true under the assumptions of Theorem 3.11 that

$$\int_{(f^\#)^{-1}(a,b)} \Phi(|\nabla f^\#|)\, dx \le \int_{f^{-1}(a,b)} \Phi(|\nabla f|)\, dx, \quad -\infty \le a < b \le \infty.$$

The deduction of the latter inequality from Theorem 3.11 is the same as the deduction of Corollary 3.8 from Theorem 3.7. Similarly, the extension argument used to deduce Corollary 3.9 from Theorem 3.7 works for Φ-Dirichlet integrals, and produces the following result: If $\Omega$ is an open subset of $\mathbb{R}^n$ with $\mathcal{L}^n(\Omega) < \infty$, $f \in \mathrm{Lip}(\Omega, \mathbb{R}^+)$, and $f = 0$ on $\partial\Omega$, then

$$\int_{\Omega^\#} \Phi(|\nabla f^\#|)\, dx \le \int_{\Omega} \Phi(|\nabla f|)\, dx.$$

When $\Phi$ is strictly concave on $\mathbb{R}^+$, the argument in §3.2 for $\Phi(t) = t^p$, $0 < p < 1$, shows again that

$$\int_{\mathbb{R}^n} \Phi(|\nabla f^\#|)\, dx > \int_{\mathbb{R}^n} \Phi(|\nabla f|)\, dx$$

when $f$ is one of the functions $f_a$, $0 < a < 1$, constructed at the beginning of §3.2. On the other hand, for decreasing rearrangements $f^*$ of functions of one variable the analogue of (3.10) is true when $\Phi$ is any increasing function. This result is included in Theorem 3.13, to be presented later in this section.

*Proof of Theorem 3.11* It suffices to prove Theorem 3.11 under the additional assumptions that $f$ is nonnegative and has compact support; the passage to general functions is accomplished in the same way as in the proof of Theorem 3.7.

Assume then that $f \in \mathrm{Lip}(\mathbb{R}^n, \mathbb{R}^+)$ and that $f$ has compact support. Let $M = \|\nabla f\|_\infty$. Then $M = \|f\|_{\mathrm{Lip}(\mathbb{R}^n,\mathbb{R})}$, by Corollary 3.4, so that $M < \infty$. By Theorem 3.6, we know that $f^\# \in \mathrm{Lip}(\mathbb{R}^n, \mathbb{R})$ with $\|\nabla f^\#\|_\infty \le M$. Thus, the function

$$g(r) \equiv f^\#(re_1)$$

is in $\mathrm{Lip}(\mathbb{R}^+, \mathbb{R}^+)$, and is decreasing. Hence $g'$ exists a.e. in $\mathbb{R}^+$, $g' \in L^\infty(\mathbb{R}^+)$, $-M \le g' \le 0$ a.e. in $\mathbb{R}^+$, and for almost every $x \in \mathbb{R}^n$ we have

$$|\nabla f^\#(x)| = -\frac{\partial f^\#}{\partial r}(x) = -g'(|x|).$$

Let $R$ be large enough so that $\operatorname{supp} f \subset \mathbb{B}^n(0,R)$. Then $g(r) = 0$ for $r \ge R$. Let $\epsilon > 0$ be given. Define $\delta$ by

$$\epsilon = \delta\, \beta_{n-1} \left[ M\Phi'(M+) + \Phi(M) + \frac{R^n}{n}\Phi'(M+) \right],$$

where $\beta_{n-1} = n\alpha_n$ is the surface measure of the sphere $\mathbb{S}^{n-1}$ (Folland, 1999, p. 80), and $\Phi'(M+)$ is the derivative from the right of $\Phi$ at $M$. The existence of $\Phi'(M+)$ is shown, for example, in Zygmund (1968, p. 21).

Define the Borel measure $\nu$ on $\mathbb{R}^+$ by

$$d\nu(r) = r^{n-1}\, dr.$$

Then

$$\int_{\mathbb{R}^n} \Phi(|\nabla f^\#|)\, dx = \beta_{n-1} \int_0^R \Phi(|g'(r)|)\, d\nu(r). \tag{3.11}$$

By Lusin's Theorem, there is a continuous function $h$ on $[0,R]$ such that $0 \le h \le M$ on $[0,R]$ and $h = -g'$ on $[0,R] \setminus E$, where $\nu(E) < \delta$.

Let $\{a_i\}_{i=0}^m$ be a sequence in $\mathbb{R}^+$ with $0 = a_0 < a_1 < \cdots < a_m = R$ such that the oscillation of $h$ over each $[a_{i-1}, a_i]$ is less than $\delta$. For $i = 1, \ldots, m$, let

$$b_i = g(a_i), \quad I_i = g^{-1}(b_{i-1}, b_i), \quad A_i = f^{-1}(b_{i-1}, b_i), \quad \mu_i = \mathcal{L}^n(A_i).$$

Then $I_i$ is an open subinterval of $(a_{i-1}, a_i)$, and $\mu_i = \beta_{n-1}\nu(I_i)$.

Jensen's inequality (Rudin, 1966, p. 61) asserts that if $\mu$ is a probability measure on a space $X$, $F\colon X \to \mathbb{R}$ is measurable, and $\Phi$ is a convex function on $h(X)$, then $\Phi(\int_X F\, d\mu) \le \int_X \Phi(F)\, d\mu$. Applied to the measure $\frac{1}{\mu_i} dx$ on $A_i$, Jensen's inequality gives

$$\frac{1}{\mu_i} \int_{A_i} \Phi(|\nabla f|)\, dx \ge \Phi\left( \frac{1}{\mu_i} \int_{A_i} |\nabla f|\, dx \right). \tag{3.12}$$

It is possible that $b_{i-1} = b_i$ for some $i$. In this case $A_i$ and $I_i$ are empty, and we interpret the expressions in (3.12), and similar ones involving averages with respect to $\nu$ over $I_i$, to be zero.

By Corollary 3.8 with $p = 1$, the integral on the right-hand side of (3.12) decreases when $f$ is replaced by $f^\#$ and $A_i$ by $(f^\#)^{-1}(b_{i-1}, b_i)$. Multiplying by $\mu_i$ and summing over $i$, we obtain

$$\int_{\mathbb{R}^n} \Phi(|\nabla f|)\, dx \ge \sum_{i=1}^m \mu_i \Phi\left( \frac{1}{\nu(I_i)} \int_{I_i} |g'|\, d\nu \right). \tag{3.13}$$

Write

$$\begin{aligned}\Phi\left(\frac{1}{\nu(I_i)}\int_{I_i}|g'|\,d\nu\right)-\frac{1}{\nu(I_i)}\int_{I_i}\Phi(|g'|)\,d\nu\\&=\Phi\left(\frac{1}{\nu(I_i)}\int_{I_i}|g'|\,d\nu\right)-\Phi\left(\frac{1}{\nu(I_i)}\int_{I_i}h\,d\nu\right)\\&\quad+\Phi\left(\frac{1}{\nu(I_i)}\int_{I_i}h\,d\nu\right)-\frac{1}{\nu(I_i)}\int_{I_i}\Phi(h)\,d\nu\\&\quad+\frac{1}{\nu(I_i)}\int_{I_i}\Phi(h)\,d\nu-\frac{1}{\nu(I_i)}\int_{I_i}\Phi(|g'|)\,d\nu\\&=s_i+t_i+u_i.\end{aligned}\tag{3.14}$$

Now $g$ is constant on $(a_{i-1},a_i)\setminus I_i$. From (3.11), (3.14), and (3.13) we obtain

$$\begin{aligned}\int_{\mathbb{R}^n}\Phi(|\nabla f^{\#}|)\,dx&=\beta_{n-1}\sum_{i=1}^{m}\int_{I_i}\Phi(|g'|)\,d\nu=\sum_{i=1}^{m}\mu_i\frac{1}{\nu(I_i)}\int_{I_i}\Phi(|g'|)\,d\nu\\&=\sum_{i=1}^{m}\mu_i\Phi\left(\frac{1}{\nu(I_i)}\int_{I_i}|g'|\,d\nu\right)-\sum_{i=1}^{m}\mu_i(s_i+t_i+u_i)\\&\leq\int_{\mathbb{R}^n}\Phi(|\nabla f|)\,dx-\sum_{i=1}^{m}\mu_i(s_i+t_i+u_i).\end{aligned}\tag{3.15}$$

Since $|g'|=h$ except on $E$, and each of $|g'|$ and $h$ is nonnegative and $\leq M$, their $\nu$-averages over each $I_i$ lie between 0 and $M$, and differ by at most $M\nu(E\cap I_i)/\nu(I_i)$. From convexity of $\Phi$, it follows that

$$|s_i|\leq\Phi'(M+)\,M\frac{\nu(E\cap I_i)}{\nu(I_i)}.$$

Similarly,

$$|u_i|\leq\Phi(M)\frac{\nu(E\cap I_i)}{\nu(I_i)}.$$

Each of the terms $\Phi\Big(\frac{1}{\nu(I_i)}\int_{I_i}h\,d\nu\Big)$ and $\frac{1}{\nu(I_i)}\int_{I_i}\Phi(h)\,d\nu$ is at most $\Phi(\sup_{I_i}h)$ and at least $\Phi(\inf_{I_i}h)$. Since $0\leq\sup_{I_i}h-\inf_{I_i}h\leq\delta$, we have

$$|t_i|\leq\Phi'(M+)\delta.$$

Since $\mu_i=\beta_{n-1}\nu(I_i)$ and $\nu(E)<\delta$, the last three estimates imply

$$\begin{aligned}&\left|\sum_{i=1}^{m}\mu_i(s_i+t_i+u_i)\right|\\&\qquad\leq\beta_{n-1}\left[\Phi'(M+)M\delta+\Phi(M)\delta+\nu([0,R])\Phi'(M+)\delta\right]=\epsilon.\end{aligned}$$

This last estimate and (3.15) complete the proof of Theorem 3.11. □

We derived our Dirichlet integral inequalities from the "main integral inequality," Theorem 2.15(a), which was itself obtained from integral inequalities involving polarization. There is another approach which bypasses the general integral inequalities and uses polarization in a different way to produce Dirichlet integral inequalities for symmetric decreasing rearrangements.

Let $f \in \mathrm{Lip}(\mathbb{R}^n, \mathbb{R})$, let $H$ be an affine hyperplane in $\mathbb{R}^n$, and select a complementary half space $H^+$ for $H$. Recall that for $x \in H^+$ the polarization $f_H$ of $f$ is defined by $f_H(x) = \max(f(x), f(\rho x))$, where $\rho$ is reflection in $H$. Since polarization decreases the modulus of continuity (Proposition 1.37), it follows that $f_H \in \mathrm{Lip}(\mathbb{R}^n, \mathbb{R})$, with $\|f_H\|_{\mathrm{Lip}(\mathbb{R}^n,\mathbb{R})} \leq \|f\|_{\mathrm{Lip}(\mathbb{R}^n,\mathbb{R})}$. The argument below will show that in fact the two norms are equal.

Let $A = \{x \in H^+ : f(x) > f(\rho x)\}$, $B = \{x \in H^+ : f(x) < f(\rho x)\}$, and $C = \{x \in H^+ : f(x) = f(\rho x)\}$. Then $|\nabla f_H(x)| = |\nabla f(x)|$ and $|\nabla f_H(\rho x)| = |\nabla f(\rho x)|$ at points of $A$ at which both $f$ and $f \circ \rho$ are differentiable, and $|\nabla f_H(x)| = |\nabla f(\rho x)|$ and $|\nabla f_H(\rho x)| = |\nabla f(x)|$ at points of $B$ at which both $f$ and $f \circ \rho$ are differentiable. Note that $\rho$ is an isometry, so $|(\nabla(f \circ \rho))(x)| = |(\nabla f)(\rho x)|$, which for short we write as $|\nabla f(\rho x)|$.

If $f_1, f_2 \in \mathrm{Lip}(\mathbb{R}^n, \mathbb{R})$, then Corollary 3.5 applied to $f_1 - f_2$ implies that $\nabla f_1 = \nabla f_2$ a.e. on $(f_1 = f_2)$. At points $x$ of $(f_1 = f_2)$ where both $f_1$ and $f_2$ are differentiable and $\nabla f_1 = \nabla f_2$, the definition of differentiable shows that the functions $\max(f_1, f_2)$ and $\min(f_1, f_2)$ are differentiable, and each has gradient $\nabla f_1(x) = \nabla f_2(x)$. Thus, at almost all points $x \in C$, each of the four functions $f, f \circ \rho, f_H$ and $f_H \circ \rho$ is differentiable, and the norm of each of the four gradients is $|\nabla f(x)|$.

These arguments show that for all functions $\Phi\colon \mathbb{R}^+ \to \mathbb{R}^+$ and almost all $x \in H^+$ we have

$$\Phi(|\nabla f(x)|) + \Phi(|\nabla f(\rho x)|) = \Phi(|\nabla f_H(x)|) + \Phi(|\nabla f_H(\rho x)|).$$

Writing $\int_{\mathbb{R}^n} \Phi(|\nabla f(x)|)\, dx = \int_{H^+} \Phi(|\nabla f(x)|)\, dx + \int_{H^+} \Phi(|\nabla f(\rho x)|)\, dx$, we obtain the following result:

**Proposition 3.12** *Let $\Phi\colon \mathbb{R}^+ \to \mathbb{R}^+$ be a nonnegative Borel function, $f \in \mathrm{Lip}(\mathbb{R}^n, \mathbb{R})$, and $H \in \mathcal{H}(\mathbb{R}^n)$. Then $f_H \in \mathrm{Lip}(\mathbb{R}^n, \mathbb{R})$, and*

$$\int_{\mathbb{R}^n} \Phi(|\nabla f_H)|)\, dx = \int_{\mathbb{R}^n} \Phi(|\nabla f|)\, dx.$$

From Proposition 1.18 in Chapter 1, it follows that Proposition 3.12 is equivalent to the statement that $|\nabla f|$ and $|\nabla f_H|$ are equidistributed on $\mathbb{R}^n$.

To deduce Theorem 3.11 from Proposition 3.12, one can adapt the proofs of Theorems 2.12 and 2.15(a). It suffices to prove Theorem 3.11 for nonnegative Lipschitz $f$ with compact support. As in the proof of Theorem 2.12, set

$$\mathcal{S} = \{F \in C_c(\mathbb{R}^n, \mathbb{R}^+) \colon \omega(\cdot, F) \le \omega(\cdot, f) \text{ on } (0, \infty),$$
$$\lambda_F = \lambda_f \text{ on } (0, \infty), \text{ and } \operatorname{diam} \operatorname{supp} F \le \operatorname{diam} \operatorname{supp} f\}.$$

Define

$$\mathcal{S}_1 = \left\{F \in \mathcal{S} \colon \int_{\mathbb{R}^n} \Phi(|\nabla F|)\,dx \le \int_{\mathbb{R}^n} \Phi(|\nabla f|)\,dx\right\}.$$

Let $d = \inf_{F \in \mathcal{S}_1} \|F - f^\#\|_{L^2(\mathbb{R}^n)}$. As in the proof of Theorem 2.12, there is a sequence $\{F_k\}$ in $\mathcal{S}_1$ which converges uniformly in $\mathbb{R}^n$ to a function $F_0 \in \mathcal{S}$ such that $\|F_0 - f^\#\|_{L^2(\mathbb{R}^n)} = d$. Since $\omega(\cdot, F_0) \le \omega(\cdot, f)$, it follows that $F_0$ is Lipschitz. If $F_0 \in \mathcal{S}_1$, then, as before, one can show that $F_0 = f^\#$, and Theorem 3.11 is proved. For $F_0$ to be in $\mathcal{S}_1$, it is sufficient that

$$\int_{\mathbb{R}^n} \Phi(|\nabla F_0|)\,dx \le \limsup_{k\to\infty} \int_{\mathbb{R}^n} \Phi(|\nabla F_k|)\,dx. \tag{3.16}$$

The functions $\nabla F_k$ are uniformly bounded in $L^\infty(\mathbb{R}^n)$ and are all supported in a fixed ball. Thus, they form a bounded sequence in $L^p(\mathbb{R}^n)$ for each $p > 0$. Fix $p \in (1, \infty)$. By the Banach–Saks Theorem, see for example Wojtaszczyk (1991, p. 101), there is a subsequence of the $F_k$ for which the arithmetic means of the corresponding $\nabla F_k$ converge in $L^p(\mathbb{R}^n)$-norm. From here, an argument which we leave to the interested reader shows that if $\Phi$ is convex then (3.16) holds.

If $\Phi$ is not convex then (3.16) need not hold.

We look briefly now at inequalities of integrals of Dirichlet type for decreasing rearrangements $f^*$ of functions $f$ of one real variable. The following result is due to Yanagihara (1993). A similar theorem was independently found by Stanoyevitch (1994).

**Theorem 3.13** *Let $f \colon [0, 1] \to \mathbb{R}$ be differentiable a.e., and let $\Phi \colon \mathbb{R}\times\mathbb{R}^+ \to \mathbb{R}^+$ be Borel measurable. If $y_2 \to \Phi(y_1, y_2)$ is an increasing function on $\mathbb{R}^+$ for each $y_1 \in \mathbb{R}$, then*

$$\int_{[0,1]} \Phi(f^*(x), |(f^*)'(x)|)\,dx \le \int_{[0,1]} \Phi(f(x), |f'(x)|)\,dx.$$

In particular, for every increasing $\Phi \colon \mathbb{R}^+ \to \mathbb{R}^+$, we have

$$\int_{[0,1]} \Phi(|(f^*)'(x)|)\,dx \le \int_{[0,1]} \Phi(|f'(x)|)\,dx. \tag{3.17}$$

Using the identity (1.5), one can give an equivalent statement of (3.17) in terms of distribution functions: $\lambda_{|(f^*)'|}(t) \le \lambda_{|f'|}(t)$, for every $t \in \mathbb{R}$. As noted at the beginning of this section, the inequality corresponding to (3.17) for *symmetric* decreasing rearrangements can fail when $\Phi$ is concave. Inequality (3.17) is due to Chong (1975); special cases had been proved earlier by Ryff and by Duff. Of course, Theorem 3.13 is true when $[0, 1]$ is replaced by $[0, A]$ for any finite positive $A$. Under appropriate hypotheses on $\lambda_f$ it is also true when $[0, 1]$ is replaced by $(0, \infty)$ in the left-hand integral and by an infinite subinterval of $\mathbb{R}$ in the right-hand integral.

*Proof of Theorem 3.13* By Ryff's Theorem, Proposition 1.26, there is a measure preserving transformation $T\colon [0, 1] \to [0, 1]$ such that $f = f^* \circ T$ a.e. Now $f'(x)$ exists a.e. by assumption, and $(f^*)'(x)$ exists a.e., since $f^*$ is decreasing. Since $T$ is measure preserving, it follows that

$$E \equiv \{x \in [0, 1] \colon f(x) = f^*(T(x)), f'(x) \text{ exists, and } (f^*)'(Tx) \text{ exists}\}$$

has $\mathcal{L}(E) = 1$. Using again that $T$ is measure preserving, one shows that for $x \in E$, $|(f^*)'(Tx)| \leq |f'(x)|$; see for example Yanagihara (1993). Thus,

$$\begin{aligned}\int_{[0,1]} \Phi(f^*(x), |(f^*)'(x)|)\, dx &= \int_{[0,1]} \Phi(f^*(Tx), |(f^*)'(Tx)|)\, dx \\ &\leq \int_{[0,1]} \Phi(f(x), |f'(x)|)\, dx,\end{aligned}$$

which is (3.17). □

The integral functionals in Theorem 3.13 may depend on both $f$ and $|f'|$, whereas in our statement of Theorem 3.11 the functionals depend only on $|\nabla f|$. In fact, one can generalize Theorem 3.11 to obtain the inequality

$$\int_{\mathbb{R}^n} \Phi(f^\#, |\nabla f^\#|)\, dx \leq \int_{\mathbb{R}^n} \Phi(f, |\nabla f|)\, dx, \tag{3.18}$$

provided $\Phi\colon \mathbb{R} \times \mathbb{R}^+ \to \mathbb{R}^+$ is a bounded Borel function such that $y_2 \to \Phi(y_1, y_2)$ is an increasing convex function on $\mathbb{R}^+$ for each $y_1 \in \mathbb{R}$ which vanishes at $y_2 = 0$, and $f$ satisfies the assumptions of Theorem 3.11. To prove (3.18), one first takes $\Phi(y_1, y_2) = \chi_{(a,b)}(y_1)\Phi_0(y_2)$, where $\Phi_0$ is convex and increasing on $\mathbb{R}^+$ with $\Phi_0(0) = 0$. Then (3.18) is the truncated version of (3.10) stated right after (3.10). To complete the proof of (3.18), one makes appropriate modifications in the standard arguments involving linear combinations and monotone convergence that permit results about integrals of characteristic functions to be extended to results about integrals of nonnegative functions. Details are left to the reader.

## 3.4 Sobolev Spaces $W^{1,p}(\mathbb{R}^n)$

This section is a brief introduction to the theory of first order Sobolev spaces on open subsets of $\mathbb{R}^n$. Good references for this theory include Evans and Gariepy

(1992), Maz'ja (1985), and Ziemer (1989). Throughout this section $\Omega$ will denote an open subset of $\mathbb{R}^n$. We shall write $L^p(\Omega)$ for the Lebesgue space of real valued functions in $L^p(\Omega, \mathcal{L}^n)$. The notation $\Omega_1 \Subset \Omega$ means that $\Omega_1$ is an open set whose closure is a compact subset of $\Omega$.

A measurable function $f\colon \Omega \to \mathbb{R}$ is said to be *locally integrable* on $\Omega$ if $f \in L^1(\Omega_1)$ for each $\Omega_1 \Subset \Omega$. The set of all locally integrable functions on $\Omega$ will be denoted by $L^1_{\mathrm{loc}}(\Omega)$. More generally, for $0 < p \le \infty$, $L^p_{\mathrm{loc}}(\Omega)$ will denote the set of measurable functions $f\colon \Omega \to \mathbb{R}$ such that $f \in L^p(\Omega_1)$ for every $\Omega_1 \Subset \Omega$. Note that if $0 < p_1 \le p_2 \le \infty$, then $L^{p_2}_{\mathrm{loc}}(\Omega) \subset L^{p_1}_{\mathrm{loc}}(\Omega)$.

Let $f \in L^1_{\mathrm{loc}}(\Omega)$ and $i \in \{1, \ldots, n\}$. A function $g_i \in L^1_{\mathrm{loc}}(\Omega)$ is said to be the *weak partial derivative* of $f$ with respect to $x_i$ if

$$\int_\Omega f\, \partial_i \phi\, dx = -\int_\Omega g_i \phi\, dx, \quad \forall\, \phi \in C^1_c(\Omega).$$

When $g_i$ exists we will write $\partial_i f = g_i$, and when $g_i$ exists for each $i$ we will write $\nabla f = (\partial_1 f, \ldots, \partial_n f)$. For $1 \le p < \infty$, define

$$W^{1,p}(\Omega) = \{f \in L^p(\Omega)\colon \text{each } \partial_i f \text{ exists in the weak sense and is in } L^p(\Omega)\}.$$

The sets $W^{1,p}(\Omega)$ are called first order Sobolev spaces. Define

$$\|f\|_{W^{1,p}(\Omega)} = \|f\|_{L^p(\Omega)} + \|\nabla f\|_{L^p(\Omega)}.$$

Then

$$W^{1,p}(\Omega) = \{f \in L^1_{\mathrm{loc}}(\Omega)\colon \text{ each } \partial_i f \text{ exists in the weak sense and } \|f\|_{W^{1,p}(\Omega)} < \infty\}.$$

Note that if $f \in W^{1,p}(\Omega)$ and $F$ is a function on $\Omega$ which agrees a.e. in $\Omega$ with $f$, then $F \in W^{1,p}(\Omega)$. It is an easy exercise to show that $\|\cdot\|_{W^{1,p}(\Omega)}$ is a norm on $W^{1,p}(\Omega)$, and that $(W^{1,p}(\Omega), \|\cdot\|_{W^{1,p}(\Omega)})$ is a Banach space.

The local Sobolev space $W^{1,p}_{\mathrm{loc}}(\Omega)$ is defined to be

$$W^{1,p}_{\mathrm{loc}}(\Omega) = \{f \in L^p_{\mathrm{loc}}(\Omega)\colon f|_{\Omega_1} \in W^{1,p}(\Omega_1),\ \forall\, \Omega_1 \Subset \Omega\}.$$

Let $K\colon \mathbb{R}^n \to \mathbb{R}$ be a function satisfying

$$K \ge 0, \quad K \in C^\infty(\mathbb{R}^n), \quad \operatorname{supp} K \subset \mathbb{B}^n(1), \quad \int_{\mathbb{R}^n} K\, dx = 1. \tag{3.19}$$

We shall refer to such functions as nonnegative smooth bump functions. For $\epsilon > 0$, define $K_\epsilon : \mathbb{R}^n \to \mathbb{R}$ and $\Omega_\epsilon \subset \Omega$ by

$$K_\epsilon(x) = \epsilon^{-n} K(x\epsilon^{-n}), \quad \Omega_\epsilon = \{x \in \Omega\colon d(x, \partial\Omega) > \epsilon\},$$

where $d$ denotes distance. For $f \in L^1_{\text{loc}}(\Omega)$, the convolution

$$f_\epsilon(x) \equiv f * K_\epsilon(x) \equiv \int_{\mathbb{R}^n} f(x-y)K_\epsilon(y)\,dy, \quad x \in \Omega_\epsilon \tag{3.20}$$

is uniquely defined in $\Omega_\epsilon$, where in the last integral $f$ may be extended from $\Omega$ to $\mathbb{R}^n$ in arbitrary fashion.

Proofs of the next two propositions may be found, for example, in Evans and Gariepy (1992, pp. 123, 125).

**Proposition 3.14** *Let $K$ satisfy* (3.19) *and $f_\epsilon$ be defined by* (3.20). *Then*

(a) *If $f \in L^1_{\text{loc}}(\Omega)$ then $f_\epsilon \in C^\infty(\Omega_\epsilon)$.*
(b) *If $f \in W^{1,p}_{loc}(\Omega)$, then $\partial_i f_\epsilon = K_\epsilon * \partial_i f$ on $\Omega_\epsilon$, for $1 \le p \le \infty$.*
(c) *If $f \in W^{1,p}_{loc}(\Omega)$ and $1 \le p < \infty$, then $\lim_{\epsilon\to 0}\|f_\epsilon - f\|_{W^{1,p}(\Omega_1)} = 0$, for every $\Omega_1 \Subset \Omega$.*

**Proposition 3.15** *Let $1 \le p < \infty$. Then for each $f \in W^{1,p}(\Omega)$, there exists $\{f_k\}_{k=1}^\infty \subset W^{1,p}(\Omega) \cap C^\infty(\Omega)$ such that $f_k \to f$ in $W^{1,p}(\Omega)$.*

From Proposition 3.15, one sees that when $1 \le p < \infty$ the Banach space $W^{1,p}(\Omega)$ may be identified with the completion of $C^\infty(\Omega)$ in the norm $\|\cdot\|_{W^{1,p}(\Omega)}$

For $1 \le p < \infty$, the space $W^{1,p}_0(\Omega)$ is defined to be the norm-closure in $W^{1,p}(\Omega)$ of $C^\infty_c(\Omega)$. Thus, for $1 \le p < \infty$,

$$W^{1,p}_0(\Omega) = \Big\{f \in W^{1,p}(\Omega)\colon\ \exists\ \{f_k\}_{k=1}^\infty \subset C^\infty_c(\Omega) \text{ such that } \lim_{k\to\infty}\|f - f_k\|_{W^{1,p}(\Omega)} = 0\Big\}.$$

For general $\Omega$, $W^{1,p}_0(\Omega)$ is a proper subspace of $W^{1,p}(\Omega)$. But when $\Omega = \mathbb{R}^n$, the two spaces coincide:

**Proposition 3.16** $W^{1,p}_0(\mathbb{R}^n) = W^{1,p}(\mathbb{R}^n), \quad 1 \le p < \infty.$

*Sketch of proof* It suffices to show that each $f \in W^{1,p}(\mathbb{R}^n)$ can be approximated in the $W^{1,p}(\mathbb{R}^n)$-norm by a sequence of functions in $W^{1,p}_0(\mathbb{R}^n)$. The reader may verify that such a sequence is given by $f_k = fg_k$, where $g_k(x) = 1$ for $0 \le |x| \le k$, $g_k(x) = 2 - \frac{|x|}{k}$ for $k \le |x| \le 2k$, $g_k(x) = 0$ for $|x| \ge 2k$. □

Proposition 3.15 is not true when $p = \infty$, that is, $W^{1,\infty}(\Omega)$ is strictly larger than the closure of $C^\infty(\Omega) \cap W^{1,\infty}(\Omega)$ in the $W^{1,\infty}$-norm. It is true that $W^{1,\infty}(\Omega)$ is the closure of $C^\infty(\Omega) \cap W^{1,\infty}(\Omega)$ in the weak-* topology carried by $W^{1,\infty}(\Omega)$ as the dual space of the Banach space $W^{-1,1}(\Omega)$. See Ziemer (1989, p. 187) for a description of the spaces $W^{-1,p}$, and discussion of the duality

$$W^{1,p} = (W^{-1,p'})^*, \quad 1 < p \le \infty,$$

where $p'$ is the Hölder conjugate exponent of $p$: $\frac{1}{p} + \frac{1}{p'} = 1$.

The $W^{1,\infty}$ spaces are essentially the same as Lipschitz spaces:

**Proposition 3.17** *Let $f \in L^1_{\mathrm{loc}}(\Omega)$, where $\Omega$ is a convex domain. Then $f \in W^{1,\infty}(\Omega)$ if and only if there exists $g \in \mathrm{Lip}(\Omega, \mathbb{R}) \cap L^\infty(\Omega)$ such that $g = f$ a.e. in $\Omega$.*

*Proof* Let $f \in W^{1,\infty}(\Omega)$, and let $K$ be a nonnegative smooth bump function. Set $f_\epsilon = f * K_\epsilon$, $\epsilon > 0$. Then $f_\epsilon \in C^\infty(\Omega_\epsilon)$ and $\nabla f_\epsilon = K_\epsilon * \nabla f$ in $\Omega_\epsilon$, by Propositions 3.14 parts (a) and (b). From the convolution equation, it follows that $\|\nabla f_\epsilon\|_{L^\infty(\Omega_\epsilon,\mathbb{R})} \le \|\nabla f\|_{L^\infty(\Omega)}$. The argument in the proof of Corollary 3.4 of §3.1 gives $\|f_\epsilon\|_{\mathrm{Lip}(\Omega_\epsilon,\mathbb{R})} \le \|\nabla f\|_{L^\infty(\Omega)}$. Since $\lim_{\epsilon\to 0} f_\epsilon = f$ a.e. in $\Omega$, it follows that there exists a set $E \subset \Omega$ with $\mathcal{L}^n(E) = 0$ such that $|f(x) - f(y)| \le \|\nabla f\|_{L^\infty(\Omega)}|x - y|$, whenever both $x$ and $y$ are in $\Omega \backslash E$. Since $\Omega\backslash E$ is dense in $\Omega$, $f$ has a continuous extension $g$ to $\Omega$ which satisfies $\|g\|_{\mathrm{Lip}(\Omega,\mathbb{R})} \le \|\nabla f\|_{L^\infty(\Omega)}$ and $g = f$ a.e. in $\Omega$. Since $\|f\|_{L^\infty(\Omega)} \le \|f\|_{W^{1,\infty}(\Omega)} < \infty$, we have also $g \in L^\infty(\Omega)$.

Conversely, suppose that $g \in \mathrm{Lip}(\Omega, \mathbb{R}) \cap L^\infty(\Omega)$. To finish the proof of Proposition 3.17, we need to show that $\nabla g$ exists in the weak sense on $\Omega$, and that this weak gradient belongs to $L^\infty(\Omega)$. By Corollaries 3.3 and 3.4 of §3.1, the $\partial_i g$ exist pointwise a.e. in $\Omega$, and belong to $L^\infty(\Omega)$. So it suffices to show that the pointwise derivatives $\partial_i g$ are also weak derivatives. Take $\phi \in C^1_c(\Omega)$. Then, for each $i \in \{1, \dots, n\}$,

$$\begin{aligned}
\int_\Omega g\,\partial_i\phi\,dx &= \lim_{\epsilon\to 0}\int_\Omega \frac{\phi(x+\epsilon e_i) - \phi(x)}{\epsilon} g(x)\,dx \\
&= \lim_{\epsilon\to 0}\int_\Omega \frac{g(x-\epsilon e_i) - g(x)}{\epsilon}\phi(x)\,dx = -\int_\Omega \partial_i g\,\phi\,dx.
\end{aligned}$$

The equalities are justified by the dominated convergence theorem and the fact that both $g$ and $\phi$ satisfy Lipschitz conditions. We have shown that the a.e. pointwise partial derivatives are indeed also weak derivatives. □

## 3.5 Weak Compactness

Our next main goal is to extend Theorem 3.7 from Lipschitz to Sobolev functions. The proof will use some results about weak compactness in $L^p$. This section contains a summary of these results. For more information, one may consult Dunford and Schwartz (1958), Wojtaszczyk (1991), or various other sources.

Let $B$ be a real Banach space (complete normed linear space), $B^*$ be its dual space, and $[x, y]$ denote the action of the linear functional $y \in B^*$ on $x \in B$. The *weak topology* on $B$ is the coarsest topology on $B$ – the one with the fewest open sets – with respect to which the maps $x \to [x, y]$ from $B$ to $\mathbb{R}$ are all continuous. The *weak* topology* on $B^*$ is the coarsest topology on $B^*$ with respect to which the maps $y \to [x, y]$ from $B^*$ to $\mathbb{R}$ are all continuous.

A set $A \subset B$ is said to be relatively weakly compact if the closure of $A$ in the weak topology is compact in the weak topology. According to the *Eberlein–Šmulyan Theorem*, $A$ is relatively weakly compact if and only if $A$ is relatively weakly sequentially compact, that is: each sequence in $A$ contains a subsequence that weakly converges in $B$ to some element of $B$. Also, according to *Alaoglu's Theorem*, each norm-bounded subset of $B^*$ is relatively compact in the weak-* topology. The converse statement is easy to prove. Thus: $A \subset B^*$ is relatively compact in the weak-* topology if and only if $A$ is norm-bounded.

If $(X, \mu)$ is a sigma-finite measure space, then for $1 \le p < \infty$, the dual of the Banach space $L^p(X, \mu)$ can be identified with $L^q(X, \mu)$, where $\frac{1}{p} + \frac{1}{q} = 1$, and $q \equiv \infty$ when $p = 1$. The identification is accomplished by means of the pairing $[f, g] = \int_X fg\, d\mu$, where $f \in L^p$ and $g \in L^q$.

If $1 < p < \infty$, then $L^p = (L^q)^*$, and the corresponding weak-* topology on $L^p$ coincides with the weak topology on $L^p$. From the theorems of Alaoglu and Eberlein–Šmulyan, it follows that:

For $1 < p < \infty$, each norm-bounded sequence $\{f_j\} \subset L^p$ contains a subsequence, also denoted $\{f_j\}$, which weakly converges to some $f \in L^p$. That is, there exists $f \in L^p$ such that

$$\lim_{j\to\infty} \int_X f_j g\, d\mu = \int_X fg\, d\mu, \quad \forall g \in L^q. \tag{3.21}$$

For $p = 1$ things are more complicated. Norm-bounded sets are not necessarily relatively weakly compact. Instead, we have the following characterization, due to Dieudonné (1951). See Dunford and Schwartz (1958, p. 371) and Wojtaszczyk (1991, p. 137) for related results. A set $A \subset L^1$ is said to be *uniformly integrable* if for each $\epsilon > 0$ there exists $\delta > 0$ such that for each measurable $E \subset X$ with $\mu(E) < \delta$ we have $\int_E |f|\, d\mu < \epsilon$ for every $f \in A$.

**Theorem 3.18** (Dieudonné) *Let $\mu$ be a Radon measure on a locally compact topological space $X$. Then a set $A \subset L^1(X, \mu)$ is relatively weakly compact if and only if the following three conditions hold:*

*(a) $A$ is norm bounded in $L^1$.*
*(b) $A$ is uniformly integrable.*

*(c) For each $\epsilon > 0$ there exists a compact set $K \subset X$ such that*

$$\int_{X\setminus K} |f|\, d\mu < \epsilon, \quad \forall f \in A.$$

Morrey (1940) was perhaps the first to apply weak compactness arguments in the context of Sobolev spaces. Radon measures are, by definition, nonnegative regular Borel measures on $X$ which are finite on compact sets. See Folland (1999). We shall refer to condition (c) as the *tail condition.*

From the theorems of Dieudonné and Eberlein–Šmulyan we see that:

If the sequence $\{f_j\} \subset L^1$ is uniformly integrable, is norm-bounded in $L^1$, and satisfies (c), then $\{f_j\}$ contains a subsequence which weakly converges in $L^1$ to some $f \in L^1$.

A further, perhaps surprising, consequence of Dieudonné's Theorem is this: $A \subset L^1$ is relatively weakly compact if and only if the set $|A| \equiv \{|f| : f \in A\}$ is relatively weakly compact. Of course, the same is true for $L^p$, $1 < p < \infty$, by Alaoglu's Theorem.

Functions $f \in L^1(X, d\mu)$ may be identified with (signed) measures $f\, d\mu$. The $L^1$ norm of $f$ equals the total variation of $f\, d\mu$. The space $M(X)$ of all Borel measures with finite total variation on $\mathbb{R}^n$ is the dual of the space $C_0(X)$ of all continuous functions on $X$ which have limit zero at $\infty$. So, if $\{f_j\}$ is a norm-bounded sequence in $L^1(X, d\mu)$, then $\{f_j\}$ need not contain a subsequence converging in the weak topology on $L^1(X, d\mu)$, but there is a subsequence of $\{f_j\, d\mu\}$ converging to some measure $\nu \in M(X)$ in the weak-* topology on $M(X)$. Then (3.21) holds for all *continuous* $g$ vanishing at $\infty$, with $d\nu$ on the right instead of $f\, d\mu$. If we assume that the supports of all the $f_j$ are contained in a single compact set, then $\nu$ is absolutely continuous with respect to $\mu$ if and only if $\{f_j\}$ is uniformly integrable.

In the next section we shall need an alternative characterization of uniform integrability. For $f \in L^1(X, \mu)$ with $X$ and $\mu$ as above, define

$$I(t, f) = \int_X (|f| - t)^+\, d\mu, \quad t \geq 0.$$

**Lemma 3.19** *Let $A \subset L^1(X, \mu)$ be norm-bounded in $L^1$. Then the following are equivalent:*

*(a) $A$ is uniformly integrable.*
*(b) $\lim_{t\to\infty} \sup_{f\in A} I(t, f) = 0$.*

*Proof* Since $A$ is uniformly integrable if and only if $|A|$ is, we may assume that all functions in $A$ are nonnegative.

Suppose that (b) is false. Then there exist $\epsilon > 0$, functions $f_j \in A$, and numbers $t_j > 0$ such that $I(t_j,f_j) \geq \epsilon$ for $j \geq 1$ and $\lim_{j\to\infty} t_j = \infty$. Let $E_j = \{f > t_j\}$. Then

$$\epsilon \leq \int_{E_j} (f_j - t_j)\, d\mu < \int_{E_j} f_j\, d\mu. \tag{3.22}$$

On the other hand, $\mu(E_j)t_j \leq \int_{E_j} f\, d\mu \leq \sup_{f\in A} \int_X f\, d\mu$. Since $A$ is norm-bounded, it follows that $\lim_{j\to\infty} \mu(E_j) = 0$. With (3.22), this shows that $A$ can not be uniformly integrable. We have proved that (a) implies (b).

Suppose (a) is false. Then there exist $\epsilon > 0$, sets $E_j$ and functions $f_j \in A$ such that $\lim_{j\to\infty} \mu(E_j) = 0$ and $\int_{E_j} f_j \geq \epsilon$, for $j \geq 1$. Let $t_j = \frac{1}{2}\epsilon(\mu(E_j))^{-1}$. Then $\lim_{j\to\infty} t_j = \infty$, and

$$\epsilon \leq \int_{E_j} [(f - t_j) + t_j]\, d\mu \leq I(t_j,f_j) + t_j\mu(E_j) = I(t_j,f_j) + \frac{\epsilon}{2}.$$

Thus $I(t_j,f_j) \geq \frac{1}{2}\epsilon$ for each $j$, so (b) is false. The proof is complete. □

## 3.6 Symmetrization Decreases the $p$-Dirichlet Integral in $W^{1,p}(\mathbb{R}^n)$

Theorem 3.7 asserts that $\|\nabla f^\#\|_{L^p(\mathbb{R}^n)} \leq \|\nabla f\|_{L^p(\mathbb{R}^n)}$ when $1 \leq p < \infty$, $f$ is Lipschitz on $\mathbb{R}^n$, and its distribution function satisfies

$$\lambda_f(t) < \infty, \quad \forall\, t > \inf f. \tag{3.23}$$

Our aim in this section is to extend this symmetrization inequality from Lipschitz functions to Sobolev functions.

**Theorem 3.20** *Let* $1 \leq p < \infty$ *and* $f \in W^{1,p}(\mathbb{R}^n, \mathbb{R}^+)$. *Then* $f^\# \in W^{1,p}(\mathbb{R}^n)$, *and*

$$\|\nabla f^\#\|_{L^p(\mathbb{R}^n)} \leq \|\nabla f\|_{L^p(\mathbb{R}^n)}. \tag{3.24}$$

The $\mathbb{R}^+$ in $W^{1,p}(\mathbb{R}^n, \mathbb{R}^+)$ means that $f$ in Theorem 3.20 is assumed to be nonnegative. By contrast, Theorem 3.7 was applicable to all real valued Lipschitz functions satisfying (3.23). The reasons for the difference are as follows: To talk about $f^\#$ for the $f$ in Theorem 3.20 we need $f$ to satisfy (3.23), with inf replaced by ess inf. If $f \in W^{1,p}(\mathbb{R}^n)$ then $f \in L^p(\mathbb{R}^n)$. As the reader may easily verify, if $f \in L^p(\mathbb{R}^n)$ for some $0 < p < \infty$, then $f$ satisfies the ess inf version of (3.23) if and only if $f \geq 0$ a.e. Thus, $f \in W^{1,p}(\mathbb{R}^n)$ satisfies the ess inf version of (3.23) if and only if $f \geq 0$ a.e.

As just noted, Theorem 3.20 carries with it the assumption that $f \in L^p(\mathbb{R}^n)$, whereas in Theorem 3.7 there is no such explicit assumption on the size of $f$. One can formulate other generalizations of Theorem 3.7, with conclusion (3.24), which involve less restriction or different restriction on the size of $f$ and different assumptions about the nature of $\nabla f$, but we will not pursue them in this book.

Theorem 3.11, the assertion that symmetrization decreases the $\Phi$-Dirichlet integral, also admits extensions from Lipschitz to Sobolev functions; see Brothers and Ziemer (1988).

The analogue of Theorem 3.20 for $p = \infty$ goes as follows: If $f \in W^{1,\infty}(\mathbb{R}^n)$, with $f$ not necessarily nonnegative, and $f$ satisfies $\lambda_f(t) < \infty \ \forall\ t >$ ess inf $f$, then

$$\|\nabla f^{\#}\|_{L^\infty(\mathbb{R}^n)} \le \|\nabla f\|_{L^\infty(\mathbb{R}^n)}.$$

This inequality follows from Proposition 3.17 and Theorem 3.6.

Here are two corollaries of Theorem 3.20.

**Corollary 3.21** *Let $1 \le p < \infty$, $f \in W^{1,p}(\mathbb{R}^n, \mathbb{R}^+)$, and $0 \le a < b \le \infty$. Then*

$$\int_{(f^{\#})^{-1}(a,b)} |\nabla f^{\#}|^p \, dx \le \int_{f^{-1}(a,b)} |\nabla f|^p \, dx.$$

*Proof* Apply Theorem 3.20 to $g = \min((f - a)^+, b - a)$. The implication $f \in W^{1,p}(\mathbb{R}^n, \mathbb{R}^+) \Longrightarrow g \in W^{1,p}(\mathbb{R}^n)$ and $\nabla g = \chi_{f^{-1}(a,b)} \nabla f$ a.e. can be proved by adapting the proof of Evans and Gariepy (1992, p. 130, Theorem 4(iii)). □

**Corollary 3.22** *Let $1 \le p < \infty$, $\Omega$ be an open subset of $\mathbb{R}^n$ and $f \in W_0^{1,p}(\Omega, \mathbb{R}^+)$. Then $f^{\#} \in W_0^{1,p}(\Omega^{\#}, \mathbb{R}^+)$, and*

$$\|\nabla f^{\#}\|_{L^p(\Omega^{\#})} \le \|\nabla f\|_{L^p(\Omega)}.$$

The proof of Corollary 3.22 will be given after the proof of Theorem 3.20.

*Proof of Theorem 3.20* For $j \ge 1$, let $F_j = K_{\epsilon_j} * f_j$, where $K$ is a smooth bump function as defined in (3.19) and $\{f_j\}$ is the sequence appearing in the proof of Proposition 3.16. Then $F_j \in C_c^\infty(\mathbb{R}^n)$, each $F_j \ge 0$, and the reader may verify that if $\epsilon_j$ goes to zero sufficiently rapidly then $F_j \to f$ in $W^{1,p}(\mathbb{R}^n)$. Each $F_j$ satisfies the hypotheses of Theorems 3.6 and 3.7, so $F_j^{\#} \in \mathrm{Lip}(\mathbb{R}^n, \mathbb{R}^+)$ and

$$\|\nabla F_j^{\#}\|_p \le \|\nabla F_j\|_p, \quad j \ge 1, \tag{3.25}$$

where here and throughout the proof we write $\|\cdot\|_p = \|\cdot\|_{L^p(\mathbb{R}^n)}$. If $g\colon \mathbb{R}^n \to \mathbb{R}^n$ is a vector field with components $g_i$, then $|g|^2 \equiv \sum_{i=1}^n g_i^2$, and we write $\|g\|_p = \|(|g|)\|_p$.

The statement $F_j \to f$ in $W^{1,p}(\mathbb{R}^n)$ means that $\|F_j - f\|_p \to 0$ and $\|\nabla F_j - \nabla f\|_p \to 0$. The second fact implies that

$$\lim_{j\to\infty} \|\nabla F_j\|_p = \|\nabla f\|_p. \tag{3.26}$$

Moreover, Corollary 2.20 implies $\|F_j^\# - f^\#\|_p \le \|F_j - f\|_p$. We conclude that $F_j^\# \to f^\#$ in $L^p(\mathbb{R}^n)$. Observe, in particular, that $f^\# \in L^p(\mathbb{R}^n)$. Also, (3.25) and (3.26) imply that the partial derivatives of $F_j^\#$ form a bounded set in $L^p(\mathbb{R}^n)$.

Suppose that $1 < p < \infty$. Then, as discussed in §3.5, there is a subsequence of $\{F_j^\#\}$, also denoted $\{F_j^\#\}$, and a vector field $g = (g_1, \ldots, g_n) \in L^p(\mathbb{R}^n, \mathbb{R}^n)$ such that for $i = 1, \ldots, n$, $\{\partial_i F_j^\#\}$ converges weakly in $L^p$ to $g_i$ as $j \to \infty$. Take $h \in L^q(\mathbb{R}^n, \mathbb{R}^n)$ with $\|h\|_q = 1$, where $q$ is the Hölder conjugate exponent of $p$. Then

$$\int_{\mathbb{R}^n} g \cdot h\, dx = \lim_{j\to\infty} \int_{\mathbb{R}^n} \nabla F_j^\# \cdot h\, dx \le \liminf_{j\to\infty} \|\nabla F_j^\#\|_p.$$

Taking the sup over all $h$, we deduce

$$\|g\|_p \le \liminf_{j\to\infty} \|\nabla F_j^\#\|_p. \tag{3.27}$$

Take $\phi \in C_c^1(\mathbb{R}^n)$ and $i \in \{1, \ldots, n\}$, then

$$\begin{aligned}\int_{\mathbb{R}^n} g_i \phi\, dx &= \lim_{j\to\infty} \int_{\mathbb{R}^n} \phi\, \partial_i F_j^\#\, dx \\ &= -\lim_{j\to\infty} \int_{\mathbb{R}^n} F_j^\# \partial_i \phi\, dx = -\int_{\mathbb{R}^n} f^\# \partial_i \phi\, dx,\end{aligned}$$

where the last equality follows from the fact noted above that $\|F_j^\# - f^\#\|_p \to 0$. Thus, the weak first order partial derivatives of $f^\#$ coincide with the $g_i$. From (3.26), (3.25), and (3.27), we deduce

$$\|\nabla f^\#\|_p = \|\nabla g\|_p \le \|\nabla f\|_p.$$

Theorem 3.20 is proved for $1 < p < \infty$.

The proof for $p = 1$ is exactly the same as for $1 < p < \infty$ once we know that for each $i \in \{1, \ldots, n\}$, the sequence $\{\partial_i F_j^\#\}$ is relatively weakly compact in $L^1(\mathbb{R}^n)$. From (3.25) and (3.26), we know that these sequences are norm bounded in $L^1$. By Dieudonné's Theorem, to prove the relative weak compactness it will suffice to show that the sequences are uniformly integrable and satisfy the tail condition (c) of Dieudonné's Theorem, Theorem 3.18.

From Lemma 3.19, it follows that the desired uniform integrability is equivalent to the condition

$$\lim_{t\to\infty} \sup_j \int_{\mathbb{R}^n} (|\nabla F_j^\#| - t)^+ \, dx = 0. \tag{3.28}$$

By Theorem 3.11, each integral on the left side of (3.28) increases when the $F_j^\#$ are replaced by $F_j$. Since $\|\nabla F_j - \nabla f\|_1 \to 0$, it follows that the sequence $\{\nabla F_j\}$ is relatively weakly compact in $L^1$, hence uniformly integrable. Invoking Lemma 3.19 again, we conclude that the limit in (3.28) equals zero when the $F_j^\#$ are replaced by $F_j$. Thus (3.28) holds as stated, and the sequences $\{\partial_i F_j^\#\}$ are uniformly integrable.

To prove that the sequences $\{\partial_i F_j^\#\}$ satisfy the tail condition, it is equivalent to show that

$$\lim_{R\to\infty} \sup_j \int_{|x|\geq R} |\nabla F_j^\#| \, dx = 0. \tag{3.29}$$

The $F_j^\#$ and $|\nabla F_j^\#|$ are functions of $|x|$. Moreover, $|\nabla F_j^\#(x)| = -\frac{\partial F_j^\#}{\partial r}$ for almost every $x \in \mathbb{R}^n$, $F_j^\#$ is Lipschitz, and $F_j^\#$ is decreasing on rays. Thus,

$$\int_{|x|\leq R} F_j^\# \, dx \geq F_j^\#(R)\alpha_n R^n.$$

Here, and below, we write $F_j^\#(R)$ instead of $F_j^\#(Re_1)$. Recall that $\alpha_n$ and $\beta_{n-1}$ denote respectively the volume of the unit ball in $\mathbb{R}^n$ and the surface measure of the unit sphere $\mathbb{S}^{n-1} \subset \mathbb{R}^n$. Using $\|F_j^\#\|_1 = \|F_j\|_1$ with the inequality above, we obtain

$$F_j^\#(R) \leq \alpha_n^{-1} R^{-n} \int_{|x|\leq R} F_j^\# \, dx \leq \alpha_n^{-1} R^{-n} \|F_j\|_1.$$

Thus,

$$\begin{aligned}
\int_{|x|\geq R} |\nabla F_j^\#| \, dx &= -\beta_{n-1} \int_R^\infty \frac{\partial F_j^\#}{\partial r}(r) r^{n-1} \, dr \\
&= \beta_{n-1} \left[ R^{n-1} F_j^\#(R) + (n-1) \int_R^\infty F_j^\#(r) r^{n-2} \, dr \right] \\
&\leq \beta_{n-1}\alpha_n^{-1} \|F_j\|_1 \left[ R^{-1} + (n-1) \int_R^\infty r^{-2} \, dr \right] \\
&= n^2 \|F_j\|_1 R^{-1},
\end{aligned} \tag{3.30}$$

where in the last line we used the fact that $\beta_{n-1} = n\alpha_n$.

Since the sequence $\{\|F_j\|_1\}$ is bounded, (3.29) follows from (3.30). Theorem 3.20 is proved completely. □

The proof of Theorem 3.20 would be easier if we knew that $F_j \to f$ in $W^{1,p}(\mathbb{R}^n)$ implies $F_j^\# \to f^\#$ in $W^{1,p}(\mathbb{R}^n)$. For then (3.26) would hold when $F_j$ and $f$ are replaced by their symmetric decreasing rearrangements, and the proof of Theorem 3.20 would be finished by application of (3.26) and (3.25). We shall see though, in the next section, that $F_j \to f$ in $W^{1,p}(\mathbb{R}^n)$ does not imply $F_j^\# \to f^\#$ in $W^{1,p}(\mathbb{R}^n)$.

*Proof of Corollary 3.22* Let $\{f_k\} \subset C_c^\infty(\Omega, \mathbb{R}^+)$ be chosen such that $\lim_{k\to\infty} \|f - f_k\|_{W^{1,p}(\Omega)} = 0$. Extend each $f_k$ to $F_k\colon \mathbb{R}^n \to \mathbb{R}$ by setting $F_k = 0$ on $\mathbb{R}^n \setminus \Omega$. Then $\{F_k\}$ is a Cauchy sequence in $W^{1,p}(\mathbb{R}^n)$. Let $F$ denote its limit function. Then, in $\Omega$, $F$ satisfies $F = f$ and $\nabla F = \nabla f$ a.e. In $\mathbb{R}^n \setminus \Omega$, the $L^p$ convergence implies $F = 0$ a.e., so that $\nabla F = 0$ a.e., by Evans and Gariepy (1992, p. 130, Theorem 4(iv)). Thus, $F \in W^{1,p}(\mathbb{R}^n, \mathbb{R}^+)$, $F = \chi_\Omega f$ a.e., and $\nabla F = \chi_\Omega \nabla f$ a.e. in $\mathbb{R}^n$. By Theorem 3.20, $F^\# \in W^{1,p}(\mathbb{R}^n)$ and $\|\nabla F^\#\|_{L^p(\mathbb{R}^n)} \le \|\nabla F\|_{L^p(\mathbb{R}^n)}$. Since $F \ge 0$ in $\Omega$ and $F = 0$ a.e in $\mathbb{R}^n \setminus \Omega$, it follows that $F^\# = \chi_{\Omega^\#} f^\#$ a.e. We conclude that $f^\# \in W^{1,p}(\Omega^\#)$, that $\nabla F^\# = \chi_{\Omega^\#} \nabla f^\#$ a.e., and that $\|\nabla f^\#\|_{L^p(\Omega^\#)} \le \|\nabla f\|_{L^p(\Omega)}$.

We still must show that $f^\# \in W_0^{1,p}(\Omega^\#)$. To do this, we use functional analysis. As in the proof of Theorem 3.20, we may assume that $\nabla F_j^\#$ converges weakly in $L^p(\mathbb{R}^n)$ to the vector field $\nabla F^\#$. By Mazur's Theorem (Wojtaszczyk, 1991, II.A.5), there is a sequence with terms of the form $\sum_{i=j}^{N(j)} \lambda_{i,j} \nabla F_i^\#$ which converges strongly in $L^p(\mathbb{R}^n)$ to $\nabla F^\#$, where the $\lambda_{i,j}$ are nonnegative with $\sum_{i=1}^{N}(j)\lambda_{i,j} = 1$, and the $N(j)$ are finite. Then $\sum_{i=j}^{N(j)} \lambda_{i,j} F_j^\#$ converges strongly in $W^{1,p}(\mathbb{R}^n)$ to $F^\#$. It is easy to see that each $F_j^\#$ has compact support in $\Omega^\#$. From the definition of $W_0^{1,p}$, we conclude that $f^\# \in W_0^{1,p}(\Omega^\#)$. □

## 3.7 Continuity and Discontinuity of the Symmetric Decreasing Rearrangement Operator

This section contains a brief discussion of some elegant and surprising results of Almgren and Lieb (1989). The results are too beautiful to omit but too technical to expound in any detail in this book. We will supply neither motivation nor proofs, for both of which we refer the reader to the superbly written article of Almgren and Lieb (1989).

In this section we will write $f^\# = \mathcal{R}f$ and will always assume that $1 \le p < \infty$. Corollary 2.20 asserts that the (nonlinear) operator $\mathcal{R}$ is a contraction on $L^p(\mathbb{R}^n, \mathbb{R}^+)$. In particular, $\mathcal{R}$ is a *strongly continuous* mapping of $L^p(\mathbb{R}^n, \mathbb{R}^+)$ into itself: If $f_k$, $k \ge 1$, and $f$ are nonnegative functions in $L^p(\mathbb{R}^n)$ such that $f_k \to f$ in the $L^p(\mathbb{R}^n)$-norm, then $\mathcal{R}f_k \to \mathcal{R}f$ in the $L^p(\mathbb{R}^n)$-norm. Moreover, if $f \in W^{1,p}(\mathbb{R}^n, \mathbb{R}^+)$, then $\|\nabla \mathcal{R} f\|_p \leq \|\nabla f\|_p$, by Theorem 3.20. Thus, as a mapping of $W^{1,p}(\mathbb{R}^n, \mathbb{R}^+)$ equipped with the $W^{1,p}(\mathbb{R}^n)$-norm into itself, $\mathcal{R}$ is strongly continuous at the zero function.

In view of this evidence, it seems plausible that $\mathcal{R}: W^{1,p}(\mathbb{R}^n, \mathbb{R}^+) \to W^{1,p}(\mathbb{R}^n, \mathbb{R}^+)$ would be strongly continuous everywhere on $W^{1,p}(\mathbb{R}^n, \mathbb{R}^+)$. For $n = 1$ this surmise was proved by Coron in 1984. But for $n \geq 2$ it was disproved by Almgren and Lieb. In fact, Almgren and Lieb discovered a condition, dubbed by them *coarea regularity*, which is necessary and sufficient for $\mathcal{R}$ to be strongly continuous at a given function. We will state their theorem, which is on p. 694 of Almgren and Lieb (1989), and then explain their definition.

**Theorem** (Almgren–Lieb) *Let $n \geq 2$ and $f \in W^{1,p}(\mathbb{R}^n, \mathbb{R}^+)$. Then $\mathcal{R}: W^{1,p}(\mathbb{R}^n, \mathbb{R}^+) \to W^{1,p}(\mathbb{R}^n, \mathbb{R}^+)$ is strongly continuous at $f$ if and only if $f$ is coarea regular.*

Take $f \in W^{1,p}(\mathbb{R}^n, \mathbb{R}^+)$. Then $\lambda_f(t) < \infty$ for each $t > 0$. Decompose $\lambda_f$ as

$$\lambda_f(t) = \mathcal{L}^n(f > t,\ \nabla f \neq 0) + \mathcal{L}^n(f > t,\ \nabla f = 0) \equiv A_f(t) + B_f(t).$$

Almgren and Lieb call $A_f$ the *coarea distribution function* and $B_f$ the *residual distribution function* of $f$. Each of $A_f$ and $B_f$ are decreasing functions on $(0, \infty)$. The measure on $(0, \infty)$ induced by $A_f$ turns out to be locally absolutely continuous with respect to Lebesgue measure. But the locally absolutely continuous and singular components of the measure $B_f$ can both be nontrivial.

**Definition** $f \in W^{1,p}(\mathbb{R}^n, \mathbb{R}^+)$ is called coarea regular if the Borel measure induced on $(0, \infty)$ by $B_f$ is purely singular with respect to $\mathcal{L}$. Otherwise, $f$ is called coarea irregular.

In Almgren and Lieb (1989, chapter 5) it is proved that when $n = 1$ every $f \in W^{1,p}(\mathbb{R}^n, \mathbb{R}^+)$ is coarea regular, but when $n \geq 2$ the subsets of $W^{1,p}(\mathbb{R}^n, \mathbb{R}^+)$ consisting of regular and of irregular functions are each norm-dense in $W^{1,p}(\mathbb{R}^n, \mathbb{R}^+)$.

By contrast, Burchard (1997) proved that the corresponding operator $\mathcal{R}$ is strongly continuous on $W^{1,p}(\mathbb{R}^n, \mathbb{R}^+)$ with respect to Steiner symmetrization.

## 3.8 Notes and Comments

Special cases of the decrease of Dirichlet integrals under symmetrization, under various regularity hypotheses, were established by Faber, Krahn, Pólya, and Szegö in the period 1923–1948 en route to the solution of famous optimization problems for fundamental frequencies, electrostatic capacity, and torsional rigidity. We'll discuss these problems in Chapter 5. The book Pólya and Szegő (1951) presents a number of such results. From a different direction, Gehring (1961), in his study of quasiconformal mappings, was led to prove that the 3-Dirichlet integral in $\mathbb{R}^3$ decreases under symmetric decreasing rearrangement and spherical symmetrization. Mostow (1968) extended Gehring's results to $n$-Dirichlet integrals in $\mathbb{R}^n$. The first comprehensive result is apparently due to Sperner (1973), who proved for smooth nonnegative functions the result stated here as Theorem 3.7, that symmetric decreasing rearrangement decreases the $p$-Dirichlet integral on $\mathbb{R}^n$ for every $p \geq 1$ and $n \geq 1$. The extension to Sobolev functions, Theorem 3.20, is due to Hildén (1976), while the extension to $\Phi$-Dirichlet integrals, Theorem 3.11, is due to Bandle (1980) for real analytic $f$ and to Brothers and Ziemer (1988) for more general functions. The proofs in these papers are derived from the isoperimetric inequality, and require careful analysis of level sets. The latter aspect became more incontestably rigorous after Federer's precise coarea formula (1959, 1969) became a working tool. In §4.5 we'll see how some of these arguments go.

The proofs presented here of Dirichlet integral inequalities were sketched in Baernstein (1994). These proofs do not use the isoperimetric inequality, but instead view the Dirichlet integral as a limiting case of integrals on $\mathbb{R}^n \times \mathbb{R}^n$ to which the inequalities of Chapter 2 apply. Thus, the Dirichlet integral inequalities appear as fairly close descendants of the two-point symmetrization result Theorem 2.8. Beckner (1992) takes a similar point of view.

The results obtained for $\mathbb{R}^n$ in this chapter go over with little change to spheres and hyperbolic spaces. See Chapter 7.

The idea of Proposition 3.12, which provides another way to obtain Dirichlet integral inequalities via polarization, is due to Dubinin (1987). He further develops this idea in several papers, such as Dubinin (1993). In this connection, see also Brock and Solynin (2000).

There also are weighted Dirichlet integral inequalities, in which one compares integrals taken with respect to more general measures than Lebesgue measure. See, for example, Epperson (1990), Schulz and Vera de Serio (1993), Talenti (1997), and Bobkov and Houdré (1997).

# 4

# Geometric Isoperimetric and Sharp Sobolev Inequalities

In its most basic form the isoperimetric inequality asserts that among all simple closed curves in $\mathbb{R}^2$ with the same length, the circle encloses maximum area. Alternatively, among all simple closed curves enclosing the fixed area, the circle has the smallest length. Still another statement: Let $L$ be the length of the curve and $A$ be the area it encloses. Then

$$L^2 \geq 4\pi A.$$

Equality holds when the curve is a circle.

A generalization of the isoperimetric inequality to $\mathbb{R}^n$ may be stated as follows: For all reasonable sets $E \subset \mathbb{R}^n$ we have

$$\text{area}(\partial E) \geq \text{area}(\partial E^{\#}) \tag{4.1}$$

where $E^{\#}$ is the open ball centered at the origin with the same $\mathcal{L}^n$- measure as $E$, area means $(n-1)$-dimensional surface area, and we write $\partial E^{\#}$ instead of the more precise $\partial(E^{\#})$.

What is the meaning of "surface area" of $\partial E$ for general sets $E$? There are many possibilities; see, for example, Federer (1969, p. 261). Here we shall consider three versions of $(n-1)$-dimensional area: The $(n-1)$-dimensional Hausdorff measure $\mathcal{H}^{n-1}(\partial E)$, the Minkowski content $\mathcal{M}^{n-1}(\partial E)$, and the perimeter $P(E)$. The "reasonable sets" $E$ will be $\mathcal{L}^n$-measurable sets of finite measure in the Hausdorff and Minkowski cases, and sets of "finite perimeter" in the perimeter case.

The main isoperimetric inequalities for $\mathbb{R}^n$ are presented in Theorem 4.10, Corollary 4.13, and Theorem 4.16. Theorem 4.10 asserts that for all $E$ with finite perimeter we have $P(E) \geq P(E^{\#})$. This result is a special case of the extension of the Dirichlet integral inequality $\|\nabla f^{\#}\|_1 \leq \|\nabla f\|_1$ to functions of "bounded variation" on $\mathbb{R}^n$. With the aid of a little geometric measure theory, we then show that Theorem 4.10 implies that (4.1) holds when "area" is taken to mean $\mathcal{H}^{n-1}$ and $E$ is an $\mathcal{L}^n$-measurable set with $\mathcal{L}^n(E) < \infty$. This is Corollary 4.13. In Theorem 4.16, we prove that for the same class of sets $E$, (4.1) holds when area is taken to mean $(n-1)$-dimensional lower Minkowski content. The Minkowski result is a simple consequence of Theorem 2.12, which asserts that symmetrization decreases the modulus of continuity.

The Sobolev Embedding Theorem asserts that for $1 \le p < n$, functions $f \in W^{1,p}(\mathbb{R}^n)$ satisfy inequalities $\|f\|_{p^*} \le C(n,p)\|\nabla f\|_p$, where $p^* = np/(n-p)$. In Theorems 4.21 and 4.23 we prove this inequality with smallest possible $C(n,p)$. Theorem 4.21 treats the case $p = 1$ and Theorem 4.23 the case $1 < p < n$. In Section 4.8 we briefly mention some other best possible inequalities in Sobolev spaces.

Tools for Chapter 4 include functions of bounded variation, the area and coarea formulas, and the Gauss–Green Theorem for transformation of integrals. These topics are briefly discussed in Sections 4.1, 4.2, and 4.5.

## 4.1 Hausdorff Measures, Area Formula, and the Gauss–Green Theorem

Hausdorff measures are the most commonly used set functions to measure the size of lower dimensional subsets of $\mathbb{R}^n$. They are defined as follows. Let $E$ be any set in $\mathbb{R}^n$. For $0 < \delta \le \infty$ and $0 \le s < \infty$ define

$$\mathcal{H}^s_\delta(E) = \inf \sum_{j=1}^{\infty} \alpha_s \frac{\operatorname{diam} C_j}{2},$$

where diam denotes diameter,

$$\alpha_s = \frac{\pi^{s/2}}{\Gamma(\frac{s}{2}+1)},$$

and the infimum is taken over all sequences of sets $\{C_j\}$ such that $\operatorname{diam} C_j \le \delta$ for each $j$ and $\bigcup_{j=1}^{\infty} C_j \supset E$. Here $\Gamma$ denotes the usual gamma function. Note that $\alpha_n$ agrees with our earlier use of $\alpha_n$ to denote $\mathcal{L}^n(\mathbb{B}^n(1))$.

Define now

$$\mathcal{H}^s(E) = \lim_{\delta \to 0} \mathcal{H}^s_\delta(E).$$

The limit exists since $\mathcal{H}^s_\delta(E)$ is an increasing function of $\delta$. The quantity $\mathcal{H}^s(E) \in [0,\infty]$ is called the $s$-dimensional *Hausdorff measure* of $E$. The set function $\mathcal{H}^s$ is countably sub-additive. In the terminology of most texts $\mathcal{H}^s$ is an outer measure on the set of all subsets of $\mathbb{R}^n$. Its restriction to the class of measurable sets is a countably additive measure. Here a set $E \subset \mathbb{R}^n$ is defined to be measurable if

$$\mathcal{H}^s(B) = \mathcal{H}^s(B \cap E) + \mathcal{H}^s(B \setminus E)$$

for every set $B \subset \mathbb{R}^n$.

The terminology in Evans and Gariepy (1992) is a bit different: for those authors, "measure" means "outer measure." We shall follow their practice. Then for arbitrary sets $E \subset \mathbb{R}^n$, $\mathcal{L}^n(E)$, denotes the Lebesgue outer measure of $E$.

**Proposition 4.1** *For every set $E \subset \mathbb{R}^n$,*

$$\mathcal{H}^n(E) = \mathcal{L}^n(E).$$

For a proof, see Evans and Gariepy (1992, p. 70). The proof there uses the isodiametric inequality, Corollary 2.14.

It is easy to see that $\mathcal{H}^s \equiv 0$ when $s > n$ and that $\mathcal{H}^0(\{x\}) = 1$ for each $x \in \mathbb{R}^n$ when $s = 0$. From the latter, it follows that

$$\mathcal{H}^0(E) = \text{number of points in } E \tag{4.2}$$

for every $E \subset \mathbb{R}^n$.

Next, we want to show that if $m$ is an integer with $1 \leq m \leq n-1$ and $E \subset \mathbb{R}^n$ is a nice $m$-dimensional set, then $\mathcal{H}^m(E)$ equals the $m$-dimensional surface area of $A$ computed in the "usual way." Suppose, for example, that $E \subset \mathbb{R}^3$ is the graph of some function $h \in C^1(\mathbb{R}^2, \mathbb{R})$ over an open set $A \subset \mathbb{R}^2$. Then, writing $x = (x_1, x_2)$,

$$E = \{(x_1, x_2, h(x)) : x \in A\}.$$

In calculus books, the surface area of $E$ is usually defined to be

$$\text{Area}(E) = \int_A (1 + |\nabla h|^2)^{1/2}\, dx.$$

To confirm that

$$\int_A (1 + |\nabla h|^2)^{1/2}\, dx = \mathcal{H}^2(E), \tag{4.3}$$

one may apply the *area formula*. To explain this formula, let us first consider a linear map $L\colon \mathbb{R}^m \to \mathbb{R}^n$, where $1 \leq m \leq n - 1$. Then $L$ has a polar decomposition

$$L = O \circ S,$$

(see Evans and Gariepy, 1992, p. 87) where $S\colon \mathbb{R}^m \to \mathbb{R}^m$ is a symmetric linear map and $O\colon \mathbb{R}^m \to \mathbb{R}^n$ is an orthogonal linear map. Orthogonal means that

$Ox \cdot Oy = x \cdot y$ for all $x, y \in \mathbb{R}^m$. For $A \subset \mathbb{R}^m$, we have $\mathcal{L}^m(SA) = |\det S|\mathcal{L}^m(A)$, and it is also true that $\mathcal{H}^m(OA) = \mathcal{L}^m(A)$. Proofs are in Evans and Gariepy (1992, p. 92). Consequently,

$$\mathcal{H}^m(LA) = |\det S|\mathcal{L}^m(A), \quad A \subset \mathbb{R}^m.$$

Now $O^* \circ O = I$, the identity map on $\mathbb{R}^m$, from which we deduce $L^*L = S^2$. The map $L^*L$ is positive semidefinite, and we have

$$|\det S| = (\det(L^*L))^{1/2}.$$

Define the Jacobian $J$ of $L$ to be the nonnegative constant

$$J = J_L = (\det(L^*L))^{1/2}.$$

We have shown that

$$\mathcal{H}^m(LA) = J_L \mathcal{L}^m(A), \quad A \subset \mathbb{R}^m. \tag{4.4}$$

Suppose now that $f\colon \mathbb{R}^m \to \mathbb{R}^n$ is any Lipschitz function. Then $f$ is differentiable a.e. For such $x$, denote by $Df(x)$ the $n \times m$ matrix whose $(i,j)$ entry is $\partial f_i/\partial x_j$, and define the Jacobian of $f$ to be

$$J(x) = J_f(x) = (\det((Df(x))^*(Df(x)))^{1/2}.$$

**Theorem 4.2** (Area Formula) *Let $f \in \mathrm{Lip}(\mathbb{R}^m, \mathbb{R}^n)$, $1 \le m \le n$ and $A$ be a $\mathcal{L}^m$-measurable set in $\mathbb{R}^m$. Then*

$$\int_A J_f \, d\mathcal{L}^m = \int_{\mathbb{R}^n} \mathcal{H}^0(A \cap f^{-1}(y)) \, d\mathcal{H}^m(y).$$

As noted in (4.2), $\mathcal{H}^0$ is the counting measure on $\mathbb{R}^n$. For $m = n$ the area formula reduces to the usual change of variable formula for mappings $f\colon \mathbb{R}^n \to \mathbb{R}^n$.

To prove the area formula, one splits the integral on the left into a large number of small pieces on which $f$ is approximately linear, then uses (4.4). For details, see Evans and Gariepy (1992, p. 96).

If $f$ is 1-1 on $A$, the area formula simplifies to

$$\int_A J_f \, dx = \mathcal{H}^m(f(A)). \tag{4.5}$$

For the rest of this section we shall confine attention to the case $m = n - 1$. If $L\colon \mathbb{R}^{n-1} \to \mathbb{R}^n$ is linear, then a special case of the Cauchy–Binet Formula (Evans and Gariepy, 1992, p. 89) tells us that

$$\det(L^*L) = \sum_{i=1}^{n} \det((P_iL)^*(P_iL)),$$

where $P_i : \mathbb{R}^n \to \mathbb{R}^{n-1}$ is the projection map defined by

$$P_i(x_1, \ldots, x_n) = (x_1, \ldots, x_{i-1}, x_{i+1}, \ldots, , x_n).$$

Set $L_i = P_i L$. Then

$$\det(L^* L) = \sum_{i=1}^{n} \det(L_i^* L_i),$$

and $L_i$ is the $(n-1) \times (n-1)$ matrix obtained by deleting the $i$th row of the matrix for $L$.

For $f = (f^1, \ldots, f^n) \in \mathrm{Lip}(\mathbb{R}^{n-1}, \mathbb{R}^n)$, it follows that

$$(J_f)^2 = \sum_{i=1}^{n} J_{g_i}^2, \tag{4.6}$$

where $g_i$ is the map of $\mathbb{R}^{n-1}$ into itself given by

$$g_i = (f^1, \ldots, f^{i-1}, f^{i+1}, \ldots, f^n).$$

Assuming that $f$ is 1–1 on the $\mathcal{L}^{n-1}$-measurable set $A \subset \mathbb{R}^{n-1}$, (4.5) and (4.6) give us the formula

$$\mathcal{H}^{n-1}(f(A)) = \int_A \sum_{i=1}^{n} J_{g_i}^2 \, dx.$$

The right side is the usual definition of $(n-1)$-dimensional surface area for the parametric hypersurface $f(A)$.

Suppose that a set $E \subset \mathbb{R}^n$ is represented by a graph:

$$E = \{(x, h(x)) : x \in A\}$$

where $A \subset \mathbb{R}^{n-1}$ and $h \in \mathrm{Lip}(\mathbb{R}^{n-1}, \mathbb{R})$. Then $E = f(A)$ where $f(x) = (x, h(x))$, and (4.6) becomes

$$J_f^2 = 1 + \sum_{i=1}^{n-1} \left( \frac{\partial h}{\partial x_i} \right)^2 .$$

Hence, we have

$$\mathcal{H}^{n-1}(E) = \int_A (1 + |\nabla h|^2)^{1/2} \, dx. \tag{4.7}$$

In particular, for $n = 3$ this confirms (4.3).

To conclude this section we put on record a statement of the Gauss–Green Theorem in a form we will use in the sequel. The Gauss–Green Theorem is also known as the *Divergence Theorem*. Our statement is taken from Evans (1998, p. 626).

**Theorem 4.3** (Gauss–Green Theorem) *Suppose that $\Omega$ is a bounded domain in $\mathbb{R}^n$ with $C^1$ boundary and that $u \in C^1(\overline{\Omega}, \mathbb{R}^n)$. Then*

$$\int_{\partial\Omega} u \cdot \nu \, d\mathcal{H}^{n-1} = \int_{\Omega} \operatorname{div} u \, dx. \tag{4.8}$$

Here $\operatorname{div} u = \sum_{i=1}^{n} u_{x_i}$ and $\nu(x)$ is the outward pointing unit normal vector to $\partial\Omega$ at the point $x \in \partial\Omega$. One says that $\Omega$ has a $C^1$ boundary if for each $x^0 \in \partial\Omega$ there exist a number $r > 0$ and a function $h \in C^1(\mathbb{R}^{n-1}, \mathbb{R})$ such that – upon relabeling and reorienting the coordinate axes if necessary – we have

$$\Omega \cap B(x^0, r) = \{x \in B(x^0, r) \colon x_n > h(x_1, \ldots, x_{n-1})\}. \tag{4.9}$$

This form of the Gauss–Green Theorem may be proved as follows. Take a finite open cover of $\overline{\Omega}$ consisting of balls $B(x^i, r_i)$ as in (4.9) with $x^i \in \partial\Omega$, and an open set $\Omega_1$ with $\overline{\Omega_1} \subset \Omega$. Take a partition of unity $\{\zeta_i\}$ subordinate to the cover. Replacing $u$ by $u\zeta_i$, the problem is reduced to proving (4.8) when $u$ is compactly supported in $\Omega$ or is compactly supported in one of the $B(x^i, r_i)$. In the first case, by integrating first in the $x_i$ variable one obtains $\int_\Omega u_{x_i} \, dx = 0$ for each $i$. Hence

$$\int_{\Omega} \operatorname{div} u \, dx = 0, \tag{4.10}$$

which gives (4.8).

In the second case, one makes a change of variables in which $B(x^i, r_i) \cap \partial\Omega$ is transformed to a set in $\mathbb{R}^{n-1}$, then does clever manipulations to establish (4.8). For $n = 2$ and $n = 3$ see, for example Folland (2002, Appendix B.7). Straightforward modifications give (4.8) for general $n$. From (4.7), it follows that the surface measure used by Folland coincides with $\mathcal{H}^{n-1}$.

By means of geometric measure theory the Gauss–Green Theorem can be extended to much more general sets and functions. See, for example, Evans and Gariepy (1992, p. 209) or Federer (1969, p. 478).

We note here that for $u \in C_c^1(\Omega, \mathbb{R}^n)$, formula (4.10) is true for every open set $\Omega \subset \mathbb{R}^n$, bounded or not, without any assumptions on $\partial\Omega$. In particular, we can take $\Omega = \mathbb{R}^n$. If $u$ belongs to the Sobolev space $W^{1,1}(\mathbb{R}^n, \mathbb{R}^n)$ the results stated in Section 3.5 imply that $u$ is the limit in $W^{1,1}(\mathbb{R}^n, \mathbb{R}^n)$ norm of a sequence of compactly supported $C^\infty$ vector fields. It follows from (4.10) that

$$\int_{\mathbb{R}^n} \operatorname{div} u \, dx = 0, \quad \forall\, u \in W^{1,1}(\mathbb{R}^n, \mathbb{R}^n). \tag{4.11}$$

## 4.2 Functions of Bounded Variation in $\mathbb{R}^n$

For $f \in L^1_{loc}(\mathbb{R}^n)$, define

$$V(f) = \sup_{\phi} \int_{\mathbb{R}^n} f \operatorname{div} \phi \, dx,$$

where the sup is taken over all $\phi \in C^1_c(\mathbb{R}^n, \mathbb{R}^n)$ with $\sup_{\mathbb{R}^n} |\phi| \leq 1$. We call $V$ the *total variation* of $f$ over $\mathbb{R}^n$. It is an interesting exercise to prove that when $n = 1$, $V(f)$ is finite if and only if $f$ coincides almost everywhere with a function of bounded variation on $\mathbb{R}$ in the usual sense, and then $V(f)$ equals the classical total variation of this function. See Evans and Gariepy (1992, p. 220) for an analogous interpretation when $n \geq 2$.

We define $f \colon \mathbb{R}^n \to \mathbb{R}$ to be a function of *bounded variation* if $f \in L^1(\mathbb{R}^n)$ and $V(f) < \infty$. The set of functions of bounded variation in $\mathbb{R}^n$ is denoted by $BV(\mathbb{R}^n)$. It is a Banach space with the norm

$$\|f\|_{BV} = \|f\|_{L^1(\mathbb{R}^n)} + V(f).$$

**Proposition 4.4** *If $f \in W^{1,1}(\mathbb{R}^n)$, then*

$$V(f) = \int_{\mathbb{R}^n} |\nabla f| \, dx. \tag{4.12}$$

*Proof* Take $\phi \in C^1_c(\mathbb{R}^n, \mathbb{R}^n)$ with $\sup_{\mathbb{R}^n} |\phi| \leq 1$. Then $f\phi \in W^{1,1}(\mathbb{R}^n, \mathbb{R}^n)$. Application of the Gauss–Green identity (4.11) to $u = f\phi$ yields

$$\int_{\mathbb{R}^n} f \operatorname{div} \phi \, dx = -\int_{\mathbb{R}^n} \nabla f \cdot \phi \, dx.$$

Taking the sup over all $\phi$, (4.12) follows. □

As a consequence of Proposition 4.4, we have $W^{1,1}(\mathbb{R}^n) \subset BV(\mathbb{R}^n)$, with $\|f\|_{W^{1,1}} = \|f\|_{BV}$.

**Proposition 4.5** *Let $f \in L^1_{loc}(\mathbb{R}^n)$. Then $f \in BV(\mathbb{R}^n)$ if and only if there exists a Radon measure $\mu$ on $\mathbb{R}^n$ and a $\mu$- measurable function $\sigma : \mathbb{R}^n \to \mathbb{R}^n$ such that $|\sigma(x)| = 1$ for $\mu$-almost every $x \in \mathbb{R}^n$ and*

$$\int_{\mathbb{R}^n} f \operatorname{div} \phi \, dx = -\int_{\mathbb{R}^n} \phi \cdot \sigma \, d\mu, \quad \phi \in C^1_c(\mathbb{R}^n, \mathbb{R}^n). \tag{4.13}$$

*Proof* Take $f \in BV(\mathbb{R}^n)$. Define a linear functional $L \colon C^1_c(\mathbb{R}^n, \mathbb{R}^n) \to \mathbb{R}$ by $L(\phi) = -\int_{\mathbb{R}^n} f \operatorname{div} \phi \, dx$. Then, we have

$$|L(\phi)| \leq V(f) \|\phi\|_{L^\infty(\mathbb{R}^n)}. \tag{4.14}$$

Take $\psi \in C_c(\mathbb{R}^n, \mathbb{R}^n)$ and a bounded open set $\Omega \subset \mathbb{R}^n$ which contains the support of $\psi$. There exists a sequence $\{\phi_k\} \subset C_c^1(\mathbb{R}^n, \mathbb{R}^n)$ such that the support of each $\phi_k$ is contained in $\Omega$ and $\{\phi_k\}$ converges uniformly to $\psi$ in $\mathbb{R}^n$. Define

$$L(\psi) = \lim_{k\to\infty} L(\phi_k).$$

By (4.14), the limit exists, is independent of the approximating sequence, and satisfies (4.14) with $\psi$ in place of $\phi$. Thus, we have extended $L$ to a linear functional defined on all of $C_c(\mathbb{R}^n, \mathbb{R}^n)$ that satisfies (4.14). The Riesz Representation Theorem for existence of vector valued measures (Evans and Gariepy, 1992, p. 49) produces a Radon measure $\mu$ and a unimodular function $\sigma : \mathbb{R}^n \to \mathbb{R}^n$ such that

$$L(\psi) = \int_{\mathbb{R}^n} \psi \cdot \sigma \, d\mu, \qquad \psi \in C_c(\mathbb{R}^n, \mathbb{R}^n).$$

Thus, (4.13) holds. Moreover, from the construction of $\mu$, or from taking suprema in (4.13), it follows that

$$\mu(\mathbb{R}^n) = V(f), \tag{4.15}$$

so that the measure $\mu$ is finite.

This proves the “only if” part of Proposition 4.5. The “if” part is obtained by taking suprema in (4.13). □

The arguments giving (4.15) show also that for every open set $\Omega \subset \mathbb{R}^n$ we have

$$\mu(\Omega) = \sup_{\phi} \int_{\mathbb{R}^n} f \operatorname{div} \phi \, dx,$$

where the sup is taken over all $\phi \in C_c^1(\mathbb{R}^n, \mathbb{R}^n)$ with $\sup_{\mathbb{R}^n} |\phi| \le 1$ and supp $\phi$ contained in $\Omega$. Thus, $\mu(\Omega)$ can be interpreted as the total variation of $f$ over $\Omega$.

The measure $\mu$ is unique, while the function $\sigma$ is uniquely determined up to a set of $\mu$-measure zero.

In Evans and Gariepy (1992), the measure we are calling $\mu$ is denoted by $\|Df\|$.

Proposition 4.5 tell us that a function $f \in L^1(\mathbb{R}^n)$ is in $BV(\mathbb{R}^n)$ if and only if its distributional partial derivatives are finite real valued (signed) measures on $\mathbb{R}^n$. The measure corresponding to $\partial f/\partial x_i$ is $\sigma \cdot e_i \, d\mu$. The $\mathbb{R}^n$-valued measure $\sigma \, d\mu$ is the distributional gradient of $f$.

If $f \in W^{1,1}(\mathbb{R}^n)$ then the Gauss–Green Theorem as applied in the proof of Proposition 4.4 implies the equation

$$\sigma \, d\mu = \nabla f \, dx. \tag{4.16}$$

It follows that $\mu$ and $\mathcal{L}^n$ are mutually absolutely continuous, with $d\mu = |\nabla f|\, dx$, and $\sigma = \frac{\nabla f}{|\nabla f|}$ at points where $|\nabla f| > 0$. The set where $\nabla f = 0$ has $\mu$-measure zero, so we can define, for example, $\sigma = 0$ at points where $\nabla f = 0$.

If $f \in BV(\mathbb{R}^n)$ has absolutely continuous $\sigma\, d\mu$ then from (4.13) it follows that its distributional gradient is a function in $L^1(\mathbb{R}^n)$, and hence $f \in W^{1,1}(\mathbb{R}^n)$. Combined with the paragraph above, we see that a function in $BV(\mathbb{R}^n)$ belongs to $W^{1,1}(\mathbb{R}^n)$ if and only if its associated measure $\sigma\, d\mu$ is absolutely continuous with respect to $\mathcal{L}^n$.

If $f \in BV(\mathbb{R}^n) \cap C^1(\mathbb{R}^n)$ then the Gauss–Green theorem implies that (4.16) holds for $f$. Thus $\sigma\, d\mu$ is absolutely continuous, and hence $f \in W^{1,1}(\mathbb{R}^n)$. This gives the inclusion

$$BV(\mathbb{R}^n) \cap C^1(\mathbb{R}^n) \subset W^{1,1}(\mathbb{R}^n). \tag{4.17}$$

The inclusion is still true if we replace $C^1$ by Lip. To see this, just confirm that $\int_{\mathbb{R}^n} div\,(f\phi)\, dx = 0$ when $f \in \mathrm{Lip}(\mathbb{R}^n)$, $\phi \in C_c^1(\mathbb{R}^n, \mathbb{R}^n)$, then reason as in $C^1$ case.

The next result asserts that the total variation enjoys a lower semicontinuity property. We do not assume in this proposition that any of the total variations are finite.

**Proposition 4.6** *Suppose that $\{f_k\} \subset L^1_{loc}(\mathbb{R}^n)$ and that $f_k \to f$ in $L^1_{loc}(\mathbb{R}^n)$. Then*

$$V(f) \leq \liminf_{k\to\infty} V(f_k).$$

*Proof* This easily follows from the definition of $V(f)$. □

**Proposition 4.7** *Suppose that $f \in BV(\mathbb{R}^n)$. There exists a sequence $\{f_k\} \subset BV(\mathbb{R}^n) \cap C^\infty(\mathbb{R}^n)$ such that $f_k \to f$ in $L^1(\mathbb{R}^n)$ and $V(f_k) \to V(f)$ as $k \to \infty$.*

*Proof* This proposition is a special case of Theorem 2 in Evans and Gariepy (1992, p. 172). We will construct an approximating sequence in our setting, and refer the reader to Evans and Gariepy for proof that this sequence possesses the requisite properties.

Let $B(r) = \mathbb{B}^n(0, r)$ denote the open ball with radius $r$ centered at the origin in $\mathbb{R}^n$. Let $\mu$ be the measure associated with $f$ by Proposition 4.5, and let $\epsilon > 0$ be given. Choose $R$ so large that $\mu(\mathbb{R}^n \setminus B(R+1)) < \epsilon$, and define open sets $W_k$ by

$$W_1 = B(R+2), \quad W_k = B(R+k+1) \setminus \overline{B}(R+k-1), \quad k \geq 2.$$

Let $\{\zeta_k\}$ be a sequence of functions with $\zeta_k \in C_c^\infty(\mathbb{R}^n)$, $0 \leq \zeta_k \leq 1$, supp $\zeta_k \subset W_k$ for $k \geq 1$, and $\sum_{k=1}^\infty \zeta_k \equiv 1$ on $\mathbb{R}^n$. Let $K\colon \mathbb{R}^n \to \mathbb{R}^+$ be a nonnegative smooth bump function with support in $B(1)$, as in §3.4. Recall that $K_\delta(x) = \delta^{-n}K(\delta^{-1}x)$ and that $*$ denotes convolution. For each $k \geq 1$, choose $\epsilon_k$ so small that the support of $K_{\epsilon_k} * (f\zeta_k)$ is contained in $W_k$ and that

$$\|K_{\epsilon_k} * (f\zeta_k) - f\zeta_k\|_{L^1(\mathbb{R}^n)} < \epsilon_k \quad \text{and} \quad \|K_{\epsilon_k} * (f\nabla\zeta_k) - f\nabla\zeta_k\|_{L^1(\mathbb{R}^n)} < \epsilon_k.$$

Define

$$f_\epsilon = \sum_{k=1}^{\infty} K_{\epsilon_k} * (f\zeta_k).$$

Then $f_\epsilon \in C^\infty(\mathbb{R}^n)$, and, as $\epsilon \to 0$, $f_\epsilon \to f$ and $V(f_\epsilon) \to V(f)$. To obtain an approximating *sequence*, as in the statement of the proposition, just take a sequence of $\epsilon$s tending to zero and consider the corresponding $f_\epsilon$. □

With our new tools in hand we return to the subject of symmetrization. According to Theorem 3.20, for nonnegative $f \in W^{1,1}(\mathbb{R}^n)$ we have $\|\nabla f^\#\|_{L^1(\mathbb{R}^n)} \leq \|\nabla f\|_{L^1(\mathbb{R}^n)}$. By Proposition 4.4, this may be restated as $V(f^\#) \leq V(f)$ for $f \in W^{1,1}(\mathbb{R}^n)$. The theorem below tells us that symmetrization in fact decreases the total variation of *all* functions of bounded variation.

**Theorem 4.8** *If $f \in BV(\mathbb{R}^n, \mathbb{R}^+)$ then $f^\# \in BV(\mathbb{R}^n)$ and*

$$V(f^\#) \leq V(f).$$

*Proof* Let $\{f_k\}$ be a sequence satisfying the conditions of Proposition 4.7. Then each $f_k \in W^{1,1}(\mathbb{R}^n)$, by (4.17). By construction, each $f_k$ is nonnegative. By Theorem 3.20, $V(f_k^\#) \leq V(f_k)$. Also, $f_k \to f$ in $L^1(\mathbb{R}^n)$, so Corollary 2.20 implies that $f_k^\# \to f^\#$ in $L^1(\mathbb{R}^n)$. From Proposition 4.6 it follows that

$$V(f^\#) \leq \liminf_{k\to\infty} V(f_k^\#) \leq \liminf_{k\to\infty} V(f_k) = V(f). \qquad \square$$

Since $\|f^\#\|_{L^1(\mathbb{R}^n)} = \|f\|_{L^1(\mathbb{R}^n)}$, Theorem 4.8 may also be stated as

$$\|f^\#\|_{BV(\mathbb{R}^n)} \leq \|f\|_{BV(\mathbb{R}^n)},$$

when $f$ is nonnegative.

## 4.3 Isoperimetric Inequalities for Perimeter and Hausdorff Measure

Throughout this section

$$E \subset \mathbb{R}^n$$

will denote a $\mathcal{L}^n$-measurable set. We shall use the notations for $BV(\mathbb{R}^n)$ functions established in §4.2. In particular, $V(f)$ equals the total variation of $f$ over $\mathbb{R}^n$, and $BV(\mathbb{R}^n)$ functions are assumed to be in $L^1(\mathbb{R}^n)$.

**Definition 4.9** $E$ has finite perimeter if $\chi_E \in BV(\mathbb{R}^n)$.

Thus, $E$ has finite perimeter if and only if $\mathcal{L}^n(E) < \infty$ and $V(\chi_E) < \infty$. We write

$$P(E) = V(\chi_E),$$

and call $P(E)$ the *perimeter* of $E$.

The perimeter in this sense was introduced by Caccioppoli (1953) and De Giorgi (1954, 1955). See Ziemer (1989) for more references. Its exact definition varies from author to author. Our definition matches the one in Evans and Gariepy (1992), Ziemer (1989), and Giusti (1984). Other relevant sources involving isoperimetry and perimeter include Talenti (1993), Chavel (2001), and Burago and Zalgaller (1988).

Suppose that $\Omega$ is a bounded domain in $\mathbb{R}^n$ with $C^1$-boundary and that $\phi \in C^1(\mathbb{R}^n, \mathbb{R}^n)$ with $\sup_{\mathbb{R}^n} |\phi| \leq 1$. Then the Gauss–Green formula (4.8) gives

$$\int_\Omega \operatorname{div} \phi \, dx = \int_{\partial\Omega} \phi \cdot \nu \, d\mathcal{H}^{n-1},$$

where $\nu$ is the unit exterior normal vector field on $\partial\Omega$. From the definition of $V(\chi_\Omega)$ it follows that $\Omega$ has finite perimeter and that $P(\Omega) \leq \mathcal{H}^{n-1}(\partial\Omega)$. On the other hand, $\nu$ is defined and continuous at every point of $\partial\Omega$. We can uniformly approximate $\nu$ on $\partial\Omega$ by restrictions to $\partial\Omega$ of vector fields $\phi$ on $\mathbb{R}^n$ satisfying $|\phi| \leq 1$. It follows that

$$P(\Omega) = \mathcal{H}^{n-1}(\partial\Omega), \quad \text{for bounded } C^1 \text{ domains } \Omega \subset \mathbb{R}^n. \tag{4.18}$$

Equality (4.18) still holds for piecewise $C^1$ domains, in particular for domains bounded by polyhedra. But if we take $\Omega = \mathbb{D} \setminus A$ where $\mathbb{D}$ is the unit disk in the complex plane and $A$ is a small closed line segment in $\overline{\mathbb{D}}$ with length $L$, then $P(\Omega) = P(\mathbb{D}) = 2\pi$ whereas $\mathcal{H}^1(\partial\Omega) = 2\pi + L$. In general, if sets $E_1$ and $E_2$ differ by a $\mathcal{L}^n$-null set then it follows from the definition of perimeter that $P(E_1) = P(E_2)$. It is also true that $P(E) \leq \mathcal{H}^{n-1}(\partial E)$ for every set $E \subset \mathbb{R}^n$ of finite perimeter. This is the content of Proposition 4.11 below.

Recall from Example 1.23 in §1.4 that $\alpha_n = \mathcal{L}^n(\mathbb{B}^n(0,1)) = \frac{\pi^{n/2}}{\Gamma(\frac{n}{2}+1)}$, and set $\beta_{n-1} = n\alpha_n$. In §4.5, with the aid of the coarea formula, we shall show that $P(\mathbb{B}^n(1)) = \beta_n$. Taking this for granted now, it follows that for the ball $\mathbb{B}^n(R)$ we have

$$P(\mathbb{B}^n(R)) = \beta_{n-1} R^{n-1} = n\alpha_n^{1/n} \mathcal{L}^n(\mathbb{B}^n(R))^{\frac{n-1}{n}}. \tag{4.19}$$

Here now is the *isoperimetric inequality for perimeter*.

**Theorem 4.10** *For every set $E \subset \mathbb{R}^n$ with finite perimeter we have*

$$P(E) \geq P(E^\#) = n\alpha_n^{1/n} \mathcal{L}^n(E)^{\frac{n-1}{n}}. \tag{4.20}$$

*Proof* To get the inequality apply Theorem 4.8 with $f = \chi_E$. The equality statement follows from (4.19). □

Our next aim is to show that (4.20) still holds when $P(E)$ is replaced by $\mathcal{H}^{n-1}(\partial E)$. Consider first the case when $E$ has finite perimeter. It will suffice then to show that $P(E) \leq \mathcal{H}^{n-1}(\partial E)$. To prove the latter inequality, it appears that one must wade a bit into the waters of geometric measure theory. We shall sketch the argument, and refer to Evans and Gariepy (1992) for details.

Since $\chi_E \in BV(\mathbb{R}^n)$, by Proposition 4.5 there exist a Radon measure $\mu$ on $\mathbb{R}^n$ and a $\mu$-measurable function $\sigma : \mathbb{R}^n \to \mathbb{R}^n$ with $|\sigma| = 1$ $\mu$-a.e. such that

$$\int_E \operatorname{div} \phi \, dx = -\int_{\mathbb{R}^n} \phi \cdot \sigma \, d\mu, \quad \phi \in C^1(\mathbb{R}^n, \mathbb{R}^n). \tag{4.21}$$

Let $x_0 \in \mathbb{R}^n \setminus \partial E$ and $B$ be be an open ball containing $x_0$ whose closure does not intersect $\partial E$. The left-hand integral in (4.21) is zero for all $\phi \in C^1(\mathbb{R}^n, \mathbb{R}^n)$ with support in $B$. Thus, the support of $\mu$ is contained in $\partial E$. This is to be expected, since $\sigma \, d\mu$ can be viewed as the gradient of $\chi_E$, which is locally constant in the complement of $\partial E$.

The *reduced boundary* $\partial^* E$ of $E$ may be defined to be the set of $x \in \partial E$ such that the limit

$$\lim_{r \to 0} \frac{1}{\mathcal{L}^n(B(x,r))} \int_{B(x,r)} \sigma(y) \, d\mu(y)$$

exists in $\mathbb{R}^n$, equals $\sigma(x)$, and has norm 1. It is clear that $\partial^* E$ is contained in the support of $\mu$. On the other hand, from the Lebesgue–Besicovitch differentiation theorem (Evans and Gariepy, 1992, p. 43), it follows that

$$\mu(\mathbb{R}^n \setminus \partial^* E) = 0.$$

This definition of $\partial^* E$ depends on the choice of $\sigma$, which is uniquely determined only up to a set of $\mu$-measure zero. Two different $\sigma$s will produce $\partial^* E$s which differ at most by a $\mu$-nullset. So properly speaking the reduced boundary of $E$ is not a single set but an equivalence class of sets.

**Proposition 4.11** *For a set $E \subset \mathbb{R}^n$ of finite perimeter, the measure $\mu$ satisfies*

$$d\mu = \chi_{\partial^* E} \, d\mathcal{H}^{n-1} \tag{4.22}$$

*as Borel measures in $\mathbb{R}^n$.*

This proposition is Evans and Gariepy (1992, Thm. 2.3 (iii)). It is the main link between perimeter and Hausdorff measure.

From (4.22) and (4.15) we see that $P(E) = V(\chi_E) = \mu(\mathbb{R}^n) = \mathcal{H}^{n-1}(\partial^* E)$. Since $\partial^* E \subset \partial E$, we have

**Corollary 4.12**

$$P(E) \leq \mathcal{H}^{n-1}(\partial E),$$

*for every set $E \subset \mathbb{R}^n$ with finite perimeter.*

Here now is the *isoperimetric inequality for Hausdorff measure*:

**Corollary 4.13** *For every $\mathcal{L}^n$-measurable set $E \subset \mathbb{R}^n$ with $\mathcal{L}^n(E) < \infty$ and every $n \geq 2$ we have*

$$\mathcal{H}^{n-1}(\partial E) \geq \mathcal{H}^{n-1}(\partial(E^{\#})) = n\alpha_n^{1/n}\mathcal{L}^n(E)^{\frac{n-1}{n}}. \tag{4.23}$$

*Proof* If $E$ has finite perimeter the result follows from Corollary 4.12 and Theorem 4.10. Let $E$ be any set satisfying the hypotheses of Corollary 4.13. If $\mathcal{H}^{n-1}(\partial E) = \infty$ we are done, so assume $\mathcal{H}^{n-1}(\partial E) < \infty$. Then $\mathcal{L}^n(\partial E) = 0$, so that $\mathcal{L}^n(E) = \mathcal{L}^n(\overline{E})$. Since also $\partial E \supset \partial \overline{E}$, it will suffice to prove (4.23) when $E$ is closed.

For $R > 0$, write $B(R) = \mathbb{B}^n(0, R)$, $S(R) = \partial(B(R))$, and $E_R = E \cap \overline{B}(R)$. Then we have

$$\partial(E_R) = (\partial E \cap B(R)) \cup (E \cap S(R)),$$

so

$$\mathcal{H}^{n-1}(\partial(E_R)) \leq \mathcal{H}^{n-1}(\partial E) + \mathcal{H}^{n-1}(E \cap S(R)). \tag{4.24}$$

Thus, $\mathcal{H}^{n-1}(\partial(E_R)) < \infty$. By a theorem of Federer (see Evans and Gariepy, 1992, p. 222), this implies that $E_R$ has "locally finite perimeter." Using the definitions, it is easy to see that a bounded set with locally finite perimeter has finite perimeter, so $E_R$ has finite perimeter. Inequalities (4.24) and (4.23) give

$$\mathcal{H}^{n-1}(\partial E) \geq n\alpha_n^{1/n}\mathcal{L}^n(E_R)^{\frac{n-1}{n}} - \mathcal{H}^{n-1}(E \cap S(R)). \tag{4.25}$$

Now $\mathcal{L}^n(E) < \infty$. From the polar coodinates expression of $\mathcal{L}^n(E) = \int_{\mathbb{R}^n} \chi_E \, d\mathcal{L}^n$, there follows existence of a sequence $\{R_k\}$ with

$$\lim_{k\to\infty} R_k = \infty \quad \text{and} \quad \lim_{k\to\infty} \mathcal{H}^{n-1}(E \cap S(R)) = 0.$$

Letting $R \to \infty$ in (4.25) through the sequence $\{R_k\}$, (4.23) follows. □

When $E$ is the complement of the unit ball we have $0 < \mathcal{H}^{n-1}(\partial E) < \infty$, $\partial(E^{\#})$ is the empty set, and $\mathcal{L}^n(E) = \infty$. Thus, the assumption in Corollary 4.13 that $E$ have finite measure cannot be dropped.

If we write $L = \mathcal{H}^1$, $A = \mathcal{H}^2$, and $V = \mathcal{H}^3$, then for $n = 2$ and $n = 3$, the isoperimetric inequality (4.23) can be written (respectively) as

$$L^2 \geq 4\pi A \quad \text{and} \quad A^3 \geq 36\pi V^2.$$

## 4.4 Isoperimetric Inequalities for Minkowski Content

Let $E \subset \mathbb{R}^n$ be a subset of $\mathbb{R}^n$, not necessarily measurable. Define

$$d(x,E) = \text{distance from } x \text{ to } E = \inf_{y\in E} |x-y|.$$

Also, for $\delta > 0$, define

$$E(\delta) = \{x \in \mathbb{R}^n : d(x,E) < \delta\}, \quad E(-\delta) = \{x \in E : d(x,E^c) \geq \delta),$$

where $E^c \equiv \mathbb{R}^n \setminus E$. Then $E(\delta)$ is open, $E(-\delta)$ is closed, and $E(-\delta) \subset E \subset E(\delta)$. It is a simple exercise to show that

$$(\partial E)(\delta) = E(\delta) \setminus E(-\delta). \tag{4.26}$$

We shall call $E(\delta)$ the $\delta$-*collar* of $E$. If, for example, $E$ is the unit circle $|z| = 1$ in the complex plane and $\delta < 1$, then $E(\delta)$ is the annulus $1-\delta < |z| < 1+\delta$. In any dimension, if $E$ is a ball with radius $R$ then $E(\delta)$ is the concentric open ball with radius $R+\delta$, while $E(-\delta)$ is the concentric closed ball of radius $R-\delta$ when $\delta \leq R$, and is empty when $\delta > R$.

Here are the main symmetrization results involving collars.

**Theorem 4.14** *Let $E \subset \mathbb{R}^n$ and $\delta > 0$. Then*

(i) $E^\#(\delta) \subset E(\delta)^\#$, *and*
(ii) $E(-\delta)^\# \subset E^\#(-\delta)$.

**Corollary 4.15** *Let $E \subset \mathbb{R}^n$ and $\delta > 0$. Then*

*(a)* $\mathcal{L}^n(E(\delta)) \geq \mathcal{L}^n(E^\#(\delta))$,
*(b)* $\mathcal{L}^n(E^\#(-\delta)) \geq \mathcal{L}^n(E(-\delta))$,
*(c)* $\mathcal{L}^n((\partial E)(\delta)) \geq \mathcal{L}^n((\partial E^\#)(\delta))$.

Parts (a) and (b) of the corollary follow right away from (a) and (b) of the theorem. Part (c) of the corollary follows from (4.26), (a) and (b).

*Proof of Theorem 4.14(a)* Since $E \subset \overline{E}$ and $E(\delta) = \overline{E}(\delta)$, we may assume that $E$ is closed. We may also assume that $\mathcal{L}^n(E(\delta)) < \infty$. Set

$$f(x) = \left(1 - \frac{d(x,E)}{\delta}\right)^+, \quad x \in \mathbb{R}^n.$$

Then $f \in C(\mathbb{R}^n, [0,1])$ and $\{x : f(x) > 0\} = E_\delta$. By the triangle inequality, the function $d(\cdot, E)$ is in $\mathrm{Lip}(\mathbb{R}^n)$ with Lipschitz constant $\leq 1$. Thus, $f \in \mathrm{Lip}(\mathbb{R}^n)$, with Lipschitz constant $\leq \delta^{-1}$. Since symmetrization decreases the modulus of continuity (Theorem 2.12), we have

$$\Omega(t,f^{\#}) \leq \Omega(t,f) \leq \frac{t}{\delta}, \quad t \in (0,\infty),$$

where $\Omega$ denotes modulus of continuity.

Take $x \in E^{\#}(\delta)$ and $y \in E^{\#}$ such that $|x - y| < \delta$. Then

$$1 - f^{\#}(x) = f^{\#}(y) - f^{\#}(x) \leq \Omega(|y - x|, f^{\#}) \leq \frac{|y - x|}{\delta} < 1,$$

so that $f^{\#}(x) > 0$.

We saw above that $f^{-1}(0,\infty) = E(\delta)$. Thus $(f^{\#})^{-1}(0,\infty) = E(\delta)^{\#}$, and hence $x \in E(\delta)^{\#}$, which proves (a). □

*Proof of Theorem 4.14(b)* We may assume that $E(-\delta)$ is nonempty. Since $E(-\delta)$ is contained in the interior $E^0$ of $E$ and $E^0(-\delta) = E(-\delta)$, we may also assume that $E$ is open.

We have $(E(-\delta))(\delta) \subset E$. Applying (a) to the set $E(-\delta)$, it follows that

$$(E(-\delta)^{\#})(\delta) \subset (E(-\delta))(\delta)^{\#} \subset E^{\#}. \tag{4.27}$$

Take $x \in E(-\delta)^{\#}$. Suppose that $x \notin E^{\#}(-\delta)$. Then $d(x, (E^{\#})^c) < \delta$, so there exists $y \in (E^{\#})^c$ such that $|x - y| < \delta$. Then $y \in E(-\delta)^{\#}(\delta)$, so from (4.27), we obtain $y \in E^{\#}$, a contradiction. We conclude that $x \in E^{\#}(-\delta)$, which proves (b). □

Next, we introduce the Minkowski content. Let $A \subset \mathbb{R}^n$ be a subset of $\mathbb{R}^n$, not necessarily measurable, and $s \in [0, n]$. Recall that

$$\alpha_s = \frac{\pi^{s/2}}{\Gamma(\frac{s}{2} + 1)}.$$

When $s$ is an integer, $\alpha_s$ is the $s$-dimensional Lebesgue measure of the unit ball in $\mathbb{R}^s$. Define

$$\mathcal{M}_*^s(A) = \liminf_{\delta \to 0} \frac{\mathcal{L}^n(A(\delta))}{\alpha_{n-s}\delta^{n-s}},$$

$$\mathcal{M}^{*s}(A) = \limsup_{\delta \to 0} \frac{\mathcal{L}^n(A(\delta))}{\alpha_{n-s}\delta^{n-s}}.$$

$\mathcal{M}_*^s(A)$ and $\mathcal{M}^{*s}(A)$ are called the $s$-dimensional lower and upper Minkowski content of $A$ respectively. When $\mathcal{M}_*^s(A) = \mathcal{M}^{*s}(A)$, the common value is called the $s$-dimensional Minkowski content of $A$ and is denoted by $\mathcal{M}^s(A)$.

If $A$ is $\mathcal{L}^n$-measurable then $\mathcal{M}^n(A) = \mathcal{L}^n(A) = \mathcal{H}^n(A)$. If $B$ is a ball of radius $R$ in $\mathbb{R}^n$, then it is easy to show from the definition of $\mathcal{M}^{n-1}$, the coarea formula (4.32) and (4.19) that

$$\mathcal{M}^{n-1}(\partial B) = \mathcal{H}^{n-1}(\partial B) = n\alpha_n R^{n-1}.$$

Much more generally, a theorem of Federer (1969, p. 275) asserts that if $s$ is an integer $m \in \{0, \dots, n\}$ and $A$ is a closed and *m-rectifiable set* in $\mathbb{R}^n$, then

$$\mathcal{M}^m(A) = \mathcal{H}^m(A).$$

The set $A \subset \mathbb{R}^n$ is defined to be $m$-rectifiable if there exists a bounded set $S \subset \mathbb{R}^m$ and a Lipschitz function $f\colon S \to \mathbb{R}^n$ such that $A \subset f(S)$. Using the Lipschitz Extension Theorem 3.1, it is easy to show that the finite union of $m$-rectifiable sets is $m$-rectifiable. In particular, if $\Omega$ is a domain in $\mathbb{R}^n$ with Lipschitz boundary, then $\partial\Omega$ is $(n-1)$-rectifiable.

For $m = n - 1$, Corollary 4.15 immediately implies the *isoperimetric inequality for Minkowski content*:

**Theorem 4.16** *For each $\mathcal{L}^n$-measurable set $E \subset \mathbb{R}^n$ with $\mathcal{L}^n(E) < \infty$, we have*

$$\mathcal{M}_*^{n-1}(\partial E) \geq \mathcal{M}^{n-1}(\partial(E^{\#})) = n\alpha_n^{1/n}\mathcal{L}^n(E)^{\frac{n-1}{n}}.$$

Next, we examine some relations between Minkowski content and perimeter. The remarks above about $(n-1)$-rectifiability, combined with (4.18), show that

$$\mathcal{M}^{n-1}(\partial\Omega) = \mathcal{H}^{n-1}(\partial\Omega) = P(\Omega), \quad \text{for bounded } C^1 \text{ domains } \Omega \subset \mathbb{R}^n,$$

where $P$ denotes perimeter in $\mathbb{R}^n$.

For general measurable sets $E \subset \mathbb{R}^n$, perimeter and Minkowski content are related as follows:

**Proposition 4.17** *For every set $E \subset \mathbb{R}^n$ of finite perimeter, we have*

$$P(E) \leq \mathcal{M}_*^{n-1}(\partial E).$$

Proposition 4.17 and Theorem 4.10, the isoperimetric inequality for perimeter, provide another proof of Theorem 4.16 for sets $E$ with finite perimeter. The set $E = B \setminus A$ with $B$ the unit disk in the complex plane and $A$ a small line segment in $\overline{B}$ of length $L$, which we considered in §4.3, satisfies

$$P(E) = \pi, \quad \mathcal{M}^{n-1}(\partial E) = \mathcal{H}^{n-1}(\partial E) = \pi + L.$$

Thus, strict inequality can occur in Proposition 4.17.

*Proof of Proposition 4.17* Since $E$ has finite perimeter, we have $\mathcal{L}^n(E) < \infty$. Also, we may assume that $\mathcal{L}^n(E) > 0$ and that $\mathcal{M}_*^{n-1}(\partial E) < \infty$. The latter implies that $\mathcal{L}^n(\partial E) = 0$, so that $P(E) = P(\overline{E})$. Also $\partial E \supset \partial\overline{E}$, so we may assume that $E$ is closed.

For $\delta > 0$, define the function $g_\delta\colon \mathbb{R}^n \to [0,2]$ by

$$g_\delta(x) = \begin{cases} \left(1 - \frac{d(x,E)}{\delta}\right)^+, & \text{if } x \in E^c, \\ \min\left\{1 + \frac{d(x,E^c)}{\delta}, 2\right\}, & \text{if } x \in E. \end{cases}$$

The reader may verify that $g_\delta$ is Lipschitz with Lipschitz constant at most $1/\delta$. Thus $g_\delta$ is differentiable $\mathcal{L}^n$-a.e., and $|\nabla g_\delta| \le 1/\delta$. Moreover, for each $\epsilon > \delta$, $\nabla g_\delta = 0$ at all points in the complement of $(\partial E)(\epsilon)$. Thus, we have

$$\int_{\mathbb{R}^n} |\nabla g_\delta|\, dx \le \delta^{-1}\mathcal{L}^n((\partial E)(\epsilon)).$$

Taking $\epsilon = \delta + \delta^2$, this inequality gives

$$\liminf_{\delta \to 0} \int_{\mathbb{R}^n} |\nabla g_\delta|\, dx \le 2\mathcal{M}_*^{n-1}(\partial E). \tag{4.28}$$

We continue to write $\epsilon = \delta + \delta^2$. The support of $g_\delta$ is contained in $E(\epsilon)$, and the finiteness of $\mathcal{M}_*^{n-1}(\partial E)$ implies $\mathcal{L}^n((\partial E)(\epsilon)) < \infty$ for all sufficiently small $\delta$. Hence

$$\mathcal{L}^n(E(\epsilon)) \le \mathcal{L}^n(E) + \mathcal{L}^n((\partial E)(\epsilon)) < \infty.$$

It follows that each $g_\delta$ belongs to $L^1(\mathbb{R}^n)$, and hence each $g_\delta \in W^{1,1}(\mathbb{R}^n)$. The integral on the left side of (4.28) equals $V(g_\delta)$. Moreover, as $\delta \to 0$, we have

$$\|g_\delta - 2\chi_E\|_{L^1(\mathbb{R}^n)} \le 2\mathcal{L}^n(E(\epsilon) \setminus E(-\epsilon)) \to 2\mathcal{L}^n(\partial E) = 0.$$

From the definition of perimeter, the lower semicontinuity of variation (Proposition 4.6), and (4.28) we obtain

$$P(E) = V(\chi_E) \le \frac{1}{2} \liminf_{\delta \to 0} V(g_\delta) \le \mathcal{M}_*^{n-1}(\partial E). \qquad \square$$

## 4.5 Coarea Formula

Throughout this section, we let $m$ and $n$ be nonnegative integers with

$$0 \le m \le n.$$

The area formula, Theorem 4.2, tells us how to change variables in integrals involving $f\colon \mathbb{R}^m \to \mathbb{R}^n$. Now we shall present a corresponding formula for $f\colon \mathbb{R}^n \to \mathbb{R}^m$.

First, consider a linear map $L\colon \mathbb{R}^n \to \mathbb{R}^m$. We saw in §4.1 that $L^*\colon \mathbb{R}^m \to \mathbb{R}^n$ has a polar decomposition $L^* = O \circ S$, where $S\colon \mathbb{R}^m \to \mathbb{R}^m$ is symmetric and $O\colon \mathbb{R}^m \to \mathbb{R}^n$ is orthogonal. Thus, $L\colon \mathbb{R}^n \to \mathbb{R}^m$ has a polar decomposition

$$L = S \circ O^*,$$

where $S\colon \mathbb{R}^m \to \mathbb{R}^m$ is symmetric and $O\colon \mathbb{R}^m \to \mathbb{R}^n$ is orthogonal.

Define the Jacobian $J$ of $L$ to be the nonnegative constant

$$J = J_L = |\det S| = (\det(LL^*))^{1/2}.$$

One can view $O^*$ as the orthogonal projection of $\mathbb{R}^n$ onto an $m$-dimensional subspace $M \subset \mathbb{R}^n$ and $S$ as a self-map of $M$. Then $J_L$ tells us about distortion of area within $M$: For sets $A \subset M$,

$$\mathcal{H}^m(LA) = J_L \mathcal{H}^m(A).$$

For "$n$-dimensional" sets $A \subset \mathbb{R}^n$ we have

**Proposition 4.18** (Coarea Formula for linear maps) *For $\mathcal{L}^n$-measurable sets $A \subset \mathbb{R}^n$,*

$$J_L \mathcal{L}^n(A) = \int_{\mathbb{R}^m} \mathcal{H}^{n-m}(A \cap L^{-1}y)\, d\,\mathcal{L}^m(y). \tag{4.29}$$

If $L$ is the orthogonal projection of $\mathbb{R}^n$ onto the subspace $\{x \in \mathbb{R}^n\colon x_{m+1} = \cdots = x_n = 0\}$ then (4.29) follows from Fubini's Theorem. The case when $L$ is the orthogonal projection onto some subspace can be reduced via rotations to the special case above. To handle the general case when $L = S \circ O^*$, first get a formula like (4.29) corresponding to $O^*$, then apply the usual $\mathbb{R}^m \to \mathbb{R}^m$ change of variable formula to $S$. For details, see Evans and Gariepy (1992, p. 104).

Suppose now that $f\colon \mathbb{R}^n \to \mathbb{R}^m$ is any Lipschitz function. For $x$ at which $f$ is differentiable, define the Jacobian of $f$ to be

$$J(x) = J_f(x) = (\det(Df(x)(Df(x))^*)^{1/2}.$$

**Theorem 4.19** (Coarea Formula) *Let $f \in \mathrm{Lip}(\mathbb{R}^n, \mathbb{R}^m)$, $1 \le m \le n$, and $A$ be a $\mathcal{L}^n$-measurable set in $\mathbb{R}^n$. Then*

$$\int_A J_f\, d\mathcal{L}^n = \int_{\mathbb{R}^m} \mathcal{H}^{n-m}(A \cap f^{-1}y)\, d\mathcal{L}^m(y).$$

As with the area formula, the Coarea Formula is deduced from the linear case by splitting the left-hand integral into a large number of small pieces. For details, see Evans and Gariepy (1992, p. 112).

The Coarea Formula leads to a more general change of variable formula (Evans and Gariepy, 1992, p. 117):

**Theorem 4.20** (Coarea change of variable formula) *If $f \in \mathrm{Lip}(\mathbb{R}^n, \mathbb{R}^m)$, $1 \leq m \leq n$, and $g \in L^1(\mathbb{R}^n, \mathbb{R})$, then the restriction of $g$ to $f^{-1}y$ is $\mathcal{H}^{n-m}$-integrable for $\mathcal{L}^n$-almost every $y \in \mathbb{R}^n$, and*

$$\int_{\mathbb{R}^n} gJ_f \, d\mathcal{L}^n = \int_{\mathbb{R}^m} d\mathcal{L}^m(y) \int_{f^{-1}y} g \, d\mathcal{H}^{n-m}.$$

Let us consider the case $m = 1$. If $f \in \mathrm{Lip}(\mathbb{R}^n, \mathbb{R})$ then $Df = \nabla f = (f_{x_1}, \ldots, f_{x_n})$, and

$$Df(Df)^* = \sum_{i=1}^{n} (f_{x_i})^2 = |\nabla f|^2.$$

Thus,

$$J_f = |\nabla f|.$$

The coarea and change of variables formulas become

$$\int_A |\nabla f| \, d\mathcal{L}^n = \int_{\mathbb{R}} \mathcal{H}^{n-1}(A \cap f^{-1}y) \, dy \tag{4.30}$$

and

$$\int_{\mathbb{R}^n} g|\nabla f| \, d\mathcal{L}^n = \int_{\mathbb{R}} dy \int_{f^{-1}y} g \, d\mathcal{H}^{n-1}. \tag{4.31}$$

### Polar Coordinates and Hausdorff Measure on Spheres

For $n \geq 1$, define $f \in \mathrm{Lip}(\mathbb{R}^n, \mathbb{R}^+)$ by $f(x) = |x|$. Then $\nabla f = \frac{x}{|x|}$ for $x \in \mathbb{R}^n \setminus \{0\}$. For $r > 0$, write $S(r) = \partial \mathbb{B}^n(r)$. The Coarea Formula (4.31) becomes, for $g \in L^1(\mathbb{R}^n)$,

$$\int_{\mathbb{R}^n} g \, d\mathcal{L}^n = \int_0^\infty dr \int_{S(r)} g(x) \, d\mathcal{H}^{n-1}(x).$$

For $E \subset S(r)$, the definition of Hausdorff measure gives

$$\mathcal{H}^{n-1}(E) = r^{n-1}\mathcal{H}^{n-1}(r^{-1}E).$$

Thus,

$$\int_{\mathbb{R}^n} g \, d\mathcal{L}^n = \int_0^\infty r^{n-1} dr \int_{\mathbb{S}^{n-1}} g(rx) \, d\mathcal{H}^{n-1}(x). \tag{4.32}$$

Folland (1999, Th 2.49) uses facts about product measures to show existence of a unique positive Borel measure $\sigma_{n-1}$ on $\mathbb{S}^{n-1}$ for which (4.32) holds for all allowable $g$. The uniqueness implies that

$$\mathcal{H}^{n-1}(E) = \sigma_{n-1}(E), \quad \forall \text{ Borel } E \subset \mathbb{S}^{n-1}.$$

Thus, $\sigma_{n-1}$ equals the restriction of $\mathcal{H}^{n-1}$ to $\mathbb{S}^{n-1}$. We shall call $\sigma_{n-1}$ the *canonical measure on* $\mathbb{S}^{n-1}$.

Note also that if $E \subset S(r)$ then

$$\mathcal{H}^{n-1}(E) = r^{n-1}\sigma_{n-1}(r^{-1}E).$$

For $n \geq 1$, set

$$\beta_{n-1} = \sigma_{n-1}(S^{n-1}).$$

Recall from Example 2 in §1.4 that $\alpha_n = \mathcal{L}^n(B^n(0,1)) = \frac{\pi^{n/2}}{\Gamma(\frac{n}{2}+1)}$. From the top line on (Folland, Gerald 2002, p. 77) we see that $\beta_{n-1} = n\alpha_n$. Since $\Gamma(x+1) = x\Gamma(x)$, it follows that

$$\beta_{n-1} = \frac{2\pi^{n/2}}{\Gamma(\frac{n}{2})}.$$

Returning now to general $f \in \mathrm{Lip}(\mathbb{R}^n, \mathbb{R})$ and taking $A = \mathbb{R}^n$ in (4.30), we obtain

$$\int_{\mathbb{R}^n} |\nabla f|\, d\mathcal{L}^n = \int_{\mathbb{R}} \mathcal{H}^{n-1}(f = t)\, dt.$$

If $f \in \mathrm{Lip}(\mathbb{R}^n, \mathbb{R})$ and ess inf $|\nabla f| > 0$, then for fixed $t \in \mathbb{R}$ we can replace the $L^1$ function $g$ in (4.31) by $g\chi_{f>t}|\nabla f|^{-1}$ and obtain

$$\int_{f>t} g\, d\mathcal{L}^n = \int_t^\infty ds \int_{f=s} g\, \frac{1}{|\nabla f|}\, d\mathcal{H}^{n-1}. \tag{4.33}$$

Recall the notation $\lambda_f(t) = \mathcal{L}^n(f > t)$. If $\lambda_f(t) < \infty$, the choice $g = \chi_{f>t}$ in (4.33) gives

$$\lambda_f(t) = \int_t^\infty ds \int_{f=s} \frac{1}{|\nabla f|}\, d\mathcal{H}^{n-1}. \tag{4.34}$$

If, for example, $f \in \mathrm{Lip}(\mathbb{R}^n)$ is nonnegative with $\lambda_f(t) < \infty$ for every $t > 0$ and $|\nabla f|$ is essentially bounded from below, then (4.34) can be differentiated to give

$$\lambda_f{}'(t) = -\int_{f=t} \frac{1}{|\nabla f|}\, d\mathcal{H}^{n-1}$$

for almost every $t \in (0, \infty)$.

**Isoperimetry and Dirichlet Integrals**

In §4.3 we established the isoperimetric inequality for perimeter as a simple consequence of the Dirichlet integral inequality $\int_{\mathbb{R}^n} |\nabla f^{\#}|\, dx \leq \int_{\mathbb{R}^n} |\nabla f|\, dx$. In previous treatments of our subject the order is usually reversed: The isoperimetric inequality, for Hausdorff measure, say, is taken as known. Then

Dirichlet integral inequalities are proved with the aid of the Coarea Formula. Here, with several loose ends, is a sketch of the argument.

Suppose that $f$ is a nonnegative function in $C_c^\infty(\mathbb{R}^n)$ and that $1 \le p < \infty$. Write $E_t = (f = t)$. By the Coarea Formula,

$$\int_{\mathbb{R}^n} |\nabla f|^p \, dx = \int_0^\infty dt \int_{E_t} |\nabla f|^{p-1} \, d\mathcal{H}^{n-1}. \tag{4.35}$$

Fix $t \in (0, \infty)$. Define a measure $\mu$ on $E_t$ and a number $k$ by

$$d\mu(x) = \chi_{E_t} \frac{d\mathcal{H}^{n-1}(x)}{|\nabla f(x)|}, \qquad k = \mu(E_t).$$

Take $p \in [1, \infty)$. By Jensen's inequality and the isoperimetric inequality for Hausdorff measure (4.23),

$$\begin{aligned} \int_{E_t} |\nabla f|^{p-1} \, d\mathcal{H}^{n-1} &= \int_{E_t} |\nabla f|^p \, d\mu \ge k \left( \frac{1}{k} \int_{E_t} |\nabla f| \, d\mu \right)^p \\ &= k^{1-p} \mathcal{H}^{n-1}(E_t) \ge k^{1-p} n \alpha_n^{1/n} \mathcal{L}^n(f \ge t). \end{aligned} \tag{4.36}$$

Since $\lambda_f = \lambda_{f^\#}$, it follows from (4.34) that $f$ and $f^\#$ have the same $k$. Also, if $f$ is replaced by $f^\#$ then the inequality in (4.36) becomes an equality. Thus, for almost every $t$,

$$\int_{E_t} |\nabla f|^{p-1} \, d\mathcal{H}^{n-1} \ge \int_{E_t} |\nabla f^\#|^{p-1} \, d\mathcal{H}^{n-1}.$$

From (4.35) follows the desired inequality

$$\int_{\mathbb{R}^n} |\nabla f|^p \ge \int_{\mathbb{R}^n} |\nabla f^\#|^p, \quad 1 \le p < \infty.$$

For careful versions of this argument, see, for example, Hildén (1976); Talenti (1976a); Aubin (1976); Sperner (1973); Brothers and Ziemer (1988).

## 4.6 Sharp Sobolev Embedding Constant for $p = 1$

Given $n \ge 1$ and $p \in [1, n)$, define

$$p^* = \frac{np}{n-p}.$$

The exponent $p^*$ is called the *Sobolev conjugate exponent* of $p$.

In the next two sections $\|f\|_p$ will denote the $L^p$-norm in $L^p(\mathbb{R}^n, \mathcal{L}^n)$. The Sobolev embedding theorem (Evans and Gariepy, 1992, p. 138) asserts the existence of constants $C(n,p)$ such that

$$\|f\|_{p^*} \leq C(n,p)\|\nabla f\|_p, \quad f \in W^{1,p}(\mathbb{R}^n), \quad 1 \leq p < n. \tag{4.37}$$

These inequalities can be succintly restated as

$$W^{1,p}(\mathbb{R}^n) \hookrightarrow L^p(\mathbb{R}^n),$$

where $\hookrightarrow$ is read "is continuously embedded in".

In this section and the next we shall prove these inequalities with the best possible, that is, the smallest, values of $C(n,p)$. This section is devoted to the case $p = 1$, the next to $1 < p < n$.

**Theorem 4.21** *For $f \in BV(\mathbb{R}^n)$, $n \geq 2$, we have*

$$\|f\|_{\frac{n}{n-1}} \leq n^{-1}\alpha_n^{-1/n}V(f). \tag{4.38}$$

*Equality holds when $f = \chi_B$ for some ball $B \subset \mathbb{R}^n$.*

Recall from §4.2 that $W^{1,1}(\mathbb{R}^n) \subset BV(\mathbb{R}^n)$ and that for the total variation $V$ we have $V(f) = \|\nabla f\|_1$ for $f \in W^{1,1}(\mathbb{R}^n)$. Thus, Theorem 4.21 is a slightly extended version of (4.37) when $p = 1$. In view of the equality statement in the theorem the constant on the right in (4.38) is certainly best possible within $BV(\mathbb{R}^n)$. This constant is still best possible in the smaller space $W^{1,1}(\mathbb{R}^n)$. To see this take $R \in (0,1)$. Let $f_R$ be the function that is 1 on $|x| \leq 1 - R$, zero on $|x| \geq 1$, and $(1 - |x|)/(1 - R)$ on $R < |x| < 1$. Then $f_R \in W^{1,1}(\mathbb{R}^n)$. As $R \to 1$, we have

$$\|f_R\|_{\frac{n}{n-1}} \to (\alpha_n)^{\frac{n-1}{n}} \quad \text{and} \quad \|\nabla f_R\|_1 \to n\alpha_n.$$

On the other hand, the proof of Theorem 4.21 will show that for $f \in W^{1,1}(\mathbb{R}^n)$ strict inequality always holds in (4.38) unless $f \equiv 0$.

The constant on the right in (4.38) is the reciprocal of the isoperimetric constant studied in Sections 4.3 and 4.4. It turns out that the sharp Sobolev inequality (4.38) is in fact equivalent to the sharp geometric isoperimetric inequality. At the end of this section we shall comment further on this phenomenon.

*Proof of Theorem 4.21* Suppose first that $f \in W^{1,1}(\mathbb{R}^n)$. Since $|\nabla f| = \nabla|f|$ a.e., we may assume that $f \geq 0$. By Theorem 3.20 or Theorem 4.8, we may also assume that $f$ is symmetric decreasing. Let

$$h(r) = f(re_1).$$

Then $h$ is a decreasing function on $[0,\infty)$. The proof of Theorem 3.20 shows that $h$ is absolutely continuous on each compact subinterval of $(0,\infty)$. Since $f \in W^{1,1}(\mathbb{R}^n) \subset L^1(\mathbb{R}^n)$, it follows that $\lim_{r\to\infty} h(r) = 0$. Thus, the nonpositive function $h'$ belongs to $L^1(r,\infty)$ for each $r \in (0,\infty)$, and we have the following formulas:

$$h(r) = -\int_r^\infty h'(s)\,ds, \quad r \in (0,\infty), \tag{4.39}$$

$$\|f\|_{\frac{n}{n-1}}^{\frac{n}{n-1}} = \beta_{n-1}\int_0^\infty |h(r)|^{\frac{n}{n-1}}\, r^{n-1}\,dr,$$

$$\|\nabla f\|_1 = \beta_{n-1}\int_0^\infty |h'(r)|\, r^{n-1}\,dr.$$

□

**Lemma 4.22** *Let* $h\colon (0,\infty) \to \mathbb{R}^+$ *be decreasing, satisfy* $\lim_{r\to\infty} h(r) = 0$, *and be absolutely continuous on each compact subinterval of* $(0,\infty)$. *Then for each* $\gamma > 1$,

$$\int_0^\infty h(r)^{\gamma'}\, r^{\gamma-1}\,dr \le \gamma^{-1}\left\{\int_0^\infty |h'(r)|\, r^{\gamma-1}\,dr\right\}^{\gamma'}. \tag{4.40}$$

*If equality holds for some* $\gamma \in (0,\infty)$, *then* $h \equiv 0$.

Here $\gamma' = \frac{\gamma}{\gamma-1}$ is the Hölder conjugate exponent of $\gamma$.

Application of Lemma 4.22 with $\gamma = n$ yields the desired inequality (4.38) for $f \in W^{1,1}(\mathbb{R}^n)$. Take $f \in BV(\mathbb{R}^n)$. By Proposition 4.7, there is a sequence $\{f_k\}$ in $W^{1,1}(\mathbb{R}^n)$ such that $V(f_k) \to V(f)$ and $f_k \to f$ in $\mathcal{L}^n(\mathbb{R}^n)$ as $k \to \infty$. Choose a subsequence, still denoted $\{f_k\}$, which converges to $f$ a.e. in $\mathbb{R}^n$. Then

$$\|f\|_{\frac{n}{n-1}} \le \liminf_{k\to\infty} \|f_k\|_{\frac{n}{n-1}} \le \liminf_{k\to\infty} n^{-1}\alpha_n^{-1/n} V(f_k) = n^{-1}\alpha_n^{-1/n} V(f).$$

The first inequality follows from Fatou's Lemma. This proves the inequality in Theorem 4.21 for $f \in BV(\mathbb{R}^n)$, modulo the Lemma. To prove the statement about equality suppose that the ball $B$ has radius $R$. Then, using $P$ to denote perimeter,

$$V(\chi_B) = P(B) = \mathcal{H}^{n-1}(\partial B) = \beta_{n-1}R^{n-1} \text{ and } \|\chi_B\|_{\frac{n}{n-1}} = (\alpha_n R^n)^{\frac{n-1}{n}}.$$

Since $\beta_{n-1} = n\alpha_n$, the equality statement in Theorem 4.21 follows.

*Proof of Lemma 4.22* Define functions $F, G,$ and $K$ from $\mathbb{R}$ into $\mathbb{R}^+$ by

$$F(x) = e^{\gamma x}|h'(e^x)|, \quad G(x) = e^{(\gamma-1)x}h(e^x), \quad K(x) = e^{(\gamma-1)x}\chi_{(-\infty,0)}(x).$$

Equation (4.39) is valid for our $h$. Setting $r = e^x$, $s = e^y$, (4.39) becomes

$$G(x) = \int_{\mathbb{R}} F(y)K(x-y)\,dy = F * K(x).$$

Also, by the definitions of $F$ and $G$,

$$\int_{\mathbb{R}} G^{\gamma'}\, dx = \int_0^\infty h(r)^{\gamma'}\, r^{\gamma-1}\, dr, \tag{4.41}$$

and

$$\|F\|_1 = \int_{\mathbb{R}} |h'(r)|\, r^{\gamma-1}\, dr.$$

By a special case of Young's inequality for convolution (Rudin, 1966, Exercise 4c, p. 148),

$$\|G\|_{\gamma'} \le \|F\|_1 \|K\|_{\gamma'},$$

with strict inequality unless $F \equiv 0$. Also, $\gamma'(\gamma - 1) = \gamma$, so

$$\int_{\mathbb{R}} K^{\gamma'}\, dx = \frac{1}{\gamma}. \tag{4.42}$$

The desired inequality (4.40) follows from (4.41)–(4.42). □

Our proofs of the isoperimetric inequality Theorem 4.10 and the sharp Sobolev inequality Theorem 4.21 are based on the result that symmetrization decreases the Dirichlet integral. Theorem 4.21 is due to Federer and Fleming (1960), who derive the sharp Sobolev inequality from the isoperimetric inequality and the Coarea Formula. Here is a sketch of their argument, as presented in Ziemer (1989).

Take a nonnegative function $f \in C_c^\infty(\mathbb{R}^n)$. Let $f_t = \min(f, t)$. Set $A_t = \{x \colon f(x) \ge t\}$ and $g(t) = \|f_t\|_{\frac{n}{n-1}}$. Then for $h > 0$, we have

$$0 \le f_t(x) \le f_{t+h}(x) \le f_t(x) + h\chi_{A_t}(x), \quad x \in \mathbb{R}^n. \tag{4.43}$$

Write $m(t) = \mathcal{L}^n(A_t)^{\frac{n}{n-1}}$. Then we get

$$g(t) \le g(t+h) \le g(t) + hm(t),$$

from (4.43) and Minkowski's inequality. Thus $g$ is Lipschitz on $[t_0, \infty)$ for each $t_0 > 0$ and for almost every $t$,

$$0 \le g'(t) \le m(t).$$

From Sard's Theorem and the Implicit Function Theorem it follows that for a.e. $t$ the set $f^{-1}(t)$ is a smooth manifold, and that $f^{-1}(t) = \partial A_t$. Assuming the isoperimetric inequality (4.23) for such sets, we have

$$m(t) \le n^{-1}\alpha_n^{-1/n}\mathcal{H}^{n-1}(f = t), \quad \text{for almost every } t.$$

Thus

$$\|f\|_{\frac{n}{n-1}} = \lim_{t\to\infty} g(t) = \int_0^\infty g'(t)\,dt \le n^{-1}\alpha_n^{-1/n} \int_0^\infty \mathcal{H}^{n-1}(f = t)\,dt$$
$$= n^{-1}\alpha_n^{-1/n} \int_{\mathbb{R}^n} |\nabla f|\,dx.$$

The last equality comes from the Coarea Formula (4.30). This establishes the sharp Sobolev inequality for smooth functions $f$ when $p = 1$.

Federer and Fleming also observe that this sharp Sobolev inequality is in fact equivalent to the isoperimetric inequality. Take a nice set $E$ and $h > 0$. Suppose that the Sobolev inequality is true, Apply it to the function

$$f(x) = \left(1 - \frac{d(x, E)}{h}\right)^+.$$

Represent $\mathcal{L}^n(0 < f < 1)$ via the Coarea Formula, then let $h \to 0$. The isoperimetric inequality for Hausdorff measure follows. For more detail, see Ziemer (1989, p. 83).

## 4.7 Sharp Sobolev Embedding Constants for $1 < p < n$

Recall that

$$p^* = \frac{np}{n-p} \quad \text{and} \quad p' = \frac{p}{p-1}.$$

For $n \ge 2$ and $p \in (1, n)$ define functions $g = g_{n,p}$ by

$$g(x) = (1 + |x|^{p'})^{-n/p^*}. \tag{4.44}$$

**Theorem 4.23** *Let $f \in W^{1,p}(\mathbb{R}^n)$, $1 < p < n$, $n \ge 2$. Then we have*

$$\|f\|_{p^*} \le (n\alpha_n^{1/n})^{-1}(p^*/p')^{1/p'} \left(\frac{p'}{n}\frac{\Gamma(n)}{\Gamma(n/p)\Gamma(n/p')}\right)^{1/n} \|\nabla f\|_p. \tag{4.45}$$

*Equality holds for $f = g_{n,p}$.*

It is easy to check that equality also holds for all functions $ag(bx + c)$, with $a, b, c$ real constants such that $b \ne 0$.

Note that

$$g(x) \sim |x|^{-\frac{n-p}{p-1}}, \quad |\nabla g| \sim |x|^{-\frac{n-1}{p-1}}, \quad x \to \infty.$$

It follows that $\nabla g \in L^p(\mathbb{R}^n)$ for $1 < p < n$ and that $g \in L^p(\mathbb{R}^n)$ for $1 < p < \sqrt{n}$, $g \notin L^p(\mathbb{R}^n)$ for $p \ge \sqrt{n}$. Thus $g \in W^{1,p}$ if and only if $1 < p < \sqrt{n}$. The ratio $\|f\|_{p^*}/\|\nabla f\|_p$ is maximized over $W^{1,p}(\mathbb{R}^n)$ by $g$ when $1 < p < \sqrt{n}$.

When $\sqrt{n} \leq p < n$ the constant on the right in (4.45) is still best possible for $f \in W^{1,p}(\mathbb{R}^n)$, as one sees by approximating $g$ by $gh_R$, where $h_R(x) = 1$ on $|x| < R$, $h_R(x) = 2 - \frac{|x|}{R}$ for $R \leq |x| \leq 2R$, $h_R(x) = 0$ for $|x| \geq 2R$ and $R$ is large.

Theorem 4.23 is due, independently, to E. Rodemich (unpublished, 1960s), Talenti (1976a), and Aubin (1976). We shall follow a fascinating and different proof, recently discovered by Cordero-Erausquin, Nazaret, and Villani (2004). Their proof appears also in Villani (2003, p. 201).

*Proof of Theorem 4.23* First we shall verify the equality statement. Let $a, b, c$ be positive numbers with $c < ab$. Then

$$\int_0^\infty (1 + r^a)^{-b} r^{c-1}\, dr = \frac{1}{a} \int_0^1 s^{\frac{c}{a}-1}(1-s)^{b-\frac{c}{a}-1}\, ds = \frac{\Gamma(\frac{c}{a})\Gamma(b-\frac{c}{a})}{a\Gamma(b)}. \tag{4.46}$$

The first equality comes from the change of variable $s = r^a/(1-r^a)$ and the second from Euler's identity for the beta function. Write

$$B = \frac{\Gamma(n/p)\Gamma(n/p')}{\Gamma(n)}.$$

Passing to polar coordinates and using (4.46), one computes

$$\int_{\mathbb{R}^n} |\nabla g|^p\, dx = \beta_{n-1} n^p (p')^{p-2} (p^*)^{-p/p'} B$$

and

$$\int_{\mathbb{R}^n} g^{p^*}\, dx = \beta_{n-1}(p')^{-1} B.$$

After some manipulations with $p, p'$, and $p^*$, one finds that $\|g\|_{p^*}/\|\nabla g\|_p$ equals the constant on the right side of (4.45). The equality statement in Theorem 4.23 is proved.

To prove inequality (4.45), take $f \in W^{1,p}(\mathbb{R}^n)$. We may assume that $f \geq 0$. Since Dirichlet integrals decrease under symmetrization (Theorem 3.20), we may also assume that $f$ is symmetric decreasing on $\mathbb{R}^n$. Since $f \in L^p(\mathbb{R}^n)$ and $p^* > p$, it follows that $\int_{|x|\geq 1} f^{p^*}\, dx < \infty$. Approximating $f$ by $\min(f, N)$ for large $N$, we may also assume that $f$ is bounded, which insures that $\int_{|x|\leq 1} f^{p^*}\, dx < \infty$. Finally, multiplying $f$ by a suitable constant, we may assume that

$$\int_{\mathbb{R}^n} f^{p^*}\, dx = \int_{\mathbb{R}^n} g^{p^*}\, dx \tag{4.47}$$

where $g = g_{n,p}$ is the function of (4.44).

Define functions $F$, $G$ on $\mathbb{R}^n$ and $F_1$, $G_1$ on $\mathbb{R}^+$ by

$$F = f^{p^*}, \qquad G = g^{p^*},$$

$$F_1(r) = \int_0^r F(se_1)s^{n-1}\,ds, \qquad G_1(r) = \int_0^r G(se_1)s^{n-1}\,ds.$$

Define also $\psi_1 : \mathbb{R}^+ \to \mathbb{R}^+$ and $T : \mathbb{R}^n \to \mathbb{R}^n$ by

$$\psi_1 = G_1^{-1} \circ F_1, \quad T(x) = \frac{x}{|x|}\psi_1(|x|).$$

Using (4.47), one sees that if $f > 0$ on $\mathbb{R}^n$ then $\psi_1$ maps $\mathbb{R}^+$ $1-1$ onto $\mathbb{R}^+$, while if $f$ has compact support then $\psi_1$ maps $[0, R_0)$ $1-1$ onto $\mathbb{R}^+$, where $R_0$ is the smalllest $R$ such that $f(Re_1) = 0$.

For $0 \leq R_1 < R_2$, the relation $F_1 = G_1 \circ \psi_1$ gives

$$\int_{R_1}^{R_2} F(re_1)r^{n-1}\,dr = \int_{\psi_1(R_1)}^{\psi_1(R_2)} G(se_1)\,s^{n-1}\,ds.$$

It follows that

$$\int_E F\,dx = \int_{T(E)} G\,dx \tag{4.48}$$

for all polar rectangles of the form $E = \{ry\colon r \in [R_1, R_2],\ y \in E'\}$, where $E'$ is a Borel set on $\mathbb{S}^{n-1}$, and this implies that (4.48) holds for all $\mathcal{L}^n$-measurable sets $E \subset \mathbb{R}^n$. We also have

$$\int_{T(E)} G(y)\,dy = \int_E G(Tx)J_T(x)\,dx,$$

where $J_T$ is the Jacobian of $T$. Combined with (4.48), we deduce that

$$F = (G \circ T)\,J_T, \quad \text{on} \quad \mathbb{R}^n, \tag{4.49}$$

and hence, for measurable functions $H\colon \mathbb{R}^n \to \mathbb{R}^+$,

$$\int_{\mathbb{R}^n} GH\,dx = \int_{\mathbb{R}^n} (H \circ T)\,F\,dx. \tag{4.50}$$

Define $\phi_1 : \mathbb{R}^+ \to \mathbb{R}^+$ and $\phi\colon \mathbb{R}^n \to \mathbb{R}^+$ by

$$\phi_1(r) = \int_0^r \psi_1(s)\,ds \quad \text{and} \quad \phi(x) = \phi_1(|x|).$$

Then

$$T = \nabla\phi, \quad \text{on} \quad \mathbb{R}^n.$$

Now $\phi_1$ is an increasing convex function on $[0, \infty)$. For $a, b \in \mathbb{R}^n$ the function $t \to |a + tb|$ is convex on $\mathbb{R}$, and hence $t \to \phi_1(|a + tb|)$ is convex on $\mathbb{R}$. Thus, $\phi$ is a convex function on $\mathbb{R}^n$.

Let $D^2\phi = (\frac{\partial^2\phi}{\partial x_i \partial x_j})_{ij}$ denote the Hessian matrix of $\phi$. Then $D^2\phi(x)$ is nonnegative and symmetric at each point $x$, hence has nonnegative eigenvalues $\lambda_1, \ldots, \lambda_n$. at each point. Thus, using the arithmetic-geometric inequality, at each point of $\mathbb{R}^n$ we have

$$J_T = \det DT = \det D^2\phi = \prod_{i=1}^{n} \lambda_i \le \left(\frac{1}{n}\sum_{i=1}^{n}\lambda_i\right)^n$$
$$= \left(\frac{1}{n}\operatorname{tr} D^2\phi\right)^n = \left(\frac{1}{n}\Delta\phi\right)^n.$$

Taking $H = G^{-1/n}$ in (4.50) and using (4.49), we obtain

$$\begin{aligned}
\int_{\mathbb{R}^n} G^{1-\frac{1}{n}}\, dx &= \int_{\mathbb{R}^n} F^{1-\frac{1}{n}} J^{1/n}\, dx \le \frac{1}{n}\int_{\mathbb{R}^n} F^{1-\frac{1}{n}}\Delta\phi\, dx \\
&= -\frac{1}{n}\int_{\mathbb{R}^n} \nabla(F^{1-\frac{1}{n}})\cdot\nabla\phi\, dx \\
&= -\frac{1}{n}\frac{n-1}{n}\int_{\mathbb{R}^n} F^{-1/n}\,\nabla F\cdot\nabla\phi\, dx \qquad (4.51)\\
&\le \frac{1}{n}\frac{n-1}{n}\int_{\mathbb{R}^n} F^{-1/n}|\nabla F|\,|\nabla\phi|\, dx \\
&= \frac{1}{n}\frac{n-1}{n}\int_{\mathbb{R}^n} F^{-1/n}|\nabla F|\,|T|\, dx.
\end{aligned}$$

From the definitions, it follows that

$$F^{-1/n}|\nabla F| = p^* f^{p^*/p'}|\nabla f| = p^* F^{1/p'}|\nabla f|.$$

So, using Hölder's inequality,

$$\begin{aligned}
\frac{1}{n}\frac{n-1}{n}\int_{\mathbb{R}^n} F^{-1/n}|\nabla F|\,|T|\, dx &= \frac{n-1}{n}\frac{p}{n-p}\int_{\mathbb{R}^n} F^{1/p'}|T||\nabla f|\, dx \\
&\le \frac{n-1}{n}\frac{p}{n-p}\left(\int_{\mathbb{R}^n} F|T|^{p'}\, dx\right)^{1/p'} \|\nabla f\|_p.
\end{aligned} \qquad (4.52)$$

Using (4.50) with $H(x) = |x|^{p'}$,

$$\int_{\mathbb{R}^n} F|T|^{p'}\, dx = \int_{\mathbb{R}^n} |x|^{p'} G\, dx. \qquad (4.53)$$

Combining (4.51)–(4.53), we obtain

$$\|\nabla f\|_p \ge \frac{n(n-p)}{p(n-1)}\left(\int_{\mathbb{R}^n} G^{\frac{n-1}{n}}\, dx\right)\left(\int_{\mathbb{R}^n} |x|^{p'} G\, dx\right)^{-1/p'}. \qquad (4.54)$$

In the derivation of (4.54) all but three of the steps were equalities. Suppose that $f = g$. Then $T(x) = x$ and $\phi(x) = \frac{1}{2}|x|^2$, so $J_T{}^{1/n} = \frac{1}{n}\Delta\phi = 1$. Also, since $G$ and $\phi$ are both radial with $G$ decreasing, $\phi$ increasing,

$$\nabla G \cdot \nabla\phi = -|\nabla G||\nabla\phi|.$$

Thus, the two inequalities in (4.51) are each equalities.

Now we examine (4.52). For equality to hold in Hölder's inequality $\int h_1 h_2 \le \|h_1\|_p \|h_2\|_{p'}$ with nonnegative $h_1$ and $h_2$, it is necessary and sufficient that $h_2 = ch_1^{p-1}$ a.e. for some constant $c$. Take

$$h_1 = |\nabla g|, \quad h_2(x) = G^{1/p'}(x)|T(x)| = |x|g^{p^*/p'}(x).$$

Calculation gives

$$|\nabla g(x)| = c|x|^{p'-1}g(x)^{1+\frac{p^*}{n}}, \quad x \in \mathbb{R}^n.$$

Thus, we have

$$|\nabla g(x)|^{p-1} = c|x|^{(p-1)(p'-1)}g(x)^{(1+\frac{p^*}{n})(p-1)}$$

where $c$ denotes a constant which can change from line to line. But

$$(p'-1)(p-1) = 1 \quad \text{and} \quad \left(1 + \frac{p^*}{n}\right)(p-1) = p^*/p'.$$

Thus $h_1^{p-1} = ch_2$, so in (4.52) Hölder's inequality holds with equality when $f = g$.

We've shown now that equality holds in (4.54) when $f = g$. Thus, recalling (4.47),

$$\|\nabla f\|_p \ge \|\nabla g\|_p, \quad \text{when} \quad \int_{\mathbb{R}^n} f^{p^*}\,dx = \int_{\mathbb{R}^n} g^{p^*}\,dx.$$

Theorem 4.23 is proved. □

Like the proof just given, the proofs of Theorem 4.23 by Aubin and Talenti employ an initial symmetrization to reduce the problem to an extremal problem for integrals of functions of one variable. They solved this extremal problem by invoking a theorem of Bliss (1930), whose proof involves calculus of variations. The proof by Cordero-Erausquin et al. (2004) does not use symmetrization. Instead, given any two nonnegative functions $f$ and $g$ with gradients in $L^p(\mathbb{R}^n)$ and with $\int_{\mathbb{R}^n} f^{p^*}\,dx = \int_{\mathbb{R}^n} g^{p^*}\,dx$, they invoke a theorem of Brenier to produce a convex $\phi\colon \mathbb{R}^n \to \mathbb{R}$ such that $T = \nabla\phi$, $F = f^{p^*}$ and $G = g^{p^*}$ satisfy (4.49) above:

$$F = G \circ T\, J_T. \tag{4.55}$$

The rest of their proof is exactly like the one presented here.

Another way to state (4.55) is as follows: Consider the measures $\mu$ and $\nu$ on $\mathbb{R}^n$ defined by $d\mu = F\,dx$, $d\nu = G\,dx$. Then $T_{\#}(\mu) = \nu$, that is,

$$\nu(E) = \mu(T^{-1}E), \quad \text{for all Borel sets} \quad E \subset \mathbb{R}^n.$$

In Villani (2003) one learns that this $T$ solves the *Monge Transportation Problem* for the given measures $\mu$ and $\nu$.

By performing the initial symmetrization as we have done here, one can immediately write down the mapping $T$. No deep theorem like Brenier's is needed.

## 4.8 More about Sobolev Spaces

In the preceding two sections we saw that $W^{1,p}(\mathbb{R}^n) \subset L^{p^*}(\mathbb{R}^n)$ when $1 \leq p < n$, where $p^* = \frac{np}{n-p}$. For $p = n$ one might expect that functions in $W^{1,n}$ are in $L^\infty(\mathbb{R}^n)$, but the function

$$f(x) = (\log^+(1/|x|))^\alpha$$

furnishes a counterexample when $0 < \alpha < \frac{n-1}{n}$.

If $n < p < \infty$ then $W^{1,p}(\mathbb{R}^n)$ is contained in $L^\infty(\mathbb{R}^n)$. In fact, a good deal more is true. Given $n \geq 2$ and $n < p < \infty$, define $\gamma$ by

$$\gamma = 1 - \frac{n}{p}.$$

**Theorem 4.24** (Morrey's Embedding Theorem) *For $n \geq 2$ and $n < p < \infty$, there exist constants $C(p, n)$ such that each equivalence class in $W^{1,p}(\mathbb{R}^n)$ contains a function $f$ such that*

$$|f(x) - f(y)| \leq C(n,p)\, \|\nabla f\|_p\, |x-y|^\gamma, \quad x, y \in \mathbb{R}^n, \tag{4.56}$$

*and*

$$\|f\|_{L^\infty(\mathbb{R}^n)} \leq C(n,p)\, \|f\|_{W^{1,p}(\mathbb{R}^n)}. \tag{4.57}$$

Functions $f$ satisfying (4.56) are said to satisfy a *Hölder condition* of order $\gamma$, or to be $\gamma$-Hölder continuous. The set of all functions which are $\gamma$-Hölder continuous in a domain $\Omega$ is denoted by $C^\gamma(\Omega)$. Thus, Morrey's Theorem can be restated as

$$W^{1,p}(\mathbb{R}^n) \hookrightarrow C^\gamma(\mathbb{R}^n) \cap L^\infty(\mathbb{R}^n), \quad n < p < \infty.$$

A proof of (4.56) is in Evans and Gariepy (1992, p. 143). To get (4.57), use (4.56) together with the assumption $f \in L^p(\mathbb{R}^n)$.

Note that $\gamma \to 1$ as $p \to \infty$. Thus, in the limit, Morrey's Theorem "converges" to the identification $W^{1,\infty}(\mathbb{R}^n) = \text{Lip}(\mathbb{R}^n) \cap L^\infty(\mathbb{R}^n)$ (Proposition 3.17).

We saw above that in the borderline case $p = n$ functions in $W^{1,n}$ can be unbounded. The following results, though, show that they cannot be very unbounded. To get a clean statement we shall leave $W^{1,n}(\mathbb{R}^n)$ and consider instead the space $W_0^{1,n}(\Omega)$, the closure of $C_c^1(\Omega)$ in the $W^{1,n}$-norm, where $\Omega$ is an open set in $\mathbb{R}^n$ with $\mathcal{L}^n(\Omega) < \infty$. Set

$$\delta_n = (n\alpha_n^{1/n})^{\frac{n}{n-1}}.$$

**Theorem 4.25** (Moser) *For each $n \geq 2$ there exists a constant $C_n$ such that for each $\Omega$ with $\mathcal{L}^n(\Omega) < \infty$ and each $f \in W_0^{n,1}(\Omega)$ with $\|\nabla f\|_n \leq 1$, we have*

$$\frac{1}{\mathcal{L}^n(\Omega)} \int_\Omega \exp[\delta_n |f|^{\frac{n}{n-1}}]\, dx \leq C_n. \tag{4.58}$$

If $\frac{n}{n-1}$ is replaced by larger number then the integral in (4.58) can diverge. If $\delta_n$ is replaced by any larger number the integral still converges, but can be made arbitrarily large by appropriate choice of $f$. The most remarkable aspect of the theorem is the uniform boundedness at the critical exponent $\delta_n$. This property has applications to curvature problems in geometry. See, for example, Chang (1996).

One may assume in Moser's theorem that $f \geq 0$. To prove his theorem, Moser begins with a symmetrization of $f$, under which the gradient integral decreases, as in Corollary 3.9, while the integral involving $f$ remains the same. The reduced problem involving one-variable integrals is difficult, but Moser (1970/71) managed to solve it. A shorter proof of the one-variable problem is in Marshall (1989).

As far as I know, the best constants on the right-hand sides of (4.56), (4.57) and (4.58) are not known. For Moser's Theorem, it is known that extremal functions exist (Carleson and Chang, 1986).

Returning to the global space $W^{1,n}(\mathbb{R}^n)$, by applying Moser's Theorem to the function $(f-t)^+$, we get

$$\int_{f \geq t} \exp(\delta_n (f-t)^{\frac{n}{n-1}})\, dx \leq C_n \mathcal{L}^n(f \geq t)$$

for $t > 0$, provided $f \in W^{1,n}(\mathbb{R}^n)$ with $f \geq 0$ and $\|\nabla f\|_n = 1$.

## Higher Order Sobolev Spaces

An $n$-tuple $\alpha = (\alpha_1, \ldots, \alpha_n)$ with the $\alpha_i$ nonnegative integers is called a multi-index. The number $|\alpha| = \sum_{i=1}^n \alpha_i$ is called the order of $\alpha$. For functions $f$, write

$$D^\alpha f = \frac{\partial^{|\alpha|} f}{\partial^{\alpha_1} x_1 \cdots \partial_n^{\alpha} x_n}.$$

Let $\Omega$ be an open set in $\mathbb{R}^n$ and $f \in L^1_{\text{loc}}(\Omega)$. For a multi-index $\alpha$, the function $g \in L^1_{\text{loc}}(\Omega)$ is said to be the weak $\alpha$th derivative of $f$ if

$$\int_\Omega f D^\alpha \phi \, dx = (-1)^{|\alpha|} \int_\Omega g\phi \, dx, \quad \forall \phi \in C_c^\infty(\Omega).$$

When $g$ exists we'll write $D^\alpha f = g$ and say that $D^\alpha f$ exists in the weak sense.

For $p \in [1, \infty)$ and integers $n \geq 1$, $k \geq 1$, define the space $W^{k,p}(\Omega)$ to be the set of all $f \in L^p(\Omega)$ such that $D^\alpha f$ exists in the weak sense and is in $L^p(\Omega)$ for each $\alpha$ with $|\alpha| \leq k$.

Define the norm in $W^{k,p}(\Omega)$ to be

$$\|f\|^p_{W^{k,p}(\Omega)} = \int_\Omega \sum_{0\leq|\alpha|\leq k} |D^\alpha f|^p \, dx.$$

The space $W^{k,p}(\Omega)$ is the closure of $C_c^\infty(\Omega)$ in this norm, hence is a Banach space. Accounts of $W^{k,p}(\Omega)$ may be found in Gilbarg and Trudinger (1983); Maz'ja (1985); Ziemer (1989).

For $k$ and $p$ given, set

$$q = \frac{np}{n - kp}, \quad 1 \leq p < \frac{n}{k}, \quad 1 \leq k \leq n - 1.$$

By iterating the result $W^{1,p} \hookrightarrow L^{p^*}$, one obtains for $W^{k,p}$ the embedding

$$W^{k,p}(\mathbb{R}^n) \hookrightarrow L^q(\mathbb{R}^n), \quad 1 \leq p < \frac{n}{k}.$$

The corresponding result for $p > \frac{n}{k}$ is

$$W^{k,p}(\mathbb{R}^n) \hookrightarrow BC^{m,\beta}(\mathbb{R}^n), \quad \frac{n}{k} < p < \infty,$$

where the integer $m$ and the number $\beta \in (0, 1]$ are determined by

$$k - \frac{n}{p} = m + \beta.$$

Here $BC^{m,\beta}$ is the set of bounded functions whose derivatives of order $\leq m$ are bounded and continuous, with $m$th order partials Hölder continuous of order $\beta$.

Take $n \geq 3$, $k = 2$, and $1 \leq p < \frac{n}{2}$. Then

$$q = \frac{np}{n - 2p}.$$

If $f \in W^{2,p}(\mathbb{R}^n)$ then $f = cK * (\Delta f)$, where $K$ is the Riesz kernel of order $n-2$. In Chapter 8, when we take up the Hardy–Littlewood–Sobolev inequality, we shall use this representation to obtain the best constant in the inequality

$$\|f\|_q \leq C\|\Delta f\|_p.$$

We shall also discuss there a generalization of Moser's inequality (4.58) due to Adams (1988), which, in particular, gives sharp exponential integrability results for the borderline cases $p = n/k$.

## 4.9 Notes and Comments

The isoperimetric inequality for length and area in the plane goes back to antiquity. See Tikhomirov (1990) for discussion of the supposed work of Queen Dido (c.800 BC) and of the Greek mathematician Eudoxus (324 BC). Good sources for the more recent past and for the present include Osserman (1978), Burago and Zalgaller (1988), Federer (1969), Chavel (2001), Talenti (1993), and the article on isoperimetry in Hazewinkel (1995).

For Hausdorff measure we have mainly followed Evans and Gariepy (1992). Other sources include Ziemer (1989), Mattila (1995) and Folland (1999). The notions of perimeter and of functions of bounded variation on $\mathbb{R}^n$ in the sense presented here grew out of work by Caccioppoli and De Georgi in the 1950s. Good introductions to the theory include Evans and Gariepy (1992), Giusti (1984), and Ziemer (1989). The isoperimetric inequality for perimeter, Theorem 4.10, is due to De Giorgi. Talenti (1993) contains a nice variational proof. Proofs of the isoperimetric inequality for Minkowski content, Theorem 4.16, can be found, for example, in Federer (1969, p. 278), Burago and Zalgaller (1988, p. 84), and Chavel (2001, p. 77). The proofs by Federer and Burago–Zalgaller are based on the Brunn–Minkowski inequality, which we will take up in Chapter 8. Chavel's is based on Steiner symmetrization, which we study in Chapter 6. Corollary 4.13 about Hausdorff measure is from Federer (1969, Theorem 4.5.9 (31)). He attributes it to Sobolev (1938).

The Coarea Formula, Theorem 4.19, in its modern generality was discovered by Federer (1959). According to Federer (1969, p. 207) the Area Formula is "classical," and the proof presented in Federer (1969, p. 243) "follows the spirit" of a proof by Kolmogorov in 1932.

The Sobolev Embedding Theorem is essentially due to Sobolev (1938). Morrey's Embedding Theorem 4.24 is in Morrey (1938). Theorem 4.21, the sharp Sobolev inequality for $p = 1$, and its equivalence to the isoperimeric inequality, are due to Federer and Fleming (1960). As discussed in the text, the sharp Sobolev inequality for $1 < p < n$ is due independently to Rodemich, Aubin and Talenti. The proof presented in this book is a hybrid using both the "classical" method of symmetrization and the recent mass transportation approach of Cordero-Erausquin et al. (2004). That paper contains a number of interesting related results and open problems.

In addition to Evans and Gariepy (1992) and Ziemer (1989), good references for Sobolev spaces include Maz'ja (1985) and Gilbarg and Trudinger (1983). The latter, though, considers only spaces of functions on bounded domains, whereas we are concerned mostly with functions defined in all of $\mathbb{R}^n$.

# 5

# Isoperimetric Inequalities for Physical Quantities

In this chapter we apply the decrease under symmetrization of the $L^2$-Dirichlet integral, proved in Chapter 3, to solve three famous extremal problems of classical mathematical physics: among all bounded open sets in the plane with the same area the disk has smallest principal frequency and largest torsional rigidity, while among all compact sets in $\mathbb{R}^3$ with the same volume the ball has smallest Newtonian capacity.

The principal frequency result had been conjectured by Lord Rayleigh in 1877, the torsional rigidity result by St. Venant in 1856, and the capacity result by Poincaré in 1887. For frequency, the first proofs were given independently by Faber and Krahn in 1923–1925, for torsional rigidity by Pólya in 1948, and for capacity by Szegő in 1930. Moreover, the mathematical problem whose solution gives the torsional result gives also the result that among all domains $\Omega \subset \mathbb{R}^n$ with fixed volume, the mean lifetime of a Brownian motion started at a random point of $\Omega$ and killed when it reaches $\partial\Omega$ is largest when $\Omega$ is a ball. We shall also obtain symmetrization inequalities involving other notions of capacity. Corollary 5.16, for example, is Carleman's result that symmetrization of ring domains in the plane increases the conformal modulus.

To formulate the problems in terms of Dirichlet integrals we shall introduce some standard analytic tools such as weak solutions of Poisson equations and approximation of general open sets by smoothly bounded open sets.

## 5.1 Weak Solutions of $\Delta u = -f$

The Laplace operator $\Delta$ on $\mathbb{R}^n$ is the operator defined by

$$\Delta u = \sum_{i=1}^{n} u_{x_i x_i},$$

where $u$ is some $C^2$ function defined in a subset of $\mathbb{R}^n$. In this section we assume that $\Omega$ is a bounded open subset of $\mathbb{R}^n$. For $u \in C^2(\Omega), v \in C^1(\Omega)$ we have

$$\operatorname{div}(v\nabla u) = \nabla u \cdot \nabla v + v\Delta u.$$

If $v$ is also compactly supported in $\Omega$, then application of the Gauss–Green formula (Theorem 4.3) to the vector field $v\nabla u$ gives

$$\int_\Omega v\,\Delta u\,dx + \int_\Omega \nabla u \cdot \nabla v\,dx = 0.$$

For $f \in C(\Omega)$, it follows that $u \in C^2(\Omega)$ is a solution of $\Delta u = -f$ if and only if

$$\int_\Omega \nabla u \cdot \nabla v\,dx = \int_\Omega f v\,dx, \qquad \forall\, v \in C_c^1(\Omega). \tag{5.1}$$

By approximation, if (5.1) holds for all $v \in C_c^1(\Omega)$ then it still holds for all $v \in W_0^{1,2}(\Omega)$. Moreover, (5.1) makes sense if we assume only that $u \in W^{1,2}(\Omega)$ and that $f \in L^2(\Omega)$. Accordingly, we define $u$ to be a *weak solution* of $\Delta u = -f$ in $\Omega$ if $u \in W^{1,2}(\Omega)$, $f \in L^2(\Omega)$, and (5.1) holds for all $v \in W_0^{1,2}(\Omega)$.

In this chapter we are often interested in solutions $u$ of $\Delta u = -f$ which vanish on $\partial\Omega$. If $u$ is continuous on $\overline{\Omega}$ this means that $u(x) \to 0$ as $x \to \partial\Omega$. The appropriate weak notion is to assume that $u \in W_0^{1,2}(\Omega)$. Thus, we arrive at the following definition.

**Definition 5.1** $u$ is a weak solution of

$$\Delta u = -f \quad \text{in} \quad \Omega, \quad u = 0 \quad \text{on} \quad \partial\Omega,$$

if $u \in W_0^{1,2}(\Omega)$, $f \in L^2(\Omega)$ and (5.1) holds for all $v \in W_0^{1,2}(\Omega)$.

Here is the basic existence and uniqueness theorem for weak solutions in bounded open sets $\Omega$.

**Proposition 5.2** *For each $f \in L^2(\Omega, \mathbb{R})$, there exists a unique function $u \in W_0^{1,2}(\Omega)$ such that $u$ is a weak solution of*

$$\Delta u = -f \quad in \quad \Omega, \quad u = 0 \quad on \quad \partial\Omega.$$

*Proof* Write $H = W_0^{1,2}(\Omega)$. For $u, v \in H$, define

$$B(u, v) = \int_\Omega \nabla u \cdot \nabla v\,dx.$$

The usual inner product on $H$ making it a real Hilbert space is

$$[u, v] \equiv \int_\Omega uv\,dx + B(u, v).$$

Letting $\|\cdot\|_H$ denote the norm associated with $[\cdot,\cdot]$, Schwarz's inequality gives

$$|B(u,v)| \leq \|u\|_H \|v\|_H. \tag{5.2}$$

Suppose that $n \geq 2$. Set $p = \frac{2n}{n+2}$. Then $1 \leq p < 2$, and the Sobolev conjugate exponent $p^* = \frac{np}{n-p}$ equals 2. For $u \in H$, set $u = 0$ in $\mathbb{R}^n \setminus \Omega$. Then the extended $u$ belongs to $W^{1,2}(\mathbb{R}^n)$. By the Sobolev embedding theorem (Theorem 4.23) and the finiteness of $\mathcal{L}^n(\Omega)$, we have

$$\|u\|_{L^2(\Omega)} \leq C_1 \|\nabla u\|_{L^p(\mathbb{R}^n)} = C_1 \|\nabla u\|_{L^p(\Omega)} \leq C_2 B(u,u)^{1/2},$$

where $C_1$ depends only on $n$, and $C_2$ on $n$ and $\mathcal{L}^n(\Omega)$. Thus

$$B(u,u) \geq C_3 \|u\|_H^2, \tag{5.3}$$

where $C_3$ depends only on $n$ and $\mathcal{L}^n(\Omega)$.

It follows from (5.2) and (5.3) that the symmetric bilinear form $B$ is itself an inner product on $H$ and that the pair $(H, B)$ is a Hilbert space. Take $f \in L^2(\Omega)$. From Schwarz's inequality and (5.3), it follows that the linear functional $v \to \int_\Omega f v\,dx$ is bounded on $(H, B)$. By the Riesz Representation Theorem for Hilbert spaces (Evans, 1998, p. 639), there exists a unique $u \in H$ such that for every $v \in H$,

$$B(u,v) = \int_\Omega f v\,dx. \tag{5.4}$$

Thus (5.1) is fulfilled, and Proposition 5.2 is proved for $n \geq 2$. The case $n = 1$ is left to the reader. □

With the situation of Proposition 5.2, define an operator $K\colon L^2(\Omega) \to H$ by $Kf = u$. Then $K$ is linear and $\Delta(Kf) = -f$, in the weak sense, for each $f \in L^2(\Omega)$. Thus, $K$ is a right inverse of $-\Delta$.

For $f \in L^2(\Omega)$ and $u = Kf$, we have

$$B(u,u) = \int_\Omega f u\,dx \leq \|f\|_{L^2(\Omega)} \|u\|_{L^2(\Omega)} \leq C \|f\|_{L^2(\Omega)} B(u,u)^{1/2},$$

where the first equality is from (5.4) and the last inequality is from (5.3). Thus $B(u,u) \leq C\|f\|_{L^2(\Omega)}$. Another application of (5.3) shows that

$$\|u\|_H \leq C \|f\|_{L^2(\Omega)},$$

where $C$ depends only on $n$ and $\mathcal{L}^n(\Omega)$. We have shown that $K$ is a *bounded* operator from $L^2(\Omega)$ into $H = W_0^{1,2}(\Omega)$.

Next, take $f, g \in L^2(\Omega)$. Write $u = Kf$, $v = Kg$. Then

$$\int_\Omega (Kf) g\,dx = \int_\Omega u g\,dx = B(u,v),$$

where we have applied (5.4) with the roles of $u$ and $v$ reversed. Since $B$ is symmetric, it follows that $K$ is *symmetric* when $L^2(\Omega)$ is equipped with its usual inner product. Moreover, $B(u, u) > 0$ unless $\nabla u = 0$ a.e. which, via (5.1), implies that $f = 0$ a.e.. Thus, $\int_\Omega (Kf)f\, dx > 0$, unless $f = 0$ in $L^2(\Omega)$, so that $K$ is a *positive* symmetric operator.

By the definition of $\|\cdot\|_H$ we have $\|u\|_{L^2(\Omega)} \le \|u\|_H$. Thus, the inclusion map $i\colon H \to L^2(\Omega)$ is bounded. The Rellich–Kondrachov Theorem (Evans, 1998, p. 274) asserts that much more is true: the operator $i$ is *compact*. Compactness means that each bounded sequence $\{u_k\}$ in $H$ contains a subsequence whose image under $i$ is norm-convergent in $L^2(\Omega)$. It follows that $i \circ K$ is a compact operator from $L^2(\Omega)$ into $L^2(\Omega)$.

Let us suppress the inclusion $i$ and regard $K$ as a map from $L^2(\Omega)$ into itself. Then the discussion above shows that $K$ is compact, symmetric, and positive. The spectral theory of such operators (see Evans, 1998, pp. 643–645) leads to the following statements. As always in this section, $\Omega$ is assumed to be a bounded open subset of $\mathbb{R}^n$.

**Proposition 5.3**

*(a) There is a countably infinite orthonormal basis $\{u_k\}_{k=1}^\infty$ of $L^2(\Omega)$ consisting of eigenfunctions of $K$.*

*(b) Let $\mu_k$ be the eigenvalue corresponding to $u_k$. Then each $\mu_k$ is positive real, and*

$$\lim_{k\to\infty} \mu_k = 0.$$

Recall that $\mu$ is an eigenvalue of $K$ with corresponding eigenfunction $u$ if $u \not\equiv 0$ and $Ku = \mu u$. A proof of Proposition 5.3 can be gleaned from Theorems 6 and 7 of Evans (1998, pp. 643–645), bearing in mind that $L^2(\Omega)$ is a separable infinite dimensional Hilbert space and that strictly positive symmetric operators can have only positive eigenvalues.

From Proposition 5.3(b), it follows that one can relabel the sequence of eigenvalues so that they lie in descending order:

$$\mu_1 \ge \mu_2 \ge \cdots > 0. \tag{5.5}$$

**Proposition 5.4** *With the situation of Proposition 5.3, the largest eigenvalue $\mu_1$ of $K$ satisfies*

$$\mu_1 = \max\left\{\int_\Omega (Kf)f\, dx\colon \|f\|_{L^2(\Omega)} = 1\right\}.$$

*Proof* For $f \in L^2(\Omega)$ and $\{u_i\}$ an orthonormal basis of eigenfunctions, one has

$$f = \sum_{i=1}^{\infty} a_i u_i,$$

where $a_i = \int_\Omega f u_i \, dx$ and the series converges in $L^2(\Omega)$. Moreover,

$$\sum_{i=1}^{\infty} a_i^2 = \|f\|_{L^2(\Omega)}^2.$$

Thus, if $\|f\|_{L^2(\Omega)} = 1$,

$$\int_\Omega (Kf) f \, dx = \sum_{i=1}^{\infty} \mu_i a_i^2 \le \mu_1 \sum_{i=1}^{\infty} a_i^2 = \mu_1,$$

with equality when $f = u_1$. □

## 5.2 Eigenvalues of the Laplacian

We continue to assume that $\Omega$ is a bounded open subset of $\mathbb{R}^n$. Let $K: L^2(\Omega) \to W_0^{1,2}(\Omega)$ be the right-inverse of $-\Delta$ defined in §5.1. Suppose that $u \in L^2(\Omega)$ is an eigenfunction of $K$ with eigenvalue $\mu$. Then $u \in W_0^{1,2}(\Omega)$ since $K$ maps $L^2(\Omega)$ into $W_0^{1,2}(\Omega)$, and $u$ is a weak solution of

$$u = -\Delta(Ku) = -\mu \Delta u.$$

Set $\lambda = 1/\mu$. Then $u$ is a weak solution of

$$\Delta u = -\lambda u \quad \text{in} \quad \Omega, \qquad u = 0 \quad \text{on} \quad \partial\Omega. \tag{5.6}$$

Solutions $u \not\equiv 0$ of (5.6) are said to be eigenfunctions of $-\Delta$ in $\Omega$ with corresponding eigenvalue $\lambda$. Thus, $u$ is an eigenfunction of $-\Delta$ if and only if $u$ is an eigenfunction of $K$, and the corresponding eigenvalues are reciprocals of each other.

A priori, eigenfunctions of $-\Delta$ belong only to $W_0^{1,2}(\Omega)$. Via elliptic regularity, one can show that they in fact belong to $C^\infty(\Omega)$. Moreover, if $\partial\Omega$ is sufficiently smooth, then eigenfunctions extend continuously to $\overline{\Omega}$ and solve the boundary value problem (5.6) in the classical sense. See, for example, Evans (1998, p. 335) or Gilbarg and Trudinger (1983).

The boundary condition $u = 0$ on $\partial\Omega$ is called the Dirichlet boundary condition, and the eigenfunctions and eigenvalues in (5.6) are called Dirichlet eigenfunctions and eigenvalues. The *Neumann* boundary value problem

$$\Delta u = -\lambda u \quad \text{in} \quad \Omega, \qquad \nabla u \cdot \nu = 0 \quad \text{on} \quad \partial\Omega,$$

is also of considerable interest, the corresponding eigenvalues and eigenfunctions being prefixed by "Neumann." Here $\nu$ is the unit exterior normal vector field on $\partial\Omega$, and one assumes that $\partial\Omega$ is sufficiently regular that $\nu$ can be suitably defined. In this book we shall consider only the Dirichlet boundary condition, and thus for us "eigenvalue" and "eigenfunction" are understood to mean Dirichlet eigenvalue or eigenfunction.

List the eigenvalues $\mu_k$ of $K$ in decreasing order, as in (5.5). Define

$$\lambda_k = \mu_k^{-1}, \quad k = 1, 2, \dots$$

Then, by Proposition 5.3,

$$0 < \lambda_1 \leq \lambda_2 \leq \cdots \tag{5.7}$$

and

$$\lim_{k\to\infty} \lambda_k = \infty.$$

In the list (5.7) each eigenvalue of $-\Delta$ is listed according to its multiplicity. That is, a number $\lambda$ appears in the list $m$ times if the nullspace of $\Delta + \lambda$ has dimension $m$. The next proposition asserts, among other things, that $\lambda_1$ is *simple,* that is, has multiplicity $m = 1$. Thus $\lambda_1 < \lambda_2$. We call $\lambda_1$ the *principal eigenvalue* of $-\Delta$ on $\Omega$.

**Proposition 5.5** *For a bounded open subset $\Omega$ of $\mathbb{R}^n$ we have*

$$\lambda_1 = \min\left\{\int_\Omega |\nabla u|^2\,dx \colon u \in W_0^{1,2}(\Omega),\ \|u\|_{L^2(\Omega)} = 1\right\}. \tag{5.8}$$

*The minimum is attained by each eigenfunction $u$ with eigenvalue $\lambda_1$ and $\|u\|_{L^2(\Omega)} = 1$. Moreover, if $\Omega$ is connected then each eigenfunction for $\lambda_1$ is either positive in $\Omega$ or negative in $\Omega$, and any two eigenfunctions for $\lambda_1$ are multiples of each other.*

Formula (5.8) is called the *variational principle*, or *Rayleigh's formula,* for the principal eigenvalue. The formula can be restated as

$$\lambda_1 = \min \frac{\int_\Omega |\nabla u|^2\,dx}{\int_\Omega u^2\,dx}, \tag{5.9}$$

where the minimum is over all $u \in W_0^{1,2}(\Omega)$ with $u \not\equiv 0$. The ratio in this formula is called the *Rayleigh quotient* of $u$.

Proposition 5.5 is taken from Theorem 2 on p. 336 of Evans (1998), to which we refer for proofs of the statements following "Moreover."

*Proof of* (5.8) *and its equality statement* Let $u_1 \in W_0^{1,2}(\Omega)$ be an eigenfunction for $\lambda_1$ with $\|u_1\|_{L^2(\Omega)} = 1$. Then $u_1 \in C^\infty(\Omega)$. By the Gauss–Green theorem,

$$\int_\Omega |\nabla u_1|^2\, dx = -\int_\Omega u_1 \Delta u_1\, dx = \lambda_1 \int_\Omega u_1^2\, dx = \lambda_1,$$

which confirms the equality statement in (5.8).

To prove the inequality statement in (5.8) we may assume that $u \in C_c^2(\Omega)$. Define $f = -\Delta u$. Then $f \in C_c(\Omega)$. Since $K$ is the right-inverse of $\Delta$ we also have $f = -\Delta(Kf)$. The uniqueness assertion in Proposition 5.2 implies that $Kf = u$.

The positive symmetric operator $K\colon L^2(\Omega) \to L^2(\Omega)$ has a positive symmetric square root $K^{1/2}$. Write $g = K^{1/2}f$. Then

$$\int_\Omega |\nabla u|^2\, dx = -\int_\Omega u \Delta u\, dx = \int_\Omega uf\, dx = \int_\Omega (Kf)f\, dx = \int_\Omega g^2\, dx,$$

and

$$\int_\Omega u^2\, dx = \int_\Omega (Kf)^2\, dx = \int_\Omega (Kg)g\, dx.$$

From these two equalities and Proposition 5.4 with $w = g/\|g\|_{L^2}$, we obtain when $\|u\|_{L^2(\Omega)} = 1$ that

$$\int_\Omega |\nabla u|^2\, dx = \frac{\int_\Omega g^2\, dx}{\int_\Omega (Kg)g\, dx} \geq \lambda_1.$$

This is the inequality statement for (5.8). □

## 5.3 Symmetrization Decreases the Principal Eigenvalue

We continue to let $\Omega$ denote a bounded open subset of $\mathbb{R}^n$, and will denote by $\lambda_1(\Omega)$ the principal eigenvalue of $-\Delta$ on $\Omega$. When $n = 2$ we can interpret $\Omega$ as a membrane or drumhead, and then $\sqrt{\lambda_1(\Omega)}$ represents the smallest natural frequency at which the drum can oscillate. Accordingly, $\lambda_1(\Omega)$ has been called the principal frequency, the fundamental frequency, the bass note, or the principal tone of the drum. We shall continue to use this language in all dimensions.

What does the size or shape of $\Omega$ tell us about $\lambda_1(\Omega)$? To begin with, let us consider dilations of $\Omega$. If $u$ is an eigenfunction on $\Omega$ with eigenvalue $\lambda$ and $\rho > 0$ is a positive constant, then the function $u_\rho(x) = u(x/\rho)$ is an eigenfunction for $-\Delta$ on the domain $\rho\Omega$ with eigenvalue $\rho^{-2}\lambda$. In particular, for the principal eigenvalues we have

$$\lambda_1(\rho\Omega) = \rho^{-2}\lambda_1(\Omega).$$

If $\rho$ becomes large then $\lambda_1(\rho\Omega)$ becomes small. This suggests that large domains tend to have small principal eigenvalues.

For another example, take positive numbers $a$ and $b$ and let $\Omega_{a,b}$ be the rectangle $(0, a\pi) \times (0, b\pi)$ in $\mathbb{R}^2$. Then

$$u(x_1, x_2) = \sin\frac{x_1}{a}\sin\frac{x_2}{b}$$

is a positive eigenfunction of $-\Delta$ on $\Omega_{a,b}$ with eigenvalue $a^{-2} + b^{-2}$. Since $u > 0$ in $\Omega_{a,b}$ and eigenfunctions corresponding to distinct eigenvalues are orthogonal, $a^{-2} + b^{-2}$ must be the principal eigenvalue of $\Omega_{a,b}$. If we specialize to $b = 1/a$, then $\mathcal{L}^2(\Omega_{a,b}) = \pi^2$ for all $a > 0$ while

$$\lambda_1(\Omega_{a,b}) = a^2 + a^{-2},$$

which is minimal for $a = 1$. So if two rectangles in $\mathbb{R}^2$ have the same area, the longer thinner one has the larger principal eigenvalue. Put another way, the more symmetric rectangle has the smaller principal eigenvalue.

Among all bounded open $\Omega \subset \mathbb{R}^n$ with the same Lebesgue measure, which ones have smallest $\lambda_1$? The rectangle example suggests that the minimizing $\Omega$, if there is one, should be as symmetric as possible, and thus is quite likely to be a ball of the prescribed measure. For $n = 2$ this result was conjectured by Lord Rayleigh in 1877 (see Rayleigh, 1945, pp. 339–340), who wrote:

> If the area of a membrane be given, there must evidently be some form of boundary for which the pitch (of the principal tone) is the gravest possible, and this form can be no other than the circle.

We shall also refer to the associated $n$-dimensional statement as Rayleigh's Conjecture.

**Conjecture** (Rayleigh) *For each bounded open set $\Omega \subset \mathbb{R}^n$ we have*

$$\lambda_1(\Omega) \geq \lambda_1(\Omega^\#). \tag{5.10}$$

Rayleigh's Conjecture was independently proved by Faber (1923) and Krahn (1925) in 2 dimensions, and by Krahn in all dimensions (see Lumiste and Peetre, 1994, pp. 139–174).

**Theorem 5.6** (Faber–Krahn) *Rayleigh's Conjecture* (5.10) *is true.*

*Proof* By Proposition 5.5 there exists a positive function $u \in W_0^{1,2}(\Omega)$ with $\|u\|_{L^2(\Omega)} = 1$ such that

$$\lambda_1(\Omega) = \int_\Omega |\nabla u|^2 \, dx. \tag{5.11}$$

By Corollary 3.22, $u^\# \in W_0^{1,2}(\Omega^\#)$ and

$$\int_{\Omega^\#} |\nabla u^\#|^2 \, dx \le \int_\Omega |\nabla u|^2 \, dx. \tag{5.12}$$

Since $u$ and $u^\#$ have the same distribution, we have $\|u^\#\|_{L^2(\Omega^\#)} = \|u\|_{L^2(\Omega)} = 1$, so by Proposition 5.5,

$$\lambda_1(\Omega^\#) \le \int_{\Omega^\#} |\nabla u^\#|^2 \, dx.$$

Inequality (5.10) now follows from (5.12) and (5.11). □

A different rearrangement proof of the Faber–Krahn theorem will be given in Example 10.11, based on an elliptic comparison theorem.

Next, we shall compute the principal eigenvalue and associated eigenfunctions for the unit ball $\mathbb{B}^n = \{x \in \mathbb{R}^n \colon |x| < 1\}$. Let $u$ be an eigenfunction for $\lambda_1(\mathbb{B}^n)$. By Proposition 5.5, $u$ is either everywhere positive or everywhere negative in $\mathbb{B}^n$. Let us assume $u$ is positive. Then, since symmetrization decreases Dirichlet integrals, $u^\#$ is also a minimizer for (5.11) and hence is also an eigenfunction for $\lambda_1(\mathbb{B}^n)$. Since any two eigenfunctions for $\lambda_1(\mathbb{B}^n)$ are multiples of each other, and $u$ and $u^\#$ have the same distribution, we must have $u = u^\#$. Thus, $u$ is positive and symmetric decreasing on $\mathbb{B}^n$. To normalize, let us take $u(0) = 1$, instead of normalizing $\|u\|_{L^2(\Omega)}$.

Committing a slight abuse of notation, we write $u(r) = u(x)$ when $r = |x|$. Then $u$ is a decreasing function on $[0, 1]$. By elliptic regularity theory, $u$ is continuous on $[0, 1]$ with $u(1) = 0$, and $u \in C^\infty(0, 1)$. By the chain rule,

$$\Delta u = u'' + (n-1)r^{-1}u',$$

where the prime denotes differentiation with respect to $r$.

By equating coefficients, one verifies that the singular boundary value problem

$$U''(r) + (n-1)r^{-1}U'(r) = -U(r) \quad \text{on} \quad (0, \infty), \qquad U(0) = 1, \tag{5.13}$$

has solution

$$U(r) = \sum_{k=0}^{\infty} \frac{(-1)^k 2^{-2k}\Gamma(n/2)}{k!\,\Gamma(k + \frac{n}{2})} r^{2k}.$$

The Bessel function $J_\nu$ of the first kind of order $\nu \in \mathbb{R}$ is defined by the series

$$J_\nu(r) = r^\nu \sum_{k=0}^{\infty} \frac{(-1)^k 2^{-2k} 2^{-\nu}}{k!\,\Gamma(k+\nu+1)} r^{2k}.$$

Take

$$\nu = \frac{n}{2} - 1, \quad n \geq 1.$$

Then the solution $U$ of (5.13) can be written as

$$U(r) = 2^\nu \Gamma(\nu+1) r^{-\nu} J_\nu(r).$$

For $\nu \geq 0$ the zeros of $J_\nu$ are symmetrically placed on $\mathbb{R}$ and there are infinitely many of them. See, e.g., Lebedev (1972). Denote by $j_\nu$ the smallest positive zero of $J_\nu$. Then the function $v(r) = U(j_\nu r)$ satisfies $v(0) = 1$, $v(r) > 0$ on $[0, 1)$, $v(1) = 0$ and, using (5.13),

$$v''(r) + (n-1)r^{-1}v'(r) = -j_\nu^2 v(r).$$

Thus, $v(x)$ is a positive eigenfunction of $-\Delta$ on $\mathbb{B}^n$ with $v = 0$ on $\partial\mathbb{B}^n$ and $v(0) = 1$. We conclude that $v = u$, and that $\lambda_1(\mathbb{B}^n) = j_\nu^2$. These facts, and a bit more, are summarized in the following proposition.

**Proposition 5.7** *Let $\nu = \frac{n}{2} - 1$. Then*

*(a)* $\lambda_1(\mathbb{B}^n) = j_\nu^2$.
*(b)* $u(x) = 2^\nu\Gamma(\nu+1)(j_\nu|x|)^{-\nu}J_\nu(j_\nu|x|)$ *is the eigenfunction associated with* $\lambda_1(\mathbb{B}^n)$*, normalized by* $u(0) = 1$.
*(c) If $B$ is an open ball in $\mathbb{R}^n$ of radius $R$, then*

$$\lambda_1(B) = R^{-2} j_\nu^2 = \alpha_n^{2/n} (\mathcal{L}^n(B))^{-2/n} j_\nu^2.$$

Taking $B = \Omega^\#$, the Faber–Krahn Theorem can be restated as follows.

**Corollary 5.8** *For each bounded open set $\Omega \subset \mathbb{R}^n$ we have*

$$\lambda_1(\Omega) \geq \left(\frac{\mathcal{L}^n(\Omega)}{\alpha_n}\right)^{-2/n} j_{\frac{n}{2}-1}^2,$$

*with equality when $\Omega$ is a ball.*

Returning to unit balls, we note that $J_{-1/2}(r) = (2/\pi r)^{1/2}\cos r$ and that $J_{1/2}(r) = (2/\pi r)^{1/2}\sin r$. Thus, for $n = 1$ we have $\lambda_1(\mathbb{B}^1) = \pi^2/4, u(x) = \cos(\frac{\pi}{2}x)$, and for $n = 3$, $\lambda_1(\mathbb{B}^3) = \pi^2, u(x) = (\pi x)^{-1}\sin(\pi x)$.

For $n = 2$,

$$j_0 \approx 2.4, \quad \lambda_1(\mathbb{B}^2) \approx 2.4^2, \quad u(x) = J_0(j_0|x|).$$

In these formulas, $u$ continues to denote the principal eigenfunction of the unit ball with $u(0) = 1$.

In addition to the Faber–Krahn Theorem, there are a number of other "isoperimetric inequalities" involving eigenvalues. For example, in 1955 Payne, Pólya and Weinberger (1955) conjectured that for bounded open sets in $\mathbb{R}^n$ the ratio

$$\lambda_2(\Omega)/\lambda_1(\Omega)$$

is maximal when $\Omega$ is a ball. This conjecture was proved by Ashbaugh and Benguria (1992). Their article from 1994 surveys related work and lists many open problems.

If we think of $\Omega$ as representing a "clamped plate" rather than a membrane, then the fundamental mode of vibration is determined by the smallest eigenvalue $\Lambda_1(\Omega)$ and corresponding eigenfunction $u$ of the problem

$$\Delta\Delta u = \Lambda u, \quad \text{in} \quad \Omega, \quad u = \nabla u = 0 \quad \text{on} \quad \partial\Omega.$$

The partial differential equation is understood to hold in the weak sense, and the boundary conditions to mean that $u \in W_0^{2,2}(\Omega)$. The variational characterization of $\Lambda_1$ is

$$\Lambda_1(\Omega) = \min\left\{\int_\Omega |\Delta u|^2\, dx \colon u \in W_0^{2,2}(\Omega),\ \|u\|_{L^2(\Omega)} = 1\right\}.$$

The minimum is achieved by eigenfunctions corresponding to $\Lambda_1(\Omega)$.

Rayleigh also conjectured that among all $\Omega \subset \mathbb{R}^2$ with fixed Lebesgue measure, $\Lambda_1(\Omega)$ is minimal when $\Omega$ is a ball (see Rayleigh, 1945, p. 382). The conjecture was finally proved by Nadirashvili (1993). Ashbaugh and Benguria (1995) proved the corresponding result when $n = 3$. For $n \geq 4$, the analogue of Rayleigh's plate conjecture remains open. See Ashbaugh, Benguria, and Laugesen (1997) for an excellent survey of pertinent results and open problems, and Chasman and Langford (2016) for the vibrating plate problem in Gauss space.

## 5.4 Domain Approximation Lemmas

Let $\Omega$ be a domain (connected open set) in $\mathbb{R}^n$. In the next section, and in some later chapters, we will need to approximate general domains by domains with smooth boundary. In this section we establish some results along these lines.

Recall that for $x \in \mathbb{R}^n$ and $E \subset \mathbb{R}^n$, $E^c \equiv \mathbb{R}^n \setminus E$ and $d(x, E) = \inf\{|x - y| \colon y \in E\}$. The notation $E_1 \Subset E_2$ means that $\overline{E_1}$ is a compact subset of $E_2$. For $\epsilon > 0$, set

$$\Omega_\epsilon = \{x \in \Omega : d(x, \partial\Omega) > \epsilon\}.$$

Then $\Omega_\epsilon$ is an open subset of $\Omega$, but is not necessarily connected.

**Lemma 5.9** (Domain Approximation Lemma 1) *Let $\Omega$ be a bounded domain in $\mathbb{R}^n$. Given $\epsilon > 0$ there exists a $C^\infty$ domain $\Omega'$ such that*

$$\Omega_\epsilon \Subset \Omega' \Subset \Omega.$$

*Proof* We may assume that $\Omega_\epsilon$ is nonempty. Fix $x_0 \in \Omega_\epsilon$. For $\delta > 0$, let $D_\delta$ be the set of all points $x \in \Omega$ for which there exists a curve $\gamma \in C([0,1], \Omega)$ such that $\gamma(0) = x_0$, $\gamma(1) = x$, and $\min_{t \in [0,1]} d(\gamma(t), \partial\Omega) > \delta$. An easy argument shows that each $D_\delta$ is open and connected. The connectedness of $\Omega$ implies that $\bigcup_\delta D_\delta = \Omega$. Also, $\Omega_\epsilon \Subset \Omega$, since $\Omega$ is bounded. The sets $D_\delta$ increase as $\delta$ decreases, and so there exists $\delta$ such that

$$\overline{\Omega_\epsilon} \subset D_\delta.$$

Furthermore, $D_\delta \subset \Omega_\delta$, so that $D_\delta \Subset \Omega$. This $\delta$ is fixed for the rest of the proof.

Let $K : \mathbb{R}^n \to \mathbb{R}^+$ be a smooth nonnegative bump function as in §3.4, and for $\eta > 0$, let $K_\eta(x) = \eta^{-n} K(\eta^{-1} x)$. Let $f = K_\eta * \chi_{\Omega_{\delta/2}}$. Then $f \in C^\infty(\mathbb{R}^n, [0,1])$. Since $D_\delta \Subset \Omega_{\delta/2} \Subset \Omega$, we can choose $\eta$ small enough so that $f = 1$ on $D_\delta$ and $f$ is compactly supported in $\Omega$.

By Sard's Theorem (see Smith, 1983, p. 342), the set of critical values of $f$ has $\mathcal{L}^1$-measure zero. Fix a regular value $t \in (0,1)$. The set $(f > t)$ is an open set compactly contained in $\Omega$. The implicit function theorem implies that the boundary of each component of $f > t$ has $C^\infty$ boundary, as defined in Evans (1998, p. 626), and that the closures of the components are disjoint. Let $\Omega'$ be the component of $f > t$ which contains $x_0$. Since $x_0 \in \Omega_\epsilon \subset D_\delta \subset (f > t)$ and $D_\delta$ is connected, we must have $D_\delta \subset \Omega'$. This $\Omega'$ satisfies the requirements of Lemma 5.9. □

The next Domain Approximation Lemma will be about unbounded domains $\Omega$. For such domains, given $\epsilon > 0$ and $R \in (0, \infty)$, set

$$\Omega_\epsilon(R) = \Omega_\epsilon \cap \mathbb{B}^n(R),$$

where $\Omega_\epsilon$ has the same meaning as above.

**Lemma 5.10** (Domain Approximation Lemma 2) *Let $\Omega$ be an unbounded domain in $\mathbb{R}^n$. Given $\epsilon > 0$ and $R \in (0, \infty)$, there exists a $C^\infty$ domain $\Omega'$ such that*

$$\Omega_\epsilon(R) \Subset \Omega' \Subset \Omega.$$

*Proof* Assuming that $\Omega_\epsilon(R)$ is nonempty, take $x_0 \in \Omega_\epsilon(R)$. For positive numbers $\delta$ and $S$, let $D_\delta(S)$ be the set of all points $x$ in $\Omega$ for which there exists a path $\gamma \in C([0,1],\Omega)$ such that $\gamma(0) = x_0$, $\gamma(1) = x$, $\min_{t\in[0,1]} d(\gamma(t), \partial\Omega) > \delta$ and $\max_{t\in[0,1]} |\gamma(t)| < S$. The $D_\delta(S)$ are open, connected, compactly contained in $\Omega$ and increase when $\delta$ decreases and $S$ increases. The union over $\delta$ and $S$ of all the $D_\delta(S)$ equals $\Omega$. The proof that some $D_\delta(S)$ is contained in suitable $\Omega'$ proceeds as in Lemma 5.9. □

By repeated application of Lemma 5.9 or 5.10, we obtain the following result, which is valid for all domains $\Omega$, bounded or unbounded.

**Lemma 5.11** (Domain Approximation Lemma 3) *Let $\Omega$ be a domain in $\mathbb{R}^n$. There exists a sequence of domains*

$$\Omega^{(1)} \Subset \Omega^{(2)} \Subset \cdots$$

*such that each $\Omega^{(j)}$ is $C^\infty$, and*

$$\bigcup_{j=1}^{\infty} \Omega^{(j)} = \Omega.$$

Such a sequence is sometimes called a smooth or a regular exhaustion of $\Omega$.

If $\Omega$ is a disconnected open set in $\mathbb{R}^n$, then by regularly exhausting each component of $\Omega$ and forming appropriate unions we can construct a regular exhaustion of $\Omega$. Of course, the sets $\Omega^{(j)}$ in this exhaustion will eventually become disconnected.

## 5.5 Symmetrization Decreases Newtonian Capacity

Let $K$ be a nonempty compact subset in $\mathbb{R}^3$. For $\mu \in M^+(K)$, the set of nonnegative Borel measures on $K$, define

$$V(\mu) = \int_{K\times K} |x-y|^{-1}\, d\mu(x)\, d\mu(y), \tag{5.14}$$

$$V(K) = \inf\{V(\mu) \colon \mu \in M^+(K) \colon \mu(K) = 1\}, \tag{5.15}$$

$$\mathrm{Cap}(K) = V(K)^{-1}. \tag{5.16}$$

When $K$ is the empty set we define $V(K) = \infty$ and $\mathrm{Cap}(K) = 0$.

The quantity $V(\mu)$ is called the *Newtonian energy* of $\mu$, $V(K)$ the *Newtonian energy* of $K$, and Cap($K$) the *Newtonian capacity* of $K$. The adjectives *electrostatic* or *gravitational* are sometimes used instead of Newtonian. In this section we will not consider any other energies or capacities, and so the adjective Newtonian will be dropped. References for this section include Lieb and Loss (1997), Hayman and Kennedy (1976), and Landkof (1972). Because of different normalizations, our value of Cap($K$) differs from Landkof's by a factor of $\pi$. The potential kernels in Hayman and Kennedy (1976) are the negative of ours. Kellogg (1967) discusses some of the physical background.

What is "energetic" about $V$, and what does the "capacity" of $K$ have to do with the ability of $K$ to hold something? To find out, we shall take a quick and informal detour through electrostatics. According to Coulomb's inverse-square law, electrical charges $q_1$ and $q_2$ at points $x_1$ and $x_2$ of $\mathbb{R}^3$ repel each other with a force of magnitude $q_1 q_2 |x_1 - x_2|^{-2}$. Here $q_1$ and $q_2$ can be either positive or negative, and if they have opposite sign the force is understood to be attractive. So if $q$ is a charge at a fixed point $a \in \mathbb{R}^3$, then the force exerted by $q$ on a unit positive charge located at $x$ is given by the vector field

$$F(x) = q\frac{x-a}{|x-a|^3}.$$

Set $u(x) = q|x-a|^{-1}$. Then $\nabla u = -F$, and we call $u$ the potential function of the charge $q$ at the point $a$. Now

$$u(x) = \int_\gamma F \cdot dx$$

where the line integral is taken over any path $\gamma$ from $\infty$ to $x$ not passing through $a$. Thus, $u(x)$ is the work required to bring a positive unit charge from $\infty$ to $x$ when the charge $q$ is fixed at $a$. Typically, $q$ is measured in coulombs and $u$ in volts.

From now on, we shall mostly consider only positive charges. Let $K$ be a nonempty compact subset of $\mathbb{R}^3$, which we imagine to be a perfect conductor with the rest of $\mathbb{R}^3$ being a vacuum. Let $\mu \in M^+(K)$ and $V(\mu)$ be defined by (5.14). Then for Borel sets $E \subset K$, $\mu(E)$ represents the amount of charge in $E$. If $\mu$ is a continuous measure, i.e. has a continuous charge density, then $V(\mu)$ is a limit of sums of the form

$$\sum \mu(K_i)\mu(K_j)|x_i - x_j|^{-1} = 2\sum_{1\le i<j\le N} \mu(K_i)\mu(K_j)|x_i - x_j|^{-1}$$

where the $K_i$ are disjoint Borel sets with union $K$, $x_i \in K_i$, and the first sum is over all pairs of indices $(i,j)$ with $1 \le i \le N$, $1 \le j \le N$, and $i \ne j$. The second sum is twice the total work required to move charges $q_1, \dots, q_N$ with strengths $q_i = \mu(K_i)$ from $\infty$ to the points $x_i$. Thus, $V(\mu)$ may be interpreted as twice the work required to distribute a unit point charge at $\infty$ onto the set $K$

so that each Borel subset $E$ of $K$ has charge $\mu(E)$. This is one reason we call $V(\mu)$ the energy of $\mu$.

By similar reasoning, one sees that the force $F(x)$ on a positive unit charge at $x \in \mathbb{R}^3$ exerted by the charge distribution $\mu$ is given formally by

$$F(x) = \int_K \frac{x-y}{|x-y|^3}\, d\mu(x).$$

The Newtonian potential $u$ of $\mu$ is defined on $\mathbb{R}^3$ by

$$u(x) = \int_K |x-y|^{-1}\, d\mu(y).$$

Then $\nabla u = -F$, and $u(x)$ is the work required to bring a unit positive charge from $\infty$ to $x$ when the charge distribution $\mu$ is fixed on $K$.

The energy $V(K)$ is defined in (5.15) to be the infimum of all energies $V(\mu)$ as $\mu$ runs over nonnegative charge distributions on $K$ with total charge one. Frostman (1935) proved that if $V(K) < \infty$ then in fact a minimizing $\mu$ exists and is unique. Moreover, if $K$ is a nice set, then the potential $u$ of the minimizing $\mu$ is constant on $K$. It follows that the electric force field $F$ created by $\mu$ is zero in the interior of $K$, and that the tangential component of $F$ on $\partial K$ is also zero. If the charges are all confined to $K$ and if the system of charges reaches the distribution $\mu$, then the charges will move no more: they have reached equilibrium. Accordingly, the energy minimizing $\mu$ is called the *equilibrium distribution* of $K$. The energy $V(K)$ of the set $K$ equals the energy of its equilibrium distribution.

We continue to assume that $K$ is a nice set with $V(K) < \infty$, and will denote now the equilibrium measure of $K$ by $\mu_K$. Let $\nu$ be any nonnegative Borel measure on $K$ with $\nu(K) > 0$, and denote its Newtonian potential by $u_\nu$. Then

$$V(\nu) = \int_K u_\nu\, d\nu \le \nu(K)\|u_\nu\|_K,$$

where $\|u\|_K \equiv \sup_K u$. Since $\nu/\nu(K)$ is a probability measure, the definition of $V(K)$ gives

$$V(K)[\nu(K)]^2 \le V(\nu).$$

Combining the two inequalities, we obtain

$$\nu(K)/\|u_\nu\|_K \le 1/V(K).$$

By definition (5.16), the right-hand side equals $\operatorname{Cap}(K)$. Moreover, the equilibrium potential $u_K \equiv u_{\mu_K}$ takes the constant value $V(K)$ on $K$, so that equality holds when $\nu = \mu_K$. Thus

$$\operatorname{Cap}(K) = \sup\{\nu(K)/\|u_\nu\|_K \colon \nu \in M^+(K), \nu(K) > 0\}.$$

The sup need be taken only over $\nu$ with $\|u_\nu\|_K = 1$, or over $\nu$ with $\|u_\nu\|_K \le 1$. We conclude that

Cap$(K)$ *equals the maximum amount of charge $K$ can hold while keeping the voltage at most* 1 *everywhere on $K$.*

In electrostatics, capacity is also called *capacitance*.

Let us return now to mathematics, and discuss a symmetrization problem for capacity. Poincaré (1887) asserted that among all compact $K \subset \mathbb{R}^3$ with the same volume, Cap$(K)$ is smallest when $K$ is a ball, and offered a partial proof. Another partial proof was given by Krahn (1925). Szegő (1930) gave the first complete proof.

**Theorem 5.12** (Szegő) *For each compact $K \subset \mathbb{R}^3$, we have*

$$\operatorname{Cap}(K) \ge \operatorname{Cap}(\overline{K^\#}). \tag{5.17}$$

Recall that for measurable sets $E \subset \mathbb{R}^n$, $E^\#$ is the *open* ball centered at the origin with the same Lebesgue measure as $E$. We have defined the capacity only for compact sets. That is why we take the closure of $K^\#$ in (5.17).

If $B$ is a closed ball in $\mathbb{R}^3$ of radius $R$, then $\operatorname{Cap}(B) = R$, by Landkof (1972, p. 163) or Lieb and Loss (1997, p. 164), and so another way to state Szegő's Theorem is

$$\operatorname{Cap}(K) \ge (\alpha_3^{-1}\mathcal{L}^3(K))^{1/3},$$

where $\alpha_3 = \frac{4\pi}{3}$ is the volume of the unit ball in $\mathbb{R}^3$.

To prove Szegő's Theorem, we need a variational characterization of Cap$(K)$ in terms of Dirichlet integrals. Let

$$\mathcal{A}(K) = \{v \in \operatorname{Lip}(\mathbb{R}^3)\colon\ 0 \le v \le 1 \text{ in } \mathbb{R}^3, \quad v = 1 \text{ on } K, \quad \lim_{x\to\infty} v(x) = 0\}.$$

**Proposition 5.13** *For compact $K \subset \mathbb{R}^3$ we have*

$$\operatorname{Cap}(K) = \inf\left\{\frac{1}{4\pi}\int_{\mathbb{R}^3} |\nabla v|^2\,dx\colon\ v \in \mathcal{A}(K)\right\}. \tag{5.18}$$

If $v \in \mathcal{A}(K)$, then $v^\# \in \mathcal{A}(\overline{K^\#})$. Thus, once we have proved Proposition 5.13, Theorem 5.12 is an immediate consequence of the fact that symmetrization decreases Dirichlet integrals (Theorem 3.6 or 3.11).

*Proof* The complement of $K$ in $\mathbb{R}^3$ is the disjoint union of countably many connected open sets, exactly one of which, call it $\Omega_0$, is unbounded. Let $K_0 = \mathbb{R}^3 \setminus \Omega_0$. Then $K_0$ equals $K$ with its holes filled in, and $\partial K_0 \subset K \subset K_0$. By Landkof (1972, p. 162), we have

$$\operatorname{Cap}(K) = \operatorname{Cap}(K_0) = \operatorname{Cap}(\partial K_0). \tag{5.19}$$

Denote by $\mathcal{D}(K)$ the right-hand side of (5.18). If $v \in \mathcal{A}(K)$, define $v_0$ by $v_0 = v$ on $\Omega_0$, $v_0 = 1$ on $K_0$. One easily checks that $v_0 \in \mathcal{A}(K_0)$ and that $\int_{\mathbb{R}^3} |\nabla v_0|^2 \, dx \le \int_{\mathbb{R}^3} |\nabla v|^2 \, dx$. Thus, $\mathcal{D}(K_0) \le \mathcal{D}(K)$. The opposite inequality is trivial, since $K \subset K_0$. Using also (5.19), we see that each side of (5.18) is unchanged when $K$ is replaced by $K_0$. Thus it suffices to prove Proposition 5.13 for compact $K$ for which $\Omega \equiv \mathbb{R}^3 \setminus K$ is connected. We may further assume that $\mathrm{Cap}(K) > 0$, since if $\mathrm{Cap}(K) = 0$ we can express $K$ as a decreasing intersection of sets with positive capacity, and then use continuity properties of Cap and $\mathcal{D}$ as in (5.24) and (5.25) below. Thus, for the rest of the proof of Proposition 5.13, we assume that:

**Assumption A1.** $\Omega \equiv \mathbb{R}^3 \setminus K$ is connected, and $\mathrm{Cap}(K) > 0$.

By Frostman's Theorem (see Hayman and Kennedy, 1976, p. 235), the infimum of $V(\mu)$ in (5.15) is attained by a unique probability measure $\mu_K$ on $K$ called the equilibrium measure or equilibrium distribution. The function

$$u_K(x) = \int_{\mathbb{R}^3} |x-y|^{-1} \, d\mu_K(y), \quad x \in \mathbb{R}^3,$$

is called the *equilibrium potential* of $K$. It is harmonic in $\Omega$, and satisfies $0 \le u_K \le V(K)$ in $\mathbb{R}^3$ with $u_K = V(K)$ on $K$ except for a set of Newtonian capacity zero. Write, for simplicity, $u = u_K$.

In addition to Assumption A1, let us assume temporarily that $\partial\Omega$ is $C^\infty$. Choose $R$ large enough that $K \subset \mathbb{B}^3(R)$, and write $\Omega(R) = \Omega \cap \mathbb{B}^3(R)$. In this case each point of $\partial\Omega(R)$ is regular for the Dirichlet problem (Gilbarg and Trudinger, 1983, pp. 24–27), and Theorem 6.14 of Gilbarg and Trudinger (1983, p. 107) shows that $u \in C^2(\overline{\Omega(R)})$.

The integral defining $u$ also shows $\nabla u$ is bounded in $|x| \ge R$, and that

$$u(x) = R^{-1} + O(R^{-2}), \qquad \frac{\partial u}{\partial r} = -R^{-2} + O(R^{-3}), \tag{5.20}$$

uniformly for $|x| = R$ with $R$ large. These facts imply that $u \in \mathrm{Lip}(\overline{\Omega})$. Also $u \equiv V(K)$ on $K$, and so it follows that $u \in \mathrm{Lip}(\mathbb{R}^3)$.

The Gauss–Green formula (4.8), applied in $\Omega(R)$ to the vector field $u\nabla u$, along with harmonicity of $u$ and (5.20), gives

$$\begin{aligned}\int_{\mathbb{B}^3(R)} |\nabla u|^2 \, dx &= \int_{\Omega(R)} |\nabla u|^2 \, dx \\ &= -\int_{\Omega(R)} u \Delta u \, dx + \int_{\partial\Omega} u \frac{\partial u}{\partial \nu} \, |dx| + \int_{|x|=R} u \frac{\partial u}{\partial r} \, |dx| \\ &= V(K) \int_{\partial\Omega} \frac{\partial u}{\partial \nu} \, |dx| + O(R^{-1}).\end{aligned}$$

Here $|dx|$ denotes the Hausdorff measure $\mathcal{H}^2$ and $\nu$ the outer unit normal. Letting $R \to \infty$, we deduce

$$\int_{\mathbb{R}^3} |\nabla u|^2 \, dx = V(K) \int_{\partial\Omega} \frac{\partial u}{\partial \nu} \, |dx|.$$

Applying Gauss–Green again in $\Omega(R)$, this time to the vector field $\nabla u$, and then recalling (5.20), we obtain

$$\int_{\partial\Omega} \frac{\partial u}{\partial \nu} \, |dx| = -\int_{|x|=R} \frac{\partial u}{\partial r} \, |dx| = 4\pi + O(R^{-1}).$$

The last two equalities give

$$\int_{\mathbb{R}^3} |\nabla u|^2 \, dx = 4\pi V(K). \tag{5.21}$$

Set now $v_K = u/V(K)$. Then $v_K \in \mathcal{A}(K)$, and by (5.21),

$$\int_{\mathbb{R}^3} |\nabla v_K|^2 \, dx = 4\pi \frac{1}{V(K)} = 4\pi \operatorname{Cap}(K). \tag{5.22}$$

Compact sets have finite capacity, so (5.22) implies that $\nabla v_K \in L^2(\mathbb{R}^3)$.

Let $v$ denote an arbitrary function in $\mathcal{A}(K)$ with $\nabla v \in L^2(\mathbb{R}^3)$. Set $\phi = v - v_K$. Then $\phi = 0$ on $K$ and at $\infty$. From Gauss–Green and (5.20), we deduce

$$\int_{\mathbb{B}^3(R)} \nabla v_K \cdot \nabla \phi \, dx = \int_{\Omega(R)} \nabla v_K \cdot \nabla \phi \, dx = \int_{|x|=R} \phi \frac{\partial v_K}{\partial r} \, |dx| = o(1)$$

as $R \to \infty$. Hence

$$\int_{\mathbb{B}^3(R)} |\nabla v|^2 \, dx = \int_{\mathbb{B}^3(R)} |\nabla v_K|^2 \, dx + \int_{\mathbb{B}^3(R)} |\nabla \phi|^2 \, dx + o(1)$$

and so by letting $R \to \infty$,

$$\int_{\mathbb{R}^3} |\nabla v|^2 \, dx \geq \int_{\mathbb{R}^3} |\nabla v_K|^2 \, dx. \tag{5.23}$$

From (5.22) and (5.23), it follows that

$$\operatorname{Cap}(K) = \frac{1}{4\pi} \int_{\mathbb{R}^3} |\nabla v_K|^2 \, dx = \frac{1}{4\pi} \inf \left\{ \int_{\mathbb{R}^3} |\nabla v|^2 \, dx \colon v \in \mathcal{A}(K) \right\}.$$

Proposition 5.13 is proved when $\partial\Omega$ is $C^\infty$.

Let now $\Omega$ be any domain satisfying assumption A1 and let $R$ be so large that $K \subset \mathbb{B}^3(R-2)$. Recall that

$$\mathcal{D}(K) \equiv \inf \left\{ \int_{\mathbb{R}^3} |\nabla v|^2 \, dx \colon v \in \mathcal{A}(K) \right\}.$$

Given $0 < \epsilon \leq 1$, the domain approximation Lemma 5.10 applied to $\Omega$ furnishes a bounded domain $\Omega' \Subset \Omega$ with $C^\infty$ boundary such that $\Omega'$ contains the sphere $|x| = R$, and $d(x, K) < \epsilon$ for each $x \in \Omega \setminus \Omega'$ with $|x| < R$. Let $\Omega'' = \Omega' \cup (|x| > R)$. Then $\Omega''$ is a domain, and $K'' \equiv \mathbb{R}^3 \setminus \Omega''$ is compact. If $x \in K''$ then $d(x, K) < \epsilon$. For positive integers $m$, let $K_m$ be the set $K''$ corresponding to $\epsilon = 1/m$.

Take $v \in \mathcal{A}(K)$. Set

$$\gamma_m = \inf_{x \in K_m} v \quad \text{and} \quad v_m = \min\left\{\frac{v}{\gamma_m}, 1\right\}.$$

Then $v_m \in \mathcal{A}(K_m)$, so

$$\mathcal{D}(K_m) \leq \int_{\mathbb{R}^3} |\nabla v_m|^2 \, dx \leq \gamma_m^{-2} \int_{\mathbb{R}^3} |\nabla v|^2 \, dx.$$

Taking the inf over $v \in \mathcal{A}(K)$ yields

$$\mathcal{D}(K_m) \leq \gamma_m^{-2} \mathcal{D}(K).$$

Since $v \in \operatorname{Lip}(\mathbb{R}^3)$, we have $\gamma_m \to 1$ as $m \to \infty$. Thus

$$\limsup_{m \to \infty} \mathcal{D}(K_m) \leq \mathcal{D}(K).$$

The reverse inequality, with lim inf in place of lim sup, is also true since $K \subset K_m$ and hence $\mathcal{D}(K) \leq \mathcal{D}(K_m)$. We conclude that

$$\lim_{m \to \infty} \mathcal{D}(K_m) = \mathcal{D}(K). \tag{5.24}$$

From the continuity of capacity (Landkof, 1972, p. 141), it follows that

$$\lim_{m \to \infty} \operatorname{Cap}(K_m) = \operatorname{Cap}(K). \tag{5.25}$$

Since, by the first part of the proof, $\operatorname{Cap}(K_m) = \frac{1}{4\pi} \mathcal{D}(K_m)$ for each $m$, (5.24) and (5.25) imply the desired inequality $\operatorname{Cap}(K) = \frac{1}{4\pi} \mathcal{D}(K)$. □

The main ingredient of the above proof is inequality (5.23), which asserts, roughly, that among all reasonable functions in $\Omega$ with boundary values 1 on $K$ and 0 at $\infty$, the minimizer $v_K$ of the Dirichlet integral is the competitor which is *harmonic* in $\Omega$. This is a special case of *Dirichlet's principle*. See Evans (1998) or Gilbarg and Trudinger (1983) for more.

## 5.6 Other Types of Capacity

The term "capacity" is applied to many different sorts of set functions. In this section we will briefly examine two families of capacities, each of which contains the Newtonian capacity studied in §5.5, and will also take a look at the logarithmic capacity in the plane.

### 5.6.1 Variational $p$-Capacity

For $n \geq 1$, let $U$ be an open subset of $\mathbb{R}^n$ and $K$ be a compact subset of $U$. The pair $(K, U)$ is called a *condenser.* In electrostatics, condensers are also called capacitors.

For $1 \leq p < \infty$, the (variational) $p$-capacity of the condenser $(K, U)$ is typically defined to be

$$\mathrm{Cap}_p(K, U) = \inf\left\{\int_U |\nabla u|^p \, dx \colon u \in \mathcal{A}(K, U)\right\}$$

where $\mathcal{A}(K, U)$ is the class of all $u \in \mathrm{Lip}(\mathbb{R}^n)$ satisfying

$$u = 1 \text{ on } K,\ \ 0 \leq u \leq 1 \text{ in } \mathbb{R}^n,\ \ u = 0 \text{ on } \mathbb{R}^n \setminus U,\ \ u(x) \to 0 \text{ as } x \to \infty.$$

As usual, we get the same infimum if the competing functions are taken over various other classes. See the treatment in Evans and Gariepy (1992) of "capacity" for an approach using Sobolev functions.

Proposition 5.13 shows that when $n = 3$ the variational 2-capacity of $(K, \mathbb{R}^3)$ coincides with the Newtonian capacity of $\mathrm{Cap}(K)$ of $K$, except for a factor of $4\pi$. We also saw in §5.5 that $\mathrm{Cap}(K)$ can be interpreted as the maximum amount of charge $K$ can hold when the voltage drop between any point of $K$ and $\infty$ is at most 1. For general open $U \subset \mathbb{R}^3$ with $K \subset U$, $\mathrm{Cap}_p(K, U)$ has a similar interpretation: Imagine that both $K$ and $\partial U$ are conductors, which are separated by some insulating material. Place a positive charge distribution $\mu$ on $K$ with total charge $Q$ and a negative charge distribution $\nu$ on $\partial U$ with total charge $-Q$. Then, apart from constants, $\mathrm{Cap}_2(K, U)$ equals the maximum $Q$ over all $\mu$ and $\nu$ for which the Newtonian potential $u$ of $\mu + \nu$ satisfies $\sup_K u - \inf_{\partial U} u \leq 1$.

Returning now to $\mathrm{Cap}_p$ for general $p$, we note that if $u \in \mathcal{A}(K, U)$ then it is easy to see that $u^\# \in \mathcal{A}(\overline{K^\#}, U^\#)$. From Theorem 3.6 or Theorem 3.11, about decrease of $p-$Dirichlet integrals under symmetrization, we obtain:

**Theorem 5.14** (Symmetrization decreases the capacity of condensers) *For $K \subset U \subset \mathbb{R}^n$ as above, we have*

$$\mathrm{Cap}_p(K, U) \geq \mathrm{Cap}_p(\overline{K^\#}, U^\#), \quad 1 \leq p < \infty, \quad n \geq 1.$$

For the energy functional $\int |\nabla u|^p \, dx$, the associated Euler–Lagrange equation is the "$p$-Laplace equation"

$$-\operatorname{div}(|\nabla u|^{p-2}\nabla u) = 0. \tag{5.26}$$

Solutions to (5.26) are called $p$-harmonic functions. For $p = 2$, (5.26) reduces to $\Delta u = 0$, and so 2-harmonic functions in open subsets of $\mathbb{R}^n$ are ordinary harmonic functions.

For an introduction to Euler–Lagrange equations, see Evans (1998, chapter 8) or Gilbarg and Trudinger (1983, §11.5). For more about $p$-harmonic functions, variational methods, and related topics, see Heinonen et al. (1993) and Iwaniec and Martin (2001).

A real function $u$ on an open set $\Omega$ is said to be a weak solution of the $p$-harmonic equation if $u \in W^{1,p}(\Omega)$ and for every $v \in W_0^{1,p}(\Omega)$,

$$\int_\Omega |\nabla u|^{p-2} \nabla u \cdot \nabla v \, dx = 0.$$

For $p = 2$, interior elliptic regularity (Evans, 1998, p. 316) insures that weak solutions of $\Delta u = 0$ are in fact in $C^\infty(\Omega)$. But when $p > 2$, the $p$-Laplace equation becomes degenerate at points where $\nabla u = 0$, and when $1 < p < 2$ it becomes singular at those points. The best one can say about regularity of solutions when $p \neq 2$ is that $u \in C^{1,\alpha}(\Omega)$ for some $\alpha < 1$ depending on $p$ and $n$. For $n = 2$ the best possible exponent $\alpha$ was found by Iwaniec and Manfredi (1989).

For $n \geq 1$, it turns out that a minimizer $u$ of the $p$-Dirichlet integral over $\mathcal{A}(K, U)$ is (weakly) $p$-harmonic in $U \setminus K$ and thus solves the Dirichlet problem for the equation (5.26) with boundary values $u = 1$ on $\partial K$, $u = 0$ on $\partial U$. Conversely, Lipschitz solutions to this Dirichlet problem also solve the minimization problem. "Lipschitz" can be replaced by "Sobolev," in which case the boundary values are to be understood in the trace sense. For existence of minimizers, see Heinonen et al. (1993) or Evans (1998, p. 448). (The $C^\infty$-Urysohn Lemma as in Folland (1999, p. 245) produces a smooth function on $\mathbb{R}^n$ which is 1 on $K$, zero on $\partial U$, which insures that the "admissible set" in Evans (1998) is not empty.)

When $p = 2$, the minimizer $u$ is harmonic in $U \setminus K$ with $u = 1$ on $\partial K$, $u = 0$ on $\partial U$. This function $u$ is called the *harmonic measure* of $\partial K$ with respect to $U \setminus K$.

Apart from $p = 2$, the most interesting $p$-capacity in $\mathbb{R}^n$ is when $p = n$. This is because the $n$-Dirichlet integral is conformally invariant: If $F$ is a conformal mapping between domains $\Omega_1$ and $\Omega_2$ in $\mathbb{R}^n$ and $u$ is a suitable function on $\Omega_2$, then

$$\int_{\Omega_2} |\nabla u|^n \, dx = \int_{\Omega_1} |\nabla (u \circ F)|^n \, dx.$$

It follows that the $n$-capacity of a condenser in $\mathbb{R}^n$ is invariant under conformal maps. $\mathrm{Cap}_n$ is therefore called the *conformal capacity*. Because of conformal invariance, the $n$-Laplace equation plays an important role in quasiconformal analysis in $\mathbb{R}^n$. See, for example, Iwaniec and Martin (2001).

### 5.6.2 Riesz $\alpha$-Capacity

Let $0 < \alpha < n$ and $K$ be a compact subset of $\mathbb{R}^n$. Define the *Riesz $\alpha$-energy* $W_\alpha(K)$ of $K$ by

$$W_\alpha(K) = \inf \int_{K\times K} |x-y|^{\alpha-n}\, d\mu(x)\, d\mu(y),$$

where the infimum is taken over all probability measures $\mu$ on $K$.

Define the *Riesz $\alpha$-capacity* $C_\alpha(K)$ of $K$ by

$$C_\alpha(K) = (W_\alpha(K))^{-1}.$$

For $\alpha = 2$ and $n = 3$, $C_2(K)$ is the Newtonian capacity studied in §5.5. In the notations for energy and capacity we have suppressed the dependence on $n$. References for $\alpha$-capacity include Landkof (1972) and Hayman and Kennedy (1976). (We have followed the notation of Landkof (1972) except for a normalizing constant. Our $C_\alpha(K)$ is the same as Hayman–Kennedy's $C_{n-\alpha}(K)^{n-\alpha}$.)

When $\alpha = 2$ and $n \geq 3$ the kernel $|x|^{2-n}$ is harmonic in $\mathbb{R}^n \setminus \{0\}$. The proof of Proposition 5.13 in §5.5 carries over without change to give an extension of that proposition to all dimensions $n \geq 3$:

$$C_2(K) = \frac{1}{(n-2)\beta_{n-1}} \operatorname{Cap}_2(K, \mathbb{R}^n), \quad n \geq 3, \tag{5.27}$$

where $\beta_{n-1}$ is the Hausdorff measure of the sphere $\mathbb{S}^{n-1}$.

By Theorem 5.14, it follows that for compact $K \subset \mathbb{R}^n$,

$$C_2(K) \geq C_2(\overline{K^\#}) \quad n \geq 3. \tag{5.28}$$

Watanabe (1983) (see also Betsakos (2004) and Méndez-Hernández (2006)) proved that the analogue of (5.28) holds when $n \geq 2$ and $0 < \alpha < 2$. It is not known if (5.28) holds when $n \geq 3$ and $2 < \alpha < n$.

The potential theory associated with $\alpha = 2$ is closely connected with the theory of Brownian motion. Similarly, the potential theory associated with $0 < \alpha < 2$ is related to the theory of symmetric $\alpha$-stable processes, also called stable Lévy processes. References include Doob (1953), Hoel et al. (1972), Betsakos (2004), Wu (2002), and Blumenthal and Getoor (1968).

### 5.6.3 Logarithmic Capacity

For compact $K \subset \mathbb{R}^n$, define the *logarithmic energy* $W_n(K)$ of $K$ to be

$$W_n(K) = \inf \int_{K \times K} \log \frac{1}{|x-y|} \, d\mu(x) d\mu(y),$$

where the infimum is over all Borel probability measures $\mu$ on $K$. The *logarithmic capacity* of $K$, which we shall denote by $\mathrm{Lcap}(K)$, is defined to be

$$\mathrm{Lcap}(K) = e^{-W_n(K)}.$$

We shall restrict the rest of our discussion to the case $n = 2$. In addition to Hayman and Kennedy (1976) and Landkof (1972), good accounts of log capacity in the plane can be found in Tsuji (1975), Ransford (1995), and Kellogg (1967).

To interpret logarithmic potential theory electrostatically, imagine a doubly infinite cylinder $\widehat{K}$ in $\mathbb{R}^3$ with axis parallel to the $x_3$-axis and base $K$ in the $x_1 x_2$ plane, with $K$ compact. For $\mu \in M^+(K)$, $\mu(K) = 1$, define a measure $\hat{\mu}$ on $\widehat{K}$ by $d\hat{\mu} = d\mu \, dx_3$. Then $\hat{\mu}$ describes a charge distribution on $\widehat{K}$ which is uniform in the $x_3$-direction and has total charge 1 per unit $x_3$-length. One can show that the Newtonian potential $u$ of $\hat{\mu}$ depends only on $x_1$ and $x_2$, and apart from a constant multiple, is given by the logarithmic potential

$$u(x) = \int_K \log \frac{1}{|x-y|} \, d\mu(y), \quad x = (x_1, x_2) \in \mathbb{R}^2.$$

Like the kernels $|x|^{2-n}$ in $\mathbb{R}^n$, the kernel $-\log|x|$ is harmonic in $\mathbb{R}^2 \setminus \{0\}$. But unlike $|x|^{2-n}$, $-\log|x|$ does not approach zero as $x \to \infty$. This causes the logarithmic potential theory in $\mathbb{R}^2$ to differ from the Newtonian theory in $\mathbb{R}^n$ for $n \geq 3$.

When $B$ is a closed disk of radius $R$ the uniform probability measure on $\partial B$ is the minimizing measure for the $W_2(B)$ problem. It follows that $W_2(B) = \log \frac{1}{R}$, and hence $\mathrm{Lcap}(B) = R$. One can also show that when $U = \mathbb{R}^2$, then

$$\mathrm{Cap}_2(K, \mathbb{R}^2) = 0$$

for every compact $K \subset \mathbb{R}^2$. Thus, in contrast to (5.27) when $n \geq 3$, $\mathrm{Lcap}(K)$ is not a function of $\mathrm{Cap}_2(K, \mathbb{R}^2)$. There is, though, a more complicated formula connecting log capacity with the variational 2-capacity:

$$\mathrm{Lcap}(K) = \lim_{R \to \infty} R \exp\left( -\frac{2\pi}{\mathrm{Cap}_2(K, \mathbb{B}^2(R))} \right), \tag{5.29}$$

whose proof we sketch below. Since, by Theorem 5.14, symmetrization decreases the capacity of condensers when $n = p = 2$, from (5.29) we obtain:

**Theorem 5.15** *For each compact $K \subset \mathbb{R}^2$, we have*

$$\mathrm{Lcap}(K) \geq \mathrm{Lcap}\left(\overline{K^\#}\right).$$

Theorem 5.15 appears in Szegő (1930), along with Theorem 5.12 about capacity in $\mathbb{R}^3$.

*Proof* To prove (5.29) we use the fact that logarithmic capacity in the plane can be expressed in terms of *Green functions* (Tsuji, 1975, p. 83), (Ahlfors, 1973, p. 27). Given compact $K \subset \mathbb{R}^2$, let $\Omega \subset \mathbb{R}^2$ be the unbounded component of the complement of $K$. Assume that $\mathrm{Lcap}(K) > 0$, and let $g(x) = g(x, \infty)$ be the Green function of the domain $\Omega \cup \{\infty\}$ on the Riemann sphere with pole at $\infty$. Then $g$ is harmonic in $\Omega$, $g(x) \to 0$ as $x$ tends to most points of $\partial\Omega$, and

$$g(x) = \log |x| + \gamma + o(1), \quad x \to \infty, \tag{5.30}$$

for some real constant $\gamma = \gamma(K)$. It turns out that $\gamma = W_2(K)$, so that

$$-\log(\mathrm{Lcap}(K)) = \gamma = \lim_{x\to\infty} [g(x, \infty) - \log |x|].$$

Continuing now with our sketch of proof of (5.29), we assume that $\Omega$ has $C^\infty$ boundary. After this case is done, we can pass to the case of arbitrary $K$ as in the proof of Proposition 5.13. For large positive values of $t$ the level set $\Gamma(t) = g^{-1}(t)$ is an analytic Jordan curve enclosing $K$ in its interior. Let $\Omega_t$ be the intersection of $\Omega$ with the inside of $\Gamma(t)$. Then $\Omega_t$ is a domain, and $1-(g/t)$ is the harmonic measure of $K$ with respect to $\Omega_t$. Set $U_t = \Omega_t \cup K$. Then

$$\begin{aligned}\mathrm{Cap}(K, U_t) &= \int_{\Omega_t} |\nabla g/t|^2 \, dx = t^{-2} \int_{\Gamma(t)} g \frac{\partial g}{dv} \, |dx| \\ &= t^{-1} \int_{\Gamma(t)} \frac{\partial g}{dv} \, |dx| = 2\pi t^{-1}.\end{aligned} \tag{5.31}$$

To obtain the last equality, use the fact that $\int_{\Gamma(t)} \frac{\partial g}{dv} \, |dx|$ does not change if we change the path of integration to $|x| = r$ for large $r$, differentiate (5.30) with respect to $r$, then let $r \to \infty$.

Next, choose a nonnegative function $\epsilon(R)$ decreasing to zero such that

$$\big| g(x) - \gamma - \log |x| \big| < \epsilon(R)$$

for every $|x| = R$ when $R$ is large. For fixed $R$, take

$$t_1 = \log R + \gamma - \epsilon(R), \quad t_2 = \log R + \gamma + \epsilon(R).$$

Then

$$\Omega_{t_1} \subset \mathbb{B}^2(R) \setminus K \subset \Omega_{t_2}.$$

The capacity of a condenser $(K, U)$ decreases if $K$ stays fixed and $U$ increases. Thus

$$\mathrm{Cap}(K, U_{t_2}) \leq \mathrm{Cap}(K, \mathbb{B}^2(R)) \leq \mathrm{Cap}(K, U_{t_1}). \tag{5.32}$$

Formula (5.29) easily follows from (5.31) and (5.32). □

Suppose now that $U$ is a simply connected domain in $\mathbb{R}^2$ and that $K$ is a compact connected subset of $U$. Then $\Omega = U \setminus K$ is called a ring domain. The ring domain is called nondegenerate if $K$ has more than one point and $U$ is not the whole plane. When the domain is nondegenerate there is a unique number $R \in (1, \infty)$ and a conformal mapping $f$ of $\Omega$ onto the annulus $A(R) = \{1 < |z| < R\}$; see Ahlfors (1973, p. 255). The (conformal) *modulus* of the ring domain $\Omega$ is defined to be

$$\mathrm{Mod}\,\Omega = \log R.$$

It is clear that the modulus of a ring domain is a conformal invariant. As discussed above, the variational 2-capacity of condensers in the plane is also a conformal invariant. For the annulus $A(R)$ the modulus is, of course,

$$\mathrm{Mod}\,A(R) = \log R.$$

The harmonic measure of $|x| = 1$ with respect to $A(R)$ is the function $\omega(x) = \log(R/|x|)\big/\log R$. We have

$$\mathrm{Cap}_2(\overline{\mathbb{B}^2(1)}, \mathbb{B}^2(R)) = \int_{A(R)} |\nabla\omega|^2\,dx = \frac{2\pi}{\log R}.$$

The first equality follows from the relation between variational 2-capacity and harmonic measure discussed in the paragraph above (5.27). The second equality is a simple computation. We conclude that for nondegenerate ring domains $U \setminus K$ the capacity of the associated condenser $(K, U)$ is reciprocal to the modulus, i.e.,

$$\mathrm{Mod}(U \setminus K) = \frac{2\pi}{\mathrm{Cap}_2(K, U)}.$$

By Theorem 5.14, symmetrization decreases the $p$-capacity when $n = p = 2$. Combining with (5.29), we obtain a theorem of Carleman (1918):

**Corollary 5.16** (Carleman) *Let $U \setminus K$ be a nondegenerate ring domain, with $K \subset U$, $K$ compact and connected, $U$ connected and simply connected. Then*

$$\mathrm{Mod}(U \setminus K) \leq \mathrm{Mod}\,(U^{\#} \setminus \overline{K^{\#}}).$$

## 5.7 Symmetrization Increases Torsional Rigidity and Mean Lifetime

Let $\Omega$ be a bounded open set in $\mathbb{R}^n$. By Proposition 5.2, the Poisson boundary value problem

$$\Delta u = -2 \quad \text{in} \quad \Omega, \qquad u = 0 \quad \text{on} \quad \partial\Omega,$$

has a unique weak solution $u \in W_0^{1,2}(\Omega)$. By interior elliptic regularity (Evans, 1998, p. 316), $u$ is in fact in $C^\infty(\Omega)$. If $\partial\Omega$ is sufficiently smooth, then $u$ is a classical solution to the Poisson boundary value problem (Evans, 1998, §6.3). Define also

$$T(\Omega) = 2\int_\Omega u\,dx.$$

When $n = 2$ and $\Omega$ is simply connected, $u$ is called the *stress function* and $T(\Omega)$ the *torsional rigidity* of $\Omega$. One imagines an infinitely long cylinder with axis parallel to the $x_3$-axis in $\mathbb{R}^3$ and with cross-section $\Omega$. When the cylinder is twisted around its axis at a rate of $\phi$ angular units per unit length, the stress vector equals $\mu\phi$ times the curl of the vector field $(0, 0, u)$, where $\mu$ is the shear modulus. The torsional rigidity turns out to equal the torque required to accomplish a unit angle of twisting per unit length, divided by the shear modulus. See Bandle (1980, p. 63), Pólya and Szegő (1951, §5.2).

For ease of reference, we shall continue to call $u$ the stress function in all dimensions and for all bounded open $\Omega$, simply connected or not. In this general context, the pde $\Delta u = -2$ is the Euler equation corresponding to the energy functional

$$I(w) = \frac{1}{2}\int_\Omega |\nabla w|^2\,dx - \int_\Omega 2w\,dx.$$

See Evans (1998, p. 435). From the Remark in Evans (1998, p. 452), it follows that the function $u$ is the unique minimizer of $I(w)$ over $w \in W_0^{1,2}(\Omega)$. Since $I(u) \geq I(|u|)$, we must have $u \geq 0$.

By Gauss–Green,

$$\int_\Omega |\nabla u|^2\,dx = -\int_\Omega u\Delta u\,dx = \int_\Omega 2u\,dx.$$

It follows that

$$I(u) = -\int_\Omega u\,dx = -\frac{1}{2}T(\Omega). \tag{5.33}$$

Let $v$ denote the solution of the symmetrized problem

$$\Delta v = -2 \quad \text{in} \quad \Omega^\#, \qquad v = 0 \quad \text{on} \quad \partial\Omega^\#.$$

The function $u^\#$ is in $W_0^{1,2}(\Omega^\#)$. Applying Corollary 3.22 (symmetrization decreases the Dirichlet integral of Sobolev functions), we obtain

$$I(u) \geq I(u^\#) \geq I(v).$$

From (5.33) we deduce:

**Theorem 5.17** (Pólya) *For bounded open sets $\Omega \subset \mathbb{R}^n$, we have*

$$T(\Omega) \leq T(\Omega^\#). \tag{5.34}$$

The phenomenon that among all simply connected plane domains of fixed area the disk has largest torsional rigidity had been suggested by St. Venant in 1856. The first proof was given by Pólya (1948). Proofs appear also in Pólya and Szegő (1951, chapter V) and Bandle (1980, p. 67), along with related results. Later in this book, Theorem 10.10 proves a pointwise inequality $u^\#(x) \leq v(x)$ (due to Talenti) that integrates over $\Omega^\#$ to give another proof of Pólya's Theorem 5.17.

Incidentally, for the ball $\mathbb{B}^n(R)$, the stress function $v$ is

$$v(x) = \frac{1}{n}(R^2 - |x|^2).$$

### Mean Lifetimes of Brownian Particles

Standard Brownian motion in $\mathbb{R}^n$ is a stochastic process $B_t$ with values in $\mathbb{R}^n$ starting at a point $B_0 = x \in \mathbb{R}^n$ that satisfies certain hypotheses. Rather than list them, we advise the non-probabilistic reader to think of Brownian motion as a continuous $n$-dimensional random walk by a particle that is equally likely to go in any direction. Let $\Omega$ be a bounded open subset of $\mathbb{R}^n$. We shall always assume our Brownian motion starts in $\Omega$. Define the *exit time* from $\Omega$ to be the random variable $\tau_\Omega$ on the Brownian probability space defined by

$$\tau_\Omega = \text{first time that the Brownian particle leaves } \Omega.$$

One can think of the particle as dying when it hits $\partial\Omega$. Then $\tau_\Omega$ is the particle's age at death.

Let $g$ be a bounded function in $C^\infty(\Omega)$. Suppose that $u$ is a solution to the Poisson boundary value problem

$$\frac{1}{2}\Delta u = -g \quad \text{in} \quad \Omega, \qquad u = 0 \quad \text{on} \quad \partial\Omega.$$

Then, as with the stress function, $u \in C^\infty(\Omega)$ and the boundary condition is classically satisfied for smooth enough $\partial\Omega$. The solution $u$ to this Poisson problem has a probabilistic representation:

$$u(x) = E_x\left(\int_0^{\tau_\Omega} g(B_t)\,dt\right), \quad x \in \Omega, \tag{5.35}$$

where $E_x Y$ denotes the mean of a random variable $Y$ on the Brownian probability space when the particle starts at $x$. This representation is an example of Dynkin's Formula in probability theory.

Here is an outline of a proof for (5.35), taken from Durrett (1984, pp. 251–252) and with details omitted. By Itô's formula, for $0 < t < \tau_\Omega$,

$$u(B_t) = u(B_0) + \int_0^t \nabla u(B_s) \cdot dB_s + \frac{1}{2}\int_0^t \Delta u(B_s)\,ds.$$

Let $t \to \tau_\Omega$. Since $u$ satisfies the Poisson equation and vanishes on $\partial\Omega$, it follows that

$$0 = u(B_0) + \int_0^{\tau_\Omega} \nabla u(B_s) \cdot dB_s - \int_0^{\tau_\Omega} g(B_s)\,ds. \tag{5.36}$$

If the particle starts at $x \in \Omega$ then $B_0 \equiv x$. Moreover, the stochastic integral involving $dB_s$ is the limit of a martingale. When we apply $E_x$ to (5.36) the stochastic integral term vanishes, and the result is exactly (5.35). □

Let's specialize now to $g \equiv 1$. The solution $u$ is then the stress function of $\Omega$ and (5.35) gives

$$u(x) = E_x(\tau_\Omega), \quad x \in \Omega.$$

The expression $E_x(\tau_\Omega)$ represents the average lifetime of a particle started at $x$. Suppose that we choose the starting point $x$ according to a uniform distribution on $\Omega$. Then

$$M(\Omega) \equiv \frac{1}{\mathcal{L}^n(\Omega)}\int_\Omega u(x)\,dx = \frac{1}{\mathcal{L}^n(\Omega)}\int_\Omega E_x(\tau_\Omega)\,dx$$

can be regarded as the average lifetime of a particle born at random somewhere in $\Omega$. From the definition of $T(\Omega)$, we see that

$$M(\Omega) = \frac{1}{2\mathcal{L}^n(\Omega)} T(\Omega).$$

Thus, Theorem 5.17 may be restated as:

**Corollary 5.18** (Symmetrization increases the mean lifetime) *For bounded open $\Omega \subset \mathbb{R}^n$, we have*

$$M(\Omega) \le M(\Omega^\#).$$

For further inequalities connecting the mean lifetime and geometric or function theoretic quantities, see Bañuelos and Carroll (1994) and Bañuelos et al. (2002).

## 5.8 Notes and Comments

In this chapter most of the history and citations are imbedded in the text. Here we shall just mention a survey article by Payne (1967) discussing problems open at the time, some still open, and a list compiled by J. Hersch of many articles related to Pólya-Szegő type inequalities. The list appears in Pólya (1984) as part of Hersch's commentary on item 177 in Pólya's list of publications. A wealth of extremal results and conjectures on eigenvalues of the Laplacian can be found in a recent book on shape optimization and spectral theory (Henrot, 2017).

As noted in §3.3, Dirichlet integrals do not change under polarization in a hyperplane $H \subset \mathbb{R}^n$. It follows that

$$\lambda_1(\Omega) \geq \lambda_1(\Omega_H),$$

for bounded open sets $\Omega$. Similar inequalities hold for the other physical quantities considered in this chapter.

# 6

# Steiner Symmetrization

## 6.1 Definition of Steiner Symmetrization

Let $n$ and $k$ be integers with $n \geq 2$ and $1 \leq k \leq n-1$. Set $m = n - k$. Decompose $x = (x_1, \ldots, x_n) \in \mathbb{R}^n$ as

$$x = (y, z) = (y_1, \ldots, y_k, z_1, \ldots, z_m).$$

Then $\mathbb{R}^n = \mathbb{R}^k \times \mathbb{R}^m$, and $\mathbb{R}^n$ is foliated into parallel affine $k$-planes $\{(y, z) \colon y \in \mathbb{R}^k\}$, indexed by $z$. For example, when $k = 1$, $\mathbb{R}^n$ is foliated into lines parallel to the $x_1$-axis, i.e., orthogonal to the hyperplane $x_1 = 0$. Such a line intersects this hyperplane at the point $(0, z)$. When $k = n - 1$, $\mathbb{R}^n$ is foliated into parallel affine hyperplanes orthogonal to the $x_n$-axis, and $z \in \mathbb{R}$ is the "height" at which the hyperplane hits the $x_n$-axis.

Let $E \subset \mathbb{R}^n$ be an $\mathcal{L}^n$-measurable set. Set

$$E(z) = \{y \in \mathbb{R}^k \colon (y, z) \in E\}, \quad z \in \mathbb{R}^m.$$

We call $E(z)$ the slice of $E$ through $z$. By Fubini's Theorem, $E(z)$ is $\mathcal{L}^k$-measurable for $\mathcal{L}^m$-almost every $z \in \mathbb{R}^m$. For measurable $E(z)$, define $E^\#(z)$ to be the symmetric decreasing rearrangement of $E(z)$ in $\mathbb{R}^k$. Then $E^\#(z)$ is the open ball centered at the origin in $\mathbb{R}^k$ with the same $\mathcal{L}^k$-measure as $E(z)$. Set $E^\#(z) = \emptyset$ if $E(z)$ is not $\mathcal{L}^k$-measurable. Note that we have $E^\#(z) = \{0\}$ when $\mathcal{L}^k(E(z)) = 0$.

**Definition 6.1** For $1 \leq k \leq n-1$, the $(k, n)$-Steiner symmetrization of $E$ is the set

$$E^\# = \{(y, z) \colon y \in E^\#(z), z \in \mathbb{R}^m\} = \bigcup_{z \in \mathbb{R}^m} (E^\#(z) \times \{z\}).$$

This notation $E^\#$ was used earlier to denote the $n$-dimensional symmetric decreasing rearrangement of $E$, which can be regarded as the $(n, n)$-Steiner symmetrization. In this chapter, though, we reserve $E^\#$ to denote $(k,n)$-Steiner symmetrization with $1 \le k \le n-1$.

The function $h(x) = h(y,z) = \alpha_k |y|^k - \mathcal{L}^k(E(z))$ is $\mathcal{L}^n$-measurable on $\mathbb{R}^n$ (the $E(z)$ term by Fubini's Theorem), and $E^\# = \{h < 0\}$. Thus, $E^\#$ is an $\mathcal{L}^n$-measurable set. Also,

$$\mathcal{L}^n(E^\#) = \int_{\mathbb{R}^m} \mathcal{L}^k(E^\#(z))\, dz = \int_{\mathbb{R}^m} \mathcal{L}^k(E(z))\, dz = \mathcal{L}^n(E). \tag{6.1}$$

To visualize the $(1,n)$-Steiner symmetrization of $E$, take a line $L$ in $\mathbb{R}^n$ which is orthogonal to the hyperplane $x_1 = 0$. Compute the 1-dimensional measure of $L \cap E$. Replace $L \cap E$ with an open interval symmetric about $x_1 = 0$ which has the same length as $L \cap E$. Then $E^\#$ is the union of these intervals as $L$ ranges over all such lines.

To visualize $(n-1,n)$-Steiner symmetrization, slice $E$ with an affine hyperplane $H$ orthogonal to the $x_n$-axis. Compute the $(n-1)$-dimensional measure of $H \cap E$. Replace $H \cap E$ with an $(n-1)$-dimensional open ball centered on the $x_n$-axis which has the same $\mathcal{L}^{n-1}$-measure as $H \cap E$. Then $E^\#$ is the union of these $(n-1)$-balls as $H$ ranges over all such hyperplanes.

As an interesting conceptual exercise, the reader is invited to visualize $(2,4)$-Steiner symmetrization in $\mathbb{R}^4$. If successful, he or she can then try $(k,n)$-symmetrization in $\mathbb{R}^n$ for general $2 \le k \le n-2$. A generic example of a bounded set $E \subset \mathbb{R}^n$ and its Steiner symmetrization is depicted in Figure 6.1.

Next, let $Z$ be an open set in $\mathbb{R}^m$. Define the open set $X \subset \mathbb{R}^n$ by

$$X = \mathbb{R}^k \times Z.$$

Let $f\colon X \to \mathbb{R}^+$ be a nonnegative $\mathcal{L}^n$-measurable function on $X$. Assume that for $\mathcal{L}^m$-almost every $z \in Z$ the slice function $f^z \equiv f(\cdot, z)$ satisfies

$$\mathcal{L}^k(f^z > t) < \infty, \qquad \text{for every } t > 0. \tag{6.2}$$

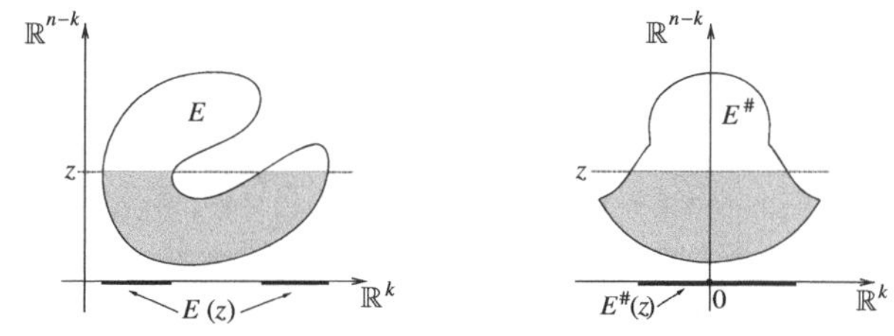


Figure 6.1 A measurable set $E \subset \mathbb{R}^n$ and its $(k,n)$-Steiner symmetrization $E^\#$. For any $z \in \mathbb{R}^{n-k}$, the $k$-dimensional slice of $E^\#$ through $z$ is the centered open ball that has the same $k$-dimensional measure as the corresponding slice of $E$.

Condition (6.2) will be called the *finiteness condition.* In the Steiner context, we do not require that $\mathcal{L}^n(f > t) < \infty$ for all $t > 0$.

The symmetric decreasing rearrangement $(f^z)^\#$ is well defined on $\mathbb{R}^k$ when (6.2) is satisfied. Set $(f^z)^\# \equiv 0$ for the $z \in \mathbb{R}^m$ at which (6.2) is not satisfied.

**Definition 6.2** For $1 \leq k \leq n-1$, the $(k,n)$-Steiner symmetrization of $f$ is the function $f^\#$ defined on $X$ by

$$f^\#(x) = f^\#(y,z) = (f^z)^\#(y), \quad y \in \mathbb{R}^k, \ z \in Z.$$

For example, to construct the $(1,n)$-Steiner symmetrization of $f$ on $X = \mathbb{R} \times Z$ with $Z \subset \mathbb{R}^{n-1}$, consider lines in $Z$ orthogonal to the hyperplane $x_1 = 0$ which intersect this hyperplane at points $(0,z)$, with $z \in Z$. On each such line, replace $f$ by its 1-dimensional symmetric decreasing rearrangement.

For $(n-1,n)$-Steiner symmetrization, $Z \subset \mathbb{R}$ is a countable union of open intervals. On each affine hyperplane $\mathbb{R}^{n-1} \times \{z\}$ with $z \in Z$, replace $f$ by its $(n-1)$-dimensional s.d.r. on that hyperplane. Thus, $f$ is a function of $n$ variables, but $f^\#$ depends on only the two variables $|y|$ and $z$. The line $\{(0,x_n) \in \mathbb{R}^n : x_n \in \mathbb{R}\}$ is the axis of symmetry for the symmetrized function.

For general $k$, the $(k,n)$-Steiner symmetrization $f^\#$ depends on $n-k+1$ variables. The author finds it useful to view $f(y,z)$ as a family of functions of $y$ depending on an $m$-dimensional parameter $z$.

According to Definition 1.29 in Chapter 1, each of the $(f^z)^\#$ is defined everywhere on $\mathbb{R}^k$. Hence, $f^\#$ is defined everywhere in $X$. Also, if $z$ is such that (6.2) holds, then the sets $\{y \in \mathbb{R}^k : f(y,z) > t\}^\#$ and $\{y \in \mathbb{R}^k : f^\#(y,z) > t\}$ are equal for each $t > 0$. Since the set of $z$ at which (6.2) fails is an $\mathcal{L}^m$-nullset, it follows that the sets $\{x \in \mathbb{R}^n : f(x) > t\}^\#$ and $\{x \in \mathbb{R}^n : f^\#(x) > t\}$ differ at most by an $\mathcal{L}^n$-null set. Hence, $f^\#$ is $\mathcal{L}^n$-measurable on $X$. Moreover, since the slice functions $f^z$ and $(f^\#)^z$ are equidistributed on $\mathbb{R}^k$ for $\mathcal{L}^m$-almost every $z$, it follows as in (6.1) that $f$ and $f^\#$ are equidistributed on $X \subset \mathbb{R}^n$.

We have defined $(k,n)$-Steiner symmetrization with respect to a certain distinguished $m$-plane, namely $x_1 = \cdots = x_k = 0$. Given any affine $m$-plane $H \subset \mathbb{R}^n$, one can define in the same way Steiner symmetrization of sets or functions with respect to $H$. For example, the $(1,n)$-Steiner symmetrization of a set $E$ with respect to an affine hyperplane $H$ is the union of line segments centered on $H$ and orthogonal to $H$, the length of each segment being equal to the 1-dimensional measure of the intersection of $E$ with the line containing that segment. Almost all the results to follow involving Steiner symmetrization with respect to $x_1 = \cdots = x_k = 0$ are still valid for Steiner symmetrization with respect to any $H$. To verify this, one can inspect the proofs, or one can reduce the case of general $H$ to that of special $H$ by means of an isometry

$T = a + S$ of $\mathbb{R}^n$ onto itself which maps $H$ onto $x_1 = \cdots = x_k = 0$. Here $a \in \mathbb{R}^n$ and $S \in O(n)$.

Chapters 1–5 contain 32 theorems, corollaries and propositions involving the symmetric decreasing rearrangement $f^\#$ on $\mathbb{R}^n$. They are listed below.

- Chapter 1: Propositions 1.30 and 1.43.
- Chapter 2: Propositions 2.10 and 2.11, Theorems 2.12 and 2.15, Corollaries 2.13, 2.16, 2.19, 2.20, 2.22, and 2.23(b).
- Chapter 3: Theorems 3.6, 3.7, 3.11, and 3.20; Corollaries 3.8, 3.9, 3.21, and 3.22.
- Chapter 4: Theorems 4.8, 4.10, 4.14, and 4.16; Corollaries 4.13 and 4.15.
- Chapter 5: Theorems 5.6, 5.12, 5.14, 5.15, and 5.17, and Corollary 5.18.

Most of these results assert that some functional increases or decreases under s.d.r. In the rest of this chapter we shall present Steiner counterparts. The reader may verify at the end of the chapter that practically all the items in the table have either been given a Steiner version or have been shown not to have one. Some Steiner results are proved just like they were in the s.d.r. case, or as consequences thereof, but in some situations, especially inequalities for Dirichlet integrals, significantly new tools must be constructed. Our plan is to start with results from Chapter 1, then move monotonically upward to end with results from Chapter 5.

In the rest of this chapter the integers $n \geq 1$ and $1 \leq k \leq n - 1$ will be fixed, and $f^\#$ will always denote the $(k, n)$-Steiner symmetrization, unless otherwise noted. We shall often invoke Fubini's Theorem, without saying so. Bear in mind that we are Steiner symmetrizing only nonnegative functions whose slice functions satisfy the finiteness condition (6.2), whereas the results for symmetric decreasing rearrangements on $\mathbb{R}^n$ often just assumed that $f$ is a real valued function satisfying

$$\lambda_f(t) = \mathcal{L}^n(f > t) < \infty, \quad \forall\, t > \operatorname{ess\,inf} f, \tag{6.3}$$

which is condition (1.1) in Chapter 1.

## 6.2 Steiner Counterparts for Results in Chapter 1

According to Proposition 1.18(a), a function and its s.d.r. have the same distribution. In the discussion after Definition 6.2, we noted that this is also true for each of the $(k, n)$-Steiner symmetrizations. Parts (b)–(e) of Proposition 1.30 are not true for Steiner symmetrization, but evident analogues hold for the slice functions $f^z$.

Now let $\{f_j\}$ be a sequence of nonnegative $\mathcal{L}^n$-measurable functions on $X = \mathbb{R}^k \times Z$, $Z$ open in $\mathbb{R}^m$, each of which satisfies (6.2).

**Proposition 6.3** *(a) If $f_j \nearrow f$ $\mathcal{L}^n$-a.e. on $X$, and $f$ satisfies* (6.2)*, then $f_j^\# \nearrow f^\#$ $\mathcal{L}^n$-a.e. on $X$.*
*(b) If $f_j \to f$ in measure on $X$ and each $f_j$ satisfies* (6.3)*, then $f_j^\# \to f^\#$ in measure on $X$.*

At the end of this section we give an example showing that (b) can fail if (6.3) does not hold.

*Proof of Proposition 6.3(a)* The hypothesis of (a) implies that $f_j^z \nearrow f^z$ $\mathcal{L}^k$-a.e for $\mathcal{L}^m$-almost every $z$. By Proposition 1.43(a), for $\mathcal{L}^m$-almost every $z$, $f_j^\#(y,z) \nearrow f^\#(y,z)$ at each $y \in \mathbb{R}^k$. Thus, $f_j^\# \nearrow f^\#$ $\mathcal{L}^n$-a.e. in $X$. □

To prove (b) we need the following lemma, in which $*$ denotes decreasing rearrangement and $X$ an arbitrary measure space.

**Lemma 6.4** *Let $f$ and $g$ be nonnegative measurable functions on a measure space $(X,\mu)$ which satisfy $\lambda_f(t) < \infty$, $\lambda_g(t) < \infty$ for every $t > 0$. Then for each $\epsilon > 0$ and $M > 0$ we have*

$$\mu(|f^* - g^*| \geq \epsilon) \leq 6M\epsilon^{-1}\mu(|f-g| \geq \epsilon/3) + \max\{\lambda_f(M), \lambda_g(M)\}.$$

*Proof* Assume first that $\max\{\|f\|_{L^\infty}, \|g\|_{L^\infty}\} \leq K < \infty$. Let $E = (|f-g| \geq \epsilon/3)$, $E^c = X \setminus E$ and $\alpha = \mu(E)$. Define

$$F = f\chi_{E^c}, \quad G = g\chi_{E^c}.$$

Then $0 \leq F \leq f$ and $0 \leq G \leq g$. Moreover, the following inequalities hold:

$$\|F - G\|_{L^\infty} \leq \epsilon/3, \tag{6.4}$$

$$f^*(x+\alpha) \leq F^*(x), \quad g^*(x+\alpha) \leq G^*(x), \quad x \in (0,\infty). \tag{6.5}$$

Inequality (6.4) is clear. As for (6.5), $f$ and $F$ agree except on the set $E$, so $(f > t) \subset (F > t) \cup E$, hence

$$\lambda_f(t) \leq \lambda_F(t) + \alpha, \quad t \geq 0. \tag{6.6}$$

Take $x \in (0,\infty)$. Let $t = F^*(x)$. Then $\lambda_F(t) = \mu(F > F^*(x)) = \mu(F^* > F^*(x)) \leq x$, so by (6.6),

$$\lambda_f(t) \leq x + \alpha.$$

If $f^*(x+\alpha)$ were strictly greater than $t$, then right continuity of $f^*$ would imply $\lambda_f(t) > x+\alpha$, contradicting the inequality above. Thus, $f^*(x+\alpha) \leq t = F^*(x)$, which is (6.5).

By Corollary 2.23 and (6.4), we have $\|F^* - G^*\|_{L^\infty} \leq \epsilon/3$, from which follows

$$\mu(|f^* - g^*| \geq \epsilon) \leq \mu(|f^* - F^*| \geq \epsilon/3) + \mu(|G^* - g^*| \geq \epsilon/3). \tag{6.7}$$

Also, using (6.5),

$$\begin{aligned}
\mu(|f^* - F^*| \geq \epsilon/3) &\leq \frac{3}{\epsilon}\int_0^\infty (f^* - F^*)\,dx \\
&\leq \frac{3}{\epsilon}\int_0^\infty (f^*(x) - f^*(x+\alpha))\,dx \\
&= \frac{3}{\epsilon}\int_0^\alpha f^*(x)\,dx \leq \frac{3}{\epsilon}K\mu(|f-g| \geq \epsilon/3).
\end{aligned}$$

The same inequality holds when $f^*$ and $F^*$ are replaced by $g^*$ and $G^*$. Combined with (6.7), we obtain

$$\mu(|f^* - g^*| \geq \epsilon) \leq 6\epsilon^{-1}K\mu(|f-g| \geq \epsilon/3). \tag{6.8}$$

Now let us drop the assumption that $f$ and $g$ are in $L^\infty(X)$, and let both $\epsilon$ and $M$ be given. Define (new) functions $F$ and $G$ by $F = \min(f, M)$, $G = \min(g, G)$. By (6.8),

$$\mu(|F^* - G^*| \geq \epsilon) \leq 6\epsilon^{-1}M\mu(|F-G| \geq \epsilon/3).$$

Now $F^* = \min(f^*, M)$. Thus $F^* = f^*$ on $[x_M, \infty)$, where $x_M$ is the smallest point such that $f^*(x_M) = M$. Then $x_M = \lambda_f(M)$. Similar considerations hold for $G^*$ and $g$. Thus,

$$\mu(|f^* - g^*| \geq \epsilon) \leq \max\{\lambda_f(M), \lambda_g(M)\} + \mu(|F^* - G^*| \geq \epsilon). \tag{6.9}$$

Combining (6.9), (6.10) and the pointwise inequality $|F - G| \leq |f - g|$, we obtain

$$\mu(|f^* - g^*| \geq \epsilon) \leq \max\{\lambda_f(M), \lambda_g(M)\} + 6\epsilon^{-1}M\mu(|f-g| \geq \epsilon/3).$$

The lemma is proved. □

*Proof of Proposition 6.3(b)* Let now $X$ resume its position as $\mathbb{R}^k \times Z$, take $\mu = \mathcal{L}^n$ and consider $f_j$ and $f$ as in Proposition 6.3(b). The finiteness of $\lambda_f(t)$ is part of Proposition 1.41. Recall that the $k$-dimensional s.d.r. $h^\#$ for $h\colon \mathbb{R}^k \to \mathbb{R}^+$ is related to the decreasing rearrangement by $h^\#(y) = h^*(\alpha_k|y|^k)$. Then using Lemma 6.4 with $f$ replaced by a slice function $f^z$ and $g$ by $f_j^z$, for fixed $\epsilon$ and $M$, we obtain

$$\begin{aligned}\mathcal{L}^n(|f^\# - f_j^\#| \geq \epsilon) &= \int_Z \mathcal{L}^k(|(f^z)^\# - (f_j^z)^\#| \geq \epsilon)\, dz \\ &\leq 6M\epsilon^{-1}\int_Z \mathcal{L}^k(|f^z - f_j^z| \geq \epsilon/3)\, dz + \int_Z [\lambda_{f^z}(M) + \lambda_{f_j^z}(M)]\, dz \\ &= 6M\epsilon^{-1}\mathcal{L}^n(|f - f_j| \geq \epsilon/3) + \lambda_f(M) + \lambda_{f_j}(M).\end{aligned}$$

Let $j \to \infty$, apply Proposition 1.40, then let $M \to \infty$. The result is

$$\lim_{n\to\infty} \mathcal{L}^n(|f^\# - f_j^\#| \geq \epsilon) = 0. \qquad \square$$

The proof just given of Proposition 6.3 obviously works when $\{f_j\}$ is a nonnegative sequence in $\mathbb{R}^n$ satisfying the (6.3) and converging in measure to $f$, with # denoting s.d.r. Thus, we have a new proof of the convergence in measure statement of Proposition 1.41.

The following example with (1, 1)-Steiner symmetrization in $\mathbb{R}^2$ shows that the finiteness assumption $\mathcal{L}^n(f_j > t) < \infty$ for each $j$ and for each $t > 0$ in Proposition 6.3(b) cannot be replaced by the weaker finiteness condition that each $f_j$ satisfy (6.2).

**Example 6.5** Define $\Phi$ on $\mathbb{R}^+$ by $\Phi(t) = 1/t$ for $0 < t < 1$, $\Phi(t) = 0$ for $t \geq 1$. For $0 < \delta < 1$, it is easily checked that

$$\Phi(t) - \Phi(t+\delta) > 1/2, \quad \text{for} \quad 0 < t < \delta^{1/2}.$$

Define $\Phi(t,\delta) = \Phi(t)$ for $\delta < t < 1$ and $\Phi(t,\delta) = 0$ for other $t \in \mathbb{R}^+$. Then $\Phi^*(t) = \Phi(t)$ and

$$\Phi^*(t,\delta) = \Phi(t+\delta), \quad t \in (0,\infty).$$

It follows that

$$\mathcal{L}(\{t \in \mathbb{R}^+ : |\Phi(t) - \Phi(t,\delta)| > 0\}) = \delta \tag{6.10}$$

and

$$\mathcal{L}(\{t \in \mathbb{R}^+ : |\Phi(t) - \Phi^*(t,\delta)| > 1/2\}) > \delta^{1/2}. \tag{6.11}$$

For $i \geq 1$ and $j \geq 1$, define $f_j(y,z)$ on $\mathbb{R}^2$ as follows: $f_j(y,z) = 0$ when $z \leq 0$ and

$$f_j(y,z) = \Phi(2|y|, i^{-2}j^{-1}), \quad i-1 < z \leq i, \quad i \geq 1.$$

Then each $f_j$ satisfies (6.2). Let $f(y,z) = \Phi(2|y|)$ for $z > 0$ and $f(y,z) = 0$ for $z \leq 0$. From (6.10) and (6.11), it easily follows that for $j \geq 1$,

$$\mathcal{L}^2(|f - f_j| > 0) \leq j^{-1}\pi^2/6$$

and

$$\mathcal{L}^2(|f^\# - f_j^\#| > 1/2) = \infty.$$

Thus, $\{f_j\}$ converges in measure to $f$, but the $(1, 1)$-Steiner symmetrizations $\{f_j^\#\}$ do not converge in measure to $f^\#$.

## 6.3 Steiner Analogues for Two Simple Polarization Results

Recall from §1.7 that $\mathcal{H}(\mathbb{R}^n)$ is the set of all affine hyperplanes in $\mathbb{R}^n$. Suppose that $H \in \mathcal{H}(\mathbb{R}^n)$ has the form

$$H = H_1 \times \mathbb{R}^m, \quad \text{where} \quad H_1 \in \mathcal{H}(\mathbb{R}^k). \tag{6.12}$$

This means that vectors in $\mathbb{R}^n$ normal to $H$ are parallel to $\mathbb{R}^k$. If (6.12) holds then for $y \in \mathbb{R}^n,\ z \in \mathbb{R}^m$,

$$\rho_H(y, z) = (\rho_{H_1} y, z), \tag{6.13}$$

where $\rho_H$ denotes reflection in $H$.

Moreover, for nonnegative measurable functions $f$ on $X$ the slices $f^z$ satisfy

$$(f^z)_{H_1} = (f_H)^z, \quad z \in Z. \tag{6.14}$$

In the two propositions to follow, the # denotes $(k, n)$-Steiner symmetrization of functions on $X = \mathbb{R}^k \times Z$.

**Proposition 6.6** *Let $f$ be a nonnegative measurable function on $X$ for which almost every slice function $f^z$, $z \in Z$, satisfies the finiteness condition $\mathcal{L}^k(f^z > t) < \infty$ for each $t > 0$. If $H \in \mathcal{H}(\mathbb{R}^n)$ satisfies (6.12), and $0 \notin H$, $0 \in H^+$, then*

$$(f^\#)_H = f^\#. \tag{6.15}$$

**Proposition 6.7** *Suppose that $f \in C_c(X)$ and that $f \neq f^\#$. Then there exists $H \in \mathcal{H}(\mathbb{R}^n)$ of the form (6.12) with $0 \notin H,\ 0 \in H^+$ such that*

$$\int_X f f^\# \, dx < \int_X f_H f^\# \, dx. \tag{6.16}$$

Proposition 6.6 follows easily from (6.13), (6.14) and Proposition 2.10, applied to the slice functions. The proposition can fail if $H \in \mathcal{H}(\mathbb{R}^n)$ does not satisfy (6.12).

Proposition 6.7 follows from the Fubini decomposition, Proposition 2.11(b), a points of density argument and continuity on $Z$ of the functions $z \mapsto \int_{\mathbb{R}^k} f^z\, (f^z)^\# \, dy$ and $z \mapsto \int_{\mathbb{R}^k} (f_H)^z\, (f^z)^\# \, dy$.

The Steiner analogue of Proposition 2.11(a), which asserts that $f \in C_c(X)$ implies $f^\# \in C_c(\mathbb{R}^n)$, also remains true in the Steiner case. It will be proved in §6.5.

## 6.4 Certain Integral Functionals Increase or Decrease under Steiner Symmetrization

Theorem 2.15, one of the main results of Chapter 2, asserts that, under appropriate assumptions, the integrals

$$\int_{\mathbb{R}^{2n}} \Psi(f(x_1), g(x_2))K(|x_1 - x_2|)\, dx_1\, dx_2$$

and

$$\int_{\mathbb{R}^n} \Psi(f(x), g(x))\, dx$$

decrease when $f$ and $g$ are replaced by their symmetric decreasing rearrangements. Theorem 2.15 spawns various other integral inequalities for s.d.r.s, embodied in Corollaries 2.16, 2.19, 2.20, and 2.22.

All of these integral inequalities remain true for Steiner symmetrization. We record them in a megatheorem, Theorem 6.8. Notation is as follows:

- $n = k + m,\ 1 \le k \le n - 1.$ $Z$ is an open set in $\mathbb{R}^m$. $X = \mathbb{R}^k \times Z$.
- $f^\#$ denotes the $(k, n)$-Steiner symmetrization of $f$. If $x \in \mathbb{R}^n$ then $x = (y, z)$ with $y \in \mathbb{R}^k,\ z \in Z$.
- $x_1$ and $x_2$ denote points of $\mathbb{R}^n$, with respective decompositions $x_i = (y_i, z_i),\ i = 1, 2$.

**Theorem 6.8** *Suppose that:*

*(i) $f$ and $g$ are nonnegative $\mathcal{L}^n$-measurable functions on $X$ satisfying the finiteness condition* (6.2) *for $\mathcal{L}^m$-almost every $z \in Z$.*
*(ii) $K\colon \mathbb{R}^+ \to \mathbb{R}^+$ is decreasing.*
*(iii) $\Psi\colon \mathbb{R}^+ \times \mathbb{R}^+ \to \mathbb{R}^+$ is in $AL_0$ (defined in §2.3).*
*(iv) $\Phi\colon \mathbb{R}^+ \to \mathbb{R}^+$ is convex and increasing, with $\Phi(0) = 0$.*

*Then:*

$$\begin{aligned}&\int_{X^2} \Psi(f(x_1), g(x_2))K(|x_1 - x_2|)\, dx_1\, dx_2\\ &\quad \le \int_{X^2} \Psi(f^\#(x_1), g^\#(x_2))K(|x_1 - x_2|)\, dx_1\, dx_2,\end{aligned} \tag{6.17}$$

$$\int_X \Psi(f(x), g(x))\, dx \le \int_X \Psi(f^\#(x), g^\#(x))\, dx, \tag{6.18}$$

$$\int_X fg\, dx \le \int_X f^\# g^\#\, dx, \tag{6.19}$$

$$\begin{aligned}&\int_{X^2} f(x_1)g(x_2)K(|x_1 - x_2|)\, dx_1\, dx_2 \\ &\quad\le \int_{X^2} f^\#(x_1)g^\#(x_2)K(|x_1 - x_2|)\, dx_1\, dx_2,\end{aligned} \tag{6.20}$$

$$\int_X \Phi(|f^\# - g^\#|)\, dx \le \int_X \Phi(|f - g|)\, dx, \tag{6.21}$$

$$\begin{aligned}&\int_{X^2} \Phi(|f^\#(x_1) - g^\#(x_2)|)K(|x_1 - x_2|)\, dx_1\, dx_2 \\ &\quad\le \int_{X^2} \Phi(|f(x_1) - g(x_2)|)K(|x_1 - x_2|)\, dx_1\, dx_2.\end{aligned} \tag{6.22}$$

*Moreover, if $f$, $g$, $K$, and $\Psi$ satisfy the hypotheses of Theorem 2.15(b), then equality holds in* (6.17) *and* (6.20) *if and only if there exists a translation $T$ of $\mathbb{R}^n$ such that $f = f^\# \circ T$ and $g = g^\# \circ T$, $\mathcal{L}^n$-a.e.*

*If $\Psi$ satisfies the hypotheses of Theorem 2.15(d), then equality holds in* (6.18) *and* (6.19) *if and only if the set*

$$A^z \equiv \{(y_1, y_2) \in \mathbb{R}^{2k} : f(y_1, z) < f(y_2, z) \quad \text{and} \quad g(y_1, z) > g(y_2, z)\}$$

*has $\mathcal{L}^{2k}(A^z) = 0$ for $\mathcal{L}^m$-almost every $z \in Z$.*

*Proof* The integral on the left side of (6.17) equals

$$\int_{Z^2} dz_1 dz_2 \int_{\mathbb{R}^{2k}} \Psi(f(y_1, z_1), g(y_2, z_2))K((|y_1 - y_2|^2 + |z_1 - z_2|^2)^{1/2})\, dy_1 dy_2.$$

For $a \ge 0$, the function $t \to K((t^2 + a^2)^{1/2})$ is nonnegative and decreasing on $[0, \infty)$. By Theorem 2.15(a), for each fixed $z_1, z_2$, the integral with respect to $dy_1\, dy_2$ increases when $f(y_1, z_1)$ and $g(y_2, z_2)$ are replaced by $f^\#(y_1, z_1)$ and $g^\#(y_2, z_2)$, where # denotes s.d.r. on $\mathbb{R}^k$. We conclude that inequality (6.17) remains true when # denotes $(k, n)$-Steiner symmetrization.

Each of the inequalities (6.18)–(6.22) can be similarly proved for Steiner symmetrization using Fubini's Theorem and the corresponding s.d.r inequalities in Chapter 2.

The equality statements can be derived from Theorems 2.15(b), 2.15(d) and point of density arguments. Details are left to the reader. □

## 6.5 Steiner Symmetrization Decreases the Modulus of Continuity

For s.d.r.s, a cornerstone of the theory expounded in Chapter 2 is Theorem 2.12, which asserts that the modulus of continuity of a continuous function on $\mathbb{R}^n$ decreases under symmetric decreasing rearrangement. This decrease was a principal ingredient in our proof of the basic inequalities for s.d.r. in Theorem 2.15. Now we will return the favor: with the aid of the integral inequalities, we shall prove that continuity moduli decrease under Steiner symmetrization.

We shall also require a small generalization of the s.d.r. modulus of continuity result which we record as the following lemma. In the lemma, we use $\tilde{g}$ to denote the $k$-dimensional s.d.r. of a nonnegative function $g$ on $\mathbb{R}^k$. Note that if $\lim_{y\to\infty} g(y) = 0$, then $g$ satisfies the finiteness condition $\mathcal{L}^k(g > t) < \infty$ for every $t > 0$, so that $\tilde{g}$ exists.

**Lemma 6.9** *Let $g$ and $h$ be nonnegative continuous functions on $\mathbb{R}^k$ with $\lim_{y\to\infty} g(y) = \lim_{y\to\infty} h(y) = 0$. Then for each $t > 0$,*

$$\begin{aligned}\sup\{|\tilde{g}(y_1) - \tilde{h}(y_2)| \colon (y_1, y_2) \in \mathbb{R}^{2k},\ |y_1 - y_2| \le t\}\\ \le \sup\{|g(y_1) - h(y_2)| \colon (y_1, y_2) \in \mathbb{R}^{2k},\ |y_1 - y_2| \le t\}.\end{aligned} \tag{6.23}$$

*Proof* Fix $t > 0$. Assume first that $g$ and $h$ have support in an open ball $B(R)$ centered at the origin. Let

$$A = \{(y_1, y_2) \in B(R+t) \times B(R+t) \colon |y_1 - y_2| \le t\}.$$

Then

$$\begin{aligned}&\sup\{|g(y_1) - h(y_2)| \colon (y_1, y_2) \in \mathbb{R}^{2k},\ |y_1 - y_2| \le t\}\\ &\quad = \sup\{|g(y_1) - h(y_2)| \colon (y_1, y_2) \in A\}\\ &\quad = \lim_{p\to\infty}\left[\int_A |g(y_1) - h(y_2)|^p \, dy_1 \, dy_2\right]^{1/p}\\ &\quad = \lim_{p\to\infty}\left[\int_{|y_1 - y_2| \le t} |g(y_1) - h(y_2)|^p \, dy_1 \, dy_2\right]^{1/p}.\end{aligned}$$

In the second equality, the limit of the $p$-norms equals the complete supremum rather than the essential supremum because the functions $g$ and $h$ are continuous on $\mathbb{R}^k$. By the definition of s.d.r., $\tilde{g}$ and $\tilde{h}$ are also supported in $B(R)$, and by Theorem 2.12, $\tilde{g}$ and $\tilde{h}$ are continuous on $\mathbb{R}^k$. Thus, the same identities hold when $g$ and $h$ are replaced by $\tilde{g}$ and $\tilde{h}$. By Corollary 2.19 with $K = \chi_{[0,\epsilon]}$, the last integral decreases when $f$ and $g$ are replaced by their s.d.r.s. Thus, (6.23) holds when $g$ and $h$ have compact support.

If $g$ and $h$ are continuous nonnegative functions tending to zero at $\infty$, then $(g-\epsilon)^+$ and $(h-\epsilon)^+$ have compact support. The s.d.r. of $(g-\epsilon)^+$ is $(\tilde{g}-\epsilon)^+$. For fixed $t$, the sup on the right-hand side of (6.23) is the increasing limit of the sups for $(g-\epsilon)^+$ and $(h-\epsilon)^+$. The corresponding statement holds for the s.d.r.s. Thus, the validity of (6.23) for each $(g-\epsilon)^+$ and $(h-\epsilon)^+$ implies its validity for $g$ and $h$. □

Let us return now to our usual Steiner setup: $X = \mathbb{R}^k \times Z$, where $1 \le k \le n-1$ and $Z$ is an open subset of $\mathbb{R}^m$. Here is the main theorem about Steiner symmetrization and the modulus of continuity $\omega$.

**Theorem 6.10** *Let $f$ be a continuous nonnegative function on $X$ such that*

$$\lim_{y\to\infty} f(y,z) = 0, \quad \forall\, z \in Z. \tag{6.24}$$

*Then*

$$\omega(t,f^\#) \le \omega(t,f), \quad 0 < t < \infty.$$

If (6.24) holds for a particular $z$ then, as noted above, the slice function $f^z$ satisfies the finiteness condition (6.2) for that $z$. In general the converse is false. But if $f$ is uniformly continuous on $X$ then (6.2) for a given $z$ does imply (6.24) for that $z$. In fact, for uniformly continuous $f$, satisfying (6.24) for all $z \in Z$ is equivalent to satisfying (6.2) for all $z \in Z$, which is equivalent to satisfying (6.2) for almost every $z \in Z$.

The hypothesis in Theorem 6.10 is a bit stronger than the hypothesis in Theorem 2.12. The nonnegativity hypothesis in Theorem 6.10 is there because we have only considered Steiner symmetrizations for nonnegative functions. Hypothesis (6.24) can probably be weakened, but we will not pursue this question now.

*Proof* Let $t > 0$ be given. Take $x_1 = (y_1, z_1)$, $x_2 = (y_2, z_2)$ with $|x_1 - x_2| \le t$ and $x_1, x_2 \in X$. Define $s$ by $s^2 = t^2 - |z_1 - z_2|^2$. Apply Lemma 6.9 with $g = f^{z_1}$, $h = f^{z_2}$ and $t$ replaced by $s$. Then

$$\begin{aligned}|f^\#(x_1) - f^\#(x_2)| &= |\tilde{g}(y_1) - \tilde{h}(y_2)| \\ &\le \sup_{|v_1 - v_2| \le s} |g(v_1) - h(v_2)| \le \sup_{|u_1 - u_2| \le t} |f(u_1) - f(u_2)| \\ &= \omega(t,f).\end{aligned}$$

Thus, $\omega(t,f^\#) \le \omega(t,f)$. □

Our next goal is to prove a Steiner version of the isodiametric inequality. Again, the proof is based on a lemma which generalizes the s.d.r. case. In the statement of the lemma, the tilde denotes s.d.r. on $\mathbb{R}^k$.

**Lemma 6.11** *Let $E_1$ and $E_2$ be nonempty $\mathcal{L}^k$-measurable subsets of $\mathbb{R}^k$. Then*

$$\sup\{|y_1 - y_2|: y_1 \in \tilde{E_1},\, y_2 \in \tilde{E_2}\} \leq \sup\{|y_1 - y_2|: y_1 \in E_1,\, y_2 \in E_2\}. \quad (6.25)$$

*Proof* If one of the sets is unbounded then the right-hand side of (6.25) is infinite and the inequality holds. So assume $E_1$ and $E_2$ are bounded. Take $0 < p < \infty$. Apply Corollary 2.19 with $f = \chi_{E_1}$, $g = \chi_{E_2}$ and $K(t) = (M^p - t^p)^+$, where $M$ is so large that $y_1 \in E_1$, $y_2 \in E_2$ imply $|y_1 - y_2| < M$ and $y_1 \in \tilde{E_1}$, $y_2 \in \tilde{E_2}$ imply $|y_1 - y_2| < M$. After cancellation of the term $M^p \mathcal{L}^k(E_1)\mathcal{L}^k(E_2)$, we obtain

$$\int_{\tilde{E_1} \times \tilde{E_2}} |y_1 - y_2|^p \, dy_1 \, dy_2 \leq \int_{E_1 \times E_2} |y_1 - y_2|^p \, dy_1 \, dy_2.$$

Taking $p$th roots and letting $p \to \infty$, we see that the essential supremum of the function $|y_1 - y_2|$ over $\tilde{E_1} \times \tilde{E_2}$ is $\leq$ its essential supremum over $E_1 \times E_2$. Since $\tilde{E_1}$ and $\tilde{E_2}$ are balls, the essential supremum of $|y_1 - y_2|$ over $\tilde{E_1} \times \tilde{E_2}$ equals the true supremum. The essential supremum over $E_1 \times E_2$ is $\leq$ the true supremum. Inequality (6.25) follows. □

Here now is the Steiner version of the isodiametric inequality. In the theorem, the # denotes $(k, n)$-Steiner symmetrization.

**Theorem 6.12** *Let $E \subset X$ be $\mathcal{L}^n$-measurable. Then*

$$\operatorname{diam} E^{\#} \leq \operatorname{diam} E. \quad (6.26)$$

*Proof* Take $x_1 = (y_1, z_1)$, $x_2 = (y_2, z_2) \in E^{\#}$. Application of Lemma 6.11 to the slice sets $E(z_1), E(z_2)$ yields

$$|y_1 - y_2| \leq \sup\{|v_1 - v_2|: v_1 \in E^{z_1},\, v_2 \in E^{z_2}\}.$$

Squaring this inequality and adding $|z_1 - z_2|^2$ to both sides, we obtain

$$|x_1 - x_2|^2 \leq \sup\{|(v_1, z_1) - (v_2, z_2)|^2: v_1 \in E^{z_1},\, v_2 \in E^{z_2}\}.$$

The points $(v_1, z_1)$ and $(v_2, z_2)$ on the right-hand side belong to $E$, so the right-hand side is $\leq (\operatorname{diam} E)^2$. Taking the sup over all $x_1$ and $x_2$, we obtain (6.26). □

**Corollary 6.13** *Suppose that $f$ is a continuous nonnegative function on $X$ for which every slice function $f^z$ satifies the finiteness condition* (6.2). *Then the Steiner symmetrization $f^{\#}$ satisfies*

$$\operatorname{diam} \operatorname{supp} f^{\#} \leq \operatorname{diam} \operatorname{supp} f. \quad (6.27)$$

*Proof* Let $E = (f > 0)$. Then $E^\# = (f^\# > 0)$. Applying (6.26) and using the fact that the diameter of a set equals the diameter of its closure, we obtain

$$\operatorname{diam} \operatorname{supp} f^\# = \operatorname{diam} E^\# \leq \operatorname{diam} E = \operatorname{diam} \operatorname{supp} f,$$

which is (6.27). □

The last result of this section asserts that Steiner symmetrization acts contractively on $L^\infty(X)$.

**Theorem 6.14** *Let $f, g\colon X \to \mathbb{R}^+$, and suppose that $f^z$ and $g^z$ satisfy the finiteness condition* (6.2) *for almost every $z \in Z$. Then*

$$\|f^\# - g^\#\|_{L^\infty(X)} \leq \|f - g\|_{L^\infty(X)}. \tag{6.28}$$

*Proof* If $f$ and $g$ have compact support then (6.28) is a limiting case of the $L^p$-contraction property, as in the proof of Corollary 2.23 in Chapter 2.

To prove (6.28) in general, we argue as follows. If $\|f^\# - g^\#\|_{L^\infty(X)} = 0$ we are done. Otherwise, take $0 < \alpha < \|f^\# - g^\#\|_{L^\infty(X)}$. Let

$$A = \{x \in X\colon |f(x) - g(x)| > \alpha\}, \quad B = \{x \in X\colon |f^\#(x) - g^\#(x)| > \alpha\}.$$

Then

$$0 < \mathcal{L}^n(B) = \int_Z \mathcal{L}^k(B^z)\, d\mathcal{L}^m(z).$$

Let

$$G_1 = \{z \in Z\colon \mathcal{L}^k(A^z) > 0\}, \quad G_2 = \{z \in Z\colon \mathcal{L}^k(B^z) > 0\}.$$

Then $\mathcal{L}^m(G_2) > 0$. If $z \in G_2$, then $\|\tilde{f^z} - \tilde{g^z}\|_{L^\infty(\mathbb{R}^k)} > \alpha$. Since s.d.r. acts contractively (Corollary 2.23(b)), we have $\|f^z - g^z\|_{L^\infty(\mathbb{R}^k)} > \alpha$, which implies that $\mathcal{L}^k(A^z) > 0$ and hence that $z \in G_1$. Thus $G_2 \subset G_1$, so that $\mathcal{L}^m(G_1) > 0$ and

$$\mathcal{L}^n(A) = \int_Z \mathcal{L}^k(A^z)\, d\mathcal{L}^m(z) \geq \int_{G_1} \mathcal{L}^k(A^z)\, d\mathcal{L}^m(z) > 0.$$

Thus, $\|f - g\|_{L^\infty(X)} > \alpha$, from which (6.28) follows. □

## 6.6 Steiner Symmetrization Decreases Dirichlet Integrals

Let

$$X = \mathbb{R}^k \times Z$$

with $Z \subset \mathbb{R}^m$ open, and let # denote $(k, n)$-Steiner symmetrization. For functions $f(x) = f(y, z)$ on $X$ having first order partial derivatives in some sense we write

$$\nabla f(x) = (\nabla_y f(x), \nabla_z f(x)),$$

where

$$\nabla_y f = \left(\frac{\partial f}{\partial y_1}, \ldots, \frac{\partial f}{\partial y_k}\right), \qquad \nabla_z f = \left(\frac{\partial f}{\partial z_1}, \ldots, \frac{\partial f}{\partial z_m}\right).$$

Suppose that $f \in \mathrm{Lip}(X, \mathbb{R}^+)$ and that

$$\lim_{y \to \infty} f(y, z) = 0, \quad \forall\, z \in Z. \tag{6.29}$$

Then each slice function $f^z$ is in $\mathrm{Lip}(\mathbb{R}^k)$ and, by (6.29), $f^z$ vanishes at $\infty$. Theorem 3.11, about decrease of integrals under s.d.r., is applicable, and gives

$$\int_{\mathbb{R}^k} \Phi(|\nabla_y f^\#(y, z)|)\, dy \le \int_{\mathbb{R}^k} \Phi(|\nabla_y f(y, z)|)\, dy, \quad z \in Z.$$

Fubini's Theorem gives

$$\int_X \Phi(|\nabla_y f^\#(x)|)\, dx \le \int_X \Phi(|\nabla_y f(x)|)\, dx.$$

In this section we will prove inequalities like this for the transverse gradient $\nabla_z f$ and the full gradient $\nabla f$. To accomplish this, we need to know how integrals involving $\frac{\partial f}{\partial z_i}(y, z)$ behave when $f$ is symmetrized in the $y$-variables. It turns out that they behave remarkably well.

Before examining the integrals we first record the Steiner version of the fact that symmetrization decreases the Lipschitz norm.

**Proposition 6.15** *Let $f \in \mathrm{Lip}(X, \mathbb{R}^+)$, and assume that the limit condition* (6.29) *holds for each $z \in Z$. Then $f^\# \in \mathrm{Lip}(X)$, and*

$$\|f^\#\|_{\mathrm{Lip}(X)} \le \|f\|_{\mathrm{Lip}(X)}, \tag{6.30}$$

$$\|\nabla f^\#\|_{L^\infty(X)} \le \|\nabla f\|_{L^\infty(X)}. \tag{6.31}$$

As noted in §6.5, functions satisfying (6.29) satisfy the finiteness condition (6.2) for every $z$, so the Steiner symmetrization $f^\#$ is well-defined.

*Proof* Using decrease of continuity modulus under Steiner symmetrization (Theorem 6.10), the proof of (6.30) is the same as in the s.d.r. case (Theorem 3.6). If $Z$ is convex then the proof of Corollary 3.4 works when $\mathbb{R}^n$ is replaced by $X$, and shows that

$$\|f\|_{\mathrm{Lip}(X)} = \|\nabla f\|_{L^\infty(X)}, \quad \|f^\#\|_{\mathrm{Lip}(X)} = \|\nabla f^\#\|_{L^\infty(X)}.$$

Thus, (6.31) holds when $Z$ is convex.

Consider now general open $Z$. If $\|\nabla f^\#\|_{L^\infty(X)} = 0$ then (6.31) holds. Suppose that $\|\nabla f^\#\|_{L^\infty(X)} > 0$. Take $0 < \alpha < \|\nabla f^\#\|_{L^\infty(X)}$. Set $E = (|\nabla f^\#| > \alpha)$

$\subset X$, and $F = \{z \in Z\colon \mathcal{L}^k(E(z)) > 0\}$. Then $\mathcal{L}^n(E) > 0$, and by Fubini's Theorem, $\mathcal{L}^m(F) > 0$. Let $z_0$ be a point of density of $F$, $B$ an open ball in $\mathbb{R}^m$ with $z_0 \in B \subset Z$, and $X_0 = \mathbb{R}^k \times B$. Then Fubini implies $\mathcal{L}^n(E \cap X_0) > 0$, which implies that $\|\nabla f^\#\|_{L^\infty(X_0)} > \alpha$. Hence, using convexity of $X_0$,

$$\begin{aligned}\alpha < \|\nabla f^\#\|_{L^\infty(X_0)} = \|f^\#\|_{\mathrm{Lip}(X_0)} \le \|f\|_{\mathrm{Lip}(X_0)} = \|f\|_{L^\infty(X_0)} \\ \le \|f\|_{L^\infty(X)}.\end{aligned}$$

Inequality (6.31) follows. □

Here now is the main result about Steiner symmetrization and Dirichlet integrals.

**Theorem 6.16** *Let $f \in \mathrm{Lip}(X, \mathbb{R}^+)$, and assume that the limit condition* (6.29) *holds for each $z \in Z$. Then there exists a set $E \subset Z$ with $\mathcal{L}^m(Z \setminus E) = 0$ such that, for each $z \in E$, $f$ and $f^\#$ are differentiable at $(y, z)$ for $\mathcal{L}^k$-a.e $y \in \mathbb{R}^k$, and for each convex increasing function $\Phi\colon \mathbb{R}^+ \to \mathbb{R}^+$ with $\Phi(0) = 0$, we have*

$$\int_{\mathbb{R}^k} \Phi(|\partial_{z_i} f^\#(y,z)|)\,dy \le \int_{\mathbb{R}^k} \Phi(|\partial_{z_i} f(y,z)|)\,dy, \quad i = 1, \ldots, m, \tag{6.32}$$

$$\int_{\mathbb{R}^k} \Phi(|\nabla_z f^\#(y,z)|)\,dy \le \int_{\mathbb{R}^k} \Phi(|\nabla_z f(y,z)|)\,dy, \tag{6.33}$$

$$\int_{\mathbb{R}^k} \Phi(|\nabla f^\#(y,z)|)\,dy \le \int_{\mathbb{R}^k} \Phi(|\nabla f(y,z)|)\,dy. \tag{6.34}$$

Note that we do not assume that $f(y, z) \to 0$ when $z$ approaches the boundary of $Z$ or when $z \to \infty$. The term "differentiable" in the statement of the theorem means that $f$ is differentiable as a function of all $n$ variables $y_1, \ldots, y_k, z_1, \ldots, z_m$.

Together with the theorems of Rademacher and Fubini, Theorem 6.16 immediately implies

**Corollary 6.17** *Let $f \in \mathrm{Lip}(X, \mathbb{R}^+)$, and assume that the limit condition* (6.29) *holds for each $z \in Z$. Then*

$$\int_X \Phi(|\partial_{z_i} f^\#(x)|)\,dx \le \int_X \Phi(|\partial_{z_i} f(x)|)\,dx, \quad i = 1, \ldots, m,$$

$$\int_X \Phi(|\nabla_z f^\#(x)|)\,dx \le \int_X \Phi(|\nabla_z f(x)|)\,dx,$$

$$\int_X \Phi(|\nabla f^\#(x)|)\,dx \le \int_X \Phi(|\nabla f(x)|)\,dx. \tag{6.35}$$

*Proof of Theorem 6.16* Assume first that $f$ also satisfies the following condition, which we'll call Condition S: For each compact $A \subset Z$, there exists $R \in (0, \infty)$ such that $f(y, z) = 0$ whenever $|y| \ge R$ and $z \in A$.

Define the set $P$ to be the set of all $z \in Z$ such that $f$ is differentiable at $(y, z)$ for $\mathcal{L}^k$-almost every $y \in \mathbb{R}^k$.

By Proposition 6.15, $f^\#$ is also Lipschitz, and it is easy to see that $f^\#$ also satisfies Condition S. Let $\overline{P}$ be obtained from $f^\#$ like $P$ is obtained from $f$.

By Rademacher's Theorem 3.2, we know that $f$ is differentiable $\mathcal{L}^n$-a.e. on $X$. From Fubini's Theorem, it follows that $\mathcal{L}^m(Z \setminus P) = 0$. The same is true for $f^\#$ and $\overline{P}$. Define

$$E = P \cap \overline{P}.$$

Next, we will show that if $f$ satisfies Condition S and $a \in E$, then (6.34) holds when $z = a$ and $\Phi(x) = x$. Let $B = \mathbb{B}^m(a, \rho)$ be an open $k$-ball centered at $a$ with compact closure in $Z$. There exists $R$ such that $f(y, z) = 0$ whenever $|y| \geq R$ and $z \in B$. Let $K = \chi_{[0,1]}$, and let $\epsilon > 0$ be small. Define

$$I(\epsilon,f) = \int_{\mathbb{R}^k \times \mathbb{R}^k \times B} |f(y, a) - f(v, w)|\, K(\epsilon^{-1}(|y - v|^2 + |a - w|^2)^{1/2})\, dy\, dv\, dw.$$

Integrate first with respect to $dy\, dv$. By Corollary 2.19, applied in $\mathbb{R}^k$ to the functions $f(\cdot, a)$ and $f(\cdot, w)$, we see that for each fixed $w \in B$, $I(\epsilon,f)$ decreases when $f$ is replaced by its $(k, n)$-Steiner symmetrization $f^\#$. Thus

$$I(\epsilon,f^\#) \leq I(\epsilon,f). \tag{6.36}$$

Let $v = y + \epsilon s$, $w = a + \epsilon t$. Then $s \in \mathbb{R}^k$, $t \in \mathbb{B}^m(0, \rho/\epsilon)$, and

$$I(\epsilon,f) = \epsilon^n \int_{\mathbb{R}^k \times \mathbb{R}^k \times \mathbb{B}^m(0,\rho/\epsilon)} |f(y, a) - f(y + \epsilon s, a + \epsilon t)| \\ K((s^2 + t^2)^{1/2})\, dy\, ds\, dt.$$

Multiply $I(\epsilon,f)$ by $\epsilon^{-n-1}$. The integrand is uniformly bounded above, and vanishes outside a compact subset of $\mathbb{R}^k \times \mathbb{R}^k \times \mathbb{R}^m$ when $\epsilon$ is sufficiently small. Since $a \in E, f$ is differentiable at $(y, a)$ for $\mathcal{L}^k$-*a.e* $y$, so as $\epsilon \to 0$ the integrand converges to $|\nabla f(y, a) \cdot (s, t)|$ for $\mathcal{L}^k$-*a.e* $y$. The dominated convergence theorem gives

$$\lim_{\epsilon \to 0} I(\epsilon)\epsilon^{-n-1} = \int_{\mathbb{R}^k \times \mathbb{R}^k \times \mathbb{R}^m} |\nabla f(y, a) \cdot (s, t)| K((s^2 + t^2)^{1/2})\, ds\, dy\, dt.$$

Putting $p = (s, t) \in \mathbb{R}^n$, as in the proof of Theorem 3.7 we have

$$\int_{\mathbb{R}^k \times Z} |\nabla f(y, a) \cdot (s, t)| K((s^2 + t^2)^{1/2})\, ds\, dt \\ = |\nabla f(y, a)| \int_{|p| \leq 1} |p \cdot e_1|\, d\mathcal{L}^n(p) = C_1 |\nabla f(y, a)|,$$

where $C_1$ depends only on $n$. Thus

$$\lim_{\epsilon\to 0} I(\epsilon)\epsilon^{-n-1} = C_1 \int_{\mathbb{R}^k} |\nabla f(y,a)|\, dy, \quad a \in E. \tag{6.37}$$

The same equality holds when $f$ is replaced by $f^{\#}$. From (6.36) and (6.37) we see that

$$\int_{\mathbb{R}^k} |\nabla f^{\#}(y,a)|\, dy \le \int_{\mathbb{R}^k} |\nabla f(y,a)|\, dy, \quad a \in E.$$

We have shown that (6.34) holds when $\Phi(x) = x$ and $f$ satisfies Condition S.

Next, we shall retain the assumption that $f$ satisfies Condition S, and will show that (6.34) is true for arbitrary $\Phi$ satisfying the stated hypotheses. At first we will follow the argument used for the symmetric decreasing rearrangement case, Theorem 3.11. Fix $a \in E$. By Condition S, the support of $f(\cdot, a)$ is contained in a $k$-ball $|y| \le R$. Let $M = \|f\|_{\mathrm{Lip}(X,\mathbb{R})}$, and let $\epsilon > 0$ be given. Define $\delta$ by

$$\epsilon = \delta\beta_{k-1}\left[M\Phi'(M+) + \Phi(M) + \frac{R^n}{n}\Phi'(M+)\right].$$

Define also

$$\begin{aligned} g(r) &= f^{\#}(re_1, a), & G(r) &= |\nabla f^{\#}(re_1, a)|, \\ I_i &= g^{-1}(b_i, b_{i-1}), & d\nu(r) &= r^{k-1}dr. \end{aligned} \tag{6.38}$$

Then $g$ is Lipschitz and decreasing, but $G$ need not be decreasing and might be defined only if $\mathcal{L}^1$-a.e. It is easy to see that $\frac{\partial f^{\#}}{\partial z_i}(\cdot, a)$ is a radial function on $\mathbb{R}^k$ for each $i = 1, \ldots, m$. So is $|\nabla_y f^{\#}(y,a)|$. Thus $|\nabla f^{\#}(\cdot, a)|$ is also radial, and $|\nabla f^{\#}(y,a)| = G(|y|)$. Moreover, using $\mathcal{L}^k$-a.e. differentiability of $f^a$ and proceeding as in the proof of Proposition 6.15, we obtain

$$\|G\|_{L^\infty[0,R]} \le \|f^{\#}\|_{\mathrm{Lip}(X_0)} \le \|f\|_{\mathrm{Lip}(X_0)} \le M,$$

where $X_0$ is as in Proposition 6.15 with $z_0$ there replaced by $a$. We also have

$$\int_{\mathbb{R}^k} \Phi(|\nabla f^{\#}(y,a)|)\, dy = \beta_{k-1} \int_0^R \Phi(G(r))\, d\nu(r). \tag{6.39}$$

By Lusin's Theorem, there is a continuous function $h$ on $[0,R]$ and a set $B \subset [0,R]$ such that $0 \le h \le M$ on $[0,R]$, $h = G$ on $[0,R] \setminus B$, and $\nu(B) < \delta$.

Let $\{a_i\}_{i=0}^j$ be a sequence in $\mathbb{R}^+$ with $0 = a_0 < a_1 < \cdots < a_j = R$ such that the oscillation of $h$ over each $[a_{i-1}, a_i]$ is less than $\delta$. Let $b_0 = g(a_0)$, and for $i = 1, \ldots, j$, set

$$b_i = g(a_i), \quad A_i = (f^a)^{-1}(b_i, b_{i-1}), \quad B_i = (f^a)^{-1}(b_i).$$

Let $A'_i$ and $B'_i$ be the sets obtained by changing $f$ to $f^\#$ in the definition of $A_i$ and $B_i$. Then

$$\int_{\mathbb{R}^k} \Phi(|\nabla f(y,a)|)\,dy = \sum_{i=1}^{j} \int_{A_i} \Phi(|\nabla f(y,a)|)\,dy + \sum_{i=0}^{j} \int_{B_i} \Phi(|\nabla f(y,a)|)\,dy, \tag{6.40}$$

and the analogous formula holds when $f$ is replaced by $f^\#$, $A_i$ by $A'_i$, and $B_i$ by $B'_i$.

Now $f^a$ is constant on $B_i$, so $\nabla_y f(y,a) = 0$ for $\mathcal{L}^k$-almost every $y \in B_i$. It is not true, though, that $\nabla_z f(y,a)$ must vanish for $\mathcal{L}^k$-a.e $y \in \mathbb{R}^k$. Thus, the integrals over the $B_i$ in (6.40) can be positive. It is this phenomenon which makes the Steiner case more difficult than the s.d.r. case proved in Theorem 3.11.

To surmount this obstacle, we shall establish a lemma.

**Lemma 6.18** *Let $f$ and $X$ be as in the statement of Theorem 6.16. Suppose that $a \in Z$ is such that $f$ and $f^\#$ are differentiable at $(y,a)$ for $\mathcal{L}^k$-almost every $y \in \mathbb{R}^k$, and that $b \in [0,\infty)$ is such that $B \equiv (f^a)^{-1}(b)$ has $\mathcal{L}^k(B) > 0$. Then the restriction of $|\nabla f(\cdot,a)|$ to $B$ has the same distribution, with respect to $\mathcal{L}^k$, as $|\nabla f^\#(\cdot,a)|$ restricted to $B'$, where $B' = ((f^\#)^a)^{-1}(b)$.*

We shall prove Lemma 6.18 in §6.7. Let us assume it for the time being.

Next, we claim that

$$\int_{A'_i} |\nabla f^\#(y,a)|\,dy \le \int_{A_i} |\nabla f(y,a)|\,dy, \quad i = 1,\dots,j. \tag{6.41}$$

To prove this, fix $i$. Take a sequence $\{\psi_j\}_{j\ge1} \subset C^1(\mathbb{R})$ such that $\psi_j(b_i) = 0$, $\psi'_j \ge 0$ on $\mathbb{R}$ and $\lim_{j\to\infty} \psi'_j(t) = \chi_{(b_i,b_{i-1})}(t)$ for every $t \in \mathbb{R}$. Then each $\psi_j \circ f \in \mathrm{Lip}(\mathbb{R}^n,\mathbb{R})$, and each point of differentiability of $f$ is a point of differentiability of $\psi_j \circ f$. The same considerations apply to $f^\#$. The first part of the proof is applicable to $\psi_j \circ f$ on the slice $z = a$, and we obtain

$$\int_{\mathbb{R}^k} |\nabla(\psi_j \circ f)^\#(y,a)|\,dy \le \int_{\mathbb{R}^k} |\nabla(\psi_j \circ f)|\,dy. \tag{6.42}$$

Furthermore, for any increasing (=nondecreasing) function $\psi$ on $[0,\infty)$ with $\psi(0+) = 0$, the functions $\psi \circ f^a$ and $\psi \circ (f^\#)^a$ have the same distribution on $\mathbb{R}^k$, hence the same s.d.r. on $\mathbb{R}^k$. The second function is already symmetric decreasing. We deduce that $(\psi_j \circ f)^\# = \psi_j \circ f^\#$, so that in (6.42) the left-hand integrand can be replaced by $|\nabla(\psi_j \circ f^\#)|$.

Now $\nabla(\psi_j \circ f) = (\psi'_j \circ f)\,\nabla f$ at points of differentiability of $f$. Letting $j \to \infty$, the dominated convergence theorem implies that the right-hand side of (6.42) converges to the right-hand side of (6.41), The same is true for the left-hand sides. We have proved the claim (6.41).

Write

$$\mu_i = \mathcal{L}^k(A_i).$$

Then

$$\begin{aligned}\int_{A_i} \Phi(|\nabla f(y,a)|\,dy &\geq \mu_i \Phi\left(\mu_i^{-1}\int_{A_i} |\nabla f(y,a)|\,dy\right)\\ &\geq \mu_i \Phi\left(\mu_i^{-1}\int_{A_i'} |\nabla f^{\#}(y,a)|\,dy\right)\\ &= \mu_i \Phi\left(\nu(I_i)^{-1}\int_{I_i} G(r)\,d\nu(r)\right)\\ &= \int_{A_i'} \Phi(|\nabla f^{\#}(y,a)|)\,dy + \mu_i(s_i+t_i+u_i),\end{aligned} \tag{6.43}$$

where the first inequality is by Jensen's inequality, the second by (6.41), and the first equality by the definitions (6.38) of $I_i$, $g$, $G$, and $\nu$. The numbers $s_i$, $t_i$ and $u_i$ are defined as in the proof of Theorem 3.11 between (6.15) and (6.16), except that here we substitute $G$ for $|g'|$ and $j$ for $m$. The argument in the proof of Theorem 3.11 carries over verbatim to give

$$\left|\sum_{i=1}^{j} \mu_i(s_i+t_i+u_i)\right| \leq \epsilon.$$

Summing in (6.43), we obtain

$$\sum_{i=1}^{j}\int_{A_i} \Phi(|\nabla f(y,a)|)\,dy \geq \sum_{i=1}^{j}\int_{A_i} \Phi(|\nabla f^{\#}(y,a)|)\,dy - \epsilon. \tag{6.44}$$

By Lemma 6.18, when the $A_i$ and $A_i'$ are replaced by $B_i$ and $B_i'$ there is equality in (6.44). From (6.40) and (6.44), we thus obtain

$$\int_{\mathbb{R}^k} \Phi(|\nabla f^{\#}(y,a)|)\,dy \geq \int_{\mathbb{R}^k} \Phi(|\nabla f(y,a)|)\,dy - \epsilon.$$

Modulo Lemma 6.18, we have now proved (6.34) for all $\Phi$ when $f$ satisfies Condition S.

Next, let us obtain (6.34) without assuming that $f$ satisfies Condition S. Suppose that $f$ satisfies the hypotheses of Theorem 6.16. Let $\{s_j\}_{j=1}^{\infty}$ be a sequence of positive numbers decreasing to zero. Define $f_j = (f - s_j)^+$. Then each $f_j$ is Lipschitz, hence is differentiable in $X$ except for an $\mathcal{L}^n$-nullset. Moreover, $(f_j)^{\#} = (f^{\#} - s_j)^+$.

Let $P_j$ be the set of all $z \in Z$ such that $f_j$ is differentiable at $(y, z)$ for $\mathcal{L}^k$-a.e $y \in \mathbb{R}^k$. As before, "differentiable" means differentiable as a function of all $n$ variables $y_1, \ldots, y_k, z_1, \ldots, z_m$. Let $P$ be the set obtained by replacing $f_j$ by $f$ in the definition of $P_j$. From Fubini's Theorem, it follows that $\mathcal{L}^m(Z \setminus P_j) = 0$ for each $j$ and that $\mathcal{L}^m(Z \setminus P) = 0$.

We claim that each $f_j$ satisfies Condition S. The claim is equivalent to saying that for each $s > 0$ and each compact $K \subset \mathbb{R}^m$ there exists $R$ such that $|f(y, z)| < s$ whenever $|y| \geq R$ and $z \in K$. If this last statement did not hold, then using the Bolzano–Weierstrass Theorem and the Lipschitzness of $f$ we could find some $z \in K$, some $s_0 > 0$ and a sequence $\{y_i\} \subset \mathbb{R}^k$ with $\lim_{i\to\infty} |y_i| = \infty$ such that $|f(y_i, z)| \geq s_0$ for every $i$. But this violates the limit hypotheses (6.29), thereby proving the claim.

Let $A = \{x \in X : f(x) = 0\}$ and $A(z)$ be the slice of $A$ through $z$. Fix $z \in \left(\bigcap_{j=1}^{\infty} P_j\right) \cap P$. Write

$$\int_{\mathbb{R}^k} \Phi(|\nabla f(y, z)|)\, dy = \int_{A(z)} \Phi(|\nabla f(y, z)|)\, dy + \int_{\mathbb{R}^k \setminus A(z)} \Phi(|\nabla f(y, z)|)\, dy.$$

The differentiability of $f$ at $\mathcal{L}^k$-almost every $y \in \mathbb{R}^k$ insures that the gradient appearing in the integrals is well-defined except for a set of $\mathcal{L}^k$-measure zero. Also, $f$ and each $f_j$ are differentiable at $(y, z)$ except for $y$ in a $\mathcal{L}^k$-nullset which is independent of $j$. Take a $(y, z)$ which is such a point of common differentiability. If $f(y, z) \leq s_j$ then $f_j$ has a local minimum in an $n$-dimensional neighborhood of $(y, z)$, so that $\nabla f_j(y, z) = 0$, while if $f(y, z) > s_j$, it is clear that $\nabla f_j(y, z) = \nabla f(y, z)$. Thus, as $j \to \infty$, $|\nabla f_j(y, z)| \nearrow |\nabla f(y, z)|$ $\mathcal{L}^k$-a.e. on the set of $y$ with $f(y, z) > 0$. By the monotone convergence theorem,

$$\lim_{j\to\infty} \int_{\mathbb{R}^k \setminus A(z)} \Phi(|\nabla f_j(y, z)|)\, dy = \int_{\mathbb{R}^k \setminus A(z)} \Phi(|\nabla f(y, z)|)\, dy. \tag{6.45}$$

At $y \in A(z)$, $f$ and each of the $f_j$ achieve the local minimum zero, so again $\nabla f(y, z) = \nabla f_j(y, z) = 0$ when $(y, z)$ is a point of differentiability for all of these functions. Since $\mathcal{L}^k$-almost every $y$ satisfies this differentiability condition, we have $\Phi(|\nabla f_j(y, z)|) = \Phi(|\nabla f(y, z)|) = 0$ for $\mathcal{L}^k$-almost every $y \in A(z)$. From this equation and (6.45), we conclude that for $z \in \left(\bigcap_{j=1}^{\infty} P_j\right) \cap P$,

$$\lim_{j\to\infty} \int_{\mathbb{R}^k} \Phi(|\nabla f_j(y, z)|)\, dy = \int_{\mathbb{R}^k} \Phi(|\nabla f(y, z)|)\, dy. \tag{6.46}$$

Let $\overline{P_j}$ and $\overline{P}$ be the sets corresponding to $P_j$ and $P$ when $f$ is changed to $f^{\#}$. The complement in $Z$ of each of these sets has $\mathcal{L}^m$-measure zero, and (6.46) remains valid when $f$ and the $f_j$ are replaced by $f^{\#}$ and $(f_j)^{\#}$. Let

$$E = \left(\bigcap_{j=1}^{\infty} P_j\right) \cap P \cap \left(\bigcap_{j=1}^{\infty} \overline{P_j}\right) \cap \overline{P}.$$

Take $z \in E$. We know that each integral on the left in (6.46) decreases when $f_j$ is changed to $(f_j)^{\#}$. Thus, applying (6.46) to the $f_j^{\#}$,

$$\int_{\mathbb{R}^k} \Phi(|\nabla f^{\#}(y,z)|)\,dy \le \int_{\mathbb{R}^k} \Phi(|\nabla f(y,z)|)\,dy.$$

We have now completely proved (6.34), modulo Lemma 6.18.

To prove (6.32) and (6.33), we replace the integrals $I(\epsilon,f)$ used in the proof of (6.34) with different integrals. Let $a \in Z$ be as above. To obtain (6.32), take, for fixed $i \in \{1,\ldots,n\}$,

$$I_1(\epsilon,f) = \int_{\mathbb{R}^k} |f(y,a+\epsilon e_i) - f(y,a)|\,dy,$$

and to obtain (6.33), take

$$I_2(\epsilon,f) = \int_{\mathbb{R}^k \times \mathbb{S}^{m-1}} |f(y,a+\epsilon z) - f(y,a)|\,dy\,d\sigma(z),$$

where $\sigma$ denotes Lebesgue surface measure on the unit sphere $\mathbb{S}^{m-1}$ in $\mathbb{R}^m$.

The arguments giving (6.37) also give

$$\lim_{\epsilon\to 0} I_1(\epsilon,f)\epsilon^{-1} = \int_{\mathbb{R}^k} |\partial_{z_i} f(y,a)|\,dy$$

and

$$\lim_{\epsilon\to 0} I_2(\epsilon,f)\epsilon^{-1} = C\int_{\mathbb{R}^k} |\nabla_z f(y,a)|\,dy, \tag{6.47}$$

where $C = \int_{\mathbb{S}^{m-1}} |e_1 \cdot z|\,d\sigma(z)$.

The inequality $I_1(\epsilon,f) \ge I_1(\epsilon,f^{\#})$ follows from the contraction property for s.d.r. (Corollary 2.20), applied in $\mathbb{R}^k$ to the two functions $f(y,a)$ and $f(y,a+te_i)$. The analogous inequality for $I_2$ also follows from Corollary 2.20, provided we first integrate in the $y$-variables. Thus, $\int_{\mathbb{R}^k} |\partial_{z_i} f(y,a)|\,dy$ and $\int_{\mathbb{R}^k} |\nabla_z f(y,a)|\,dy$ each decrease when $f$ is changed to $f^{\#}$.

Lemma 6.18 is still true when the full gradient is replaced by $\partial_{z_i}$ or $\nabla_z$. The function $G$ introduced before (6.39) should be replaced by $|\partial_{z_i} f^{\#}(re_1,a)|$ or $|\nabla_z f^{\#}(re_1,a)|$. With these changes, the proofs of (6.32) and (6.33) can be accomplished by repeating the arguments used to prove (6.34). Theorem 6.16 is now completely proved, modulo Lemma 6.18. □

By further use of functionals like $I(\epsilon,f)$ one can establish still more rearrangement inequalities involving $\nabla_y f$ and the derivatives $\partial_{z_i} f$.

## 6.7 Proof of Lemma 6.18

If $b = 0$, the discussion between (6.42) and (6.43) shows that $\nabla f(y,a) = \nabla f^{\#}(y,a) = 0$ $\mathcal{L}^k$-a.e on $B$ and $B'$, respectively, and we are done.

Suppose $b > 0$. Then replacing $f$ by $(f - \frac{1}{2}b)^+$, if necessary, we may assume that $f$ satisfies Condition S of §6.6. We may also assume that $\|f\|_{\mathrm{Lip}(X)} \leq 1$. For small $\epsilon > 0$ and points $w \in \mathbb{S}^{m-1}$, define sets $B_\epsilon$ and $B_{\epsilon,w} \subset \mathbb{R}^k$ by

$$B_\epsilon = (b - 2\epsilon < f^a < b + 2\epsilon), \quad B_{\epsilon,w} = (b - \epsilon < f^{a+\epsilon w} < b + \epsilon).$$

From the Lipschitz condition on $f$ and fact that $f(y,a) = b$ for $y \in B$, it follows that

$$B \subset B_{\epsilon,w} \subset B_\epsilon. \tag{6.48}$$

Take $0 < p < \infty$. The arguments giving (6.47) also give

$$\begin{aligned}\lim_{\epsilon\to 0} \epsilon^{-p} \int_{\mathbb{S}^{m-1}} d\sigma(w) \int_B |f(y, a + \epsilon w) - b|^p \, dy \\ = C_p \int_B |\nabla_z f(y,a)|^p \, dy = C_p \int_B |\nabla f(y,a)|^p \, dy,\end{aligned} \tag{6.49}$$

where $C_p = \int_{\mathbb{S}^{m-1}} |e_1 \cdot z|^p \, d\sigma(z)$. The equality $|\nabla f(y,a)| = |\nabla_z f(y,a)|$ holds for $\mathcal{L}^k$-a.e $y \in B$, since $f(\cdot, a)$ is Lipschitz on $\mathbb{R}^k$ and constant on $B$.

For each $w \in \mathbb{S}^{m-1}$ we also have

$$\lim_{\epsilon\to 0} \int_{B_\epsilon \setminus B} \epsilon^{-p} |f(y, a + \epsilon w) - b|^p \, dy = 0. \tag{6.50}$$

In fact, since $f$ has Lipschitz norm $\leq 1$ and $|f(y,a) - b| < 2\epsilon$, for each $\epsilon$ the integral on the left-hand side is $\leq 3^p \mathcal{L}^k(B_\epsilon \setminus B)$. The set $B$ is the decreasing intersection of the sets $B_\epsilon$, and since $\lambda_{f^a}(t) < \infty$ for $t > 0$, the $B_\epsilon$ have finite measure for small $\epsilon$. Thus, in (6.50), the limit as $\epsilon \to 0$ is zero. Since the bound $3^p \mathcal{L}^k(B_\epsilon \setminus B)$ on the integrals holds simultaneously for all $w \in \mathbb{S}^{m-1}$, if we replace $B$ by $B_\epsilon \setminus B$ on the left side of (6.49) the dominated convergence theorem implies that the limit in (6.49) is zero. Consequently, (6.49) remains valid if we replace $B$ on the left by $B_\epsilon$. Then (6.48) implies we can replace $B_\epsilon$ by $B_{\epsilon,w}$. We conclude that

$$\lim_{\epsilon\to 0} \epsilon^{-p} \int_{\mathbb{S}^{m-1}} d\sigma(w) \int_{B_{\epsilon,w}} |f(y, a + \epsilon w) - b|^p \, dy = C_p \int_B |\nabla f(y,a)|^p \, dy. \tag{6.51}$$

Let $B'_{\epsilon,w}$ be formed from $f^{\#}$ like $B_{\epsilon,w}$ was formed from $f$. Then (6.51) is true when $f$ is changed to $f^{\#}$ and $B_{\epsilon,w}$ to $B'_{\epsilon,w}$. The distribution with respect to $\mathcal{L}^k$ of $f^{a+\epsilon w}$ on $B_{\epsilon,w}$ is the same as the distribution of $(f^{\#})^{a+\epsilon w}$ on $B'_{\epsilon,w}$. Thus, for small $\epsilon$,

$$\int_{B_{\epsilon,w}} |f(y, a + \epsilon w) - b|^p \, dy = \int_{B'_{\epsilon,w}} |f^{\#}(y, a + \epsilon w) - b|^p \, dy. \tag{6.52}$$

From (6.51) and (6.52), we obtain

$$\int_B |\nabla f(y, a)|^p \, dy = \int_{B'} |\nabla f^{\#}(y, a)|^p \, dy, \quad 0 < p < \infty.$$

Since $f$ satisfies Condition S, for fixed $a \in Z$ the function $|\nabla f(y, a)|$ has compact support in $\mathbb{R}^k$. To complete the proof of Lemma 6.18, it suffices to prove the following claim.

**Claim** *If $g$ and $h$ are nonnegative $L^\infty$ functions on finite measure spaces $(X, \mu)$ and $(Y, \nu)$, respectively, if $\mu(X) = \nu(Y)$, and if*

$$\int_X g^p \, d\mu = \int_Y h^p \, d\nu, \quad \forall\, p \in (0, \infty), \tag{6.53}$$

*then the distribution of $g$ with respect to $\mu$ on $X$ is the same as the distribution of $h$ with respect to $\nu$ on $Y$.*

*Proof* Take $M \geq \max\{\operatorname{ess\,sup} g, \operatorname{ess\,sup} h\}$. By Corollary 1.17, (6.53) is the same as

$$\int_0^M t^{p-1} \lambda_g(t) \, dt = \int_0^M t^{p-1} \lambda_h(t) \, dt.$$

Let $Q = \lambda_g - \lambda_h$. Then $Q \in L^\infty[0, M]$. Taking $p = 1, 2, \ldots$ it follows that $Q$ is orthogonal to all polynomials on $[0, M]$. By the Weierstrass approximation theorem, $Q$ is orthogonal to all continuous functions, hence $Q = 0$ $\mathcal{L}$-a.e. on $[0, M]$. Since $Q$ is right continuous, it follows that $Q = 0$ at all points of $[0, M]$. Thus $\lambda_g(t) = \lambda_h(t)$ for every $t \in [0, M]$.

The claim is proved and, with it, Lemma 6.18 and Theorem 6.16. □

## 6.8 Steiner Symmetrization Decreases $p$-Dirichlet Integrals in $W^{1,p}(\mathbb{R}^n)$

For brevity, we confine attention to the most basic theorem about Steiner symmetrization of Sobolev functions. There are many possible refinements, variants and extensions. In this section # denotes $(k, n)$-Steiner symmetrization. The set $Z$ will be all of $\mathbb{R}^m$, so that $X = \mathbb{R}^k \times \mathbb{R}^m = \mathbb{R}^n$. We continue with the notation that points $x \in \mathbb{R}^n$ are written as $x = (y, z) = (y_1, \ldots, y_k, z_1, \ldots, z_m)$, $\nabla f$ is the $n$-dimensional gradient of functions on $\mathbb{R}^n$, and $\nabla_y f$, $\nabla_z f$ are the gradients with respect to the $y$ and $z$ variables, respectively.

**Theorem 6.19** *Let $f \in W^{1,p}(\mathbb{R}^n, \mathbb{R}^+)$ and $1 \le p < \infty$. Then $f^\# \in W^{1,p}(\mathbb{R}^n)$, and*

$$\int_{\mathbb{R}^n} |\partial_{z_i} f^\#(x)|^p \, dx \le \int_{\mathbb{R}^n} |\partial_{z_i} f(x)|^p \, dx, \quad i = 1, \dots, m, \tag{6.54}$$

$$\int_{\mathbb{R}^n} |\nabla_z f^\#(x)|^p \, dx \le \int_{\mathbb{R}^n} |\nabla_z f(x)|^p \, dx, \tag{6.55}$$

$$\int_{\mathbb{R}^n} |\nabla f^\#(x)|^p \, dx \le \int_{\mathbb{R}^n} |\nabla f(x)|^p \, dx. \tag{6.56}$$

*Proof* As in the proof of the s.d.r. case (Theorem 3.20), take a nonnegative sequence $\{f_j\} \subset C_c^\infty(\mathbb{R}^n)$ such that $\|f_j - f\|_{W^{1,p}(\mathbb{R}^n)} \to 0$. Each $f_j$ is Lipschitz and has compact support in $\mathbb{R}^n$, so for each $z \in \mathbb{R}^m$ the slice function $f_j^z$ is Lipschitz and has compact support in $\mathbb{R}^k$. Corollary 6.17 is applicable, and shows that inequalities (6.54)–(6.56) hold for each $f_j$. When $1 < p < \infty$ the proof of (6.54)–(6.56) for $f$ now proceeds exactly as it did in Theorem 3.20.

To finish the proof for $p = 1$, we just need to show that the set $\{\nabla f_j^\#\}$ is relatively weakly compact in $L^1(\mathbb{R}^n)$. As in §3.6, this is equivalent to showing that $\{\nabla f_j^\#\}$ is norm-bounded in $L^1(\mathbb{R}^n)$, is uniformly integrable, and satisfies the tail condition

$$\lim_{R\to\infty} \sup_j \int_{|x|\ge R} |\nabla f_j^\#| \, dx = 0. \tag{6.57}$$

The norm boundedness follows from Corollary 6.17. Taking $\Phi(s) = (s-t)^+$ in Corollary 6.17, we deduce uniformly integrability just as in the proof of Theorem 3.20. Thus, to finish the proof of Theorem 6.19, we just need to verify (6.57). It will suffice to verify it separately for $|\nabla_y f_j^\#|$ and $|\nabla_z f_j^\#|$.

Inequality (3.30) is valid for each $(f_j^\#)^z$, and gives

$$\int_{|y|\ge r} |\nabla_y f_j^\#(y,z)| \, dy \le r^{-1} k^2 \int_{\mathbb{R}^k} f_j^\#(y,z) \, dy, \quad r > 0. \tag{6.58}$$

For fixed $R > 0$, let

$$E_1 = \left\{x\colon |z| \ge \frac{1}{2}R\right\}, \quad E_2 = \left\{x\colon |z| \le \frac{1}{2}R,\ |y|^2 \ge R^2 - |z|^2\right\}.$$

Then $\{|x| \ge R\} \subset E_1 \cup E_2$. From (6.58) we deduce

$$\begin{aligned}\int_{E_2} |\nabla_y f_j^\#| \, dx &= \int_{|z|\le\frac{1}{2}R} dz \int_{\mathbb{R}^k} |\nabla_y f_j^\#| \, dy \\ &\le k^2 \int_{|z|\le\frac{1}{2}R} (R^2 - |z|^2)^{-1/2} \, dz \int_{\mathbb{R}^k} f_j^\#(y,z) \, dy \\ &\le 2k^2 R^{-1} \int_{\mathbb{R}^n} f_j^\# \, dx.\end{aligned}$$

Thus,

$$\int_{|x|\geq R} |\nabla_y f_j^{\#}|\, dx \leq \int_{E_1} + \int_{E_2} \leq \int_{|z|\geq \frac{1}{2}R} |\nabla_y f_j^{\#}|\, dx + 2k^2 R^{-1} \int_{\mathbb{R}^n} f_j^{\#}\, dx.$$

By Theorem 6.16 and Fubini's Theorem, the integral over $|z| \geq \frac{1}{2}R$ increases when $f_j^{\#}$ is replaced by $f_j$. The integral over $\mathbb{R}^n$ does not change. So

$$\int_{|x|\geq R} |\nabla_y f_j^{\#}|\, dx \leq \int_{|z|\geq \frac{1}{2}R} |\nabla_y f_j|\, dx + 2k^2 R^{-1} \int_{\mathbb{R}^n} f_j\, dx. \tag{6.59}$$

Now $\{f_j\}$ converges to $f$ in $W^{1,1}$-norm. Thus, the $f_j$ are norm-bounded in $L^1(\mathbb{R}^n)$, and the $\nabla f_j$ are relatively weakly compact in $L^1(\mathbb{R}^n)$, hence satisfy the tail condition. From (6.59), it follows that (6.57) is true when $\nabla$ is replaced by $\nabla_y$.

To analyze $\int_{|x|\geq R} |\nabla_z f^{\#}|\, dx$, we again split the integral into parts over $E_1$ and $E_2$. The integral over $E_1$ satisfies the tail condition, by the same argument as above. But the integral over $E_2$ requires different arguments, since estimate (6.58) stems from the fact that for s.d.r.s the norm of the gradient equals the negative of the radial derivative. Nothing like this is true for $\nabla_z$.

Set

$$D(R) = \{x \in \mathbb{R}^n\colon\ |y| \geq R, |z| \leq R\}.$$

Then the integrals of $|\nabla_z f_j^{\#}|$ over $E_2$ satisfy the tail condition if and only if the integrals over $D(R)$ do.

Assume that the integrals over $D(R)$ do not satisfy the tail condition. Then there exist $\eta > 0$, a sequence $R_j \to \infty$ and a subsequence of the original $\{f_j\}$, also denoted $\{f_j\}$, such that

$$\int_{D(R_j)} |\nabla_z f_j^{\#}|\, dx \geq \eta, \quad j = 1, 2, \ldots \tag{6.60}$$

Fix $j \geq 1$, $z \in \mathbb{R}^m$ and write $g = f_j$. For $y \in \mathbb{R}^k$ with $|y| = r$, write also $g^{\#}(r,z) = g^{\#}(y,z)$. Since $|y| \geq R$ implies $g^{\#}(y,z) \leq g^{\#}(R,z)$, we have

$$\int_{|y|\geq R} |\nabla_z g^{\#}(y,z)|\, dy \leq \int_{g^{\#}\leq g^{\#}(R,z)} |\nabla_z g^{\#}(y,z)|\, dy. \tag{6.61}$$

For $b \in [0,\infty)$, write

$$\int_{g^{\#}\leq b} |\nabla_z g^{\#}(y,z)|\, dy = \int_{g^{\#}< b} |\nabla_z g^{\#}(y,z)|\, dy + \int_{g^{\#}= b} |\nabla_z g^{\#}(y,z)|\, dy. \tag{6.62}$$

The function $g$ is Lipschitz and, having compact support, satisfies Condition S of §6.6. Let us suppose that $z$ belongs to the set $E$ described in Theorem 6.16. If $\mathcal{L}^k(\{y\colon g(y,a) = b\}) > 0$, then $g$ satisfies the hypotheses of Lemma 6.18, and the lemma implies that

$$\int_{g^\#=b} |\nabla_z g^\#(y,z)|\, dy = \int_{g=b} |\nabla_z g(y,z)|\, dy. \tag{6.63}$$

Actually, Lemma 6.18 states equality for the full gradients rather than the $z$-gradients. Since the $y$-gradients vanish $\mathcal{L}^k$-a.e on $g^\# = b$ and $g = b$, (6.63) is also correct.

Next, the argument used to get (6.41) is valid with $\nabla_z$ in place of $\nabla$, and produces

$$\int_{g^\#<b} |\nabla_z g^\#(y,z)|\, dy \le \int_{g<b} |\nabla_z g(y,z)|\, dy.$$

With (6.62) and (6.63), this gives

$$\int_{g^\#\le b} |\nabla_z g^\#(y,z)|\, dy \le \int_{g\le b} |\nabla_z g(y,z)|\, dy. \tag{6.64}$$

We also have the inequality

$$\begin{aligned} g^\#(R,z) \le \alpha_k^{-1} R^{-k} \int_{|y|\le R} g^\#(y,z)\, dy &\le \alpha_k^{-1} R^{-k} \int_{\mathbb{R}^k} g^\#(y,z)\, dy \\ &= \alpha_k^{-1} R^{-k} \int_{\mathbb{R}^k} g(y,z)\, dy. \end{aligned} \tag{6.65}$$

Define

$$A_j(R) = \left\{ (y,z) \in \mathbb{R}^n : f_j(y,z) \le \alpha_k^{-1} R^{-k} \int_{\mathbb{R}^k} f_j(v,z)\, dv \right\} \cap \{|z| \le R\}.$$

Taking $b = g^\#(R,z)$ and recalling that $g = f_j$, (6.61), (6.64), (6.65) imply that for almost every $z \in \mathbb{R}^m$ with $|z| \le R$, we have for each $j \ge 1$,

$$\int_{|y|\ge R} |\nabla_z f_j^\#(y,z)|\, dy \le \int_{\mathbb{R}^k} |\nabla_z f_j(y,z)|\, \chi_{A_j(R)}(y,z)\, dy. \tag{6.66}$$

Integration of (6.66) over $|z| \le R$ gives

$$\begin{aligned} \int_{D(R)} |\nabla_z f_j^\#|\, dx = \int_{|z|\le R} dz \int_{|y|\ge R} |\nabla_z f_j^\#|\, dy &\le \int_{\mathbb{R}^n} |\nabla_z f_j|\, \chi_{A_j(R)}\, dx \\ &\le \int_{\mathbb{R}^n} |\nabla_z f - \nabla_z f_j|\, dx + \int_{\mathbb{R}^n} |\nabla_z f|\, \chi_{A_j(R)}\, dx \\ &= \int_{\mathbb{R}^n} |\nabla_z f - \nabla_z f_j|\, dx + \int_{\mathbb{R}^n} |\nabla_z f|\, \chi_{A_j(R)}\, \chi_{f>0}\, dx. \end{aligned} \tag{6.67}$$

Equality in the last line follows from the fact that $\nabla f = 0$ $\mathcal{L}^n$-a.e. on the set where $f = 0$. Next, recall that $\nabla_z f_j \to \nabla_z f$ and $f_j \to f$ in the $L^1(\mathbb{R}^n)$-norm. If a sequence converges in $L^1$ then some subsequence converges a.e. Choose a subsequence, still denoted $\{f_j\}$, such that $f_j \to f$ $\mathcal{L}^n$-a.e. in $\mathbb{R}^n$ and

$$\lim_{j\to\infty} \int_{\mathbb{R}^k} |f(v,z) - f_j(v,z)|\, dv = 0 \tag{6.68}$$

for almost every $z \in \mathbb{R}^m$. The existence of a subsequence satisfying the latter condition follows from the fact that the functions

$$p_j(z) \equiv \int_{\mathbb{R}^k} |f(v,z) - f_j(v,z)|\, dv$$

converge to zero in the $L^1(\mathbb{R}^m)$ norm when $j \to \infty$.

Define

$$q_j(z) = \int_{\mathbb{R}^k} f_j(v,z)\, dv, \quad j \geq 1.$$

If $z$ satisfies (6.68) and also $\int_{\mathbb{R}^k} f(v,z)\, dv < \infty$, then the sequence $\{q_j(z)\}$ is convergent, hence bounded. Thus, $\{q_j(z)\}_{j\geq 1}$ is bounded for $\mathcal{L}^m$-almost every $z \in \mathbb{R}^m$.

A point $x = (y,z) \in \mathbb{R}^n$ belongs to $A_j(R_j)$ if and only if $|z| \leq R_j$ and

$$f_j(x) \leq \alpha_k^{-1} R_j^{-k} q_j(z). \tag{6.69}$$

Suppose that $x$ is a point such that $f(x) > 0$, $\lim_{j\to\infty} f_j(x) = f(x)$, $z$ satisfies (6.68), and $\int_{\mathbb{R}^k} f(v,z)\, dv < \infty$. Since $R_j \to \infty$, (6.69) shows that $x$ fails to be in $A_j(R_j)$ for all sufficiently large $j$. We conclude that

$$\lim_{j\to\infty} \chi_{A_j(R_j)}\, \chi_{f>0}\,(x) = 0$$

for $\mathcal{L}^n$-almost every $x \in \mathbb{R}^n$.

Taking $R = R_j$ in (6.67) and applying the dominated convergence theorem to the second integral on the far right, we obtain

$$\lim_{j\to\infty} \int_{D(R_j)} |\nabla_z f_j^{\#}|\, dx = 0.$$

This contradicts (6.60). We conclude that the integrals of $|\nabla_z f_j^{\#}|$ over $E_2$ indeed do satisfy the tail condition (6.57). The proof of Theorem 6.19 is complete. □

The next two corollaries are Steiner versions of the s.d.r. results Corollaries 3.21 and 3.22.

**Corollary 6.20** *Let* $1 \leq p < \infty$, $f \in W^{1,p}(\mathbb{R}^n, \mathbb{R}^+)$, *and* $0 \leq a < b \leq \infty$. *Then*

$$\int_{(f^{\#})^{-1}(a,b)} |\partial_{z_i} f^{\#}(x)|^p\, dx \leq \int_{f^{-1}(a,b)} |\partial_{z_i} f(x)|^p\, dx, \quad i = 1, \ldots, m,$$

$$\int_{(f^{\#})^{-1}(a,b)} |\nabla_z f^{\#}(x)|^p\, dx \leq \int_{f^{-1}(a,b)} |\nabla_z f(x)|^p\, dx,$$

$$\int_{(f^{\#})^{-1}(a,b)} |\nabla f^{\#}|^p\, dx \leq \int_{f^{-1}(a,b)} |\nabla f|^p\, dx.$$

*Proof* Let $\psi_j$ be the $C^1$ functions with $\psi_j' \to \chi_{(a,b)}$ used to prove (6.41). Apply Theorem 6.19 to the $\psi_j \circ f$, then pass to the limit. See Evans and Gariepy (1992) for validity of the chain rule in the Sobolev context, and for other technical points. □

Since $\nabla f = 0$ $\mathcal{L}^n$-a.e on sets $f = c$ (Evans and Gariepy, 1992, p. 130), the open interval $(a, b)$ in Corollary 6.20 can be replaced by $[a, b)$, $(a, b]$, or $[a, b]$.

**Corollary 6.21** *Let $1 \leq p < \infty$, $\Omega$ be an open subset of $\mathbb{R}^n$ and $f \in W_0^{1,p}(\Omega, \mathbb{R}^+)$. Then $f^\# \in W^{1,p}(\Omega^\#, \mathbb{R}^+)$, and*

$$\int_{\Omega^\#} |\partial_{z_i} f^\#(x)|^p \, dx \leq \int_{\Omega} |\partial_{z_i} f(x)|^p \, dx, \quad i = 1, \ldots, m,$$

$$\int_{\Omega^\#} |\nabla_z f^\#(x)|^p \, dx \leq \int_{\Omega} |\nabla_z f(x)|^p \, dx,$$

$$\int_{\Omega^\#} |\nabla f^\#|^p \, dx \leq \int_{\Omega} |\nabla f|^p \, dx.$$

*Moreover, $f^\# \in W_0^{1,p}(\Omega^\#)$.*

*Proof* The proof is exactly the same as for Corollary 3.22. The only new detail perhaps requiring explication is the fact that if $h$ is a nonnegative function with compact support in $\Omega$, then $h^\#$ has compact support in $\Omega^\#$. This point will be discussed in §6.9. □

## 6.9 Steiner Symmetrization Decreases Surface Area

We continue with the setting of §6.8:

$$X = \mathbb{R}^k \times \mathbb{R}^m,$$

where $1 \leq k \leq n - 1$, and $m = n - k$. The superscript # will denote the $(k, n)$-Steiner symmetrization of a set or function.

The isoperimetric inequalities of Chapter 4 state that passing from a suitable set $E \subset \mathbb{R}^n$ to its s.d.r. decreases the surface area of $\partial E$. The s.d.r. is a ball of the same volume, and "surface area" was interpreted to be the perimeter $P(E)$, the $(n-1)$-dimensional Hausdorff measure $\mathcal{H}^{n-1}(\partial E)$, or the $(n-1)$-dimensional Minkowski content $\mathcal{M}^{n-1}(\partial E)$. In the section at hand, we will present corresponding results involving the $(k, n)$-Steiner symmetrization $E^\#$ of $E$.

In §4.3, $P(E)$ was defined in terms of functions of bounded variation on $\mathbb{R}^n$. Theorem 4.8 asserts that for $f \in BV(\mathbb{R}^n, \mathbb{R}^+)$ the total variation $V(f)$ decreases under s.d.r. Our first order of business provides the Steiner counterpart.

**Theorem 6.22** *If $f \in BV(\mathbb{R}^n, \mathbb{R}^+)$ then $f^\# \in BV(\mathbb{R}^n, \mathbb{R}^+)$ and*

$$V(f^\#) \leq V(f).$$

*Proof* The same as the proof of Theorem 4.8, except we use Theorem 6.19(c) instead of Theorem 3.20 to obtain for the approximating functions

$$V(f_j^\#) = \|\nabla f_j^\#\|_{L^1(\mathbb{R}^n)} \leq \|\nabla f_j\|_{L^1(\mathbb{R}^n)} = V(f_j). \qquad \square$$

Let $E$ be an $\mathcal{L}^n$-measurable set in $\mathbb{R}^n$. Recall that $E$ has finite perimeter if $\chi_E \in BV(\mathbb{R}^n)$, that is:

$$\mathcal{L}^n(E) < \infty \quad \text{and} \quad V(\chi_E) < \infty,$$

in which case we define

$$P(E) = \chi_E.$$

Applying Theorem 6.22 to $\chi_E$ we obtain

**Theorem 6.23** *For every set $E \subset \mathbb{R}^n$ with finite perimeter, we have*

$$P(E^\#) \leq P(E). \tag{6.70}$$

Thus, *Steiner symmetrization decreases the perimeter.*

According to Corollary 4.12, for sets $E$ of finite perimeter we have $P(E) \leq \mathcal{H}^{n-1}(\partial E)$. Equality holds when $E$ is a ball. Thus, when $\mathcal{L}^n(E) < \infty$, the s.d.r. case of (6.70) implies that s.d.r. also reduces $\mathcal{H}^{n-1}(\partial E)$. But for Steiner symmetric sets, the Hausdorff measure of the boundary can be strictly larger than the perimeter. Take, for example, $E$ to be the square $(-1,1) \times (-1,1)$ in $\mathbb{R}^2$ from which two horizontal slits $\{(x_1, a_2) \colon a_1 \leq |x_1| < 1\}$ have been removed. Here $0 < a_1 < 1$, $|a_2| < 1$. Then $E$ is $(1,1)$-Steiner symmetric, with $P(E) = 8$ and $\mathcal{H}^1(E) = 8 + 2(1 - a_1)$.

The s.d.r results about Minkowski content also readily extend to the Steiner setting. We review the definitions: $E$ denotes an arbitrary subset of $\mathbb{R}^n$; $d(x, E)$ is the distance in $\mathbb{R}^n$ from $x$ to $E$. For $\delta > 0$, the *$\delta$-collar* $E(\delta)$ and *$\delta$-core* $E(-\delta)$ of $E$ are defined by

$$E(\delta) = \{x \in \mathbb{R}^n \colon d(x, E) < \delta\}, \quad E(-\delta) = \{x \in E \colon d(x, E) \geq \delta\}.$$

The lower $(n-1)$-dimensional Minkowski content $\mathcal{M}_*^{n-1}(A)$ of a set $A \subset \mathbb{R}^n$ is defined by

$$\mathcal{M}_*^{n-1} A = \liminf_{\delta \to 0} \frac{\mathcal{L}^n(A(\delta))}{2\delta}.$$

The upper $(n-1)$-dimensional Minkowski content $\mathcal{M}^{*(n-1)}(A)$ of $A$ is defined the same way, but with lim sup in place of lim inf. If $\mathcal{M}_*^{n-1}(A) = \mathcal{M}^{*(n-1)}(A)$, the common value is called the $(n-1)$-dimensional Minkowski content of $A$ and is denoted $\mathcal{M}^{n-1}(A)$.

**Theorem 6.24** *Let $E \subset \mathbb{R}^n$ and $\delta > 0$. Then*

*(a)* $E^{\#}(\delta) \subset E(\delta)^{\#}$;
*(b)* $E(-\delta)^{\#} \subset E^{\#}(-\delta)$.

**Corollary 6.25** *Let $E \subset \mathbb{R}^n$ and $\delta > 0$. Then*

*(a)* $\mathcal{L}^n(E(\delta)) \geq \mathcal{L}^n(E^{\#}(\delta))$;
*(b)* $\mathcal{L}^n(E^{\#}(-\delta)) \geq \mathcal{L}^n(E(-\delta))$;
*(c)* $\mathcal{L}^n((\partial E)(\delta)) \geq \mathcal{L}^n((\partial E^{\#})(\delta))$.

**Theorem 6.26** *For each $\mathcal{L}^n$-measurable set $E \subset \mathbb{R}^n$ with $\mathcal{L}^n(E) < \infty$, we have*

$$\mathcal{M}^{n-1}(\partial(E^{\#})) \leq \mathcal{M}_*^{n-1}(\partial E).$$

*The same inequality holds when the lower contents are replaced by the upper contents.*

These are the Steiner versions of Theorem 4.14, Corollary 4.15, and Theorem 4.16. The statements are identical, with one exception: in Theorem 6.26 $E^{\#}$ is no longer a ball, so we cannot state a formula for the Minkowski content of its boundary.

*Proofs of 6.24–6.26* Consider Theorem 6.24(a). If $E$ is unbounded then $\mathcal{L}^n(E(\delta)) = \infty$ and (a) is true. Thus, we may assume that $E$ is bounded, and also, as in the proof of Theorem 4.14, that $E$ is closed. Set

$$f(x) = \left(1 - \frac{d(x, E)}{\delta}\right), \qquad x \in \mathbb{R}^n.$$

By Theorem 6.10, $\omega(t,f^{\#}) \leq \omega(t,f)$ for all $t > 0$. The proof of Theorem 4.14(a) carries over to prove (a). The proofs of Theorem 6.24(b), Corollary 6.25 and Theorem 6.26 are now accomplished in exactly the same way as the corresponding results in Chapter 4. □

According to Theorem 6.26, *Steiner symmetrization decreases the $(n-1)$-dimensional Minkowski contents.*

To complete the parallel with s.d.r., we would like to prove that

$$\mathcal{H}^{n-1}(\partial E^{\#}) \leq \mathcal{H}^{n-1}(\partial E) \tag{6.71}$$

for all sets $E \subset \mathbb{R}^n$ of finite measure. If $E^{\#}$ is a bounded domain with Lipschitz boundary, then $\partial E^{\#}$ coincides with its measure theoretic boundary $\partial_* E^{\#}$. Thus,

$$\mathcal{H}^{n-1}(\partial E^{\#}) = P(\partial E^{\#}) \leq P(E) \leq \mathcal{H}^{n-1}(\partial E),$$

where the first equality follows from Evans and Gariepy (1992, Theorem 5.7.2 and Lemma 5.8.1(ii)), the first inequality holds by (6.70), and the second inequality by Corollary 4.12. So, (6.71) is true when $E^\#$ is a bounded domain with Lipschitz boundary. If $E$ itself is a bounded domain with Lipschitz boundary, it is not hard to show that the same is true for $E^\#$, so it is true at least that Steiner symmetrization reduces the Hausdorff measure of the boundary of Lipschitz domains.

For rough sets $E$ it appears to be not known if (6.71) is always true or not. A partial result, due to Sperner (1979), shows that (6.71) holds when $k = n-1$ and $E$ satisfies some extra conditions.

Theorem 6.24 enables us to take care of a loose end in the proof of Corollary 6.21.

**Corollary 6.27** *Let $\Omega$ be an open set in $\mathbb{R}^n$ and $f \in C_c(\Omega)$. Then $f^\# \in C_c(\Omega^\#)$.*

*Proof* Extend $f$ to be zero in $\mathbb{R}^n \setminus \Omega$. Then Theorem 6.10 implies continuity of $f^\#$ on $\Omega^\#$. To show that the support of $f^\#$ is a compact subset of $\Omega^\#$, set $A = \{x \in \Omega : f(x) > 0\}$. Then $\operatorname{supp} f = \overline{A}$. Take $0 < \delta < d(A, \partial\Omega)$. Then $A \subset \Omega(-\delta)$. Hence

$$A^\# \subset \Omega(-\delta)^\# \subset \Omega^\#(-\delta), \tag{6.72}$$

where the second inclusion is by Theorem 6.24(b). But $A^\# = \{x \in \Omega^\# : f^\#(x) > 0\}$. From (6.72), it follows that

$$d(\operatorname{supp} f^\#, \partial\Omega) = d(A^\#, \partial\Omega) \geq \delta$$

so that $\operatorname{supp} f^\#$ is a compact subset of $\Omega^\#$. □

## 6.10 Steiner Symmetrization Increases or Decreases Physical Quantities

We continue with the setting of the previous two sections

$$X = \mathbb{R}^k \times \mathbb{R}^m,$$

where $1 \leq k \leq n-1$, and $m = n-k$. The superscript # will denote the $(k,n)$-Steiner symmetrization of a set or function.

In Chapter 5 we showed that under symmetric decreasing rearrangement of the domain, the principal eigenvalue decreases, various capacities decrease, the torsional rigidity increases, and the mean lifetime of a Brownian particle increases. All of this is still true for Steiner symmetrization.

**Theorem 6.28** $\lambda_1(\Omega^{\#}) \leq \lambda_1(\Omega)$.

**Theorem 6.29** $\mathrm{Cap}_p(\overline{K^{\#}}, U^{\#}) \leq \mathrm{Cap}_p(K, U), \quad 1 \leq p < \infty$.

**Theorem 6.30** $T(\Omega) \leq T(\Omega^{\#})$ *and* $M(\Omega) \leq M(\Omega^{\#})$.

In the first and third theorems $\Omega$ is a bounded domain in $\mathbb{R}^n$. In the second theorem $U$ is an open subset of $\mathbb{R}^n$ and $K$ a compact subset of $U$.

For definitions, see Theorems 5.6, 5.14, 5.17, and Corollary 5.18. Recall that $T$ in Theorem 6.30 stands for torsional rigidity defined in §5.7.

For reasons explained in §5.6, Theorem 6.29 implies that Szegő's theorems about decrease of logarithmic and Newtonian capacity (Theorems 5.14 and 5.15) remain valid for Steiner symmetrization. So does Carleman's theorem about ring domains (Corollary 5.16), which is a consequence of the logarithmic capacity decrease.

*Proofs of Theorems 6.28–6.30* For the arguments proving Theorems 5.6–5.17 and Corollary 5.18 to carry over to their Steiner counterparts Theorems 6.28 and 6.30, we just need to know that Steiner symmetrization preserves the space $W_0^{1,2}(\Omega)$ and decreases the $L^2$-norm of the gradient. This is true by Corollary 6.21.

For the proof of Theorem 5.14 to carry over to Theorem 6.29, we just need to know that if $u \in \mathrm{Lip}(\mathbb{R}^n)$ satisfies

$$K \subset (u = 1) \subset (u > 0) \subset V, \quad 0 \leq u(x) \leq 1 \quad \forall\, x \in \mathbb{R}^n, \tag{6.73}$$

and

$$\lim_{x \to \infty} u(x) = 0, \tag{6.74}$$

then $u^{\#} \in \mathrm{Lip}(\mathbb{R}^n)$, $\|\nabla u^{\#}\|_{L^2(\mathbb{R}^n)} \leq \|\nabla u\|_{L^2(\mathbb{R}^n)}$, and (6.73)–(6.74) are satisfied when $u, K$ and $U$ are replaced by $u^{\#}, K^{\#}$ and $U^{\#}$.

The Lipschitz property of $u^{\#}$ and decrease of norm of gradient follow from Corollary 6.17. Relations (6.73) are true because $u$ and $u^{\#}$ have the same distribution on $\mathbb{R}^n$. As for (6.74), for each $\epsilon > 0$ there exists $R < \infty$ such that $(u > \epsilon) \subset \mathbb{B}^n(0, R)$. Since the ball $\mathbb{B}^n(0, R)$ is Steiner symmetric, it follows that $(u^{\#} > \epsilon) \subset \mathbb{B}^n(0, R)$. This implies $\lim_{x \to \infty} u^{\#}(x) = 0$. □

## 6.11 Notes and Comments

As often in symmetrization theory, our terminology in this chapter is far from universal. For example, Burago and Zalgaller (1988), Sperner (1979), and many other authors restrict the term Steiner symmetrization to mean our

$(1,n)$-Steiner symmetrization. Burago and Zalgaller call $(n-1,n)$-Steiner symmetrization Schwarz symmetrization, while Sperner calls it Blaschke symmetrization. For Chavel (2001), Schwarz symmetrization is our symmetric decreasing rearrangement.

Steiner invented $(1,2)$-Steiner symmetrization in 1838 as a device for proving the isoperimetric inequality in the plane, but he was unable to give a rigorous proof. See pp. 75–91 in Volume 2 of his collected works Steiner (1882). A main sticking point was to prove that symmetric decreasing rearrangement can be realized as the appropriate limit of a sequence of Steiner symmetrizations having various axes of symmetry. Fortified with more powerful mathematical tools such as the Hausdorff metric, rigorous versions of his arguments appeared in the twentieth century. See, for example Brascamp et al. (1974) and Talenti (1993).

Schwarz (1884) used $(2,3)$-Steiner symmetrization in his research on the isoperimetric inequality in $\mathbb{R}^3$. The article of Stammbach (2012) gives some historical context of the work of Schwarz and Steiner.

In Talenti's proof (1993) of the isoperimetric inequality for perimeter in $\mathbb{R}^n$, the main ingredient is Theorem 6.23, the fact that perimeter decreases under $(1,n)$-Steiner symmetrization.

Sarvas (1972) may be among the first to consider $(k,n)$-Steiner symmetrization for the full range $k \in \{1,\ldots,n\}$. He proves Theorem 6.29, that these symmetrizations decrease variational $p$-capacities.

Pólya and Szegő (1951, p. 154) give a semi-rigorous proof of the Dirichlet integral inequality (6.35) and the capacity inequalities in Theorem 6.23 for $(1,2)$-, $(1,3)$- and $(2,3)$-Steiner symmetrizations. See also Pólya and Szegő (1945). The first modern proof of these inequalities is due to Gehring (1961) for $n = 3$. See also Sarvas (1972) for a simpler proof.

Sperner (1979) proved Theorems 6.10 and 6.26, along with a number of other interesting properties of $(k,n)$-Steiner symmetrization.

Brock and Solynin (2000) carry out a careful study showing how $(k,n)$-Steiner symmetrizations $f^{\#}$ can be approximated by polarizations.

Burchard (1997) showed that $(1,n)$-Steiner is distinctive, in that the operator which sends $f$ to its $(k,n)$-Steiner symmetrization is strongly continuous in $W^{1,p}(\mathbb{R}^n)$ when $k = 1$. The Almgren–Lieb Theorem, see §3.7, implies this is false for $k \geq 2$.

Most studies of Steiner symmetrization operate in $\mathbb{R}^n$ rather than our space $X = \mathbb{R}^k \times Z$. The introduction of $Z$ and the "almost every slice" version of inequalities such as (6.34) is implicit in Sperner (1979). I have not seen before inequalities like (6.32) and (6.33) involving $\partial_{z_i}$ and $\nabla_z$. Nor have I seen anything like Lemma 6.18.

# 7

# Symmetrization on Spheres, and Hyperbolic and Gauss Spaces

So far our principal object of study has been the symmetric decreasing rearrangement of functions on $\mathbb{R}^n$. In this chapter we shall study symmetric decreasing rearrangements on spheres $\mathbb{S}^n$ and hyperbolic spaces $\mathbb{H}^n$. We shall also introduce "$(k, n)$-cap symmetrization" in $\mathbb{R}^n$, which is like Steiner symmetrization except that the rearranging is done on copies of $\mathbb{S}^k$ instead of $\mathbb{R}^k$. Just about everything we proved for s.d.r. and Steiner symmetrization is still true for these new symmetrizations, with the same or simpler proofs. All we need are suitable replacements on $\mathbb{S}^n$ and $\mathbb{H}^n$ for the distance function $|x-y|$ and measure $\mathcal{L}^n$, a good understanding of metric balls, rich isometry groups, and plenty of hyperplanes in which to polarize.

As an application of cap symmetrization, in §7.8 we shall prove a landmark theorem of Gehring about distortion of quasiconformal maps in $\mathbb{R}^n$. Also, in §7.7 we shall give a short discussion of a method of symmetrization on $\mathbb{R}^n$ in which the rearranging is done with respect to the Gauss measure $d\mu = (2\pi)^{-n/2} e^{-\frac{1}{2}|x|^2}\, dx$.

## 7.1 The Sphere $\mathbb{S}^n$

The unit sphere $\mathbb{S}^n \subset \mathbb{R}^{n+1}$ is

$$\mathbb{S}^n = \{x \in \mathbb{R}^{n+1} \colon |x| = 1\},$$

where $|\cdot|$ denotes the Euclidean norm. Note that $\mathbb{S}^n$ contains the $(n+1)$-dimensional coordinate vectors $e_1, \dots, e_{n+1}$. We shall think of the east pole $e_1$ as being the origin of $\mathbb{S}^n$.

Each orthogonal linear transformation $T \in O(n+1)$ maps $\mathbb{S}^n$ onto itself. By linear algebra, given $x, y \in \mathbb{S}^n$ with $x \neq \pm y$ there exists $T \in O(n+1)$ such that $Tx = e_1$ and $Ty = c_1 e_1 + c_2 e_2$ with $c_1 \in (-1, 1)$, $c_2 > 0$, and $c_1^2 + c_2^2 = 1$.

Define the *distance* $d(x, y)$ between $x$ and $y$ to be the shorter arc length distance between $Tx$ and $Ty$ on the unit circle in the plane spanned by $e_1$ and $e_2$. Put another way: there is a unique circle of radius 1 centered at $0 \in \mathbb{R}^{n+1}$ which contains $x$ and $y$. Then $d(x, y)$ is the shorter arc length distance on that circle between $x$ and $y$. To complete the definition of $d$, if $x = y$ we define $d(x, y) = 0$ and if $x = -y$ we define $d(x, y) = \pi$.

The group $O(n + 1)$ acts isometrically on $(\mathbb{S}^n, d)$. That is,

$$d(x, y) = d(Tx, Ty), \quad T \in O(n + 1).$$

Accordingly, $d$ is called the *invariant distance* on $\mathbb{S}^n$. It is also called the *canonical distance*, the *intrinsic distance*, or the *geodesic distance* on $\mathbb{S}^n$. "Intrinsic" signifies that the distance is computed without leaving $\mathbb{S}^n$, in contrast to the "chordal distance" $|x - y|$, which is computed relative to $\mathbb{R}^{n+1}$. As for "geodesic," if we endow $\mathbb{S}^n$ with the Riemannian metric inherited from $\mathbb{R}^{n+1}$ with its usual metric, then

$$d(x, y) = \min L(\gamma),$$

where $L(\gamma)$ is the Riemannian length of the path $\gamma$, and the minimum is taken over all paths on $\mathbb{S}^n$ connecting $x$ and $y$. The minima are achieved by arcs of great circles, so that these circles are geodesics for the spherical Riemannian metric.

We shall usually refer to $d$ as the canonical distance on $\mathbb{S}^n$.

Since the arc length distance between two points on a circle of radius 1 equals the angle $\theta \in [0, \pi]$ between them, and the dot product $x \cdot y$ in $\mathbb{R}^{n+1}$ equals $|x||y| \cos \theta$, we obtain

$$\cos d(x, y) = x \cdot y, \quad x, y \in \mathbb{S}^n.$$

The chordal distance on $\mathbb{S}^n$ is related to the canonical distance by

$$|x - y|^2 = 2 - 2 \cos d(x, y).$$

In §4.5 we defined the *canonical measure* $\sigma_n$ on $\mathbb{S}^n$ to be the unique nonnegative Borel measure for which the polar coordinate decomposition $d\mathcal{L}^{n+1} = r^n \, d\sigma_n \, dr$ holds. With the aid of the coarea formula, we showed that

$$\sigma_n = \mathcal{H}^n \quad \text{restricted to} \quad \mathbb{S}^n.$$

From the definition of Hausdorff measure, it is clear that $\sigma_n$ enjoys the invariance property

$$\sigma_n(TE) = \sigma_n(E), \quad E \in \mathcal{B}(\mathbb{S}^n).$$

For $\theta \in [0, \pi]$, set

$$\mathcal{K}(\theta) = \{x \in \mathbb{S}^n : d(x, e_1) < \theta\} = \{x \in \mathbb{S}^n : x \cdot e_1 > \cos \theta\}.$$

Then $\mathcal{K}(\theta)$ is the open unit ball in $\mathbb{S}^n$ centered at $e_1$ with radius $\theta$ with respect to the canonical distance function $d$. As a subset of $\mathbb{R}^{n+1}$, $\mathcal{K}(\theta)$ is an open spherical cap on $\mathbb{S}^n$ centered at $e_1$. In the next section, we shall see that

$$\sigma_n(\mathcal{K}(\theta)) = \beta_{n-1} \int_0^\theta \sin^{n-1} t\, dt, \quad 0 \leq \theta \leq \pi. \tag{7.1}$$

We shall say that a set $E \subset \mathbb{S}^n$ is $\sigma_n$-measurable if it belongs to the completion with respect to $\sigma_n$ of the Borel $\sigma$-algebra $\mathcal{B}(\mathbb{S}^n)$.

In §1.7 we defined $\mathcal{H}(\mathbb{R}^n)$ to be the set of all affine hyperplanes in $\mathbb{R}^n$. Define now $\mathcal{H}_0(\mathbb{R}^n) = \{H_0 \in \mathcal{H}(\mathbb{R}^n) \colon 0 \in H_0\}$, and define

$$\mathcal{H}(\mathbb{S}^n) = \{H_0 \cap \mathbb{S}^n \colon H_0 \in \mathcal{H}_0(\mathbb{R}^{n+1})\}.$$

Let us call elements of $\mathcal{H}(\mathbb{S}^n)$ hyperplanes on $\mathbb{S}^n$, or s-hyperplanes. Thus, s-hyperplanes are intersections of $\mathbb{S}^n$ with $n$-dimensional subspaces of $\mathbb{R}^{n+1}$. Each s-hyperplane lives on $\mathbb{S}^n$, and is an $(n-1)$-dimensional sphere in $\mathbb{R}^{n+1}$ of radius 1, center $0 \in \mathbb{R}^{n+1}$. The complement of $H$ in $\mathbb{S}^n$ is the union of two hemispheres; denote one of them by $H^+$, the other by $H^-$.

For $H = H_0 \cap \mathbb{S}^n$, let $\rho = \rho_H$ denote reflection in $\mathbb{R}^{n+1}$ with respect to $H_0$. Then $\rho_H$ is an involutive isometry of $\mathbb{S}^n$ which maps $H^+$ one-one onto $H^-$ and is the identity on $H$. Given $f \colon \mathbb{S}^n \to \mathbb{R}$ and $H \in \mathcal{H}(\mathbb{S}^n)$, define the *polarization* $f_H \colon \mathbb{S}^n \to \mathbb{R}$ of $f$ with respect to $H$ by

$$f_H(x) = \begin{cases} \max(f(x), f(\rho x)), & x \in H^+, \\ \min(f(x), f(\rho x)), & x \in H^-, \\ f(x), & x \in H. \end{cases}$$

The reader may verify that the results in §1.7 of Chapter 1 about hyperplanes and polarizations on $\mathbb{R}^n$ remain true for hyperplanes and polarizations on $\mathbb{S}^n$. All one need do is change $|x-y|$ to $d(x,y)$ and $\mathcal{L}^n$ to $\sigma_n$.

Our next job is to define symmetric decreasing rearrangement for sets and functions on $\mathbb{S}^n$. For a $\sigma_n$-measurable $E \subset \mathbb{S}^n$ we define its *symmetric decreasing rearrangement* $E^\#$ as follows:

- If $0 < \sigma(E) \leq \beta_n$ and $E \neq \mathbb{S}^n$, then $E^\#$ is the open spherical cap $\mathcal{K}(\theta)$ with $\sigma_n(E) = \sigma_n(\mathcal{K}(\theta))$.
- If $\sigma_n(E) = 0$ then $E^\#$ is empty.
- If $E = \mathbb{S}^n$ then $E^\# = \mathbb{S}^n$.

The last specification turns out to be convenient later.

Next, define a map $T \colon \mathbb{S}^n \to [0, \beta_n]$ by

$$T(x) = \sigma_n(\mathcal{K}(\theta))$$

where $\theta = d(x, e_1)$. Then $x$ lies on the boundary, relative to $\mathbb{S}^n$, of the cap $\mathcal{K}(\theta)$. One shows, as in Example 1.23 of §1.4, that $T$ is a measure preserving map from the completion of $(\mathbb{S}^n, \mathcal{B}(\mathbb{S}^n), \sigma_n)$ onto the completion of $([0, \beta_n], \mathcal{B}([0, \beta_n]), \mathcal{L})$. If $f\colon \mathbb{S}^n \to \mathbb{R}$ is $\sigma_n$-measurable, we define its *symmetric decreasing rearrangement* $f^\#$ by

$$f^\# = f^* \circ T.$$

Here $f^*$ is the decreasing rearrangement on $f$. As noted in §1.2, when the measure space on which $f$ is defined is finite, then the finiteness condition (1.1) of Chapter 1 is automatically fulfilled. Thus, in contrast to the $\mathbb{R}^n$ case, every a.e.-defined real valued $f$ on $\mathbb{S}^n$ has an s.d.r.

The s.d.r. $f^\#$ is constant on the boundaries of the caps $\mathcal{K}(\theta)$, and decreases as $\theta$ increases. All the other features of s.d.r. on $\mathbb{R}^n$ proved in §1.6 are also true for s.d.r. on $\mathbb{S}^n$, with obvious modifications, as the reader may check.

## 7.2 Spherical Coordinates on $\mathbb{S}^n$

In calculus, we learn to parameterize $\mathbb{S}^1$ by

$$x_1 = \cos\theta, \quad x_2 = \sin\theta,$$

with $0 \le \theta \le 2\pi$, and parameterize $\mathbb{S}^2$ by

$$x_1 = \cos\theta_1, \quad x_2 = \sin\theta_1 \cos\theta_2, \quad x_3 = \sin\theta_1 \sin\theta_2,$$

with $0 \le \theta_1 \le \pi, 0 \le \theta_2 \le 2\pi$. On $\mathbb{S}^2$, the coordinates have been chosen so that $\theta_1$ is the angle between the point $x$ and the east pole $e_1$.

We describe now $n$-dimensional versions of these coordinatizations. Let $n \ge 2$, and define the open set $V$ in $\mathbb{R}^n$ by

$$V = (0, \pi)^{n-1} \times (0, 2\pi).$$

Define $F\colon V \to \mathbb{R}^{n+1}$ by $F(\vec{\theta}) = x$, where $\vec{\theta} = (\theta_1, \ldots, \theta_n)$, $x = (x_1, \ldots, x_{n+1})$, and

$$x_1 = \cos\theta_1, \quad x_{n+1} = \prod_{i=1}^{n} \sin\theta_i, \quad x_j = \left(\prod_{i=1}^{j-1} \sin\theta_i\right) \cos\theta_j, \quad 2 \le j \le n. \tag{7.2}$$

The reader may check that $F(V) \subset \mathbb{S}^n$, that $F$ is 1–1 on $V$, and that

$$\mathbb{S}^n \setminus F(V) = \{x \in \mathbb{S}^n\colon x_{n+1} = 0,\ x_n \ge 0\}. \tag{7.3}$$

The last two statements can be checked by an inductive argument starting with $n = 2$.

For $x \in \mathbb{S}^n$ the coordinate $\theta_1$ tells us the angle that $x$ makes with $e_1$ in $\mathbb{R}^{n+1}$. The other spherical coordinates specify the position of $x$ within the $(n-1)$-sphere $\partial\mathcal{K}(\theta_1)$.

From (7.3), we see that $\sigma_n(\mathbb{S}^n \setminus V) = 0$. The area formula (4.5) gives for $\sigma_n$-measurable $A \subset \mathbb{S}^n$,

$$\int_{F^{-1}(A)} J_F\, dx = \sigma_n(A), \tag{7.4}$$

where

$$J_F^2 = \det[(DF)^*(DF)].$$

Here $DF$ is the $(n+1) \times n$ matrix whose $(i, j)$ entry is $\frac{\partial x_i}{\partial \theta_j}$. Denote the entries of $(DF)^*(DF)$ by $g_{ij}$. Then $g_{ij} = \frac{\partial F}{\partial x_i} \cdot \frac{\partial F}{\partial x_j}$. By an exercise for the reader we have,

$$g_{ij} = 0 \quad \text{if} \quad i \neq j, \quad g_{11} = 1, \quad g_{ii} = \prod_{j=1}^{i-1} \sin^2 \theta_j, \quad 2 \leq i \leq n.$$

Thus, we get

$$J_F = \sin^{n-1} \theta_1 \sin^{n-2} \theta_2 \cdots \sin \theta_{n-1}. \tag{7.5}$$

Let $h$ be a nonnegative $\sigma_n$-measurable function on $\mathbb{S}^n$. As in Evans and Gariepy (1992, p. 99), formula (7.4) can be generalized to the change of variable formula

$$\int_{\mathbb{S}^n} h\, d\sigma_n = \int_V (h \circ F)\, J_F\, d\mathcal{L}^n.$$

From (7.5) it follows that

$$\begin{aligned}\int_{\mathbb{S}^n} h\, d\sigma_n = &\int_0^\pi \sin^{n-1} \theta_1\, d\theta_1 \int_0^\pi \sin^{n-2} \theta_2\, d\theta_2 \cdots \\ &\cdots \int_0^\pi \sin \theta_{n-1}\, d\theta_{n-1} \int_0^{2\pi} h(x_1, \ldots, x_{n+1})\, d\theta_n,\end{aligned} \tag{7.6}$$

where the $x_j$ are related to the $\theta_j$ by (7.2).

If $h$ is radial on $\mathbb{S}^n$, that is, depends only on the angle $\theta$ between a point on $\mathbb{S}^n$ and $e_1$, then (7.6) simplifies to

$$\int_{\mathbb{S}^n} h\, d\sigma_n = \beta_{n-1} \int_0^\pi h(\theta) \sin^{n-1} \theta\, d\theta,$$

where we have employed the usual abuse of notation equating $h$ with $h \circ F$. In particular, taking $h = \chi_{\mathcal{K}(\theta)}$, we obtain

$$\sigma_n(\mathcal{K}(\theta)) = \beta_{n-1} \int_0^\theta \sin^{n-1} t\, dt, \quad 0 \le \theta \le \pi,$$

which confirms formula (7.1).

Next, we consider spherical coordinates in $\mathbb{R}^n$. Assume $n \ge 2$. Then each $x \in \mathbb{R}^n \setminus \{0\}$ can be written in exactly one way as $x = r\,y$, with $r \in (0, \infty)$, $y \in \mathbb{S}^{n-1}$. Let $W = (0, \infty) \times (0, \pi)^{n-2} \times (0, 2\pi)$, where the middle factor set is to be omitted when $n = 2$. Define $G\colon W \to \mathbb{R}^n$ by $G(\vec{\theta}) = x$, where $\vec{\theta} = (r, \theta_1, \ldots, \theta_{n-1})$, $x = (x_1, \ldots, x_n)$, and

$$x_1 = r \cos\theta_1,\; x_n = r \prod_{i=1}^{n-1} \sin\theta_i,\; x_j = r\cos\theta_j \prod_{i=1}^{j-1} \sin\theta_i, \quad 2 \le j \le n-1. \tag{7.7}$$

Then $G$ is 1–1 on $W$, and $\mathbb{R}^n \setminus G(W) = \{x \in \mathbb{R}^n : x_n = 0,\ x_{n-1} \ge 0\}$ has $\mathcal{L}^n$-measure zero.

If $h$ is a nonnegative measurable function on $\mathbb{R}^n$ then

$$\begin{aligned}
\int_{\mathbb{R}^n} h\, d\mathcal{L}^n &= \int_0^\infty r^{n-1}\, dr \int_{\mathbb{S}^{n-1}} h(ry)\, d\sigma_{n-1}(y) \\
&= \int_0^\infty r^{n-1}\, dr \int_0^\pi \sin^{n-2}\theta_1\, d\theta_1 \int_0^\pi \sin^{n-3}\theta_2\, d\theta_2 \cdots \\
&\quad \cdots \int_0^\pi \sin\theta_{n-2}\, d\theta_{n-2} \int_0^{2\pi} h(x_1, \ldots, x_n)\, d\theta_{n-1},
\end{aligned}$$

where the $x_j$ are related to the $r$ and $\theta_j$ by (7.7). The first equality was established in §4.5 and the second equality follows from (7.6). The equality of the first and third quantities can be succinctly expressed as

$$dx_1\, dx_2 \cdots dx_n = r^{n-1} \sin^{n-2}\theta_1 \sin^{n-3}\theta_2 \cdots \sin\theta_{n-2}\, dr\, d\theta_1 \cdots d\theta_{n-1}.$$

Given a domain $\Omega \subset \mathbb{R}^n \setminus G(W)$ and a function $u \in C^2(\Omega)$, let $v = u \circ G$ in $\Omega' \equiv G^{-1}(\Omega) \subset V$. Then for $n \ge 3$ we have the following formula for $\Delta u$ in spherical coordinates, see Vilenkin (1968, p. 493) or Atkinson and Han (2012, p. 94):

$$\begin{aligned}
\Delta u &= r^{1-n} \partial_r (r^{n-1} v_r) + r^{-2} \sin^{2-n}\theta_1\, \partial_{\theta_1} \left( \sin^{n-2}\theta_1\, \partial_{\theta_1} v \right) \\
&\quad + r^{-2} \sum_{j=2}^{n-1} \sin^{j+1-n}\theta_j \left( \prod_{i=1}^{j-1} \sin\theta_i \right)^{-2} \frac{\partial}{\partial\theta_j} \left( \sin^{n-j-1}\theta_j\, \partial_{\theta_j} v \right).
\end{aligned} \tag{7.8}$$

Here $\Delta u$ is evaluated at a point $x \in \Omega$ and the functions on the right side are evaluated at $y = G^{-1}(x)$. When $j = n-1$, the term $\sin^{n-j-1}\theta_j$ is understood to be 1. When some $\theta_j$ are equal to 0, formula (7.8) remains true if suitably rewritten: expand $\frac{\partial}{\partial\theta_j}\left(\sin^{n-j-1}\theta_j\,\partial_{\theta_j}v\right)$ using the product formula, and simplify.

Often, we will not distinguish between $u$ and $v$. For example, if $u$ is a radial function, that is, depends only on $r = |x|$, then (7.8) can be written as

$$\Delta u = r^{1-n}\partial_r(r^{n-1}\partial_r u) = u_{rr} + \frac{n-1}{r}u_r, \tag{7.9}$$

while if $u$ depends only on $r$ and $\theta_1$ then writing $\theta = \theta_1$, we have

$$\begin{aligned}\Delta u &= r^{1-n}\partial_r(r^{n-1}\partial_r u) + r^{-2}\sin^{2-n}\theta\,\partial_\theta(\sin^{n-2}\theta\,\partial_\theta u)\\ &= u_{rr} + \frac{n-1}{r}u_r + r^{-2}u_{\theta\theta} + r^{-2}(n-2)\cot\theta\,u_\theta.\end{aligned} \tag{7.10}$$

Identities (7.9) and (7.10) remain true when $n = 2$.

Returning now to functions on $\mathbb{S}^n$, assume that $\Omega$ is an open subset of $\mathbb{S}^n$, that $n \ge 2$ and that $f \in C^2(\Omega)$. Take $R_1, R_2$ with $0 < R_1 < 1 < R_2 < \infty$, and let $\widehat{\Omega} = \{rx\colon r \in (R_1,R_2),\ x \in \Omega\}$. Extend $f$ to a function $\tilde{f}\colon \widehat{\Omega} \to \mathbb{R}$ by $\tilde{f}(rx) = f(x)$. Then $\tilde{f} \in C^2(\widehat{\Omega})$. Define the function $\Delta_s f$ on $\mathbb{S}^n$ by

$$\Delta_s f(x) = \Delta\tilde{f}(x), \quad x \in \Omega,$$

where $\Delta$ denotes the Laplace operator in $\mathbb{R}^{n+1}$.

The operator $\Delta_s$ is called the *spherical Laplace operator* on $\mathbb{S}^n$. It coincides with the Laplace–Beltrami operator when $\mathbb{S}^n$ is equipped with the Riemannian metric inherited from the Euclidean metric on $\mathbb{R}^{n+1}$. In spherical coordinates on $\mathbb{S}^n$, the volume element of this metric is

$$d\sigma_n = \sin^{n-1}\theta_1\,\sin^{n-2}\theta_2\cdots\sin\theta_{n-1}\,d\theta_1\,d\theta_2\cdots d\theta_n.$$

See (7.6). From (7.8), with $n$ replaced by $n+1$, we obtain, when $n \ge 2$, the coordinate representation

$$\begin{aligned}\Delta_s f &= \sin^{1-n}\theta_1\,\partial_{\theta_1}(\sin^{n-1}\theta_1\,\partial_{\theta_1}f)\\ &\quad + \sum_{j=2}^{n}\sin^{j-n}\theta_j\left(\prod_{i=1}^{j-1}\sin\theta_i\right)^{-2}\frac{\partial}{\partial\theta_j}\left(\sin^{n-j}\theta_j\partial_{\theta_j}f\right).\end{aligned} \tag{7.11}$$

When $n = 1$, (7.11) simplifies to $\Delta f = f_{\theta\theta}$.

Let again $\Omega \subset \mathbb{S}^n$ and $\widehat{\Omega}$ be as above. Given a function $f$ on $\Omega$, extend it to $\tilde{f}$ on $\widehat{\Omega}$. If $f$ is differentiable at a point $x \in \Omega$ then $\tilde{f}$ is also differentiable at $x$, as a function on $\mathbb{R}^{n+1}$. We define the *spherical gradient* of $f$ at $x$ to be

$$\nabla_s f(x) = \nabla\tilde{f}(x).$$

Then $\nabla_s f(x)$ is a vector in $\mathbb{R}^{n+1}$. From the nondependence of $\tilde{f}$ on $r$, it follows that $\nabla_s f(x)$ is tangent to $\mathbb{S}^n$ at $x$. That is, $\nabla_s f(x)\cdot x = 0$.

Finally, suppose that we again have $\Omega \subset \mathbb{S}^n$. This time, let $f: \Omega \to \mathbb{R}^{n+1}$ be a tangential vector field on $\mathbb{S}^n$. Thus, we have $f(x) \cdot x = 0$ for each $x \in \Omega$. Letting again $\tilde{f}(rx) = f(x)$, define the *spherical divergence* of $f$ to be the scalar function

$$\mathrm{div}_s f(x) = \mathrm{div} \tilde{f}(x), \quad x \in \Omega.$$

If $\Omega \subset \mathbb{S}^n$ has $C^1$ boundary and $f \in C^1(\overline{\Omega})$, then by passing to $\hat{\Omega}$ the reader may check that we still have a *Gauss–Green Theorem*:

$$\int_{\partial\Omega} f \cdot \nu \, d\mathcal{H}^{n-1} = \int_{\Omega} \mathrm{div}_s f \, d\sigma_n,$$

where $\partial\Omega$ denotes boundary relative to $\mathbb{S}^n$, and $\nu(x)$ is the outward pointing unit normal vector to $\partial\Omega$ at $x$ which lies in the tangent space of $\mathbb{S}^n$ at $x$. If $\Omega = \mathbb{S}^n$, the formula becomes

$$\int_{\mathbb{S}^n} \mathrm{div}_s f \, d\sigma_n = 0.$$

Like $\Delta_s$, the operators $\nabla_s$ and $\mathrm{div}_s$ coincide with the Riemannian gradient and Riemannian divergence when $\mathbb{S}^n$ is endowed with the Riemannian metric inherited from $R^{n+1}$. See, e.g. Spivak (1965), Chavel (1993), and Bérard (1986) for discussion of calculus on manifolds. Here, we shall just write down the coordinate expressions for these operators on $\mathbb{S}^n$, and will observe that the formula

$$\Delta_s f = \mathrm{div}_s(\nabla_s f)$$

is valid.

## 7.3 Inequalities for Spherical Symmetrization, Part 1

All of the inequalities involving polarization and s.d.r. on $\mathbb{R}^n$ which were proved in Chapters 1–5 remain true for polarization and s.d.r. on $\mathbb{S}^n$, often with fewer restrictions on the data. In this section we shall elaborate on this statement for Chapters 1 and 2. Chapters 3–5 are treated in the next section.

In this section, # will denote s.d.r. on the sphere $\mathbb{S}^n$, and $f_H$ will denote the polarization of $f: \mathbb{S}^n \to \mathbb{R}$ with respect to an s-hyperplane $H \in \mathcal{H}(\mathbb{S}^n)$.

As noted in §7.1, the pertinent results in Sections 1.5 and 1.6 remain valid on $\mathbb{S}^n$. The same is true for the convergence in measure result, Proposition 1.43(b). Indeed, in the spherical case we have $\sigma_n(\mathbb{S}^n) < \infty$. From Proposition 1.41, it follows that $f_n \to f$ in measure implies $f_n^\# \to f^\#$ in measure for every sequence of real valued $\sigma_n$- measurable functions, even those with ess inf $= -\infty$.

After changing $\mathbb{R}^n$ to $\mathbb{S}^n$, $dx$ to $d\sigma_n(x)$, $|x-y|$ to $d(x,y)$, and $0 \in \mathbb{R}^n$ to $e_1 \in \mathbb{S}^n$, all of the following results from Chapter 2 and their proofs, sometimes simplified, remain valid for polarization and s.d.r on $\mathbb{S}^n$: Theorem 2.9, Propositions 2.10, 2.11, Theorem 2.12, Corollary 2.13, Theorem 2.15, Corollaries 2.16, 2.19, 2.20, 2.22, 2.23(b).

Conditions like $\lambda_f(t) < \infty$ are automatically satisfied on $\mathbb{S}^n$. The spherical version of Proposition 2.11 is true for all continuous $f$ on $\mathbb{S}^n$, not just nonnegative $f$. In Theorem 2.15(b), the map $T$, which was a translation in the $\mathbb{R}^n$ case, becomes an element of $O(n+1)$ in the $\mathbb{S}^n$ case.

For future reference, we state the spherical version of Corollary 2.19 explicitly.

**Corollary 7.1** *Let $f$ and $g$ be nonnegative Lebesgue measurable functions on $\mathbb{S}^n$ and let $K\colon \mathbb{R}^+ \to \mathbb{R}^+$ be decreasing. Then*

$$\int_{\mathbb{S}^n\times\mathbb{S}^n} f(x)g(y)K(d(x,y))\,dx\,dy \le \int_{\mathbb{S}^n\times\mathbb{S}^n} f^\#(x)g^\#(y)K(d(x,y))\,dx\,dy.$$

The $\mathbb{R}^n$ versions of the main integral inequalities, Theorems 2.9 for polarization and Theorem 2.15 for symmetrization, were stated and proved for nonnegative integrands. In the spherical case, not only do the nonnegative versions remain true, but also the finiteness of the measure $\sigma_n$ makes feasible formulations in which the integrands can change sign. Two such results are stated below, as Theorems 7.2 and 7.3.

We write $\sigma = \sigma_n$ and $L^p(\mathbb{S}^n) = L^p(\mathbb{S}^n, \sigma)$.

### Hypotheses and Definitions

Let $H \in \mathcal{H}(\mathbb{S}^n)$ be an s-hyperplane, and $\rho = \rho_H$ be the reflection in $H$. The complementary open hemispheres of $H$ are denoted by $H^+$ and $H^-$. Assume that

$$f \text{ and } g \text{ are } \sigma_n\text{-measurable real functions on } \mathbb{S}^n, \tag{7.12}$$

$$K\colon [0,\pi] \to \mathbb{R} \text{ is decreasing, and } K(d(e_1,x)) \in L^1(\mathbb{S}^n), \tag{7.13}$$

$$\Psi\colon \mathbb{R}\times\mathbb{R} \to \mathbb{R} \text{ is in } AL(\mathbb{R}\times\mathbb{R}), \tag{7.14}$$

$$\Psi(f(x),g(y))K(d(x,y)),\ \Psi(f_H(x),g_H(y))K(d(x,y)) \in L^1(\mathbb{S}^n\times\mathbb{S}^n), \tag{7.15}$$

and

$$\Psi(f(x),g(x)),\ \Psi(f_H(x),g_H(x)) \in L^1(\mathbb{S}^n). \tag{7.16}$$

Let

$$Q(f,g) = Q(f,g,K,\Psi) \equiv \int_{\mathbb{S}^n\times\mathbb{S}^n} \Psi(f(x),g(y))K(d(x,y))\,d\sigma(x)\,d\sigma(y),$$

$$A \equiv \{(x,y) \in (H^+)^2 : \quad \text{either} \quad f(x) < f(\rho x) \quad \text{and} \quad g(y) > g(\rho y), \\ \text{or} \quad f(x) > f(\rho x) \quad \text{and} \quad g(y) < g(\rho y)\},$$

and $A_0 \equiv \{x \in H^+ : (x,x) \in A\}$.

**Theorem 7.2** *Let assumptions* (7.12)–(7.16) *be satisfied. Then*

*(a)* $Q(f,g) \le Q(f_H, g_H)$.
*(b) Suppose that neither $f$ nor $g$ is constant, that $K$ is strictly decreasing on $\mathbb{R}^+$, and that $\Psi \in SAL(\mathbb{R} \times \mathbb{R})$. Then equality holds in (a) if and only if* $(\sigma \times \sigma)(A) = 0$.
*(c)* $\int_{\mathbb{S}^n} \Psi(f(x), g(x))\, d\sigma(x) \le \int_{\mathbb{S}^n} \Psi(f_H(x), g_H(x))\, d\sigma(x)$.
*(d) If $\Psi \in SAL(\mathbb{R} \times \mathbb{R})$ then equality holds in (c) if and only if* $\sigma(A_0) = 0$.

*Proof* The result is derived from the two-point symmetrization inequality Theorem 2.8, just as Theorem 2.9 was. Note that hypothesis (7.15) insures that $Q(f,g)$ and $Q(f_H, g_H)$ are well defined and are finite. □

**Theorem 7.3** *Suppose that assumptions* (7.12), (7.13), *and* (7.14) *are satisfied, that* (7.15) *and* (7.16) *are satisfied with $f^\#$, $g^\#$ replacing $f_H$, $g_H$, that*

$$A_1 \equiv \{(x,y) \in \mathbb{S}^n \times \mathbb{S}^n : f(x) < f(y) \text{ and } g(x) > g(y)\},$$

*and that*

$$\Psi(f(x), 0) \text{ and } \Psi(0, g(y)) \text{ are in } L^1(\mathbb{S}^n).$$

*Then*

*(a)* $Q(f,g) \le Q(f^\#, g^\#)$.
*(b) Under additional assumptions that neither $f$ nor $g$ is constant, that $K$ is strictly decreasing on $[0,\pi]$, and that $\Psi \in SAL(\mathbb{R} \times \mathbb{R})$, equality holds in (a) if and only if there exists $T \in O(n+1)$ such that*

$$f = f^\# \circ T \quad \text{and} \quad g = g^\# \circ T, \quad \text{a.e. on} \quad \mathbb{S}^n.$$

*(c)* $\int_{\mathbb{S}^n} \Psi(f(x), g(x))\, d\sigma(x) \le \int_{\mathbb{S}^n} \Psi(f^\#(x), g^\#(x))\, d\sigma(x)$.
*(d) If $\Psi \in SAL(\mathbb{R} \times \mathbb{R})$ then equality holds in (c) if and only if* $\sigma(A_1) = 0$.

*Proof of Theorem 7.3(a)* We sketch the proof when $\Psi \in C(\mathbb{R} \times \mathbb{R})$. The passage to discontinuous $\Psi$ is done as in the proof of Theorem 2.15.

Define

$$\Psi_1(s,t) = \Psi(s,t) - \Psi(s,0) - \Psi(0,t) + \Psi(0,0), \quad (s,t) \in \mathbb{R}^2.$$

Note that $\int_{\mathbb{S}^n} K(d(x,y))\,d\sigma(y)$ is the same for all $x$. So, integrating first with respect to $y$, we obtain

$$\int_{\mathbb{S}^n\times\mathbb{S}^n} |\Psi(f(x),0)K(d(x,y))|\,d\sigma(x)\,d\sigma(y)$$
$$= \int_{\mathbb{S}^n} |\Psi(f(x),0)|\,d\sigma(x) \int_{\mathbb{S}^n} |K(d(e_1,x))|\,d\sigma(x) < \infty.$$

Similarly, $\Psi(0,g(y))K(d(x,y))$ belongs to $L^1(\mathbb{S}^n\times\mathbb{S}^n)$. We conclude that $\Psi_1(f(x),g(y))K(d(x,y)) \in L^1(\mathbb{S}^n\times\mathbb{S}^n)$, and likewise when $f$ and $g$ are replaced by $f^\#$ and $g^\#$. Furthermore, the integrals $\int_{\mathbb{S}^n}\Psi(f(x),0)\,d\sigma(x)$ and $\int_{\mathbb{S}^n}\Psi(0,g(y))\,d\sigma(y)$ do not change when $f$ and $g$ are replaced by $f^\#$ and $g^\#$. Thus, to prove the theorem, it suffices to prove it when $\Psi$ is replaced by $\Psi_1$. In the argument to follow we shall use $Q(f,g)$ to denote $Q(f,g,K,\Psi_1)$.

We have $\Psi_1 \in AL(\mathbb{R}\times\mathbb{R})$, and $\Psi_1$ vanishes on the coordinate axes. From the definition of $AL$, it follows that $\Psi_1$ is nonnegative in the first and third quadrants, nonpositive in the second and fourth quadrants. Furthermore, within each quadrant, $|\Psi_1(s,t)|$ increases as $|s|$ increases with $t$ fixed or as $|t|$ increases with $s$ fixed.

Assume now that $f$ and $g$ are continuous and that $K$ is bounded on $[0,\pi]$. Define

$$\begin{aligned}
\mathcal{S}(f) &= \{F\in C(\mathbb{S}^n,\mathbb{R})\colon \Omega(\cdot,F)\le\Omega(\cdot,f) \text{ on } (0,\infty),\ \lambda_F=\lambda_f \text{ on } \mathbb{R}\},\\
\mathcal{S} &= \mathcal{S}(f,g) = \{(F,G)\in\mathcal{S}(f)\times\mathcal{S}(g)\colon Q(f,g)\le Q(F,G)\}, \text{and}\\
d^2 &= \inf_{(F,G)\in\mathcal{S}} \|f^\# - F\|_2^2 + \|g^\# - G\|_2^2,
\end{aligned}$$

where $\|\cdot\|_2$ is the norm in $L^2(\mathbb{S}^n)$.

Let $\{f_k\}$ and $\{g_k\}$ be sequences in $\mathcal{S}(f)$ and $\mathcal{S}(g)$ respectively such that

$$d^2 = \lim_{k\to\infty} \|f^\# - f_k\|_2^2 + \|g^\# - g_k\|_2^2.$$

By the Arzelà–Ascoli Theorem, there is a subsequence of $\{(f_k,g_k)\}$, also denoted $\{(f_k,g_k)\}$, such that $\{f_k\}$ and $\{g_k\}$ respectively converge uniformly on $\mathbb{S}^n$ to continuous functions $F_0$ and $G_0$. Since the sequences are uniformly bounded and the measure $\sigma$ is finite, the dominated converges theorem yields

$$d^2 = \|F_0 - f^\#\|_2^2 + \|G_0 - g^\#\|_2^2.$$

We have $F_0\in\mathcal{S}(f)$, $G_0\in\mathcal{S}(g)$. There exists $M$ such that

$$|\Psi_1(f_k(x),g_k(y))K(d(x,y))| \le M, \quad \forall\, x\in\mathbb{S}^n,\ y\in\mathbb{S}^n,\ k\ge 1.$$

By the dominated convergence theorem, $\lim_{k\to\infty} Q(f_k,g_k) = Q(F_0,G_0)$. Thus $Q(f,g)\le Q(F_0,G_0)$, so that $(F_0,G_0)\in\mathcal{S}$. Via Theorem 7.2, and taking $H$

to be an appropriate s-hyperplane with $e_1 \in H^+$, one shows as in the proof of Theorem 2.15 that $F_0 = f^\#$ and $G_0 = g^\#$, so that $(f^\#, g^\#) \in \mathcal{S}$, hence $Q(f,g) \leq Q(f^\#, g^\#)$. Conclusion (a) is proved when $f$ and $g$ are continuous and $K$ is bounded.

Next, we continue to suppose that $K$ is bounded, and will assume that $f$ and $g$ are in $L^\infty(\mathbb{S}^n)$. Take sequences $\{f_k\}$ and $\{g_k\}$ of functions in $C(\mathbb{S}^n, \mathbb{R})$ such that $f_k \to f$ and $g_k \to g$ in $L^2$-norm and a.e., and also $\|f_k\|_{L^\infty} \leq \|f\|_{L^\infty}$, $\|g_k\|_{L^\infty} \leq \|g\|_{L^\infty}$ for each $k$. Some subsequence also has the corresponding convergence properties for $\{f_k^\#\}$ and $\{g_k^\#\}$. An argument with the dominated convergence theorem gives $Q(f,g) \leq Q(f^\#, g^\#)$, as desired.

Let now $f, g$, and $K$ be any functions satisfying the hypotheses of Theorem 7.3. For $m \geq 1$, define $f_m(x) = f(x)$ if $x \in [-m, m]$, $f_m(x) = m$ if $f(x) \geq m$, $f_m(x) = -m$ if $f(x) \leq -m$. Define $g_m$ and $K_m$ in the same way. Then for $m \geq 1$ we have

$$\begin{aligned}\int_{\mathbb{S}^n \times \mathbb{S}^n} & \Psi_1(f_m(x), g_m(y)) K_m(d(x,y))\, d\sigma(x)\, d\sigma(y) \\ & \leq \int_{\mathbb{S}^n \times \mathbb{S}^n} \Psi_1(f_m^\#(x), g_m^\#(y)) K_m(d(x,y))\, d\sigma(x)\, d\sigma(y).\end{aligned} \tag{7.17}$$

We noted above that $|\Psi_1|$ increases on horizontal and vertical half lines within a quadrant when we walk along the half line toward infinity. It follows that, for almost every $(x, y)$ and every $m \geq 1$, we have

$$|\Psi_1(f_m(x), g_m(y)) K_m(d(x,y))| \leq |\Psi_1(f(x), g(y)) K(d(x,y))|.$$

By hypothesis, the function on the right is in $L^1(\mathbb{S}^n \times \mathbb{S}^n)$. Thus, as $m \to \infty$, the dominated convergence theorem implies that the sequence on the left in (7.17) converges to $Q(f,g)$. Since $f_m^\# = (f^\#)_m$, the same reasoning shows that the sequence on the right (7.17) converges to $Q(f^\#, g^\#)$. Part (a) of Theorem 7.3 is proved for continuous $\Psi \in AL(\mathbb{R} \times \mathbb{R})$. □

*Proof of Theorem 7.3(b), for continuous $\Psi$* Since $Q(f,g) = Q(f \circ T, g \circ T)$, the "if" part is clear. To obtain the "only if" part, define the center of mass $c(E)$ of a $\sigma$-measurable set $E \subset \mathbb{S}^n$ with $\sigma(E) > 0$ to be

$$c(E) = \frac{1}{\sigma(E)} \int_E x\, d\sigma(x).$$

Then $c(E) \in \mathbb{B}^{n+1}(0,1)$, the open unit ball in $\mathbb{R}^{n+1}$. For $T \in O(n+1)$, the linearity of $T$ gives $c(TE) = Tc(E)$. In particular, if $c(E) = 0$ then $c(TE) = 0$ for every $T \in O(n+1)$. Also,

$$c(E) \cdot e_j = \frac{1}{\sigma(E)} \int_E x \cdot e_j\, d\sigma, \quad 1 \leq j \leq n+1.$$

Set $E(t,f) = \{x \in \mathbb{S}^n : f(x) > t\}$ and $I_f = (\operatorname{ess\,inf} f, \operatorname{ess\,sup} f)$. Then $0 < \sigma(E(t,f)) < \sigma(\mathbb{S}^n)$ for $t \in I_f$. Define

$$c_1(t) = c(E(t,f)), \quad c_2(t) = c(E(t,g)),$$

where $t \in I_f$ in the definition of $c_1$ and $t \in I_g$ in the definition of $c_2$.

We now proceed by cases and subcases.

**Case 1.** Assume there exists $t \in I_f$ such that $c_1(t) = 0$ or there exists $s \in I_g$ such that $c_2(s) = 0$. Say the latter. Take such an $s \in I_g$ and take any $t \in I_f$. Write

$$E_1 = E(t,f), \quad E_2 = E(s,g), \quad E_i^c = \mathbb{S}^n \setminus E_i, \quad i = 1,2.$$

Then $c(E_2) = 0$, and also $c(E_2^c) = 0$.

Let $a$ be a point of density of $E_1$ and $b$ be a point of density of $E_1^c$. Let $H$ be the bisecting s-hyperplane of $a$ and $b$, and $H^+$ be the complementary hemisphere containing $a$. If $B$ is a sufficiently small ball centered at $a$ then we have

$$\sigma[\{x \in B : x \in E_1, \rho_H x \notin E_1\}] > 0. \tag{7.18}$$

Moreover, (7.18) still holds, with the same $B$, when $H$ is replaced by any s-hyperplane $\tilde{H}$ which is sufficiently close to $H$. Here we may define the distance $d_1$ between $H$ and $\tilde{H}$ to be $\theta$, where $\theta \in [0, \pi/2]$ is the angle between two corresponding normal vectors.

Assuming, as we may, that $B \subset H^+$, it follows that there exists $\epsilon > 0$ such that

$$\sigma[\{x \in (\tilde{H})^+ : f(x) > t, \, f(\rho x) \le t\}] > 0, \tag{7.19}$$

where $\rho = \rho_{\tilde{H}} x$ and $\tilde{H}$ is any s-hyperplane with $0 \le d_1(H, \tilde{H}) < \epsilon$.

Next, with $H$ and $H^+$ as above, we examine $E_2$.

*Subcase 1.* Suppose that the set

$$\rho(E_2^c \cap H^+) \cap E_2 \tag{7.20}$$

has positive measure. Then

$$\sigma[\{y \in H^+ : \, g(y) \le s, \; g(\rho_H y) > s\}] > 0. \tag{7.21}$$

Combining (7.21) with (7.19), then using Theorems 7.1(b) and 7.2(a), we obtain $Q(f,g) < Q(f_H, g_H) \le Q(f^\#, g^\#)$, as desired.

*Subcase 2.* Suppose that the set in (7.20) has measure zero. Write $\rho = \rho_H$. Then, except for a null set, $\rho(E_2^c \cap H^+) \subset E_2^c \cap H^-$. If the difference between the two sets had positive measure then $c(E_2^c)$ could not be zero. Thus, except for a null set we get $\rho(E_2^c \cap H^+) = E_2^c \cap H^-$. It follows that the sets $E_2$ and $E_2^c$

are essentially symmetric with respect to $\rho$, i.e., $\rho(E_2) = E_2$ except for a null set, and likewise for $E_2^c$.

We know that $E_2 \cap H^+$ and $E_2^c \cap H^+$ each have positive measure. By a real variable exercise, given $\delta > 0$ there exist $a',\ b' \in H^+$ such that $a'$ is a point of density of $E_2$, $b'$ is a point of density of $E_2^c$, and $d(a', b') < \delta$. By the essential symmetry of $E_2$, $\rho_H(a')$ is also a point of density of $E_2$. Let $\tilde{H}$ be the plane bisecting $\rho_H(a')$ and $b'$. Then, as in the proof of (7.19), (7.21) holds, with $H$ replaced by $\tilde{H}$. For small enough $\delta$ we have $d_1(H, \tilde{H}) < \epsilon$. Then (7.19) also holds, and we conclude as before that $Q(f, g) < Q(f^\#, g^\#)$. Case 1 of Theorem 7.3(b) is proved.

**Case 2.** If no $c_1(t)$ or $c_2(s)$ is ever zero for any $t \in I_f$ or $s \in I_g$, then there are two subcases: $c_1(t)$ and $c_2(s)$ are distinct nonzero numbers for some $t$ and $s$, or $c_1(t) = c_2(s) =$ nonzero constant for all relevant $t$. The proofs of each subcase can be accomplished by modifying the proof of Theorem 2.15(b). In addition to the usual spherical adjustments, the hyperplane $H(w)$ appearing in that proof should be replaced by the s-hyperplane bisecting the points $e_1$ and $w \in \mathbb{S}^n$.

This completes the proof of Theorem 7.3(b) when $\Psi$ is continuous. Verification of parts (c) and (d) of Theorem 7.3, and the extension to discontinuous $\Psi$, is left to the reader. □

In Chapter 2, each of Corollaries 2.16, 2.19, 2.20, 2.22 and 2.23(b) states integral inequalities involving nonnegative integrands. As observed at the beginning of this section, these results remain valid in the spherical case. Theorem 7.3 permits us to prove versions of these corollaries involving integrands that are not necessarily nonnegative. Here is just one example: the inequality

$$\int_{\mathbb{S}^n} f(x)g(x)\,d\sigma(x) \le \int_{\mathbb{S}^n} f^\#(x)g^\#(x)\,d\sigma(x)$$

is true when $f \in L^p(\mathbb{S}^n),\ g \in L^{p'}(\mathbb{S}^n)$, where $p$ and $p'$ are conjugate Hölder exponents.

## 7.4 Inequalities for Spherical Symmetrization, Part 2

For $f \in \mathrm{Lip}(\mathbb{S}^n, \mathbb{R})$, we define the Lipschitz norm by

$$\|f\|_{\mathrm{Lip}(\mathbb{S}^n)} = \sup_{x,y\in\mathbb{S}^n} \frac{|f(y) - f(x)|}{d(x, y)},$$

where $d$ is the canonical distance on $\mathbb{S}^n$.

If $\|f\|_{\text{Lip}(\mathbb{S}^n)} < \infty$, then application of Rademacher's Theorem 3.2 to the extension $\tilde{f}$, studied in §7.2, shows that $f$ is differentiable $\sigma_n$-a.e on $\mathbb{S}^n$. Thus, the spherical gradient $\nabla_s f$ exists a.e, as a vector in $\mathbb{R}^{n+1}$ which is tangent vector to $\mathbb{S}^n$. Denote by $|\nabla_s f|$ the length of this vector in $\mathbb{R}^{n+1}$.

Here is the main theorem about the decrease of spherical Dirichlet integrals under spherical symmetrization. Throughout this section, $f^\#$ denotes the spherical s.d.r. of $f$. Also, set

$$\mathcal{K}(x,\epsilon) = \{y \in \mathbb{S}^n : d(x,y) < \epsilon\}.$$

**Theorem 7.4** *Let* $f \in \text{Lip}(\mathbb{S}^n, \mathbb{R})$ *and* $\Phi: \mathbb{R}^+ \to \mathbb{R}^+$ *be convex and increasing, with* $\Phi(0) = 0$. *Then*

$$\int_{\mathbb{S}^n} \Phi(|\nabla_s f^\#|)\, d\sigma_n \leq \int_{\mathbb{S}^n} \Phi(|\nabla_s f|)\, d\sigma_n. \tag{7.22}$$

*Proof* For $\epsilon > 0$, $\sigma = \sigma_n$, and $K_\epsilon = \chi_{[0,\epsilon]}$, define

$$\begin{aligned} I(\epsilon,f) &= \int_{\mathbb{S}^n \times \mathbb{S}^n} |f(y) - f(x)|\, K_\epsilon(d(x,y))\, d\sigma(x)\, d\sigma(y) \\ &= \int_{\mathbb{S}^n} d\sigma(x) \int_{\mathcal{K}(x,\epsilon)} |f(y) - f(x)|\, d\sigma(y). \end{aligned}$$

Then, by the spherical version of Corollary 2.19 we get

$$I(\epsilon,f^\#) \leq I(\epsilon,f).$$

We claim that for all $f \in \text{Lip}(\mathbb{S}^n, \mathbb{R})$,

$$\lim_{\epsilon \to 0} \epsilon^{-n-1} I(\epsilon,f) = C_n \int_{\mathbb{S}^n} |\nabla_s f|\, d\sigma(x), \tag{7.23}$$

where $C_n$ is a constant depending only on $n$. In fact,

$$C_n = \frac{\beta_n}{\pi(n+1)} = \frac{1}{\pi}\alpha_{n+1}, \quad n \geq 1. \tag{7.24}$$

Once (7.23) is proved, we will have (7.22) for $\Phi(x) = |x|$. The passage from $|x|$ to general $\Phi$ can then be accomplished as in the proof of Theorem 3.11. □

*Proof of* (7.23) At points of differentiability $x \in \mathbb{S}^n$ of $f$, and each $y \in \mathbb{S}^n$, define $R(x,y)$ by

$$f(y) - f(x) = \nabla_s f(x) \cdot (y - x) + R(x,y),$$

where $\cdot$ is the inner product in $\mathbb{R}^{n+1}$. Then, for almost every $x$, we have

$$|R(x,y)| \leq 2\|f\|_{\text{Lip}(\mathbb{S}^n)} |y - x|$$

and

$$\lim_{y \to x} R(x,y)|x-y|^{-1} = 0.$$

Thus we can write

$$I(\epsilon, f) = \int_{\mathbb{S}^n \times \mathbb{S}^n} |\nabla_{\mathbb{S}} f(x) \cdot (y-x)| \, K_\epsilon(d(x,y)) \, d\sigma(x) \, d\sigma(y) + P_\epsilon, \tag{7.25}$$

where

$$|P_\epsilon| \leq \int_{\mathbb{S}^n \times \mathbb{S}^n} |R(x,y)| K_\epsilon(d(x,y)) \, d\sigma(x) \, d\sigma(y).$$

Integrating first with respect to $y$, and using $\sigma(\mathcal{K}(x,\epsilon)) \leq C_n \epsilon^n$, we obtain, for some constant $C_n$, not necessarily the one in (7.24),

$$\epsilon^{-n-1}|P_\epsilon| \leq C_n \int_{\mathbb{S}^n} \sup_{y \in \mathcal{K}(x,\epsilon)} \epsilon^{-1}|R(x,y)| \, d\sigma(x).$$

The integrand in the last integral is bounded, and approaches zero as $\epsilon \to 0$ for almost every $x$. By the dominated convergence theorem, we see that

$$\lim_{\epsilon \to 0} \epsilon^{-n-1} P_\epsilon = 0. \tag{7.26}$$

To analyze the first term on the right in (7.25), take a point of differentiability $x \in \mathbb{S}^n$, assume $\nabla_{\mathbb{S}} f(x) \neq 0$, and define $a \in \mathbb{S}^n$ by

$$\nabla_{\mathbb{S}} f(x) = a |\nabla_{\mathbb{S}} f(x)|.$$

Since $\nabla_{\mathbb{S}} f(x)$ is tangent to $\mathbb{S}^n$ at $x$, we have $a \cdot x = 0$, and hence

$$\int_{\mathbb{S}^n} |\nabla_{\mathbb{S}} f(x) \cdot (y-x)| \, K_\epsilon(d(x,y)) \, d\sigma(y) = |\nabla_{\mathbb{S}} f(x)| \int_{\mathcal{K}(x,\epsilon)} |a \cdot y| \, d\sigma(y).$$

Since $a \in \mathbb{S}^n$ and $a \cdot x = 0$, there exists $T \in O(n+1)$ such that $Te_1 = x$ and $Te_2 = a$. Letting $y = Tz$ in the integral on the right, we obtain

$$\begin{aligned} \int_{\mathbb{S}^n \times \mathbb{S}^n} & |\nabla_{\mathbb{S}} f(x) \cdot (y-x)| \, K_\epsilon(d(x,y)) \, d\sigma(x) \, d\sigma(y) \\ &= C_n(\epsilon) \int_{\mathbb{S}^n} |\nabla_{\mathbb{S}} f(x)| \, d\sigma(x) \end{aligned} \tag{7.27}$$

where $C_n(\epsilon) = \int_{\mathcal{K}(\epsilon)} |e_2 \cdot z| \, d\sigma(z)$. Integrating in spherical coordinates, one can show that

$$\lim_{\epsilon \to 0} \epsilon^{-n-1} C_n(\epsilon) = \pi^{-1}(n+1)^{-1} \beta_n.$$

Then (7.23) and (7.24) follow from (7.25), (7.26) and (7.27). This completes our proof of Theorem 7.4. □

Theorem 7.4 is the spherical version of Theorems 3.7 and 3.11. It implies that the spherical versions of Corollaries 3.8 and 3.9 are true. The reader may verify also that the inequality

$$\|\nabla_s f^\#\|_{L^\infty(\mathbb{S}^n)} \leq \|\nabla_s f\|_{L^\infty(\mathbb{S}^n)}$$

is valid for $f \in \mathrm{Lip}(\mathbb{S}^n, \mathbb{R})$. This is the spherical version of Theorem 3.6.

One may form Sobolev spaces $W^{1,p}(\mathbb{S}^n, \mathbb{R})$ of functions on spheres. Theorem 3.20 and its Corollaries 3.21 and 3.22 remain valid. In spherical versions of Theorem 3.20 and Corollary 3.21 one can drop the nonnegativity assumption on $f$.

In the discussion after the proof of Corollary 3.9 we remarked that for $n = 1$ the nonnegativity assumption on $f$ in that result could be removed. To see this, suppose that $\Omega \subset \mathbb{R}$ is open, with $\mathcal{L}(\Omega) < \infty$, and that $f \in \mathrm{Lip}(\overline{\Omega}, \mathbb{R})$ with $f = 0$ on $\partial\Omega$. By running all the component intervals together, we may assume that $\Omega$ is a finite open interval, then by translation and dilation we may assume that $\Omega = (-\pi, \pi)$. Now $f(-\pi) = f(\pi) = 0$, so we may identify $f$ with a Lipschitz function $f_0$ on the unit circle. The validity of Theorem 7.4 for $f_0$ implies the validity of the inequality

$$\int_\Omega \Phi(|(f^\#)'|)\, dx \leq \int_\Omega \Phi(|f'|)\, dx$$

for all convex increasing $\Phi$ with $\Phi(0) = 0$, where $f^\#$ denotes s.d.r. on $\Omega \subset \mathbb{R}$.

By straightforward modification of the $\mathbb{R}^n$ case, one can define spherical versions of the total variation of a function, the space $BV(\mathbb{S}^n, \mathbb{R})$, and the perimeter $P(E)$ of a set $E \subset \mathbb{S}^n$. For example, for $f \in L^1(\mathbb{S}^n, \sigma_n)$,

$$V(f) \equiv \sup_\phi \int_{\mathbb{S}^n} f \ \mathrm{div}_s\, \phi \, d\sigma_n,$$

where $\phi$ runs through all functions $\phi \in C^1(\mathbb{S}^n, \mathbb{R}^n)$ with $\sup_{\mathbb{S}^n} |\phi| \leq 1$, and $\mathrm{div}_s$ denotes the spherical divergence defined in §7.2.

One may check that the spherical versions of the following symmetrization results are true: Theorems 4.8, 4.10, 4.14, 4.16, Corollary 4.15.

The gist of these results is that s.d.r. on $\mathbb{S}^n$ reduces the total variation of functions, the perimeter of sets, and the $(n-1)$-dimensional spherical lower Minkowski content of the boundary. Corollary 4.13 asserts that s.d.r. on $\mathbb{R}^n$ reduces the $(n-1)$-Hausdorff measure of the boundary. The analogue for spherical s.d.r. is probably true, but I have not thoroughly checked it.

Theorems 4.10 and 4.16 contain simple formulas for the measure of the boundary of Euclidean balls in terms of their volume. Except in low dimensions, there are no straightforward formulas for the measure of the boundary of spherical caps in terms of their spherical volumes. Since $\partial\mathcal{K}(\theta)$ is the $(n-1)$-sphere in $\mathbb{R}^{n+1}$ with center $(\cos\theta)\,e_1$, radius $\sin\theta$, in terms of radii, we do have the simple formula

$$\mathcal{H}^{n-1}(\partial\mathcal{K}(\theta)) = \beta_{n-1}\sin^{n-1}\theta.$$

Much of the early work on isoperimetry on spheres was done in the context of Minkowski content. For emphasis, the spherical version of Theorem 4.16 is stated as Theorem 7.5.

**Theorem 7.5** *For each $\sigma_n$-measurable set $E \subset \mathbb{S}^n$, we have*

$$\mathcal{M}_*^{n-1}(\partial E) \geq \mathcal{M}^{n-1}(\partial(E^{\#})).$$

Here, # denotes s.d.r. on $\mathbb{S}^n$, and $\mathcal{M}$ is computed like its counterpart in $\mathbb{R}^n$, except that the collar $E(\delta)$ and the core $E(-\delta)$ are computed using the canonical distance $d$ on $\mathbb{S}^n$. For example, for $E \subset \mathbb{S}^n$ we have

$$E(\delta) = \{x \in \mathbb{S}^n : d(x, E) < \delta\}.$$

Sobolev embedding theorems exist in the $\mathbb{S}^n$ setting. Here is one, which is more often called the Poincaré inequality: For $f \in L^1(\mathbb{S}^n, \sigma_n)$ with distributional derivatives in $L^p(\mathbb{S}^n, \sigma_n)$,

$$\|f - f_{\rm av}\|_{p^*} \leq C_p\|\nabla f\|_p, \quad 1 \leq p < n,$$

where $p^* = \frac{np}{n-p}$ and $\|\cdot\|_p$ denotes the norm in $L^p(\mathbb{S}^n, \sigma_n)$, and $f_{\rm av} = \frac{1}{\beta_n}\int_{\mathbb{S}^n} f\, d\sigma_n$.

Apparently, the best constants $C_p$ are not known.

For a proper open subset $\Omega \subset \mathbb{S}^n$ the analysis in §§5.1–5.3 of eigenvalues of the Laplacian with Dirichlet boundary conditions remains valid. We just have to use the spherical Laplacian $\Delta_s$ instead of the Euclidean Laplacian $\Delta$. Thus, there is a smallest nonnegative number $\lambda_1 = \lambda_\Omega$ for which the p.d.e. $\Delta_s u = -\lambda_1 u$ has a solution $u \in W_0^{1,2}(\Omega)$ which is not identically zero. One such eigenfunction, call it $u_1$, is the the nonnegative minimizer of $\int_\Omega |\nabla_s u|^2 d\sigma_n$ over all $u \in W_0^{1,2}(\Omega)$ with $\int_\Omega u^2\, d\sigma_n = 1$. By the consequences of Theorem 7.4, we see that the Faber–Krahn Theorem 5.6 is still valid on $\mathbb{S}^n$:

$$\lambda_1(\Omega^{\#}) \leq \lambda_1(\Omega), \quad \text{for Dirichlet eigenvalues,}$$

where $^{\#}$ denotes spherical s.d.r.

If $U$ is a proper open set in $\mathbb{S}^n$, $K$ is a compact subset of $U$ and $1 \leq p < \infty$, we can define the *variational p-capacity* of the spherical condenser $(K, U)$ to be

$$\mathrm{Cap}_p(K, U) = \inf\left\{\int_U |\nabla_s u|^2\, d\sigma_n\right\},$$

where the infimum is taken over all $u \in \mathrm{Lip}(\mathbb{S}^n)$ with $u = 1$ on $K$, $u = 0$ on $\mathbb{S}^n \setminus U$ and $0 \leq u \leq 1$ in $\mathbb{S}^n$. It follows again from Theorem 7.4 that

$$\mathrm{Cap}_p(\overline{K^\#}, U) \leq \mathrm{Cap}_p(K, U).$$

Let again $\Omega$ be a proper open subset of $\mathbb{S}^n$. Define the torsional rigidity of $\Omega$ to be

$$T(\Omega) = 2 \int_\Omega u \, d\sigma_n,$$

where $u \in W_0^{1,2}(\Omega)$ is the solution of the Poisson problem $\Delta_s u = -2$ in $\Omega$. Then Pólya's Theorem 5.17 remains true in the spherical setting:

$$T(\Omega) \leq T(\Omega^\#).$$

This can be proved using Dirichlet integrals, as in the Euclidean case.

## 7.5 Cap Symmetrizations

Let $n \geq 2$. For $x = (x_1, \ldots, x_n) \in \mathbb{R}^n \setminus \{0\}$ write $x = ry$, where $r = |x|$ and $y = x/r \in \mathbb{S}^{n-1}$. For an $\mathcal{L}^n$-measurable set $E \subset \mathbb{R}^n$ and $r > 0$, set

$$E(r) = \{y \in \mathbb{S}^{n-1} : ry \in E\}.$$

Then $E(r)$ is $\sigma_{n-1}$-measurable for $\mathcal{L}$-almost every $r$. For such $r$, let $E^\#(r) \subset \mathbb{S}^{n-1}$ denote the spherical symmetrization of $E(r)$ with respect to the origin $e_1 \in \mathbb{S}^{n-1} \subset \mathbb{R}^n$. If $E(r)$ is not measurable, set $E^\#(r) = \emptyset$. Also, define $E^\#(0) = \{0\}$ if $0 \in E$, $E^\#(0) = \emptyset$ if $0 \notin E$.

The $(n-1, n)$-*cap symmetrization* of $E$ is defined to be the subset $E^\#$ of $\mathbb{R}^n$ given by

$$E^\# = \bigcup_{r \in \mathbb{R}^+} r E^\#(r).$$

Thus, for almost every $r > 0$, the intersection of $E^\#$ with the sphere $|x| = r$ is all of $|x| = r$ if $E$ contains $|x| = r$, is empty if $E$ intersects $|x| = r$ in a set of measure zero, and is an open spherical cap centered on the positive $x_1$-axis with the same spherical measure as $(|x| = r) \cap E$ if the intersection has positive measure and is not all of $|x| = r$. Furthermore, $0 \in E^\#$ if and only if $0 \in E$. By the same argument as for Steiner symmetrization, $E^\#$ is $\mathcal{L}^n$-measurable. The symmetrized set $E^\#$ has the positive $x_1$-axis as ray of symmetry. Of course, one can perform $(n-1, n)$-cap symmetrizations with respect to any half-line in $\mathbb{R}^n$.

Next, let $A$ be an open subset of $\mathbb{R}^+$. Define the set $X$ by

$$X = \{ry : r \in A, \ y \in \mathbb{S}^{n-1}\}.$$

Then $X$ is an open subset of $\mathbb{R}^n$ which may or may not contain 0. Suppose that $f: X \to \mathbb{R}$ is $\mathcal{L}^n$-measurable. Then for $\mathcal{L}$-almost every $r \in A$, the slice function $f^r$ defined on $\mathbb{S}^{n-1}$ by $f^r(y) = f(ry)$ is $\sigma_{n-1}$-measurable, so it has an $(n-1)$-dimensional spherical symmetric decreasing rearrangement $(f^r)^\#$. Define the *$(n-1, n)$-cap symmetrization* $f^\#$ of $f$ to be the real valued function defined $\mathcal{L}^n$-a.e. on $X$ by

$$f^\#(x) = f^\#(ry) = (f^r)^\#(y)$$

when $r > 0$ and $f^r$ is measurable. If $0 \in X$, define $f^\#(0) = f(0)$. If $f^r$ is not measurable, leave $f^\#$ undefined on $|x| = r$.

In the plane, *circular symmetrization* is the name commonly used for $(1, 2)$-cap symmetrization.

As with Steiner symmetrization, the $(n-1, n)$-cap symmetrization $f^\#$ is $\mathcal{L}^n$-measurable. The restriction of $f^\#$ to spheres $|x| = r$, $r \in A$, is symmetric decreasing, with maximum at $x = re_1$. By passing to polar coordinates and use of Fubini's theorem, one finds that $f^\#$ and $f$ are equidistributed on $X$, as well as on each sphere within $X$.

Next we define $(k, n)$-cap symmetrization. Suppose $n \geq 3$ and that $k$ is an integer with $1 \leq k \leq n-2$. Write $m = n - k$. Decompose $x = (x_1, \ldots, x_n) \in \mathbb{R}^n$ as

$$x = (w, z) = (w_1, \ldots, w_{k+1}, z_1, \ldots, z_{m-1}).$$

Then $\mathbb{R}^n = \mathbb{R}^{k+1} \times \mathbb{R}^{m-1}$, and $\mathbb{R}^n$ is foliated into parallel affine $(k+1)$-planes $\{(w, z) \colon w \in \mathbb{R}^{k+1}\}$ indexed by $z$.

Let $E \subset \mathbb{R}^n$ be an $\mathcal{L}^n$-measurable set. Define $E(z)$, the slice of $E$ through $z$, by

$$E(z) = \{w \in \mathbb{R}^{k+1} \colon (w, z) \in E\}, \quad z \in \mathbb{R}^{m-1}.$$

Then $E(z)$ is $\mathcal{L}^{k+1}$-measurable for $\mathcal{L}^{m-1}$-almost every $z \in \mathbb{R}^{m-1}$. For measurable $E(z)$, define $E^\#(z)$ to be the $(k, k+1)$-cap symmetrization of $E(z)$ (with respect to the positive $x_1$-axis in $\mathbb{R}^{k+1}$). Define the *$(k, n)$-cap symmetrization* $E^\#$ of $E$ by

$$E^\# = \bigcup \{(w, z) \colon w \in E^\#(z)\},$$

where the union is over all $z \in \mathbb{R}^{m-1}$ for which $E^\#(z)$ is measurable.

Let us say that a set $X \subset \mathbb{R}^n$ is *ring-type* if $X$ is open and each slice $X(z) \subset \mathbb{R}^{k+1}$ is has the form $X(z) = \{ry \colon r \in A(z),\ y \in \mathbb{S}^k\}$, where $A(z)$ is an open subset of $\mathbb{R}^+$. If we express points $x \in \mathbb{R}^n$ as $x = (ry, z)$ with $r \in \mathbb{R}^+, y \in \mathbb{S}^k$, $z \in \mathbb{R}^{m-1}$, then $X$ being ring-type means that if $(ry, z) \in X$ for some $y$ then $(ry, z) \in X$ for all $y$. Set

$$Z = \{z \in \mathbb{R}^{m-1} \colon X(z) \text{ is not empty}\}.$$

Let $f: X \to \mathbb{R}$, with $X$ ring-type and $f$ $\mathcal{L}^n$-measurable. Define the *$(k,n)$-cap symmetrization of $f$* to be the real valued function $f^\#$ defined $\mathcal{L}^n$-a.e. on $X$ by

$$f^\#(x) = f^\#(ry, z) = (f^z)^\#(ry), \quad z \in Z,$$

where $(f^z)^\#$ is the $(k, k+1)$-cap symmetrization of $f^z$.

Then $E^\#$ and $f^\#$ are $\mathcal{L}^n$-measurable, $E^\#$ and $E$ have the same measure, and $f^\#$ and $f$ have the same distribution. The same is true for each appropriate slice.

We have defined now $(k,n)$-cap symmetrization for each $1 \le k \le n-1$. By far, the most frequently encountered case is $(n-1, n)$. The simplest case with $k \le n-2$ is $(1,3)$, which is associated with cylindrical coordinates in $\mathbb{R}^3$. In the decomposition $x = (ry, z)$, $z = x_3$ is the "height" of $x$, $r \in \mathbb{R}^+$ is the distance from $x$ to the $x_3$-axis, and when $r > 0$, $y$ is the point $\frac{1}{r}(x_1, x_2) \in \mathbb{S}^1$. For $E \subset \mathbb{R}^3$, the set $E^\#$ is the union of circular arcs centered on the $x_3$-axis, parallel to the $x_1x_2$-plane, symmetric with respect to the positive $x_1$-axis, and containing the point $(r e_1, z)$ when nonempty.

Chapter 6 contains 27 theorems, corollaries, propositions and lemmas involving Steiner symmetrization. All remain true for cap symmetrization, provided the statements are adjusted to account for the fact that the basic symmetrization process is carried out on spheres rather than Euclidean spaces. To illustrate, we will state and outline proofs of cap versions of Theorem 6.8 (general integral inequalities) and Theorem 6.16 (Dirichlet integral inequalities). The reader, we hope, will be able to supply appropriate statements for the other results as needed. A systematic study of cap symmetrization can be found in Van Schaftingen (2006).

Let $X \subset \mathbb{R}^n$ be ring-type. Then $x \in X$ can be written as $x = (ry, z)$, with $r \in A(z) \subset \mathbb{R}^+$, $y \in \mathbb{S}^k$, and $z \in Z \subset \mathbb{R}^{m-1}$, as above. If $f$ is an integrable or nonnegative $\mathcal{L}^n$-measurable function on $X$, we can write $dx = r^k dr\, d\sigma_k(y)\, dz$. Fubini's Theorem gives

$$\int_{\mathbb{R}^n} f(x)\, dx = \int_Z dz \int_{A(z)} r^k\, dr \int_{\mathbb{S}^k} f(ry, z)\, d\sigma_k(y).$$

Here and later, *when $k = n-1$ the terms involving $z$ are to be omitted* and $X$ will reduce to the form $X = \{ry\colon r \in A,\ y \in \mathbb{S}^k\}$, for some open set $A \subset \mathbb{R}^+$.

Next, let $f$ and $g$ be nonnegative $\mathcal{L}^n$-measurable functions on $X$, $K: \mathbb{R}^+ \to \mathbb{R}^+$ be decreasing, and $\Psi \in AL_0$. Let $x_1$ and $x_2$ denote points of $X$, with respective decompositions $x_i = (r_i y_i, z_i)$. Define

$$Q(f,g) = \int_{X^2} \Psi(f(x_1), g(x_2)) K(|x_1 - x_2|)\, dx_1\, dx_2.$$

Then $Q(f,g)$ equals

$$\int_{Z^2} dz_1\, dz_2 \int_{A(z_1)\times A(z_2)} r_1^k r_2^k\, dr_1\, dr_2$$
$$\int_{\mathbb{S}^k\times\mathbb{S}^k} \Psi(f(r_1y_1,z_1),g(r_2y_2,z_2))K(|x_1-x_2|)d\sigma_k(y_1)d\sigma_k(y_2),$$

with

$$|x_1-x_2|^2=(r_2-r_1)^2+r_1r_2|y_1-y_2|^2+|z_1-z_2|^2.$$

By the nonnegative version of Theorem 7.3, applied with $K$ replaced by a suitable function $K_1$ depending on $r_1, r_2, z_1$ and $z_2$, it follows that each of the integrals over $\mathbb{S}^k\times\mathbb{S}^k$ increases when $f$ and $g$ are replaced by their $(k,n)$-cap symmetrizations $f^\#$ and $g^\#$. Thus, we have

$$Q(f,g)\le Q(f^\#,g^\#).$$

This shows that conclusion (7.22) of Theorem 6.8 holds with cap symmetrization in place of Steiner symmetrization. By similar arguments, the other inequalities asserted by Theorem 6.8 are also valid for cap symmetrization. Here is a formal statement.

**Theorem 7.6** *Let $X$ be a ring-type set in $\mathbb{R}^n$, $f$ and $g$ be nonnegative $\mathcal{L}^n$-measurable functions on $X$, $K\colon \mathbb{R}^+\to\mathbb{R}^+$ be decreasing, $\Psi\in AL_0$, and $\Phi\colon\mathbb{R}^+\to\mathbb{R}^+$ be convex and increasing with $\Phi(0)=0$. Then inequalities (6.17)–(6.22) of Theorem 6.8 hold, when $^\#$ denotes $(k,n)$-cap symmetrization in $\mathbb{R}^n$, $1\le k\le n-1$.*

The corresponding equality statements are like those in Theorem 6.8. For (6.17) and (6.20) of Theorem 6.8, we just have to change $T$ from a translation of $\mathbb{R}^k$ to an orthogonal transformation of $\mathbb{R}^{k+1}$ such that $f(ry,z)=f^\#(rTy,z)$ and $g(ry,z)=g^\#(rTy,z)$ for $\mathcal{L}^n$-almost every $(ry,z)\in X$. For (6.18) and (6.19), we just have to replace the sets $A^z$ by sets

$$A^{(r,z)}=\{(y_1,y_2)\in\mathbb{S}^k\times\mathbb{S}^k\colon f(ry_1,z)<f(ry_2,z)$$
$$\text{and } g(ry_1,z))>g(ry_2,z)\},$$

and require that $(\sigma_k\times\sigma_k)(A^{(r,z)})=0$ for $\mathcal{L}^m$-a.e $(r,z)$ with $z\in Z$ and $r\in A(z)$.

In Theorem 7.6 we assumed that all functions appearing inside integrals were nonnegative. There is another version of Theorem 7.6 derived from Theorem 7.3 and including necessary and sufficient equality statements, in which $f, g, K, \Psi$ and $\Phi$ need not be nonnegative, but instead must satisfy integrability assumptions. We will forgo a detailed statement of this alternate version, but will use particular cases as the need arises.

Next, we take up the behavior of Dirichlet integrals under cap symmetrization. The notations $X$, $x = (ry, z)$, etc. have the same meaning as previously in this section. Define also

$$Z' = \{(r, z) \in \mathbb{R}^+ \times Z \colon r \in A(z),\ z \in Z\}.$$

$f^\#$ will denote the $(k, n)$-cap symmetrization of $f$, with $1 \leq k \leq n-1$, it being understood that when $k = n-1$ all terms involving $Z$ are to be omitted. We shall also write $\sigma$ instead of $\sigma_k$.

**Theorem 7.7** *Let $f \in \mathrm{Lip}(X, \mathbb{R})$. Then there exists a set $E \subset Z'$ with $\mathcal{L}^m(Z' \setminus E) = 0$ such that, for each $(r, z) \in E$, $f$ and $f^\#$ are differentiable at $(r, y, z)$ for $\sigma_k$-a.e $y \in \mathbb{S}^k$, and for each convex increasing function $\Phi \colon \mathbb{R}^+ \to \mathbb{R}^+$ with $\Phi(0) = 0$ we have*

$$\int_{\mathbb{S}^k} \Phi(|\partial_{z_i} f^\#(ry, z)|)\, d\sigma(y) \leq \int_{\mathbb{S}^k} \Phi(|\partial_{z_i} f(ry, z)|)\, d\sigma(y), \quad 1 \leq i \leq m-1, \tag{7.28}$$

$$\int_{\mathbb{S}^k} \Phi(|\nabla_z f^\#(ry, z)|)\, d\sigma(y) \leq \int_{\mathbb{S}^k} \Phi(|\nabla_z f(ry, z)|)\, d\sigma(y), \tag{7.29}$$

$$\int_{\mathbb{S}^k} \Phi(|\partial_r f^\#(ry, z)|)\, d\sigma(y) \leq \int_{\mathbb{S}^k} \Phi(|\partial_r f(ry, z)|)\, d\sigma(y), \tag{7.30}$$

$$\int_{\mathbb{S}^k} \Phi(|\nabla f^\#(ry, z)|)\, d\sigma(y) \leq \int_{\mathbb{S}^k} \Phi(|\nabla f(ry, z)|)\, d\sigma(y). \tag{7.31}$$

In (7.28) we have $|\nabla_z f|^2 = \sum_{i=1}^{m-1} |\partial z_i f|^2$. In (7.31), $\nabla$ denotes the full gradient in $\mathbb{R}^n$. "Differentiable" means differentiable in all $n$ variables. There are lots of other inequalities like (7.28)–(7.31). For example, if we define $\nabla_s f(r, y, z)$ to be the spherical gradient of the function $f(r, \cdot, z)$, then by Theorem 7.4, (7.31) holds when $\nabla f$ is replaced by $\nabla_s f$. We have the identity

$$|\nabla f|^2 = |\partial_r f|^2 + r^{-2} |\nabla_s f|^2 + |\nabla_z f|^2. \tag{7.32}$$

The proof we sketch below can be modified to give (7.31) when $|\nabla f|$ is replaced by the square root of the sum of any two summands on the right-hand side of (7.32).

Of course, each of the Dirichlet integral inequalities in (7.28)–(7.31) can be integrated, to produce inequalities such as

$$\int_X \Phi(|\nabla f^\#|)\, dx \leq \int_X \Phi(|\nabla f|)\, dx.$$

*Sketch of proof of Theorem 7.7* The proof follows the proof of the Steiner case, Theorem 6.16, but is made a bit simpler by the compactness of $\mathbb{S}^k$. For example, we do not need to introduce a condition like Condition S.

In essence, all we need do is represent the various $L^1$-Dirichlet integrals as limits of integrals that decrease under cap symmetrization. Let us illustrate by examining the full gradient case (7.31) when $k = n-1$. Then $X = \{ry\colon r \in A,\ y \in \mathbb{S}^{n-1}\}$ where $A$ is an open set in $\mathbb{R}^+$. There is a set $E \subset A$ with $\mathcal{L}(A \setminus E) = 0$ such that $f$ and $f^\#$ are differentiable, as functions of $n$ variables, at $\sigma_{n-1}$-almost every point $ry$ when $r \in A$. We may exclude the point $r = 0$ from $E$.

Fix $a \in E$. Let $\mathcal{A}$ be an open annulus in $X$ which contains $a$. For small $\epsilon > 0$, define

$$I(\epsilon, f) = \int_{\mathbb{S}^{n-1}} d\sigma(y) \int_{\mathcal{A}} |f(u) - f(ay)| K(\epsilon^{-1}|ay - u|)\, du.$$

Here $K = \chi_{[0,1]}$, $u$ denotes a point of $\mathbb{R}^n$, $\sigma = \sigma_{n-1}$ and $du$ denotes $n$-dimensional Lebesgue measure. Let $\epsilon$ be so small that $u \in \mathbb{B}^n(ay, \epsilon)$ implies $u \in \mathcal{A}$, where $\mathbb{B}^n$ denotes an open ball in $\mathbb{R}^n$. Then

$$\begin{aligned} I(\epsilon, f) &= \int_{\mathbb{S}^{n-1}} d\sigma(y) \int_{\mathbb{B}^n(ay,\epsilon)} |f(u) - f(ay)|\, du \\ &= \epsilon^n \int_{\mathbb{S}^{n-1}} d\sigma(y) \int_{\mathbb{B}^n(0,1)} |f(ay + \epsilon s) - f(ay)|\, ds, \end{aligned}$$

so that

$$\epsilon^{-n-1} I(\epsilon, f) = \int_{\mathbb{S}^{n-1}} d\sigma(y) \int_{\mathbb{B}^n(0,1)} \epsilon^{-1} |f(ay + \epsilon s) - f(ay)|\, ds.$$

Since $f$ is Lipschitz, the integrand in the last integral is uniformly bounded for $y \in \mathbb{S}^{n-1}$. At points $ay$ of differentiability of $f$, as $\epsilon \to 0$ the integrand converges to $|\nabla f(ay) \cdot s|$. Since $a \in E$, $\sigma_{n-1}$-almost all points $ay$ are points of differentiability. As in the proof of Theorems 6.16 and 3.7, we obtain

$$\lim_{\epsilon \to 0} \epsilon^{-n-1} I(\epsilon, f) = C_1 \int_{\mathbb{S}^{n-1}} |\nabla f(ay)|\, d\sigma(y), \tag{7.33}$$

where $C_1 = \int_{\mathbb{B}^n(0,1)} |e_1 \cdot s|\, ds$.

As in the Steiner case, the integral inequalities in Theorem 7.6 imply that cap symmetrization decreases moduli of continuity. Thus, $f^\#$ is Lipschitz when $f$ is, and (7.33) holds when $f$ is replaced by $f^\#$.

We can write

$$I(\epsilon, f) = \int_{\mathcal{A}} r^{n-1}\, dr \int_{\mathbb{S}^{n-1} \times \mathbb{S}^{n-1}} |f(ay) - f(rv)| K(\epsilon^{-1}|ay - rv|)\, d\sigma(y)\, d\sigma(v).$$

Using $|ay - rv|^2 = (a-r)^2 + |y - v|^2$ and the $\mathbb{S}^{n-1}$-version of Corollary 2.22, we conclude that each of the integrals over $\mathbb{S}^{n-1} \times \mathbb{S}^{n-1}$ decreases when $f$ is replaced by $f^\#$. Thus, $I(\epsilon, f^\#) \le I(\epsilon, f)$ for each small $\epsilon$, so

$$\int_{\mathbb{S}^{n-1}} |\nabla f^{\#}(a\,y)|\,d\sigma(y) \leq \int_{\mathbb{S}^{n-1}} |\nabla f(a\,y)|\,d\sigma(y)$$

for almost every $a \in A$. Our proof of (7.31) for $(n-1,n)$-Steiner symmetrization and $\Phi(t) = t$ is complete.

To go from $\Phi(t) = t$ to general convex $\Phi$ one can follow the model of Theorem 6.16, which includes proving a cap version of Lemma 6.18. But if all one wants is the inequality for $p$-Dirichlet integrals, $p \geq 1$, then things are much easier: in the argument just given, simply substitute $|f(u) - f(a\,y)|^p$ for $|f(u) - f(a\,v)|$.

Let us remain for a moment with $(n-1,n)$-cap symmetrization. Take $a \in E$, as before, and set

$$I_1(\epsilon,f) = \int_{\mathbb{S}^{n-1}} |f((a+\epsilon)y) - f(a\,y)|\,d\sigma(y).$$

For Lipschitz $f$ on $X$, we obtain as above

$$\lim_{\epsilon \to 0} \epsilon^{-1} I_1(\epsilon,f) = \int_{\mathbb{S}^{n-1}} |\partial_r f(a\,y)|\,d\sigma(y).$$

The $L^p$-contraction property, Corollary 2.20, holds for s.d.r. on $\mathbb{S}^{n-1}$. Thus, $I_1(f^{\#},\epsilon) \leq I_1(\epsilon,f)$ for each small $\epsilon > 0$. We conclude that

$$\int_{\mathbb{S}^{n-1}} |\partial_r f^{\#}(a\,y)|\,d\sigma(y) \leq \int_{\mathbb{S}^{n-1}} |\partial_r f(a\,y)|\,d\sigma(y), \qquad a \in E.$$

This proves that (7.30) holds when $\Phi(t) = t$. The passage to general $\Phi$ is like that in (7.31).

Inequalities (7.28) and (7.29) are vacuous for $(n-1,n)$-cap symmetrization. To prove (7.28)–(7.31) for $(k,n)$-cap symmetrization when $1 \leq k \leq n-2$, we just need to find appropriate $I$-functionals. The proofs of Theorem 6.16 and of the $(n-1,n)$-cap case should furnish the so-inclined reader with sufficient clues.

This concludes our outline of the proof of Theorem 7.7. □

## 7.6 Hyperbolic Symmetrization

Let $M$ be a simply connected Riemannian manifold of dimension $n \geq 2$. Suppose that $M$ has constant sectional curvature $\kappa$. If $\kappa > 0$ then $M$ is isometric to the sphere $|x| = \kappa^{-1/2}$ with the metric induced by the usual metric on $\mathbb{R}^{n+1}$. If $\kappa = 0$ then $M$ is isometric to $\mathbb{R}^n$ with the usual metric. If $\kappa < 0$ then $M$ is said to be a model for hyperbolic $n$-space, or a hyperbolic space form, with curvature $\kappa$. Any two models with the same curvature are isometric. See Chavel (1993, §2.3) and Chavel (2001) for differential geometric background.

By $\mathbb{H}^n$, we will mean any such $M$ equipped with the pertinent metric, having $\kappa = -1$. We will examine three models for $\mathbb{H}^n$.

**Ball model** $M = \mathbb{B}^n$, the open unit ball in $\mathbb{R}^n$. The hyperbolic metric is given infinitesimally by

$$ds = \frac{2}{(1-|x|^2)}\,|dx|, \quad x \in \mathbb{B}^n, \tag{7.34}$$

where $|dx|$ denotes the usual line element in $\mathbb{R}^n$.

**Halfspace model** $M = \{x \in \mathbb{R}^n : x_n > 0\}$, where $x_n = x \cdot e_n$. The hyperbolic metric is given by

$$ds = x_n^{-1}|dx|, \quad x \in M.$$

The ball and halfspace models are also called the Poincaré ball and Poincaré halfspace.

**Hyperboloid model** (see Vilenkin and Klimyk, 1991)

$$M = \left\{ x \in \mathbb{R}^{n+1} : x_1^2 = 1 + \sum_{i=2}^{n+1} x_i^2 \right\}.$$

The metric is induced by the bilinear form

$$B(x,y) = -x_1y_1 + \sum_{i=2}^{n+1} x_iy_i,$$

which is positive definite on the tangent space of $M$.

In the hyperboloid model, if one takes the origin to be the point $e_1$, then one can construct a symmetrization theory in which the symmetrized sets $E^{\#}$ are spherical caps on the hyperboloid with center at $e_1$. Thus, the theory looks something like the symmetrization theory on $\mathbb{S}^n$. On the whole, though, I think it more convenient to develop hyperbolic symmetrization theory in the context of the ball model. So from now on in this section we will work with $\mathbb{B}^n$, equipped with the metric (7.34).

For $x, y \in \mathbb{B}^n$, define

$$d(x,y) = \inf L(\gamma),$$

where the infimum is taken over all Lipschitz curves $\gamma : [a,b] \to \mathbb{B}^n$ connecting $x$ and $y$, and

$$L(\gamma) = \int_a^b 2(1-|\gamma(t)|^2)^{-1}|\gamma'(t)|\,dt.$$

For $\mathcal{L}^n$-measurable $E \subset \mathbb{B}^n$, define the measure $\tau = \tau_n$ by

$$\tau(E) = \int_E 2^n(1 - |x|^2)^{-n}\, d\mathcal{L}^n(x).$$

Then $d$ and $\tau$ are called the *canonical hyperbolic distance* and *canonical hyperbolic measure* on $\mathbb{B}^n$, or on $\mathbb{H}^n$.

For $a \in \mathbb{R}^n$, $R > 0$, define $S(a,R) = S = \partial\mathbb{B}^n(a,R)$. Define also $\overline{\mathbb{R}^n} = \mathbb{R}^n \cup \{\infty\}$. Define $\rho = \rho_S \colon \overline{\mathbb{R}^n} \colon \to \overline{\mathbb{R}^n}$, the reflection in $S$, to be

$$\rho(x) = a + \frac{R^2(x-a)}{|x-a|^2}.$$

Then $\rho$ is the identity on $S$, $\rho$ is 1–1 on $\overline{\mathbb{R}^n}$, and $\rho$ maps the set inside $S$ onto the set outside $S$. Let $G(\mathbb{R}^n)$ be the group of homeomorphisms of $\overline{\mathbb{R}^n}$ generated by the orthogonal group $O(n)$, all reflections in affine hyperplanes $H \in \mathcal{H}(\mathbb{R}^n)$, and all reflections in spheres $S(a,R)$ as $a$ ranges over $\mathbb{R}^n$ and $R$ over $(0,\infty)$. When $n = 2$, $G(\mathbb{R}^2)$ is the set of all Möbius transformations and their complex conjugates. For dimensions $n \geq 3$ the maps in $G(\mathbb{R}^n)$ are still called Möbius transformations, or conformal transformations. Their theory is worked out in Ahlfors (1981), Reshetnyak (1989), Iwaniec and Martin (2001), and in various books on hyperbolic geometry. $G(\mathbb{R}^n)$ is often called the *conformal group* of $\mathbb{R}^n$.

Define

$$G(\mathbb{B}^n) = \{T \in G(\mathbb{R}^n) \colon T\mathbb{B}^n = \mathbb{B}^n\},$$

and define a set $H \subset \mathbb{B}^n$ to be a *hyperbolic hyperplane* if $H = \mathbb{B}^n \cap S(a,R)$ for some $a, R$ for which $S(a,R) \cap \mathbb{S}^{n-1}$ has dimension $n-2$ and intersects $\mathbb{S}^{n-1}$ orthogonally, or if $H = \mathbb{B}^n \cap H_0$ for some $H_0 \in \mathcal{H}(\mathbb{R}^n)$ for which $H_0 \cap \mathbb{S}^{n-1}$ has dimension $n-2$ and intersects $\mathbb{S}^{n-1}$ orthogonally. Set

$$\mathcal{H}_h(\mathbb{B}^n) = \text{the set of all hyperbolic hyperplanes in } \mathbb{B}^n.$$

**Proposition 7.8** *The hyperbolic distance $d$ and measure $\tau$ are $G(\mathbb{B}^n)$-invariant. That is,*

$$d(Tx, Ty) = d(x,y) \quad \textit{and} \quad \tau(TE) = \tau(E),$$

*for every $T \in G(\mathbb{B}^n)$, $x, y \in \mathbb{B}^n$, and $\mathcal{L}^n$-measurable $E \subset \mathbb{B}^n$.*

For a proof, see Ahlfors (1981, p. 18).

**Proposition 7.9**

*(a) For each $x \in \mathbb{B}^n$ there exists $T \in G(\mathbb{B}^n)$ such that $Tx = 0$.*

*(b) For each $H \in \mathcal{H}_h(\mathbb{B}^n)$, $\rho_H \in G(\mathbb{B}^n)$.*

*(c) For each $x, y \in \mathbb{B}^n$ with $x \neq y$ there exists $H \in \mathcal{H}_h(\mathbb{B}^n)$ such that $\rho_H(x) = y$.*

The proof is left to the reader. The reader may also show that the geodesic arc from 0 to a point $x \in \mathbb{B}^n$ is the ray from 0 to $x$, so that

$$d(0,x) = \log \frac{1+|x|}{1-|x|}.$$

Another computation shows that for the ball $\mathbb{B}^n(0,R)$ with Euclidean radius $R$,

$$\tau(\mathbb{B}^n(0,R)) = 2^n \beta_{n-1} \int_0^R (1-r^2)^{-n} r^{n-1}\, dr, \quad 0 < R < 1. \tag{7.35}$$

Set $p_n(R) = \int_0^R (1-r^2)^{-n} r^{n-1}\, dr$, and, for fixed $n$, define for $x \in \mathbb{B}^n$, $s \in (0,\infty)$, $B_h(x,s) = \{y \in \mathbb{B}^n : d(y,x) < s\}$. Then $B_h(x,s)$ is the hyperbolic ball centered at $x$ with hyperbolic radius $s$. If $s$ and $R$ are related by $s = \log \frac{1+R}{1-R}$, then $B_h(0,s) = \mathbb{B}^n(0,R)$, and

$$\tau(B_h(x,s)) = \tau(B_h(0,s)) = 2^n \beta_{n-1} p_n(R). \tag{7.36}$$

When $n = 2$ the integral defining $p_2$ can be evaluated, and we find that

$$\tau(B_h^2(x,s)) = 2\pi \cosh^2 \frac{s}{2}.$$

For general $n \geq 2$, it is true that

$$\tau(B_h^n(x,s)) \sim C_n e^{(n-1)s}, \quad s \to \infty,$$

for a constant $C_n$ depending on $n$.

Define $U : \mathbb{B}^n \to \mathbb{R}^+$ by $U(x) = \tau(\mathbb{B}^n(0,|x|))$. Then $U$ is a measure preserving map of $(\mathbb{B}^n, \tau)$ onto $(\mathbb{R}^+, \mathcal{L})$.

For $\mathcal{L}^n$-measurable $E \subset \mathbb{B}^n$, define $E^\#$, the hyperbolic s.d.r. of $E$, to be the ball $B_h(0,s)$ with the same $\tau$-measure as $E$. Then $s$ can be obtained from $\tau(E)$ via (7.35) and (7.36). For $\mathcal{L}^n$-measurable $f : \mathbb{B}^n \to \mathbb{R}$ satisfying the finiteness condition

$$\tau(f > t) < \infty, \quad \text{for every } t > \operatorname{ess\,inf} f, \tag{7.37}$$

define $f^\#$, the hyperbolic symmetric decreasing rearrangement of $f$, to be the function on $\mathbb{B}^n$ given by

$$f^\#(x) = f^*(Ux)$$

where $f^*$ is the decreasing rearrangement of $f$. Here, of course, the distribution function of $f$ is computed using the measure $\tau$.

As in the Euclidean and spherical cases, $f^\#$ and $f$ have the same distribution, with respect to $\tau$. The function $f^\#(x)$ depends only on the hyperbolic distance from $x$ to 0, and $f^\#(x)$ decreases as $d(0,x)$ increases.

In §6.1 we gave a list of all the results about Euclidean s.d.r. which were proved in Chapters 1–5. With only a few exceptions, *all of the results in that list remain valid for hyperbolic s.d.r., and can be obtained by suitable modification of the Euclidean proofs*. We will not provide any proof of this claim, but instead will leave verification of particular results to the interested reader. Because of noncompactness, the hyperbolic symmetrization theory is usually more akin to the Euclidean theory than to the spherical theory. Occasionally, though, the Euclidean proofs do not carry over. To see what to do then, the reader may find the spherical proofs in §7.3 and §7.4 good sources for non-Euclidean substitutes.

The hyperbolic analogue of Theorems 3.11 and 7.4 asserts that

$$\int_{\mathbb{B}^n} \Phi(|\nabla_h f^{\#}|)\, d\tau \leq \int_{\mathbb{B}^n} \Phi(|\nabla_h f|)\, d\tau,$$

where $\Phi$ is increasing and convex with $\Phi(0) = 0$, $f$ is hyperbolically Lipschitz and satisfies the finiteness condition (7.37), $f^{\#}$ denotes hyperbolic s.d.r., and $\nabla_h f$, the hyperbolic gradient of $f$, may be defined by

$$\nabla_h f(x) = \frac{1}{2}(1 - |x|^2)\nabla f(x), \quad x \in \mathbb{B}^n,$$

with $\nabla f$ the Euclidean gradient. At points $x$ of differentiability of $f$ we have the formula

$$|\nabla_h f(x)| = \limsup_{y \to x} \frac{|f(y) - f(x)|}{d(x, y)},$$

where $d$ is the canonical distance on $\mathbb{B}^n$.

One more time, we want to stress that the theory of symmetrization constructed in this book works essentially the same way in the Euclidean, spherical and hyperbolic cases, and is based upon properties of polarization with respect to hyperplanes in the various geometries. Propositions 7.8 and 7.9 tell us much of what we need to know about polarization in the hyperbolic case. But we also need a few additional properties. For example, we need to know that if $x$ and $y$ in $\mathbb{B}^n$ belong to the same complementary component of some $H \in \mathcal{H}_h(\mathbb{B}^n)$, then

$$d(x, y) < d(x, \rho_H y).$$

To verify this, we may assume that $H$ is the intersection of $\mathbb{B}^n$ with the hyperplane $x_n = 0$, and that $x$ and $y$ have positive $n$th components. Let $\gamma_1$, $\gamma_2$ be the geodesics connecting $x$ to $y$ and $x$ to $\rho y$, respectively, where $\rho = \rho_H$. Let $\gamma_3$ be the curve defined by $\gamma_3(t) = \gamma_2(t)$ if $\gamma_2(t) \cdot e_n \geq 0$ and $\gamma_3(t) = \rho\gamma_2(t)$ if $\gamma_2(t) \cdot e_n \leq 0$. Then, with $L$ denoting hyperbolic length,

$$d(x, \rho y) = L(\gamma_2) = L(\gamma_3) \geq d(x, y), \tag{7.38}$$

with equality if and only if $\gamma_3$ is a reparameterization of $\gamma_1$. But it is not hard to show that complementary components of hyperbolic hyperplanes are geodesically convex. Since $\gamma_3$ passes through $H$, it cannot be a reparameterization of $\gamma_1$. Thus, strict inequality holds in (7.38).

## 7.7 Gauss Space Symmetrization

Write $x = (x_1, \ldots, x_n)$ for points $x \in \mathbb{R}^n$. Define the standard Gauss measure $\gamma_n$ on $\mathbb{R}^n$ by

$$\gamma_n(E) = (2\pi)^{-n/2} \int_E e^{-\frac{1}{2}|x|^2}\, dx,$$

where $dx = d\mathcal{L}^n(x)$ and $E$ is an $\mathcal{L}^n$-measurable set. Then $\gamma_n$ is the product measure of $n$ copies of the 1-dimensional Gauss measure $\gamma_1$, and $\gamma_n(\mathbb{R}^n) = 1$. Denote the cumulative distribution function of $\gamma_1$ by $\Phi$. Then

$$\Phi(t) = \gamma_1(-\infty, t) = (2\pi)^{-1/2} \int_0^t e^{-\frac{1}{2}s^2}\, ds, \quad t \in \mathbb{R}.$$

The set $\mathbb{R}^n$, equipped with the measure $\gamma_n$, is called $n$-dimensional Gauss space. For $a \in \mathbb{R}$, define

$$G(a) = \{x \in \mathbb{R}^n \colon x_1 \le a\}.$$

The $G(a)$ are closed halfspaces in $\mathbb{R}^n$. Define also $G(-\infty) = \emptyset$, $G(\infty) = \mathbb{R}^n$. Given an $\mathcal{L}^n$-measurable set $E \subset \mathbb{R}^n$, define $\tilde{E}$ to be the halfspace $G(a)$ such that $\gamma_n(E) = \gamma_n(G(a))$. Then

$$\gamma_n(E) = \gamma_n(\tilde{E}) = \Phi(a), \quad \text{when} \quad \tilde{E} = \{x \in \mathbb{R}^n \colon x_1 \le a\}.$$

We could, if we like, take the symmetrized sets to be half spaces of the form $\{x \in \mathbb{R}^n \colon x \cdot v \le a\}$, where $v$ is any fixed unit vector in $\mathbb{R}^n$. But we will stick to the case $v = e_1$.

Define $U \colon \mathbb{R}^n \to (0, 1)$ by

$$U(x) = \gamma_n(G(x_1)) = \Phi(x_1).$$

Then $U$ is a measure preserving transformation of $(\mathbb{R}^n, \gamma_n)$ onto $((0, 1), \mathcal{L})$. Given an $\mathcal{L}^n$-measurable function $f \colon \mathbb{R}^n \to \mathbb{R}$, define a new function $\tilde{f} \colon \mathbb{R}^n \to \mathbb{R}$ by

$$\tilde{f} = f^* \circ U,$$

where $f^*$ is the decreasing rearrangement of $f$, and the distribution of $f$ is computed using the measure $\gamma_n$. Then $f$ and $\tilde{f}$ have the same distribution with respect to $\gamma_n$. The function $\tilde{f}$ is a function of the first component $x_1$ which decreases as $x_1$ increases on $\mathbb{R}$. We call $\tilde{f}$ the Gauss symmetrization of $f$ and $\tilde{E}$ the Gauss symmetrization of $E$.

To me, Euclidean, spherical and hyperbolic symmetrization all look alike. But Gauss symmetrization looks different. Since the Gauss measure is not translation invariant, there is no simple rearrangement analogous to polarization. Accordingly, I shall refrain from calling Gauss symmetrization a s.d.r., and will denote it with a tilde rather than a #.

It turns out that Gauss space may be realized as a limit of spheres of increasing radius with dimension going to infinity. This fact is often attributed to Poincaré, but Stroock (1993, p. 77) points out that it was already known to Mehler in 1866. We shall call it the Mehler–Poincaré formula. A proof appears in McKean (1973). Here is the result.

**Proposition 7.10** *Fix $n \geq 1$. Let $E \subset \mathbb{R}^n$ be $\mathcal{L}^n$-measurable. For $N \geq n$, set*

$$E_N = \{x \in \mathbb{R}^N \colon |x| = \sqrt{N},\ (x_1, \ldots, x_n) \in E\}.$$

*Then*

$$\lim_{N \to \infty} \lambda_N(E_N) = \gamma_n(E),$$

*where $\lambda_N$ is the uniform probability measure on the sphere $\partial\mathbb{B}^N(0, N^{1/2}) \subset \mathbb{R}^N$.*

For $E \subset \mathbb{R}^n$ and $\delta > 0$, recall from §4.4 that the $\delta$-collar of $E$ is the set

$$E(\delta) = \{x \in \mathbb{R}^n \colon d(x, E) < \delta\},$$

where $d$ denotes Euclidean distance.

**Theorem 7.11**

$$\gamma_n(E(\delta)) \geq \gamma_n(\tilde{E}(\delta)), \qquad \forall\ E \text{ and } \delta.$$

**Corollary 7.12** *For each $\mathcal{L}^n$-measurable set $E \subset \mathbb{R}^n$, we have*

$$\liminf_{\delta \to 0} \frac{1}{\delta}[\gamma_n(E(\delta)) - \gamma_n(E)] \geq \lim_{\delta \to 0} \frac{1}{\delta}[\gamma_n(\tilde{E}(\delta)) - \gamma_n(\tilde{E})]. \tag{7.39}$$

This theorem and corollary are among the main results for Gauss symmetrization. They are due to Borell (1975). The expression on the left side of (7.39) may be called the lower $(n-1)$-dimensional Gaussian Minkowski content of $\partial E$. It is the most frequently used version of surface area in Gauss space. Accordingly, (7.39) is often called the isoperimetric inequality for Gauss space. As in the Euclidean, spherical, and hyperbolic settings, the isoperimetric inequality implies that Gauss symmetrization decreases the modulus of continuity. It is also closely linked to inequalities for Dirichlet-like integrals.

Borell's proof of Theorem 7.11 in Borell (1975) made use of the Mehler–Poincaré formula and of Theorem 7.5, the analogue of Theorem 7.11 on spheres. A different proof was found later by Ehrhard (1983b, 1984). Borell (1985) gave still another proof, based on comparison theorems for parabolic PDE involving the Ornstein–Uhlenbeck operator $\Delta - x \cdot \nabla$. (See the endnotes to Chapter 10 for references to such comparison theorems.) In Chapter 8 we shall explore related topics such as hypercontractivity of semigroups and logarithmic Sobolev inequalities.

The Gaussian Hardy–Littlewood type inequality

$$\int_{\mathbb{R}^n} fg \, d\gamma_n \leq \int_{\mathbb{R}^n} \tilde{f}\tilde{g} \, d\gamma_n$$

for nonnegative $f$ and $g$ is easily established: by layer cake representations, the problem reduces to the case $f = \chi_A$, $g = \chi_B$. Then the inequality follows from the fact that one of $\tilde{A}$, $\tilde{B}$ contains the other. However, Theorem 2.15 does not carry over from the Euclidean setting.

In addition to the works of Borell and Ehrhard cited above, references for Gauss symmetrization and related topics include Ledoux (1994), Epperson (1990), González (2000), and Bobkov (1996).

A well-known result in Gauss space for which symmetrization might be relevant is the Gaussian correlation inequality. Let $A$ and $B$ be closed convex sets in $\mathbb{R}^n$ which are *balanced*, in the sense that if a point $x$ is in one of the sets, then $-x$ is also in that set. The correlation inequality says that

$$\gamma_n(A \cap B) \geq \gamma_n(A)\gamma_n(B).$$

For $n = 1$ the result is trivial, since the only sets satisfying the assumptions are closed intervals centered at the origin. For $n = 2$ the inequality was proved by Pitt (1977). Partial results were obtained by Schechtman et al. (1998) among others. The inequality was proved in all dimensions by Royen (2014), who handled also a larger class of measures, the gamma distributions. Royen's methods do not involve symmetrization.

## 7.8 Hölder Continuity of Quasiconformal Mappings

For $n \times n$ real matrices $A$ denote by $|A|$ the operator norm of $A$:

$$|A| = \sup\{|Ax| : x \in \mathbb{R}^n, \ |x| = 1\}.$$

For $n \geq 2$, let $f$ be an orientation preserving $C^1$-homeomorphism of a domain $\Omega \subset \mathbb{R}^n$ onto another domain $\Omega' \subset \mathbb{R}^n$. Denote by $Df(x)$ the $n \times n$-matrix of first order partial derivatives of $f$ at $x$ and by $J(x) = J_f(x)$ the determinant of $Df(x)$. Then for $x \in \Omega$,

$$|Df(x)| = \lim_{r \to 0} r^{-1} \sup_{|y|=r} |f(x+ry) - f(x)|.$$

Accordingly, $|Df(x)|$ is called the *maximal stretch* of $f$ at $x$. The maximal stretch equals the largest singular value of $Df(x)$. We also have

$$J(x) = \lim_{r \to 0} \alpha_n^{-1} r^{-n} \mathcal{L}^n(f(\mathbb{B}^n(x,r))),$$

so that $J(x)$ measures the infinitesimal distortion of volume near $x$. Since $J(x)$ also equals the product of the singular values of $Df(x)$, we have *Hadamard's inequality*:

$$J_f(x) \leq |Df(x)|^n, \quad x \in \Omega.$$

If Hadamard's inequality partially reverses for $f$ and its inverse $f^{-1}$, we say that $f$ is *quasiconformal*. More precisely, let $K$ be a real number with $K \geq 1$. We say that the orientation preserving $C^1$-homeomorphism $f\colon \Omega \to \Omega'$ is $K$-quasiconformal in $\Omega$ if the inequalities

$$|Df(x)|^n \leq K^{n-1} J_f(x) \quad \text{and} \quad |D(f^{-1})(x)|^n \leq K^{n-1} J_{f^{-1}}(x) \tag{7.40}$$

hold for every $x \in \Omega$ and $x \in \Omega'$, respectively. Via calculations with singular values, (7.40) implies that

$$\lambda_1(x) \leq K^{2-\frac{2}{n}} \lambda_n(x), \quad x \in \Omega,$$

where $\lambda_1(x)$ is the maximal stretch of $f$ at $x$, and $\lambda_n(x)$ the minimal stretch. Thus, quasiconformal maps map infinitesimal balls onto infinitesimal ellipsoids for which the ratio of the largest to the smallest semiaxis is uniformly bounded above in $\Omega$.

There are interesting homeomorphisms which are limits of smooth $K$-quasiconformal maps in some natural sense but fail to be in $C^1$. Thus, one seeks definitions of $K$-quasiconformality in which the smoothness requirement is relaxed. Many definitions have been introduced. Here, we shall consider two types of definitions, the *analytic* and the *geometric*.

**Definition 7.13** (Analytic definition) The homeomorphism $f\colon \Omega \to \Omega'$ is $K$-quasiconformal if $f \in W^{1,n}_{\text{loc}}(\Omega, \mathbb{R}^n)$, $f^{-1} \in W^{1,n}_{\text{loc}}(\Omega', \mathbb{R}^n)$ and (7.40) holds a.e. in $\Omega$ and $\Omega'$.

Using Evans and Gariepy (1992, p. 150) and some facts about homeomorphisms, one can show that the $W^{1,n}$ hypotheses can be replaced by the requirement that $f$ and $f^{-1}$ each be locally absolutely continuous on almost all lines in $\Omega$, respectively $\Omega'$, which are parallel to one of the coordinate axes.

**Examples**

For $n \geq 2$ and $K \geq 1$, set $f(x) = x|x|^{K-1}$ and $g(x) = x|x|^{\frac{1}{K}-1}$. Then $f$ and $g$ are homeomorphisms of $\mathbb{R}^n$ onto itself. In fact, they are "radial stretch maps," which map rays through the origin onto themselves. At each $x \in \mathbb{R}^n \setminus \{0\}$, the map $f$ has one eigenvalue of size $Kr^{K-1}$, corresponding to the radial derivative, and $n-1$ eigenvalues of size $r^{K-1}$, corresponding to the tangential derivatives, where $r = |x|$. The map $g = f^{-1}$ has $n-1$ eigenvalues of size $r^{\frac{1}{K}-1}$ and one eigenvalue of size $\frac{1}{K}r^{\frac{1}{K}-1}$. Thus, a.e., $f$ and $g$ satisfy both inequalities in (7.40). More precisely, $f$ satisfies the first with equality, while $g$ satisfies the second with equality. In particular, $f$ and $g$ are $L$-quasiconformal maps of $\mathbb{R}^n$ onto itself for all $L \geq K$, but for no $L < K$.

For $n = 2$ the 1-quasiconformal homeomorphisms coincide with the (univalent) conformal mappings. But for $n \geq 3$ the updating of a theorem of Liouville tells us that a 1-quasiconformal map $f\colon \Omega \to \mathbb{R}^n$ must be the restriction to $\Omega$ of a Möbius transformation $F$ of $\overline{\mathbb{R}^n}$ onto $\overline{\mathbb{R}^n}$.

Let $U$ be a domain in $\mathbb{R}^n$ such that $\overline{\mathbb{R}^n} \setminus U$ is connected in $\overline{\mathbb{R}^n}$, and let $S$ be a compact connected subset of $U$. Set $R = U \setminus S$. Then $R$ is connected and open. We call $R$ a *ring domain* in $\mathbb{R}^n$. In §5.6 we considered ring domains in $\mathbb{R}^2$. For general $n \geq 2$, define $\operatorname{Mod} R$, the modulus of the ring domain $R$, to be

$$\operatorname{Mod} R = [\beta_{n-1} / \operatorname{Cap}_n(S, U)]^{-1/(n-1)},$$

where $\operatorname{Cap}_n$ denotes the variational $n$-capacity of the condenser $(S, U)$, defined in §5.6.

If $R(a, b) = \{a < |x| < b\}$ is a spherical shell in $\mathbb{R}^n$, then, as with $n = 2$, the minimizing function $u$ in the extremal problem defining $\operatorname{Cap}_n(S, U)$ turns out to be

$$u(x) = [\log(b/a)]^{-1} \log(b/|x|), \quad x \in R,$$

from which we deduce, as with $n = 2$, that

$$\operatorname{Mod} R(a, b) = \log(b/a).$$

**Definition 7.14** (Geometric definition) The homeomorphism $f\colon \Omega \to \Omega'$ is $K$-quasiconformal if

$$\frac{1}{K} \operatorname{Mod} R \leq \operatorname{Mod} f(R) \leq K \operatorname{Mod} R$$

for every ring domain $R$ such that $\overline{R} \subset \Omega$.

The $K$-quasiconformal stretch maps $f(x) = x|x|^{K-1}$, $g(x) = x|x|^{\frac{1}{K}-1}$ give equality in the second, respectively first inequality when $R = R(a, b)$.

**Theorem 7.15** *A homeomorphism $f$ satisfies the analytic definition if and only if it satisfies the geometric definition.*

For $n = 3$ this theorem is due to Gehring (1962, Theorem 4). The proof is valid for all $n \geq 2$. To prove that analytic implies geometric, take a ring domain $R$ with $\overline{R} \subset \Omega$. Take a function $v \in \mathrm{Lip}(\mathbb{R}^n, \mathbb{R})$ with $v = 0$ on $\mathbb{R}^n \backslash f(U)$, $v = 1$ on $f(S)$, and $0 \leq v \leq 1$ in $f(R)$. Let $u = v \circ f$. Then, ignoring smoothness questions, the obvious extension of $u$ to $\mathbb{R}^n$ is a competing function for the problem of determining $\mathrm{Cap}_n(S, U)$. From the chain rule, it follows that

$$|\nabla u| \leq |(\nabla v) \circ f| \, |Df|,$$

then by (7.40),

$$\int_R |\nabla u|^n \, dx \leq \int_R |(\nabla v) \circ f|^n K^{n-1} J_f \, dx = K^{n-1} \int_{f(R)} |\nabla v|^n \, dx.$$

It follows that $\mathrm{Cap}_n(S, U) \leq K^{n-1} \, \mathrm{Cap}_n(f(S), f(U))$, whence

$$\mathrm{Mod} f(R) \leq K \, \mathrm{Mod} \, R.$$

The left-hand inequality in the geometric definition is proved by applying the right-hand inequality to $f^{-1}$. For more detail, and the proof that geometric implies analytic, we refer the reader to Gehring (1962) or Iwaniec and Martin (2001).

By assumption, quasiconformal maps are continuous. Conformal maps are infinitely differentiable, but the stretch map $f(x) = x|x|^{\frac{1}{K}-1}$ show that $K$-quasiconformal maps need not even be Lipschitz when $K > 1$. Is there anything at all, beyond continuity, that one can say about the regularity of quasiconformal maps? The theorem and corollary below show that the answer is emphatically positive.

**Theorem 7.16** *Let $f$ be a $K$-quasiconformal homeomorphism of $\mathbb{B}^n$ onto $\Omega' \subset \mathbb{R}^n$, with $f(0) = 0$ and $d(0, \partial\Omega') \leq 1$. Then*

$$|f(x)| \leq C_{n,K} |x|^{1/K}, \quad \forall \, |x| \leq 1/2.$$

Here $\mathbb{B}^n$ is the open unit ball of $\mathbb{R}^n$, $d$ denotes Euclidean distance, and $C_{n.K}$ denotes a constant depending on $n$ and $K$, which can change from line to line.

**Corollary 7.17** *Let $f$ be a $K$-quasiconformal homeomorphism of $\Omega \subset \mathbb{R}^n$ onto $\Omega' \subset \mathbb{R}^n$, and let $S$ be a compact subset of $\Omega$. Then*

$$|f(x) - f(y)| \leq \eta \, C_{n,K} |x - y|^{1/K}, \quad \forall x, y \in S,$$

*where* $\eta = (a+b)c^{-1/K}$ *with*

$$a = d(f(S), \partial\Omega'), \quad b = \operatorname{diam} f(S), \quad c = d(S, \partial\Omega).$$

Theorem 7.16 is an analogue for quasiconformal maps of Schwarz's Lemma and the Koebe one-quarter theorem. Corollary 7.17 says thst $K$-quasiconformal maps are *Hölder continuous* with exponent $K^{-1}$ on compact subsets of $\Omega$, with constant depending only on $n$, $K$, the distances of $S$ and $f(S)$ to the boundaries of their containing domains, and the diameter of $f(S)$. The stretch map $f(x) = x|x|^{\frac{1}{K}-1}$ shows that $1/K$ can be replaced by no larger exponent.

Our proof of Theorem 7.16 and Corollary 7.17 follows the approach of Gehring (1962). It involves some special ring domains, the *Grötzsch rings* $R_G(a)$ and the *Teichmüller rings* $R_T(b)$. For $a \in (0,\infty)$, define $R_G(a)$ to be the exterior of the unit ball in $\mathbb{R}^n$ from which the half-line $x_1 \geq a$ has been removed. For $b \in (0,\infty)$, define $R_T(b)$ to be $\mathbb{R}^n$ with the two half-lines $x_1 \leq -1$ and $x_1 \geq b$ removed. Define functions $\phi$ and $\psi$ by

$$\log \phi(a) = \operatorname{Mod} R_G(a), \quad 1 < a < \infty,$$
$$\log \psi(b) = \operatorname{Mod} R_T(b), \quad 0 < b < \infty.$$

**Theorem 7.18**

*(a)* $\psi(b) = \phi^2((1+b)^{1/2})$.
*(b)* $a^{-1}\phi(a)$ *strictly increases for* $a \in (1,\infty)$, *with*

$$\lim_{a\to 1} a^{-1}\phi(a) = 1, \qquad \lim_{a\to\infty} a^{-1}\phi(a) = C_n,$$

*where the constant* $C_n$ *is finite.*

In dimension $n = 2$ Theorem 7.18 may be expeditiously proved by conformal mapping. See Lehto and Virtanen (1973) or Ahlfors (1966). For $n \geq 3$, see Lemmas 6 and 8 of Gehring (1961), or Iwaniec and Martin (2001), or Anderson et al. (1997).

*Proof of Theorem 7.16* Take $x \in \mathbb{B}^n$ with $|x| \leq 1/2$, set $y = f(x)$, let $R$ be the ring domain obtained by removing the closed line segment $[0,x]$ from $\mathbb{B}^n$, and set $R' = f(R) = \Omega' \setminus f([0,x])$. Then each circle of radius $\leq |y|$ intersects the image of $f([0,|x|)$, and each circle of radius $\geq 1$ intersects the complement of $\Omega'$. Let $u$ be a competing function for the problem of finding $\operatorname{Cap}_n(f([0,x]), \Omega')$. Then, we have

$$\sup_{|z|=r} u(z) = 1, \quad 0 \leq r \leq |y|, \qquad \inf_{|z|=r} u(z) = 0, \quad 1 \leq r < \infty. \tag{7.41}$$

Let $u^\#$ denote the $(n-1,n)$-cap symmetrization of $u$. Then $u^\#$ is Lipschitz in $\mathbb{R}^n$ since $u$ is. Let $R''$ be the ring domain obtained by removing line segments $L_1$ and $L_2$ from $\mathbb{R}^n$, where $L_1$ is $-\infty < z_1 \leq -1$ and $L_2$ is $0 \leq z_1 \leq |y|$. From (7.41) it follows that $u^\#$ is a competitor for the problem of finding $\operatorname{Cap}_n(L_2, \mathbb{R}^n \backslash L_1)$. Applying conclusion (7.31) of Theorem 7.7, with $\Phi(t) = t^n$, we deduce that $\operatorname{Mod} R' \leq \operatorname{Mod} R''$. Thus,

$$\frac{1}{K} \operatorname{Mod} R \leq \operatorname{Mod} R' \leq \operatorname{Mod} R''. \tag{7.42}$$

Moduli of ring domains are preserved by Möbius transformations, in particular inversions such as $z \to z|z|^{-2}$. Thus $\operatorname{Mod} R = \operatorname{Mod} R_G(|x|^{-1})$ and $\operatorname{Mod} R'' = \operatorname{Mod} R_T(|y|^{-1})$. From (7.42), and setting $s = |x|^{-1}$, $t = |y|^{-1}$, we conclude that

$$\phi^{1/K}(s) \leq \psi(t). \tag{7.43}$$

With Theorem 7.18(b), (7.43) implies that $\psi(t) \geq s^{1/K} \geq 2^{1/K}$. Set $t_K = \psi^{-1}(2^{1/K})$. Then $t \geq t_K$, and from Theorem 7.18 it follows that

$$t^{-1}\psi(t) \leq C_n \frac{1+t_K}{t_K} \equiv C_{n,K},$$

where $C_n$ is the constant in Theorem 7.18(b). Thus $s^{1/K} \leq C_{n,K}\, t$, which is

$$|f(x)| \leq C_{n,K}\, |x|^{1/K}. \qquad \square$$

*Proof of Corollary 7.17* Take $x_0, y_0 \in S$ with $|x_0 - y_0| \leq c/2$. Let

$$g(x) = \frac{1}{a+b}[f(x_0 + cx) - f(x_0)], \quad x \in \mathbb{B}^n.$$

Then $g$ satisfies the hypotheses of Theorem 7.16, with $x = c^{-1}(y_0 - x_0)$. Theorem 7.16 implies the conclusion of Corollary 7.17 holds with the same $C_{n,k}$ as in Theorem 7.16.

If $x_0, y_0 \in S$ satisfy $|x_0 - y_0| \geq c/2$, then one easily shows that the desired conclusion holds with $C = 2$. Thus, the desired conclusion holds for all $x_0, y_0 \in S$ if we replace $C_{n,K}$ by $\max(C_{n,K}, 2)$. $\square$

Functions $f$ that satisfy the defining dilatation inequalities for quasiconformal maps but are not necessarily one-to-one are called *quasiregular mappings*, or *mappings of bounded distortion*. It turns out that quasiregular mappings are also Hölder continuous. For the theory of these maps, see Iwaniec and Martin (2001), Reshetnyak (1989) or Anderson et al. (1997).

## 7.9 Notes and Comments

Schmidt (1943) proved the isoperimetric inequality in $\mathbb{S}^n$ and $\mathbb{H}^n$ using the $(2,n)$-symmetrization. Sperner (1973) showed that the Dirichlet integral does not increase under decreases under symmetric decreasing rearrangement on $\mathbb{S}^n$. Pólya (1950) introduced the circular symmetrization, which is $(1,2)$-cap symmetrization in our notation. Pólya and Szegő use both the circular and spherical ($(2,3)$-cap) symmetrization in their book (1951). Hayman (1950) gave a number of applications of circular symmetrization to complex function theory, and provided detailed proofs of its properties in the book (1994). More can be found in Chapter 11.

Sarvas (1972) proved that $(k,n)$-Steiner and $(k,n)$-cap symmetrization does not increase capacity of ring domains. Symmetrization in the Gauss space was introduced and developed, together with its $(k,n)$-variant, in a series of papers by Ehrhard (1983b; 1983a; 1984).

For $n = 2$, the Hölder continuity of quasiconformal maps (Corollary 7.17) was known since early stages of the development of the theory. Ahlfors (1954) proved it, remarking that it "was undoubtedly known to Teichmüller, if not already to Grötzsch." Morrey (1938) gave a proof based on the analytic definition of quasiconformality. Mori (1956) used a symmetrization argument to prove the sharp inequality $|f(x)-f(y)| \leq 16|x-y|^{1/K}$ for quasiconformal self-maps of the disk fixing the origin. While 16 is the best constant independent of $K$, it remains unknown (Bhayo and Vuorinen, 2011) whether it can be improved to $16^{1-1/K}$. Gehring (1962) proved the Hölder continuity for $n = 3$, and his proof extends to all dimensions. A proof by a different method was given by Väisälä (1961).

The theory of quasiconformal mapping has been originally developed in dimension 2: see Ahlfors (1966), Lehto and Virtanen (1973), Astala et al. (2009). Its extension to higher dimensions is presented, e.g., in Iwaniec and Martin (2001), Anderson et al. (1997), and Baernstein and Manfredi (1983). The books Reshetnyak (1989) and Rickman (1993) develop the theory of quasiregular mappings, which are not required to be homeomorphisms. The extension of the quasiconformal theory beyond $\mathbb{R}^n$ was initiated by Mostow (1968) who used it to prove a strong rigidity theorem: two compact Riemannian manifolds of constant negative curvature and dimension $n \geq 3$ that are diffeomorphic must be conformally equivalent.

# 8

# Convolution and Beyond

The Riesz–Sobolev convolution theorem asserts that for nonnegative functions $f, g, h$ on $\mathbb{R}^n$ the integral

$$\int_{\mathbb{R}^n \times \mathbb{R}^n} f(-x)g(y)h(x-y)\,dx\,dy = f * g * h(0)$$

increases when $f, g,$ and $h$ are replaced by their symmetric decreasing rearrangements. When at least one of $f, g, h$ is already symmetric decreasing, we proved this inequality in Corollary 2.19.

In this chapter we begin by proving the analogue of the general Riesz–Sobolev inequality for convolution on the unit circle $\mathbb{S}^1$. The special case when one of the functions is symmetric decreasing is Corollary 7.1. The general circle version easily implies the general version on $\mathbb{R}$, which in turn implies the general version on $\mathbb{R}^n$ for $(1, n)$-Steiner symmetrization, which, we show, implies the general version for s.d.r. and $(k, n)$-Steiner symmetrization on $\mathbb{R}^n$.

In §8.4 we use the Riesz–Sobolev theorem to obtain the Brunn–Minkowski inequality about the volume radius of the vector sum $A + B$ of sets $A$ and $B$ in $\mathbb{R}^n$. Then, using a consequence of Brunn–Minkowski, we prove a major generalization of the Riesz–Sobolev inequality, due to Brascamp, Lieb, and Luttinger, in which there are any number $p \geq 2$ of input functions. The Brascamp–Lieb–Luttinger theorem and Trotter's product formula for the solution of the heat equation enable us to deduce a theorem of Luttinger, which asserts that symmetrization increases the trace of the Dirichlet heat kernel of domains in $\mathbb{R}^n$.

In §8.7 we give a proof of Lieb's sharp Hardy–Littlewood–Sobolev inequality $f * g * K_\lambda(0) \leq C_{\lambda,n}\|f\|_p\|g\|_p$, where $K_\lambda(x) = |x|^{-\lambda}$, $p \in (1, 2)$, $\lambda = 2n/p'$, and $C_{\lambda,n}$ is an explicit constant. In addition to the Riesz–Sobolev inequality, the proof uses conformal invariance properties of the HLS integral to transform the problem to an equivalent problem on the sphere $\mathbb{S}^n$. Following Beckner, we then show that if we rephrase Lieb's inequality in terms of spherical harmonic multiplier operators, and then let $p \to 2$, we obtain $\mathbb{S}^n$ versions of Gross's logarithmic Sobolev inequalities and Nelson's hypercontractivity theorem. When $p \to 1$, we obtain sharp bounds on $\int_{\mathbb{S}^n} e^F$ in terms of Dirichlet-like integrals, which for $n = 1$ and $n = 2$ reduce to inequalities of Lebedev–Milin and Onofri, respectively.

## 8.1 A Riesz-Type Convolution Inequality on $\mathbb{S}^1$

For real-valued functions $f, g$ on the unit circle $\mathbb{S}^1 = \{x \in \mathbb{C}\colon |x| = 1\}$, define their *convolution* $f * g\colon \mathbb{S}^1 \to \mathbb{R}$ to be the function

$$f * g(e^{i\theta}) = \int_{-\pi}^{\pi} f(e^{i\phi})g(e^{i(\theta-\phi)})\, d\phi, \quad \theta \in \mathbb{R},$$

whenever the integral exists. One easily shows that convolution is commutative and associative, that is, $f*g = g*f$ and $f*(g*h) = (f*g)*h$. The function in the latter identity will be denoted $f * g * h$. An alternate notation for convolution is

$$f * g(x) = \int_{\mathbb{S}^1} f(y)g(x\bar{y})\, d\sigma_1(y), \quad x \in \mathbb{S}^1,$$

where the bar denotes complex conjugation.

Given real-valued functions $f, g, h$ on $\mathbb{S}^1$, define

$$J(f, g, h) = \int_{-\pi}^{\pi} \int_{-\pi}^{\pi} f(e^{-i\phi})g(e^{i\theta})h(e^{i(\phi-\theta)})\, d\theta\, d\phi. \tag{8.1}$$

Then

$$J(f, g, h) = f * g * h(1),$$

from which we see that $J$ does not depend on the order of its arguments.

**Theorem 8.1** *[Baernstein, 1989a] Suppose that $f, g$, and $h$ are either nonnegative measurable functions on $\mathbb{S}^1$, or that $f, g, h \in L^1(\mathbb{S}^1)$ and the product $f^\#(e^{-i\phi})g^\#(e^{i\theta})h^\#(e^{i(\phi-\theta)}) \in L^1(\mathbb{S}^1 \times \mathbb{S}^1)$. Then*

$$J(f, g, h) \le J(f^\#, g^\#, h^\#), \tag{8.2}$$

*where # denotes symmetric decreasing rearrangement on $\mathbb{S}^1$.*

Since the convolution of two $L^1$ functions is in $L^1$, for the product $f(e^{-i\phi})g(e^{i\theta})h(e^{i(\phi-\theta)})$ to be integrable, it suffices that two of the functions be in $L^1$ and the third be in $L^\infty$. Also, according to Young's inequality (Zygmund,

1968, p. 37 or Folland, 1999) it suffices for integrability that $f \in L^p$, $g \in L^q$, $h \in L^r$, where $p, q, r \in (1, \infty)$ with $\frac{1}{p} + \frac{1}{q} + \frac{1}{r} = 2$.

If one of $f, g$, or $h$ is already symmetric decreasing, then $J(f, g, h)$ increases under polarization (see Corollary 7.1). But in the general case this need not be so. For example, let $f = g = h = \chi_A$, where $A$ is the image of $[-\epsilon, \epsilon] \cup [\pi - \epsilon, \pi + \epsilon]$ under the exponential map $\phi \mapsto e^{i\phi}$, with $0 < \epsilon < \pi/4$. Take $H$ to be the intersection of $\mathbb{S}^1$ with the line $y = -x$. Then

$$J(f, g, h) = 12\epsilon^2 \quad \text{and} \quad J(f_H, g_H, h_H) = 9\epsilon^2.$$

Furthermore, the usual proofs of the $\mathbb{R}$-version of (8.2) do not work in the $\mathbb{S}^1$ case. So to prove Theorem 8.1 we need a different kind of argument. The proof below is suggested by considerations involving the star function $f^\star$, which we will study in Chapters 9–11.

*Proof of Theorem 8.1* Assume first that $f, g$, and $h$ are nonnegative. Expressing $f, g$, and $h$ by their layer cake representations (§1.6, Proposition 1.32), we see that if (8.2) holds whenever $f, g$, and $h$ are characteristic functions, then it holds for all triples of nonnegative functions. Let $\mathcal{A}_n$ be the collection of all sets $E \subset \mathbb{S}^1$ such that $E$ is the union of at most $n$ disjoint closed intervals on $\mathbb{S}^1$. For each measurable $E \subset \mathbb{S}^1$ the function $\chi_E$ is the a.e. limit of a sequence $\chi_{A_n}$ with $A_n \in \mathcal{A}_n$. Thus, to prove (8.2) for all characteristic functions, it suffices to prove that for each fixed $n \geq 1$, (8.2) holds when $f = \chi_{E_1}$, $g = \chi_{E_2}$, $h = \chi_{E_3}$, with $E_1, E_2, E_3 \in \mathcal{A}_n$.

Let

$$Q = [0, \pi] \times [0, \pi] \times [0, \pi]$$

and fix $n$. Define $F\colon Q \to \mathbb{R}^+$ by

$$F(\theta_1, \theta_2, \theta_2) = \sup J(E_1, E_2, E_3) \tag{8.3}$$

where the supremum is taken over all sets $E_1, E_2, E_3$ with each $E_i \in \mathcal{A}_n$ and $\sigma_1(E_i) = 2\theta_i$ for $i = 1, 2, 3$. It is straightforward to show that the supremum is in fact achieved by some triple of competing sets, and that $F$ is continuous on $Q$.

Define also $G\colon Q \to \mathbb{R}^+$ by

$$G(\theta_1, \theta_2, \theta_3) = J(\chi_{I_1}, \chi_{I_2}, \chi_{I_3}),$$

where $I_i = [-\theta_i, \theta_i]$, $i = 1, 2, 3$, and we follow the usual practice of identifying sets $\{e^{i\phi} : \phi \in A\} \subset \mathbb{S}^1$ with sets $A \subset \mathbb{R} \bmod 2\pi$. Then $(\chi_{E_i})^\# = \chi_{I_i}$.

Denote the interior of $Q$ by $Q^o$, and denote the coordinate vectors in $\mathbb{R}^3$ by $e_1, e_2, e_3$.

**Claim 8.2** *Suppose that $\vec{\theta} = (\theta_1, \theta_2, \theta_2) \in Q^o$. Then for at least one $i \in \{1, 2, 3\}$,*

$$\liminf_{t\to 0} t^{-2}[F(\vec{\theta} + te_i) + F(\vec{\theta} - te_i) - 2F(\vec{\theta})] \geq -2. \tag{8.4}$$

If $F$ is twice differentiable at $\vec{\theta}$ then the expression on the left equals the partial derivative $F_{\theta_i\theta_i}(\vec{\theta})$.

Assume for now that the claim is true. Fix $\epsilon > 0$. Define $H\colon Q \to \mathbb{R}$ by

$$H(\vec{\theta}) = F(\vec{\theta}) - G(\vec{\theta}) - \epsilon \sum_{i=1}^{3} \theta_i(\pi - \theta_i).$$

Since $F$ and $G$ are continuous, $H$ achieves a maximum value $M$ in $Q$. We shall show that $M \leq 0$.

Take $\vec{\theta}_0 \in Q$ such that $H(\vec{\theta}_0) = M$. Write $\vec{\theta}_0 = (\theta_1, \theta_2, \theta_3)$, and let $E_1, E_2, E_3$ be competing sets for which the supremum in (8.3) is achieved. Then

$$F(\vec{\theta}_0) = \int_{-\pi}^{\pi} \int_{-\pi}^{\pi} \chi_{E_1}(e^{-i\phi})\chi_{E_2}(e^{i\theta})\chi_{E_3}(e^{i(\phi-\theta)})\, d\theta\, d\phi\,. \tag{8.5}$$

Consider now four cases.

**Case 1.** $\vec{\theta}_0 \in \partial Q$. Then some $\theta_i$ is equal to either 0 or $\pi$. Say $\theta_1 = 0$. Then $E_1$ is a single point, and $F(\vec{\theta}_0) = G(\vec{\theta}_0) = 0$. If $\theta_1 = \pi$, then $E_1 = \mathbb{S}^1$, and $F(\vec{\theta}_0) = G(\vec{\theta}_0) = 4\theta_2\theta_3$. Either way, $M = H(\vec{\theta}_0) = 0$.

**Case 2.** $\vec{\theta}_0 \in Q^o$, and some $\theta_i$ is greater than or equal to the sum of the other two. Say $\theta_1 \geq \theta_2 + \theta_3$. Using the bound $\chi_{E_1} \leq 1$ in (8.5), we have $F(\vec{\theta}_0) \leq 4\theta_2\theta_3$. Also,

$$G(\vec{\theta}_0) = J(\chi_{I_3}, \chi_{I_1}, \chi_{I_2}) = \int_{-\theta_3}^{\theta_3} \chi_{I_1} * \chi_{I_2}(e^{i\phi})\, d\phi,$$

and

$$\chi_{I_1} * \chi_{I_2}(e^{i\phi}) = \int_{-\theta_2}^{\theta_2} \chi_1(e^{i(\phi-\theta)})\, d\theta.$$

If $|\phi| \leq \theta_3$ and $|\theta| \leq \theta_2$ then $|\phi - \theta| \leq \theta_1$. Thus

$$\chi_{I_1} * \chi_{I_2}(e^{i\phi}) = 2\theta_2, \quad |\phi| \leq \theta_3,$$

hence $G(\vec{\theta}_0) = 4\theta_2\theta_3$, and we again have $M \leq 0$.

**Case 3.** $\vec{\theta}_0 \in Q^o$ and $\theta_1 + \theta_2 + \theta_3 \geq 2\pi$. Note that the definition of $Q$ forces each $\theta_i$ to be less than the sum of the other two.

For functions $f \in L^1(\mathbb{S}^1)$, write $m(f) = \int_{-\pi}^{\pi} f\, d\theta$. Then $f * 1$ is the constant function $m(f)$. From this, we deduce that if $f, g, h$ are three functions for which the integrand in (8.1) is integrable, then

$$J(1-f, 1-g, h) = m(h) - m(f)m(h) - m(g)m(h) + J(f, g, h).$$

Since $1 - \chi_E = \chi_{\mathbb{S}^1 \setminus E}$, this implies the symmetry relation

$$F(\pi - \theta_1, \pi - \theta_2, \theta_3) = F(\theta_1, \theta_2, \theta_3) + 4\theta_3(\pi - \theta_1 - \theta_2), \quad \forall\, (\theta_1, \theta_2, \theta_3) \in Q.$$

Translating both variables in the defining integral by 1, we see that the same relation holds for $G$. We conclude that

$$H(\theta_1, \theta_2, \theta_3) = H(\pi - \theta_1, \pi - \theta_2, \theta_3), \quad \forall\, (\theta_1, \theta_2, \theta_3) \in Q. \tag{8.6}$$

Assuming now that $(\theta_1, \theta_2, \theta_3) = \vec{\theta}_0$ is a maximizing point for $H$, we see that $(\pi - \theta_1, \pi - \theta_2, \theta_3)$ is also a maximizing point. By hypothesis, $\pi - \theta_1 + \pi - \theta_2 \le \theta_3$. Thus, by Case 2, $M = H(\pi - \theta_1, \pi - \theta_2, \theta_3) \le 0$.

**Case 4.** $Q$, the interior of $Q$, $\theta_1 + \theta_2 + \theta_3 < 2\pi$ and each $\theta_i$ is strictly smaller than the sum of the other two. Assume that (8.4) is satisfied with $i = 3$. By the symmetry relation in (8.6), we may assume that $\theta_1 + \theta_2 \le \pi$. Suppose also, without loss of generality, that $\theta_1 \ge \theta_2$. We have

$$G(\theta_1, \theta_2, s) = \int_{-s}^{s} \chi_{I_1} * \chi_{I_2}(e^{i\phi})\, d\phi, \qquad s \in [0, \pi].$$

Thus $G$ is a differentiable function of $s$, with partial derivative

$$G_s(\theta_1, \theta_2, s) = 2\chi_{I_1} * \chi_{I_2}(e^{is}).$$

The function $\chi_{I_1} * \chi_{I_2}(e^{is})$ is an even function of $s$. Since $\theta_1 + \theta_2 \le \pi$, this function is constant for $0 \le s \le \theta_1 - \theta_2$, zero for $\theta_1 + \theta_2 \le s \le \pi$, and linear with slope $-1$ for $\theta_1 - \theta_2 \le s \le \theta_1 + \theta_2$. Our hypothesis implies that $\theta_1 - \theta_2 < \theta_3 < \theta_1 + \theta_2$. It follows that $G(\theta_1, \theta_2, s)$ is a twice continuously differentiable function of $s$ in some neighborhood of $\theta_3$. Its derivative satisfies

$$G_{\theta_3\theta_3}(\vec{\theta}_0) = -2$$

and hence, by Claim 8.2,

$$\liminf_{t \to 0} t^{-2}[H(\vec{\theta}_0 + te_3) + H(\vec{\theta}_0 - te_3) - H(\vec{\theta}_0) \ge 2\epsilon.$$

Such a $\vec{\theta}_0$ cannot maximize $H$. We conclude that Case 4 cannot occur. Since Cases 1–4 exhaust all possibilities, it follows that $H \le 0$ in $Q$. Since $\epsilon$ was arbitrary, we must have $F \le G$ in $Q$. As explained at the beginning of the proof, this inequality implies that $J(f, g, h) \le J(f^\#, g^\#, h^\#)$ holds for all nonnegative

$f, g, h$. We have proved Theorem 8.1 for nonnegative functions, modulo the claim.

*Proof of Claim 8.2* Let $E_1, E_2$, and $\overline{E_3}$ be extremal sets for $\vec{\theta}$, where $\overline{E_3}$ is the complex conjugate of the set $E_3$. Then $\sigma_1(\overline{E_3}) = \sigma(E_3) = 2\theta_3$. Set $P(e^{i\phi}) = \chi_{E_1} * \chi_{E_2}(e^{i\phi})$. Then $P$ is a periodic piecewise linear function of $\phi$, and

$$F(\vec{\theta}) = \int_{E_3} P(e^{i\phi})\, d\phi. \tag{8.7}$$

Each $E_i$ is the union of $k_i$ disjoint closed nondegenerate intervals, with $1 \le k_i \le n$. Say $k_1 \le k_2 \le k_3$. We will prove that the claim holds with $i = 3$.

For $m \in \mathbb{R}$, $\delta > 0$ write $I(m, \delta) = [m - \delta, m + \delta]$. Then $E_3 = \cup_{i=1}^{k_3} I(m_i, \delta_i)$ for some $m_i$ and $\delta_i$. For $t > 0$, define

$$E_3^+(t) = \cup_{i=1}^{k_3} I(m_i, \delta_i + t), \quad E_3^-(t) = \cup_{i=1}^{k_3} I(m_i, \delta_i - t).$$

We assume that $t$ is small enough so that the intervals defining $E_3^+(t)$ are disjoint and the intervals defining $E_3^-(t)$ are nondegenerate. Set $b_i = m_i + \delta_i$, $a_i = m_i - \delta_i$. Then

$$\begin{aligned}&\int_{E_3^+(t)} P\, d\phi + \int_{E_3^-(t)} P\, d\phi - 2\int_{E_3} P\, d\phi \\ &= \sum_{i=1}^{k_3} \left\{ \int_{b_i}^{b_i+t} - \int_{b_i-t}^{b_i} - \int_{a_i}^{a_i+t} + \int_{a_i-t}^{a_i} P\, d\phi \right\}.\end{aligned}$$

For $i = 1, \ldots, k_3$, let $\lambda_i^+$ be the derivative of $P$ from the right at $b_i$ and $\lambda_i^-$ be the derivative from the left at $b_i$. Let $\mu_i^+$ and $\mu_i^-$ be the analogous derivatives from the right and left respectively of $P$ at $a_i$. Then each of these slopes has absolute value $\le k_1$. To see this, first examine the case $k_1 = 1$ and verify that the slopes are either $\pm 1$ or zero. For general $k_1$, observe that $P$ is the sum of $k_1$ functions, each of which has slopes of absolute value $\le 1$.

Calculation gives

$$\int_{b_i}^{b_i+t} P\, d\phi = tP(e^{ib_i}) + \frac{1}{2}\lambda_i^+ t^2, \quad \int_{b_i-t}^{b_i} P\, d\phi = tP(e^{ib_i}) - \frac{1}{2}\lambda_i^- t^2,$$
$$\int_{a_i}^{a_i+t} P\, d\phi = tP(e^{ia_i}) + \frac{1}{2}\mu_i^+ t^2, \quad \int_{a_i-t}^{a_i} P\, d\phi = tP(e^{ia_i}) - \frac{1}{2}\mu_i^- t^2,$$

so that

$$\begin{aligned}\int_{E_3^+(t)} P\, d\phi + \int_{E_3^-(t)} P\, d\phi - 2\int_{E_3} P\, d\phi &= \frac{1}{2}t^2 \sum_{i=1}^{k_3} (\lambda_i^+ + \lambda_i^- - \mu_i^+ - \mu_i^-) \\ &\ge -2t^2 k_1 k_3.\end{aligned}$$

The first term on the left is at most $F(\theta_1,\theta_2,\theta_3+k_3t)$ and the second is at most $F(\theta_1,\theta_2,\theta_3-k_3t)$. Combining with (8.7), we obtain

$$F(\theta_1,\theta_2,\theta_3+k_3t)+F(\theta_1,\theta_2,\theta_3-k_3t) \quad - \quad 2F(\theta_1,\theta_2,\theta_3)\geq -2t^2k_1k_3.$$

Replacing $t$ by $t/k_3$ and recalling that $k_1 \leq k_3$, we obtain (8.4). The claim is proved. □

It remains to complete the proof of Theorem 8.1. Suppose finally that $f,g,h\in L^1(\mathbb{S}^1)$ and that $f^\#(e^{-i\phi})g^\#(e^{i\theta})h^\#(e^{i(\phi-\theta)})$ is in $L^1(\mathbb{S}^1\times\mathbb{S}^1)$. Let $f_n$ be the truncation defined by $f_n=f$ if $|f|\leq n$, $f_n=-n$ if $f\leq -n$, $f_n=n$ if $f\geq n$. Define $g_n$ and $h_n$ similarly. Applying (8.2) to the functions $|f|$, $|g|$ and $|h|$, we see that $f(e^{-i\phi})g(e^{i\theta})h(e^{i(\phi-\theta)})\in L^1(\mathbb{S}^1\times\mathbb{S}^1)$. The dominated convergence theorem gives $\lim_{n\to\infty}J(f_n,g_n,h_n)=J(f,g,h)$, and likewise for the s.d.r. of these functions. Thus, to prove (8.2) in general, we just need to prove it for $f$, $g$ and $h$ bounded.

Take $c\in\mathbb{R}^+$ such that $|f|$, $|g|$ and $|h|$ are $\leq c$. Then $f+c$, $g+c$, and $h+c$ are nonnegative. Denoting the mean value of $f$ by $\mathfrak{m}(f)$, we expand

$$\begin{aligned}J(f+c,g+c,h+c)=J(f,g,h)&+c[\mathfrak{m}(f)\mathfrak{m}(g)+\mathfrak{m}(g)\mathfrak{m}(h)+\mathfrak{m}(h)\mathfrak{m}(f)]\\&+c^2[\mathfrak{m}(f)+\mathfrak{m}(g)+\mathfrak{m}(h)]+c^3,\end{aligned}$$

and apply (8.2) to the left-hand side. Since the terms involving mean values do not change under rearrangement, it follows that (8.2) holds for $f$, $g$, and $h$, and the proof of Theorem 8.1 is complete. □

## 8.2 Riesz's Convolution Inequality on $\mathbb{R}$

For real-valued functions $f,g$ on $\mathbb{R}$, the convolution $f*g\colon \mathbb{R}\to\mathbb{R}$ is defined by

$$f*g(x)=\int_{\mathbb{R}} f(y)g(x-y)\,dy, \qquad x\in\mathbb{R},$$

whenever the integral exists. As on the circle, convolution on $\mathbb{R}$ is commutative and associative.

Given real-valued functions $f,g,h$ on $\mathbb{R}$, define

$$J(f,g,h)=\int_{\mathbb{R}\times\mathbb{R}} f(-x)g(y)h(x-y)\,dx\,dy=f*g*h(0).$$

Recall that for functions $f$ on $\mathbb{R}^n$, the distribution function $\lambda_f(t)$ is defined by $\lambda_f(t)=\mathcal{L}^n(f>t)$. Our usual finiteness condition for nonnegative functions is

$$\lambda_f(t)<\infty, \quad \forall\, t>0. \tag{8.8}$$

**Theorem 8.3** (F. Riesz, 1930) *Suppose that $f, g$, and $h$ are nonnegative measurable functions on $\mathbb{R}$ satisfying the finiteness condition* (8.8). *Then*

$$J(f,g,h) \leq J(f^{\#}, g^{\#}, h^{\#}), \tag{8.9}$$

*where # denotes symmetric decreasing rearrangement on $\mathbb{R}$.*

*Proof* Assume first that $f, g$, and $h$ are all supported in an interval $[-A, A]$ with $0 < A < \pi/3$. Then $f * g * h$ is supported in $[-3A, 3A] \subset (-\pi, \pi)$. Define $f_1, g_1, h_1$ on $\mathbb{S}^1$ by $f_1(e^{i\theta}) = f(\theta)$, etc., where $|\theta| \leq \pi$. Then

$$f * g * h(\theta) = f_1 * g_1 * h_1(e^{i\theta}), \quad |\theta| \leq \pi.$$

It is also true that $f^{\#}(\theta) = f_1^{\#}(e^{i\theta})$, where $f_1^{\#}$ denotes the s.d.r. of $f_1$ on $\mathbb{S}^1$. Taking $\theta = 0$ and applying Theorem 8.1, inequality (8.9) follows.

Next, assume that $f, g$ and $h$ are all supported in some interval $[-A, A]$ with $0 < A < \infty$. Take $\alpha > 3A/\pi$. This time, define $f_1, g_1, h_1$ on $\mathbb{R}$ by $f_1(x) = f(\alpha x)$, etc. Then $f_1, g_1, h_1$ are compactly supported in $(-\pi/3, \pi/3)$, and

$$f * g * h * (\alpha x) = \alpha^2 f_1 * g_1 * h_1(x).$$

Also, we have the equation $f^{\#}(\alpha x) = f_1^{\#}(x)$. Thus, the validity of (8.9) for $f, g, h$ follows from its validity for $f_1, g_1, h_1$.

Let now $f, g, h$ be any three functions satisfying the hypotheses. For $m \geq 1$, define $f_m = f\chi_{[-m,m]}$, etc. Then $J(f_m, g_m, h_m) \leq J(f_m^{\#}, g_m^{\#}, h_m^{\#})$. By monotone convergence, $J(f_m, g_m, h_m) \to J(f, g, h)$ as $m \to \infty$, and by Proposition 1.39 and monotone convergence, $J(f_m^{\#}, g_m^{\#}, h_m^{\#}) \to J(f^{\#}, g^{\#}, h^{\#})$. Thus, (8.9) holds for $f, g, h$. □

Burchard and Hajaiej (2006, p. 16) give a simple example for which (8.9) is false when $f, g, h$ are replaced by their polarizations instead of their symmetric decreasing rearrangements.

## 8.3 The Riesz–Sobolev Inequality

The convolution $f * g$ of two real-valued functions on $\mathbb{R}^n$ is defined by the same formula as when $n = 1$, except that now $x$ and $y$ denote points of $\mathbb{R}^n$. Given real-valued functions $f, g, h$ on $\mathbb{R}^n$, define

$$J(f,g,h) = \int_{\mathbb{R}^n \times \mathbb{R}^n} f(-x)g(y)h(x-y)\,dx\,dy = f * g * h(0).$$

**Theorem 8.4** (Sobolev, 1938) *Suppose that $f, g$, and $h$ are nonnegative measurable functions on $\mathbb{R}^n$ satisfying the finiteness condition* (8.8). *Then*

$$J(f,g,h) \leq J(f^{\#}, g^{\#}, h^{\#}), \tag{8.10}$$

*where # denotes symmetric decreasing rearrangement on $\mathbb{R}^n$ or $(k,n)$- Steiner symmetrization on $\mathbb{R}^n$, $1 \le k \le n$.*

*Proof* **Step 1.** For $x, y \in \mathbb{R}^n$ write $x = (x_1, u),\ y = (y_1, v)$ where $u, v \in \mathbb{R}^{n-1}$. For nonnegative $f, g, h$ on $\mathbb{R}^n$ we can write

$$J(f,g,h) = \int_{\mathbb{R}^{n-1}\times\mathbb{R}^{n-1}} \left( \int_{\mathbb{R}\times\mathbb{R}} f(-x_1, -u) g(y_1, v) h(x_1 - y_1, u - v)\, dx_1\, dy_1 \right) du dv.$$

Let $f^s$ denote the $(1,n)$-Steiner symmetrization of $f$ (with respect to the hyperplane $x_1 = 0$). Then $f^s$ is obtained by changing each function $f(\cdot, u)$ to its symmetric decreasing rearrangement on $\mathbb{R}$. By Theorem 8.3, each of the $dx_1\, dy_1$ integrals increases when we change $f, g$ and $h$ to $f^s$, $g^s$ and $h^s$. Thus,

$$J(f,g,h) \le J(f^s, g^s, h^s). \tag{8.11}$$

Since $J$ is invariant under simultaneous rotation of the three functions, it also decreases under Steiner symmetrization with respect any other hyperplane.

**Step 2.** Assume that $f, g, h$ are in $C_c(\mathbb{R}^n, \mathbb{R}^+)$. Let $R(f)$ the radius of the smallest centered ball that contains the support of $f$. As in (2.12) of §2.4, define a class of functions $\mathcal{S}(f)$ by

$$\begin{aligned}\mathcal{S}(f) = \{F \in C_c(\mathbb{R}^n, \mathbb{R}^+) \colon\ &\omega(\cdot, F) \le \omega(\cdot, f) \text{ on } (0,\infty),\\ &\lambda_F = \lambda_f \text{ on } (0,\infty), \text{ and } R(F) \le R(f)\}.\end{aligned}$$

Here $\omega$ denotes modulus of continuity defined in §1.7, and $\lambda_F = \lambda_f$ says that $f$ and $F$ have the same distribution on $\mathbb{R}^n$. As in §2.5, define

$$\begin{aligned}\mathcal{S} &= \mathcal{S}(f,g,h)\\ &= \{(F,G,H) \in \mathcal{S}(f) \times \mathcal{S}(g) \times \mathcal{S}(h) \colon J(f,g,h) \le J(F,G,H)\}\end{aligned}$$

and

$$d^2 = \inf_{(F,G,H)\in\mathcal{S}} \left( \|F - f^{\#}\|_2^2 + \|G - g^{\#}\|_2^2 + \|H - h^{\#}\|_2^2 \right),$$

where $\|\cdot\|_2$ is the norm in $L^2(\mathbb{R}^n, \mathcal{L}^n)$ and # denotes symmetric decreasing rearrangement on $\mathbb{R}^n$.

By Theorem 6.10 and Corollary 6.13, the modulus of continuity and the diameter of the support of functions in $C_c(\mathbb{R}^n)$ decrease under Steiner symmetrization. Thus, $F \in \mathcal{S}(f)$ implies $F^s \in \mathcal{S}(f)$. Combined with (8.11), we see that $(F,G,H) \in \mathcal{S}$ implies $(F^s, G^s, H^s) \in \mathcal{S}$. Also, by applying Eq. (6.21) of Theorem 6.8 with $\Phi(t) = t^2$,

$$\|f^s - f^{\#}\|_2 \le \|f - f^{\#}\|_2.$$

We can now follow the argument in the proof of Theorem 2.15(a). Let $\{(f_k, g_k, h_k)\}$ be a sequence in $\mathcal{S}$ such that

$$d^2 = \lim_{k\to\infty} \|f_k - f^\#\|_2^2 + \|g_k - g^\#\|_2^2 + \|h_k - h^\#\|_2.$$

All of these functions are supported in the ball of radius $R$, where $R$ is the maximum of $R(f), R(g), R(h)$. We can apply the Arzelà–Ascoli theorem to get functions $(F_0, G_0, H_0) \in \mathcal{S}$ with

$$d^2 = \|F_0 - f^\#\|_2^2 + \|G_0 - g^\#\|_2^2 + \|H_0 - h^\#\|_2^2. \tag{8.12}$$

If $d^2 = 0$ then $(f^\#, g^\#, h^\#) = (F_0, G_0, H_0) \in \mathcal{S}$ and we are done. Suppose that at least one of the three equalities fails, say $F_0 \neq f^\#$. For simplicity, write $F$ instead of $F_0$. Take $a \in \mathbb{R}^n$ such that $F(a) \neq f^\#(a)$. Say $F(a) < f^\#(a)$. Take $t$ with $F(a) < t < f^\#(a)$. Let $E_1 = (F > t)$, $E_2 = (f^\# > t)$. By continuity of $F$ and $f^\#$, there is a neighborhood $U$ of $a$ such that $U \subset E_2 \setminus E_1$. Thus, $\mathcal{L}^n(E_2 \setminus E_1) > 0$. Since $\mathcal{L}^n(E_1) = \mathcal{L}^n(E_2)$, the set $E_1 \setminus E_2$ must have positive measure. Now $E_2$ is an open ball, so its boundary has measure zero. We conclude that there is a point $b \in \mathbb{R}^n$ and an open neighborhood $V$ of $b$ such that $V \subset E_1 \setminus E_2$. It follows that

$$F(x) < F(y) \quad \text{and} \quad f^\#(x) > f^\#(y) \quad \forall\ (x, y) \in U \times V. \tag{8.13}$$

If $F(a) > f^\#(a)$ we repeat the argument above but interchange the roles of $E_1$ and $E_2$. In any event, whenever $F \neq f^\#$, we have shown there exist open neighborhoods $U$ and $V$ of the points $a$ and $b$ respectively for which (8.13) holds.

By construction, the points $a$ and $b$ are distinct. After a rotation, we may assume that $a$ and $b$ have the form $a = (a_1, z^0)$, $b = (b_1, z^0)$ for some $z^0 \in \mathbb{R}^{n-1}$.

For $z \in \mathbb{R}^{n-1}$, define sets $A^z \subset \mathbb{R}^2$ by

$$A^z = \{(x_1, y_1) \in \mathbb{R}^2 \colon F(x_1, z) < F(y_1, z) \quad \text{and} \quad f^\#(x_1, z) > f^\#(y_1, z)\}.$$

Then, by (8.13),

$$A^z \supset \{(x_1, y_1) \in \mathbb{R}^2 \colon (x_1, z) \in U,\ (y_1, z) \in V\}.$$

For $z$ sufficiently close to $z^0$ in $\mathbb{R}^{n-1}$ the set on the right contains a neighborhood of $(a_1, b_1)$ in $\mathbb{R}^2$, thus has positive $\mathcal{L}^2$ measure. Thus $\mathcal{L}^2(A^z) > 0$ on a neighborhood of $z^0$ in $\mathbb{R}^{n-1}$, and hence on a set of $z$ with positive $\mathcal{L}^{n-1}$ measure.

By the equality statement in Theorem 6.8, we have the strict inequality

$$\|F_0^s - f^\#\|_2 < \|F_0 - f^\#\|.$$

It is also true that $\|G_0^s - g^\#\| \le \|G_0 - g^\#\|_2$ and likewise for $H_0$ and $h^\#$. Thus, the left-hand side of (8.12) strictly decreases when $F_0, G_0, H_0$ are replaced by $F_0^s, G_0^s, H_0^s$. Since $(F_0, G_0, H_0) \in \mathcal{S}$ implies $(F_0^s, G_0^s, H_0^s) \in \mathcal{S}$, we have reached a contradiction to the definition of $d$. It must be true, then, that $(F_0, G_0, H_0) = (f^\#, g^\#, h^\#)$. We have proved that

$$J(f, g, h) \le J(f^\#, g^\#, h^\#)$$

for all $f, g, h \in C_c(\mathbb{R}^n)$.

**Step 3.** The arguments in the proof of Parts 2 and 3 of Theorem 2.15(a) carry over, and produce inequality (8.10) for s.d.r. on $\mathbb{R}^n$, first for $f, g, h$ nonnegative, compactly supported and in $L^\infty(\mathbb{R}^n)$, then for all nonnegative measurable $f, g, h$ on $\mathbb{R}^n$.

**Step 4.** Take $1 \le k < n$ and set $m = n-k$. Then, with self-explanatory notation

$$J(f, g, h) = \int_{\mathbb{R}^m \times \mathbb{R}^m} du\, dv \int_{\mathbb{R}^k \times \mathbb{R}^k} f(-x', u) g(y', v) h(x' - y', u - v)\, dx'\, dy',$$

where $x = (x', u) \in \mathbb{R}^n$, $y = (y', v) \in \mathbb{R}^n$. Let # denote $(k, n)$-Steiner symmetrization. Then $f^\#$ is obtained by changing each function $f(\cdot, u)$ to its s.d.r. on $\mathbb{R}^k$. By the first part of the proof, each of the $dx'\, dy'$ integrals increases when we change $f, g$ and $h$ to $f^\#, g^\#$ and $h^\#$. Thus, (8.10) holds for $(k, n)$-Steiner symmetrization and all nonnegative measurable $f, g, h$.

The proof of Theorem 8.4 is complete. □

## 8.4 The Brunn–Minkowski Inequality

Let $A$ and $B$ be nonempty Lebesgue measurable subsets of $\mathbb{R}^n$, $n \ge 1$. Unless otherwise stated, we will always assume in this section that $A$ and $B$ have finite measure. Define sets $A + B$ and $C$ by

$$A + B = \{x + y \colon x \in A,\ y \in B\} = \{x \in \mathbb{R}^n \colon (x - A) \cap B \ne \emptyset\},$$

$$C = \{x \in \mathbb{R}^n \colon \mathcal{L}^n((x - A) \cap B) > 0\} = \{x \in \mathbb{R}^n \colon \chi_A * \chi_B(x) > 0\}.$$

The set $A + B$ is called the Minkowski sum of $A$ and $B$. It is not $\mathcal{L}^n$-measurable in general (Sierpiński, 1920), but if $A$ and $B$ are Borel, then $A + B$ is analytic and therefore $\mathcal{L}^n$-measurable (Gardner, 2002). In Brascamp and Lieb (1976), $C$ is called the essential sum of $A$ and $B$. It is clear that

$$C \subset A + B.$$

Note also that $C$ is open, since $\chi_A * \chi_B$ is continuous. The reader may show that if $A + B$ is open, then $C = A + B$.

**Theorem 8.5** *(Brunn-Minkowski Inequality) For $\mathcal{L}^n$-measurable sets $A, B \subset \mathbb{R}^n$ of positive measure and $C$ defined as above, we have*

$$\mathcal{L}^n(A)^{1/n} + \mathcal{L}^n(B)^{1/n} \leq \mathcal{L}^n(C)^{1/n} \leq \mathcal{L}^n(A+B)^{1/n}, \tag{8.14}$$

*where the second inequality holds whenever $A + B$ is $\mathcal{L}^n$-measurable.*

Note that the left inequality fails if $A$ has measure zero and $B$ has positive measure, since then $C$ is empty. If $A$ and $B$ are open balls of radii $R_1$ and $R_2$, then $C = A + B$ is an open ball of radius $R_1 + R_2$, and equality holds in (8.14). By a theorem of Hadwiger and Ohmann (1956), the necessary and sufficient condition for equality to hold in both inequalities of (8.14) is that $A = K_1 \setminus L_1$, $B = K_2 \setminus L_2$, where $L_1$ and $L_2$ are $\mathcal{L}^n$ nullsets, $K_1$ is convex, and $K_2 = aK_1 + b$ for some $a \in \mathbb{R}$, $b \in \mathbb{R}^n$.

When $A$ is open and $B = \mathbb{B}^n(0, \epsilon)$, the open ball of radius $\epsilon$ centered at the origin, then $C = A + B = A(\epsilon)$, the $\epsilon$-collar of $A$ as defined in §4.4, and inequality (8.14) coincides with Corollary 4.15(a).

Write $R_A$ for the volume radius of $A$, that is, $\mathcal{L}^n(A) = \alpha_n R_A^n$. Then inequality (8.14) can be written as

$$R_A + R_B \leq R_C \leq R_{A+B},$$

with equality when $A$ and $B$ are balls. Still another equivalent formulation is

$$A^\# + B^\# \subset C^\# \subset (A+B)^\#, \tag{8.15}$$

where # denotes symmetric decreasing rearrangement on $\mathbb{R}^n$. Indeed, each set in (8.15) is an open ball centered at the origin, and by (8.14), we see that $A^\# + B^\#$ has volume radius at most that of $C^\#$. The same argument shows that (8.15) implies (8.14).

*Proof of Theorem 8.5* By definition of $C$,

$$\int_C \chi_A * \chi_B \, dx = \int_{\mathbb{R}^n} \chi_A * \chi_B \, dx = \mathcal{L}^n(A)\mathcal{L}^n(B). \tag{8.16}$$

Suppose that the first inclusion relation in (8.15) is false. Then the open ball $A^\# + B^\#$ contains the closure of the open ball $C^\#$. So

$$\begin{aligned} \int_C \chi_A * \chi_B \, dx &\leq \int_{C^\#} \chi_{A^\#} * \chi_{B^\#} \, dx < \int_{A^\# + B^\#} \chi_{A^\#} * \chi_{B^\#} \, dx \\ &= \mathcal{L}^n(A^\#)\mathcal{L}^n(B^\#) = \mathcal{L}^n(A)\mathcal{L}^n(B). \end{aligned} \tag{8.17}$$

The first inequality is by the Riesz–Sobolev Theorem 8.4. The second inequality follows from the fact that the set where $\chi_{A^\#} * \chi_{B^\#} > 0$ is precisely the open ball $A^\# + B^\#$. Since (8.16) and (8.17) are contradictory, we conclude that $A^\# + B^\# \subset C^\#$. Further, $C \subset A + B$ implies that $C^\# \subset (A+B)^\#$, completing the proof of (8.16) and Theorem 8.5. □

We deduced the Brunn–Minkowski inequality (8.14) from the Riesz–Sobolev convolution inequality. Conversely, the Brunn–Minkowski inequality is used in Riesz' original proof of the convolution inequality in 1 dimension. The inequality in higher dimensions plays a key role in the proof of the Brascamp–Lieb–Luttinger inequality: see Section 8.5.

Take $\lambda \in [0, 1]$. Applying Theorem 8.5 to the sets $\lambda A$ and $(1-\lambda)B$, we find the following concavity relation for the volume radius.

**Corollary 8.6** *Let $\lambda \in [0, 1]$ and $A, B \subset \mathbb{R}^n$ be $\mathcal{L}^n$ measurable sets of finite positive measure. Then*

$$\lambda \mathcal{L}^n(A)^{1/n} + (1-\lambda)\mathcal{L}^n(B)^{1/n} \le \mathcal{L}^n(\lambda A + (1-\lambda)B)^{1/n}.$$

It follows that the volume radius is a concave function on the class of convex sets. Indeed, for convex sets $A, B$, positive numbers $a_0, b_0, a_1, b_1$ and $\lambda \in [0, 1]$, we have

$$((1-\lambda)a_0 + \lambda a_1)A + ((1-\lambda)b_0 + \lambda b_1)B = (1-\lambda)(a_0 A + b_0 B) + \lambda(a_1 A + b_1 B),$$

so we can apply Corollary 8.6 on the right. In particular, the volume radius of the $\epsilon$-collar $A(\epsilon)$ of a convex set $A$ is concave in $\epsilon$. Note that the concavity does not extend to the class of measurable sets.

Next, we derive another concavity property, which will be needed in the next section. Let $K$ be a convex subset of $\mathbb{R}^{n+1}$. Define

$$K(t) = \{x \in \mathbb{R}^n : (x, t) \in K\}, \quad \psi(t) = \mathcal{L}^n(K(t)), \quad t \in \mathbb{R}.$$

Thus, $K(t)$ is the slice of $K$ through the hyperplane $\{x_{n+1} = t\}$, and $\psi(t)$ is the measure of the slice. Define

$$U = \{t \in \mathbb{R} : \psi(t) > 0\}.$$

**Corollary 8.7** *Let $K$, $\psi$, and $U$ be as specified. Assume that $U$ contains at least two points. Then $U$ is an interval, and $\psi^{1/n}$ is a concave function on $U$.*

**Example** If $K$ is the open unit ball $\mathbb{B}^{n+1}$, then $U = (-1, 1)$, and $\psi(t) = \alpha_n(1-t^2)^{n/2}$ on $U$.

*Proof of Corollary 8.7* Each of the sets $K(t)$ with $t \in U$ is a convex set with positive $\mathcal{L}^n$ measure. Thus, each such $K(t)$ has interior points. From the convexity of $K$, we infer that if $t_1, t_2 \in U$ then each $t$ in between $t_1$ and $t_2$ also belongs to $U$. Thus, $U$ is an interval.

Take $t_1 < t_2 < t_3$ in $U$. Write $t_2 = \lambda t_1 + (1-\lambda)t_3$, where $\lambda = \frac{t_3-t_2}{t_3-t_1}$. Take $z \in \lambda K(t_1) + (1-\lambda)K(t_3)$. Then $z = \lambda x + (1-\lambda)y$ for some $x \in K(t_1)$, $y \in K(t_3)$. So

$$(z,t_2) = \lambda(x,t_2) + (1-\lambda)(y,t_2) = \lambda(x,t_1) + (1-\lambda)(y,t_3).$$

The last expression is a convex combination of points in $K$, hence belongs to $K$. We conclude that $z \in K(t_2)$, and hence that

$$\lambda K(t_1) + (1-\lambda)K(t_3) \subset K(t_2).$$

Thus,

$$\lambda\psi^{1/n}(t_1) + (1-\lambda)\psi^{1/n}(t_3) \le \mathcal{L}^n(\lambda K(t_1) + (1-\lambda)K(t_3))^{1/n} \le \psi^{1/n}(t_2),$$

where the first inequality is by Corollary 8.6. This shows that $\psi^{1/n}$ is concave on $U$. □

Corollary 8.7 appears in §11.48 of Bonnesen and Fenchel (1987).

## 8.5 The Brascamp–Lieb–Luttinger Inequality

Motivated by problems from mathematical physics, Brascamp, Lieb, and Luttinger (1974) obtained a significant extension of the Riesz–Sobolev inequality in which there can be more than three input functions. Here is the setup:

- $f_1,\dots,f_p$ are $\mathcal{L}^n$ measurable nonnegative functions on $\mathbb{R}^n$.
- $k \ge 2$ and $1 \le q \le p-1$ are integers.
- $A = (a_{j\ell}),\ 1 \le j \le p,\ 1 \le \ell \le q$, is a $p \times q$ matrix with real entries.
- $x = (x_1,\dots,x_q) \in \mathbb{R}^{nq}$, where each $x_j \in \mathbb{R}^n$.
- $L_j x = L_j(x) = \sum_{\ell=1}^{q} a_{j\ell}x_\ell, \quad 1 \le j \le p$.
- $J(f_1,\dots,f_p) \equiv \int_{\mathbb{R}^{nq}} \prod_{j=1}^{p} f_j(L_j x)\, dx_1 \dots dx_q$.

**Theorem 8.8** *With the setup just specified,*

$$J(f_1,\dots,f_p) \le J(f_1^{\#},\dots,f_p^{\#}), \tag{8.18}$$

*where # denotes symmetric decreasing rearrangement on $\mathbb{R}^n$ or $(k,n)$-Steiner symmetrization on $\mathbb{R}^n$, $1 \le k \le n-1$.*

For $p = 3$, $q = 2$ and $L_1 x = -x_1,\ L_2 x = x_2,\ L_3 x = x_1 - x_2$, we have

$$J(f_1,f_2,f_3) = \int_{\mathbb{R}^{2n}} f_1(-x_1)f_2(x_2)f_3(x_1 - x_2)\, dx_1\, dx_2,$$

and thus recover the Riesz–Sobolev inequality in $\mathbb{R}^n$.

Why the restriction $q < p$? To find out, let us look at the case $p = q = 2$, $n = 1$ and assume that the matrix $A$ is nonsingular. Then

$$\begin{aligned} J(f_1,f_2) &= \int_{\mathbb{R}^2} f_1(a_{11}x_1 + a_{12}x_2)f_2(a_{21}x_1 + a_{22}x_2)\,dx_1\,dx_2 \\ &= \det(A^{-1}) \int_{\mathbb{R}^2} f_1(y_1)f_2(y_2)\,dy_1\,dy_2 \\ &= \det(A^{-1}) \int_{\mathbb{R}} f_1\,dx_1 \int_{\mathbb{R}} f_2\,dx_2, \end{aligned}$$

where in the second equality we made an evident change of variable. For general $n$ and $p, q$ with $p = q$ the corresponding result is

$$J(f_1,\ldots,f_p) = (\det A)^{-n} \prod_{j=1}^{p} \int_{\mathbb{R}^n} f_j\,dx_j.$$

In any event, we see that when $p = q$ the value of $J$ does not change when all the $f_j$ are changed to $f_j^{\#}$. If $p < q$, then the same sort of argument shows that $J$ equals 0 or $\infty$, and if $J(f_1,\ldots,f_n) = \infty$ then $J(f_1^{\#},\ldots,f_p^{\#}) = \infty$.

*Proof of Theorem 8.8* Suppose we have proved (8.18) when $n = 1$, # is s.d.r. on $\mathbb{R}$ and each $f_j$ is the characteristic function of a set that is the union of finitely many closed disjoint intervals. Then, by the approximation argument in the proof of Theorem 8.1, (8.18) holds when the $f_j$ are characteristic functions of any measurable subsets of $\mathbb{R}$.

Next, for arbitrary nonnegative measurable functions $f_j$ on $\mathbb{R}$, the layer cake representation (Proposition 1.32) gives

$$J(f_1,\ldots,f_p) = \int_{(\mathbb{R}^+)^p} J(E_1(t_1),\ldots E_p(t_p))\,dt_1\ldots,dt_p,$$

where $E_j(t_j)$ is an abbreviation for the function $\chi_{(f_j>t_j)}$. Each term in the integrand increases under s.d.r., and so it follows that (8.18) holds for all nonnegative $f_j$ on $\mathbb{R}$. The extension of (8.18) from $n = 1$ to general $n$ is accomplished just like Theorem 8.4 was deduced from Theorem 8.3.

The proof of (8.18) when $n = 1$ and each $f_j = \chi_{E_j}$ with each $E_j$ the disjoint union of finitely many closed intervals is carried out in two steps.

**Step 1.** Let $I_1,\ldots,I_p$ be nondegenerate closed intervals in $\mathbb{R}$. Write $I_j = [b_j - \delta_j,\ b_j + \delta_j] = I_j^{\#} + b_j$, where # denotes s.d.r. on $\mathbb{R}$. For $t \in \mathbb{R}$, define

$$I_j(t) = I_j^{\#} + tb_j = [tb_j - \delta_j,\ tb_j + \delta_j].$$

Then $I_j(0) = I_j^{\#}$ and $I_j(1) = I_j$. All $I_j(t)$ have the same length. As $t$ increases, the intervals $I_j(t)$ move with speed equal to the absolute value of the midpoint of $I_j$.

Define a set $K \subset \mathbb{R}^{q+1}$ by

$$K = \{(x,t) : x \in \bigcap_{j=1}^{p} L_j^{-1}(I_j(t))\},$$

where the linear form $L_j$ defined by $L_j x = \sum_{\ell=1}^{q} a_{j\ell} x_\ell$. This is a convex set, since each strip $L_j^{-1}(I_j)$ is convex. Let

$$K(t) = \bigcap_{j=1}^{p} L_j^{-1}(I_j(t))$$

be the slice of $K$ through the hyperplane $\{x_{q+1} = t\}$, as discussed just before Corollary 8.7, and let $\psi(t) = \mathcal{L}^q(K(t))$. Then

$$\psi(t) = J(\chi_{I_1(t)}, \ldots, \chi_{I_p(t)}). \tag{8.19}$$

Since $I_j(-t) = -I_j(t)$ for each $j$ and $L_j(-x) = -L_j(x)$, it follows that

$$K(-t) = -K(t),$$

and hence $\psi$ satisfies

$$\psi(t) = \psi(-t), \quad t \in \mathbb{R}. \tag{8.20}$$

If $\psi$ is not identically zero, then $U = \{t \in \mathbb{R} : \psi(t) > 0\}$ is an open interval symmetric about 0. From Corollary 8.7, it follows that $\psi^{1/q}$ is concave on $U$. Along with the symmetry relation (8.20), the concavity implies that $\psi$ is *decreasing* in $|t|$. With (8.19), we deduce that

$$J(\chi_{I_1(t)}, \ldots, \chi_{I_p(t)}) \searrow \quad \text{as} \quad t \nearrow \quad \text{on } [0,1]. \tag{8.21}$$

**Step 2.** Suppose that

$$E_j = \bigcup_{k=1}^{N_j} I_{jk}, \quad 1 \leq j \leq p,$$

where the $I_{jk}$ are nondegenerate closed intervals which for each fixed $j$ are disjoint, and the $N_j$ are positive integers. We need to show that

$$J(\chi_{E_1}, \ldots, \chi_{E_p}) \leq J(\chi_{E_1^\#}, \ldots, \chi_{E_p^\#}). \tag{8.22}$$

Let $N = N(E_1, \ldots, E_p) = \prod_{j=1}^{p} N_j$. We shall prove (8.22) by induction on $N$. If $N = 1$ then each $E_j$ is a single interval, call it $I_j$. Let $I_j(t)$ be as in Step 1. Then $I_j(1) = E_j$, $I_j(0) = E_j^\#$, and (8.22) follows from (8.21).

Suppose (8.22) has been proved for all $p$-tuples $\{E_j\}$ such that the product $N(E_1, \ldots, E_p)$ is smaller than some $N \geq 4$. Let $\{E_j\}$ be a $p$-tuple with $N(E_1, \ldots, E_p) = N$. Substituting $\chi_{E_j} = \sum_{k=1}^{N_j} \chi_{I_{jk}}$ in the definition of $J$, we obtain

$$J(\chi_{E_1}, \ldots, \chi_{E_p}) = \sum J(\chi_{I_{1,k_1}}, \ldots, \chi_{I_{p,k_p}}), \tag{8.23}$$

where the sum is over all multi-indices $(k_1, \ldots, k_p)$ with $1 \le k_i \le N_j$, $j = 1, \ldots, p$.

For each of the intervals $I_{jk}$ let $I_{jk}(t)$ be formed from $I_{jk}$ as $I_j(t)$ was formed from $I_j$ in Step 1. Let $t_0$ be the largest value of $t$ such that $I_{jk}(t_0)$ and $I_{jm}$ share an endpoint for some $j, k, m$ with $k \neq m$. Then $0 < t_0 < 1$, because $I_{jk}(1) = I_{jk}$ and $I_{jm}(1) = I_{jm}$ are disjoint while $I_{jk}(0) = I_{jk}^\#$ and $I_{jm}(0) = I_{jm}^\#$ intersect at 0. Set $E_j(t) = \bigcup_{k=1}^{N_j} I_{jk}(t)$. Then $E_j(1) = E_j$. By (8.21), each term in the sum (8.23) increases when all the $I_{j,k_j}$ are replaced by $I_{j,k_j}(t_0)$. Thus,

$$J(\chi_{E_1}, \ldots, \chi_{E_p}) \le J(\chi_{E_1(t_0)}, \ldots, \chi_{E_p(t_0)}). \tag{8.24}$$

Since $N(E_1(t_0), \ldots, E_p(t_0)) < N$, the induction hypothesis implies that the right-hand side of (8.24) increases when we change the $E_j(t_0)$ to $E_j(t_0)^\#$. But the choice of $t_0$ insures that $E_j(t_0)^\# = E_j^\#$. This completes the proof of (8.22), and with it the proof of Theorem 8.8. □

## 8.6 Symmetrization Increases the Trace of the Heat Kernel

Let $f \in L^p(\mathbb{R}^n)$ for some $p \in [1, \infty]$. Define

$$u(x,t) = (2\pi t)^{-n/2} \int_{\mathbb{R}^n} f(y) e^{-\frac{|x-y|^2}{2t}}\, dy, \quad x \in \mathbb{R}^n,\ t \in (0, \infty). \tag{8.25}$$

Then $u$ satisfies the p.d.e.

$$u_t = \frac{1}{2} \Delta u \quad \text{in } \mathbb{R}^n \times (0, \infty), \tag{8.26}$$

where the subscript $t$ denotes differentiation and $\Delta$ is the Laplacian operator considered in Chapter 5. Moreover, $u$ satisfies

$$\lim_{t \to 0} u(x,t) = f(x), \quad x \in \mathbb{R}^n,$$

where the limit exists in $L^p$ for $p < \infty$, and also at points of continuity of $f$. We will abbreviate this statement to $u(x, 0) = f(x)$. For more information about the heat equation, we refer to Evans (1998).

Under appropriate physical conditions, $u(x,t)$ is the temperature at time $t$ at the point $x$ if at time zero the temperature is $f(x)$ at every point $x$. Accordingly, we say that $u$ solves the heat equation in (8.26) $\mathbb{R}^n \times (0, \infty)$ with initial values $f$ on $\mathbb{R}^n$. The function $K(x,y,t) = (2\pi t)^{-n/2} \exp(-|x-y|^2/(2t))$ is called the heat kernel on $\mathbb{R}^n$. A useful fact is the *semigroup property*

$$K(x, y, t+s) = \int_{\mathbb{R}^n} K(x,z,s)\, K(z,y,t)\, dz\,.$$

The factor $\frac{1}{2}$ in (8.26) is called the probabilists' one-half, the reason being that $\frac{1}{2}\Delta$ is the infinitesimal generator of standard Brownian motion in $\mathbb{R}^n$. Even for non-probabilists, an advantage to inserting the one-half is that formulas like (8.25) turn out to be more symmetric when the one-half is present.

Next, let $\Omega$ be a bounded open set in $\mathbb{R}^n$. As discussed in §5.1, the operator $-\frac{1}{2}\Delta$ with Dirichlet boundary conditions has eigenvalues $\lambda_1 < \lambda_2 \leq \lambda_3 \leq \dots$ with $\lambda_1 > 0$. There is a corresponding sequence of eigenfunctions $w_j$ defined by the property that

$$\frac{1}{2}\Delta w_j = -\lambda_j w_j \quad \text{in } \Omega, \qquad w_j = 0 \quad \text{on } \partial\Omega$$

for each $j \geq 1$. The functions $w_j$ can be normalized to form an orthonormal basis for $L^2(\Omega)$. For each $f \in L^2(\Omega)$ we have $f = \sum_{j=1}^{\infty}(f, w_j)w_j$, where the parentheses denote inner product in $L^2(\Omega)$ and the series converges in $L^2(\Omega)$.

Set

$$u(x,t) = \sum_{j=1}^{\infty} e^{-\lambda_j t}(f, w_j)w_j(x). \tag{8.27}$$

Then $u \in C^{\infty}(\Omega \times (0,\infty))$, and $u$ satisfies

$$u_t = \frac{1}{2}\Delta u \quad \text{in } \Omega \times (0,\infty).$$

If $\Omega$ has sufficiently smooth boundary, then $u$ also satisfies the boundary and initial conditions

$$u(x,t) = 0 \text{ on } \partial\Omega \times (0,\infty), \qquad u(x,0) = f(x) \text{ for } x \in \Omega.$$

If $\Omega$ is a region in space whose boundary is kept at zero temperature, and the initial temperature is given by $f(x)$, then the temperature at time $t$ and point $x$ will be given by $u(x,t)$. We say that $u$ solves the heat equation in $\Omega \times (0,\infty)$ with Dirichlet boundary conditions and initial values $f$.

The operator that carries $f$ to $u(\cdot,t)$ is the operator exponential $e^{\frac{1}{2}t\Delta}$ in the sense of spectral theory, where $\Delta$ here is the Laplacian with Dirichlet boundary conditions on $\Omega$. The family $\{e^{\frac{1}{2}t\Delta} \mid t \geq 0\}$ is a 1-parameter contraction semigroup under composition whose infinitesimal generator is $\frac{1}{2}\Delta$. For background on spectral theory and operator semigroups, see Evans (1998), Davies (1989), or Reed and Simon (1972)–(1975).

Define the *Dirichlet heat kernel* of $\Omega$ by

$$K(x,y,t,\Omega) = \sum_{j=1}^{\infty} e^{-\lambda_j t}w_j(x)w_j(y), \quad (x,y,t) \in \Omega \times \Omega \times (0,\infty). \tag{8.28}$$

Then

$$u(x,t) = \int_{\Omega} K(x,y,t,\Omega)f(y)\,dy, \quad (x,t) \in \Omega \times (0,\infty).$$

If we take $f$ to be $\delta_y$, a unit mass at $y$, then the corresponding solution $u(x,t)$ is $K(x,y,t,\Omega)$. Thus, we can view $K(x,y,t,\Omega)$ as the temperature at time $t$ and point $x$, if the boundary temperature is kept at zero and initially there is no heat except for a heat source of unit strength at $y$.

The Dirichlet heat kernel is symmetric in the spatial variables, that is, $K(x,y,t,\Omega) = K(y,x,t,\Omega)$, and has the semigroup property that

$$K(x,y,t+s,\Omega) = \int_{\Omega} K(x,z,s,\Omega)\,K(z,y,t,\Omega)\,dz\,.$$

There is also a Neumann heat kernel, but we will rarely consider it, and therefore will sometimes omit the adjective "Dirichlet."

The *trace of the heat kernel* of $\Omega$ is defined as the sum of the eigenvalues of the operator $e^{\frac{1}{2}t\Delta}$,

$$\mathrm{Tr}(t,\Omega) = \sum_{j=1}^{\infty} e^{-t\lambda_j}, \quad 0 < t < \infty. \tag{8.29}$$

By (8.28) and the orthonormality of the eigenfunctions,

$$\mathrm{Tr}(t,\Omega) = \int_{\Omega} K(x,x,t,\Omega)\,dx, \quad 0 < t < \infty.$$

The following theorem is due to Luttinger (1973a; 1973b).

**Theorem 8.9** *If $\Omega \subset \mathbb{R}^n$ is bounded and has sufficiently smooth boundary, then*

$$\mathrm{Tr}(t,\Omega) \leq \mathrm{Tr}(t,\Omega^{\#}), \quad 0 < t < \infty, \tag{8.30}$$

*where # may be either s.d.r. on $\mathbb{R}^n$ or $(k,n)$-Steiner symmetrization, $1 \leq k \leq n$.*

Letting $t \to \infty$, we see from (8.29) and (8.30) that

$$\lambda_1(\Omega) \geq \lambda_1(\Omega^{\#}).$$

If $\partial\Omega$ is bounded and sufficiently smooth, there is an asymptotic expansion (McKean and Singer, 1967; Davies, 1989)

$$(2\pi t)^{n/2}\,\mathrm{Tr}(t,\Omega) = \mathcal{L}^n(\Omega) - \frac{1}{4}(2\pi t)^{1/2}\mathcal{H}^{n-1}(\partial\Omega) + O(t), \quad t \to 0.$$

Thus, letting $t \to 0$, we see from (8.30) that

$$\mathcal{H}^{n-1}(\partial\Omega) \geq \mathcal{H}^{n-1}(\partial\Omega^{\#}).$$

Thus, Luttinger's Theorem implies both the Faber–Krahn Theorem and the isoperimetric inequality, for sufficiently regular $\Omega$.

For the proof of the theorem, we will write $\mathrm{Tr}(t, \Omega)$ as a limit of multiple integrals which satisfy the assumptions of the Brascamp–Lieb–Luttinger inequality in Theorem 8.8.

*Proof of Theorem 8.9* Fix $t > 0$. For a positive integer $q \geq 2$, define

$$K_q(x, y, t, \Omega) = \left(\frac{q}{2\pi t}\right)^{\frac{nq}{2}} \int_{\Omega^{q-1}} \exp\left(-\frac{q}{2t}\sum_{j=1}^{q} |x_{j-1} - x_j|^2\right) dx_1 \ldots dx_{q-1},$$

where $x_0 = x$ and $x_q = y$. By extending the domain of integration to $(\mathbb{R}^n)^{q-1}$ and using the semigroup property of the heat kernel on $\mathbb{R}^n$, we see that the functions $K_q$ are uniformly bounded on $\Omega \times \Omega$ by

$$0 \leq K_q(x, y, t, \Omega) \leq (2\pi t)^{-n/2} e^{-\frac{|x-y|^2}{2t}} \leq (2\pi t)^{-n/2}.$$

By the same reasoning,

$$K_{2q}(x, y, t, \Omega) \leq K_q(x, y, t, \Omega)$$

for all $x, y \in \Omega \times \Omega$. It follows that the subsequence $K_{2^j}$ is monotonically decreasing in $j$.

The Trotter product formula for semigroups, a version of which is stated below, will yield the following representation for the trace of the heat kernel:

$$\mathrm{Tr}(t, \Omega) = \lim_{q \to \infty} \int_{\Omega} K_q(x, x, t, \Omega)\, dx. \tag{8.31}$$

The integrals on the right-hand side of (8.31) satisfy the hypotheses of Theorem 8.8, with $p = 2q$, $f_j = \chi_\Omega$ for $1 \leq j \leq q$, $f_j(z) = e^{-|z|^2/2}$ for $q + 1 \leq j \leq 2q$. The $f_j$ with $j > q$ are symmetric decreasing on $\mathbb{R}^n$. By Theorem 8.8,

$$\int_{\Omega} K_q(x, x, t, \Omega)\, dx \leq \int_{\Omega^\#} K_q(x, x, t, \Omega^\#)\, dx,$$

and (8.30) follows by taking $q \to \infty$. To complete the proof of the theorem, it remains to establish (8.31) along the monotone subsequence indexed by $q = 2^j$.

Let $\{P^t\}$ and $\{P^t_\Omega\}$ denote the heat semigroups on $\mathbb{R}^n$ and $\Omega$ respectively. For $t > 0$ let $Q^t$ be the multiplication operator on $L^2(\mathbb{R}^n)$ defined by $Q^t f = \chi_\Omega f$, and let $Q^0 f = f$. Since $\chi_\Omega^2 = \chi_\Omega$, this defines a semigroup $\{Q^t\}$, which is obviously contractive. By Kato's version (1978) of Trotter's formula, the $q$-fold composition $(Q^{t/q} P^{t/q})^q$ converges in the strong $L^2$-sense to a contraction semigroup as $q \to \infty$. This means that $(\chi_\Omega P^{t/q})^q f$ converges in $L^2$ for each $f \in L^2$. Using Herbst and Zhao (1988) or Kac (1949), one shows that

$$P^t_\Omega f = \lim_{q\to\infty} (\chi_\Omega P^{t/q})^q f$$

in $L^2$ for each $f \in L^2$. (See also Simon (1979).) Since

$$(\chi_\Omega P^{t/q})^q f(x) = \int_\Omega K_q(x, y, t, \Omega) f(y)\, dy,$$

we have proved that

$$u(x, t) = \lim_{q\to\infty} \int_\Omega K_q(x, y, t, \Omega) f(y)\, dy \tag{8.32}$$

in $L^2(\Omega)$. This uniquely determines the function $K(x, y, t, \Omega)$ for every fixed $t$ almost everywhere on $\Omega \times \Omega$. In particular, by monotonicity,

$$K(x, y, t, \Omega) = \lim_{j\to\infty} K_{2^j}(x, y, t, \Omega)$$

for almost every $x, y \in \Omega$.

For every fixed $t$, the operators $P^t_\Omega$, and $(\chi_\Omega P^{t/q})^q \chi_\Omega$ are real self-adjoint Hilbert–Schmidt operators. Using the semigroup property once more, we see that the trace of the heat kernel is precisely the Hilbert–Schmidt norm of $P^{t/2}_\Omega$,

$$\mathrm{Tr}(t, \Omega) = \int_{\Omega\times\Omega} \big(K(x, y, t/2, \Omega)\big)^2\, dxdy.$$

By monotone convergence,

$$\int_{\Omega\times\Omega} \big(K(x, y, t/2, \Omega)\big)^2\, dxdy = \lim_{j\to\infty} \int_{\Omega\times\Omega} (K_{2^j}(x, y, t/2, \Omega)^2\, dxdy\,.$$

Since

$$\int_{\Omega\times\Omega} \big(K_q(x, y, t/2, \Omega)\big)^2\, dxdy = \int_\Omega K_{2q}(x, x, t, \Omega)\, dx$$

by the definition of $K_q$, it follows that

$$\mathrm{Tr}(t, \Omega) = \lim_{j\to\infty} K_{2^j}(x, x, t, \Omega)\, dx.$$

This completes the proof of Theorem 8.9. □

Burchard and Schmuckenschläger (2001) and independently Morpurgo (2002) proved that Luttinger's trace inequality (8.30) also holds for sufficiently regular domains on spheres $\mathbb{S}^n$ and hyperbolic spaces $\mathbb{H}^n$. The main ingredient of these proofs is a partial extension of the Brascamp–Lieb–Luttinger inequality to integrals of the form

$$\int_{X^q} \prod_{i=1}^{q} f_i(x_i) \prod_{1 \le i < j \le q} k_{ij}(d(x_i, x_j))\, dx_1 \ldots dx_q,$$

where $X$ denotes $\mathbb{R}^n$, $\mathbb{S}^n$ or $\mathbb{H}^n$, the $k_{ij}$ are nonnegative decreasing functions on $\mathbb{R}^+$, and $d(x, y)$, $dx_i$ are the canonical distance and canonical measure on $X$. The key observation is that such integrals increase under polarization. In particular, Theorem 8.9 is valid also when # is polarization.

Morpurgo's integral inequality in fact applies to considerably more general integrands, and leads to trace inequalities for a broad class of operators. Both papers work with a definition of trace in terms of Wiener measure which permit extension of the results to arbitrary domains and even to the case when $\Omega$ is an arbitrary Borel set. Burchard and Schmuckenschläger use an argument of Ledoux (1994) to give a formula for the perimeter of $\Omega$ in terms of *exit distributions*

$$u_\Omega(x, t) = \int_\Omega K(x, y, t, \Omega)\, dy = P_x(\tau_\Omega > t).$$

As in §5.7, $\tau_\Omega$ is the first time a Brownian motion in $\mathbb{R}^n$ leaves $\Omega$, and $P_x$ denotes probability starting from $x$. The second equation is the special case $f = \chi_\Omega$ of (8.35) below. They prove a comparison theorem for $u_\Omega$, which leads to a version of the isoperimetric inequality in $X$.

The solution (8.32) of the heat equation in $\Omega \subset \mathbb{R}^n$ via Trotter's formula has a probabilistic interpretation. We briefly discuss this interpretation in the more general context of Schrödinger equations. Let $V$ be a continuous nonnegative function on $\mathbb{R}^n$. The Schrödinger equation on $\mathbb{R}^n$ with potential $V$ is

$$u_t = \frac{1}{2}\Delta u - Vu, \quad (x, t) \in \mathbb{R}^n \times (0, \infty), \tag{8.33}$$

where $V$ acts on $u$ by pointwise multiplication.

Suppose an initial function $f$ is given. Let $\{B_t\}$ denote standard Brownian motion in $\mathbb{R}^n$, as in §5.7. Then, under appropriate assumptions (see Karatzas and Shreve, 1991, p. 270), we have:

**Theorem 8.10** (Feynman–Kac formula) *The function*

$$u(x, t) = E_x\Big(f(B_t) \exp(-\int_0^t V(B_s)\, ds)\Big) \tag{8.34}$$

*satisfies the p.d.e.* (8.33) *and the initial condition* $u(x, 0) = f(x)$, $x \in \mathbb{R}^n$.

In particular, taking $V \equiv 0$, we see that the function (8.25) has probabilistic representation

$$u(x,t) = E_x(f(B_t)).$$

The choice $V = 0$ on $\Omega$, $V = \infty$ on $\mathbb{R}^n \setminus \Omega$, results in $e^{-V} = \chi_\Omega$, which suggests that (8.27) might have representation

$$u(x,t) = E_x(f(B_t), \tau > t), \quad (x,t) \in \Omega \times (0,\infty), \tag{8.35}$$

where $\tau = \tau_\Omega$ is the first exit time from $\Omega$. Under appropriate assumptions on $\Omega$ and $f$, (8.35) indeed does hold.

For general $V$ in (8.34), take a positive integer $q$, fix $0 < t < \infty$, and approximate the integral by the Riemann sum $\frac{t}{q}\sum_{j=1}^{q} V(B_{tj/q})$. Set $X_j = B_{tj/q}$. Then by (8.34),

$$u(x,t) \approx E_x\left( f(X_q)\prod_{j=1}^{q} e^{-\frac{t}{q}V(X_j)}\right). \tag{8.36}$$

For Brownian motion started at $x \in \mathbb{R}^n$, the joint probability density of $X_j = B_{tj/q}, j = 1, \ldots, q$ is the Gaussian

$$\left(\frac{q}{2\pi t}\right)^{nq/2} \exp\left( -\frac{q}{2t}\sum_{j=1}^{q} |x_{j-1} - x_j|^2\right),$$

where $x_0 = x$. Taking $g_j(x_j) = e^{-\frac{t}{q}V(x_j)}$ for $1 \le j < q$ and $g_q(x_q) = f(x_q)e^{-\frac{t}{q}V(x_q)}$, we obtain

$$\begin{aligned} &E_x\left( f(X_q)\prod_{j=1}^{q} e^{-\frac{t}{q}V(X_j)}\right) \\ &= \left(\frac{q}{2\pi t}\right)^{nq/2} \int_{\mathbb{R}^{nq}} f(x_q)e^{-t/q\sum_{j=1}^{q} V(x_j)} e^{-\frac{q}{2t}\sum_{j=1}^{q}|x_{j-1}-x_j|^2}\, dx_1 \ldots dx_q. \end{aligned}$$

If we are lucky, then it will follow from (8.36) that the solution of the Schrödinger equation is

$$\begin{aligned} u(x,t) = {} & \lim_{q\to\infty} \left(\frac{q}{2\pi t}\right)^{nq/2} \\ & \times \int_{\mathbb{R}^{nq}} f(x_q)e^{-t/q\sum_{j=1}^{q} V(x_j)} e^{-\frac{q}{2t}\sum_{j=1}^{q}|x_{j-1}-x_j|^2}\, dx_1, \ldots dx_q. \end{aligned} \tag{8.37}$$

But this is precisely the Trotter product $(P^{t/q}Q^{t/q})^q f$, where $P^t$ is the heat semigroup and $Q^t$ is the contraction semigroup which acts by pointwise multiplication with $e^{-tV(x)}$.

Returning to the heat equation on $\Omega$, take a sequence of continuous nonnegative potentials $V_k$ in $\mathbb{R}^n$ which are zero on $\Omega$ and converge monotonically to $\infty$ in $\mathbb{R}^n \setminus \Omega$. As $k \to \infty$, we expect that the functions $u(x,t)$ constructed in (8.37), should converge to (8.35), which represents the solution of the heat equation on $\Omega$ from (8.32). Proving this requires to interchange the limit $q \to \infty$ with

$$\lim_{k\to\infty} e^{-t/q\sum_{j=1}^{q} V(x_j)} = \prod_{j=1}^{q} \chi_\Omega(x_j)\,.$$

Thus, the Feynman–Kac formula furnishes another plausibility argument for the Trotter formula solution to the heat equation with Dirichlet boundary condition.

## 8.7 The Sharp Hardy–Littlewood–Sobolev Inequality

Let $p > 1, q > 1$ and $0 < \lambda < n$ be related by

$$\frac{1}{p} + \frac{1}{q} + \frac{\lambda}{n} = 2\,.$$

The Hardy–Littlewood–Sobolev inequality states the existence of a constant $C$ depending on $p$, $q$, and $\lambda$ such that

$$\int_{\mathbb{R}^{2n}} f(x)g(y)|x-y|^{-\lambda}\,dx\,dy \le C\|f\|_p\|g\|_q \tag{8.38}$$

for all $f \in L^p(\mathbb{R}^n, \mathcal{L}^n)$ and $g \in L^q(\mathbb{R}^n, \mathcal{L}^n)$. In this section, we prove a theorem of Lieb that determines the sharp constants in (8.38) for certain special values of the parameters.

For the rest of the section, we make the convention that all functions are nonnegative and $\mathcal{L}^n$ or $\sigma_n$ measurable, and write $\|\cdot\|_p$ for the norm in $L^p(\mathbb{R}^n, \mathcal{L}^n)$ or $L^p(\mathbb{S}^n, \sigma_n)$. Also, # shall denote symmetric decreasing rearrangement on $\mathbb{R}^n$.

Write

$$K_\lambda(x) = |x|^{-\lambda}, \quad x \in \mathbb{R}^n.$$

The left side of (8.38) equals $\check{f} * g * K_\lambda(0)$, where $\check{f}(x) = f(-x)$. Note that $f$ and $\check{f}$ have the same distribution, so that $\|\check{f}\|_p = \|f\|_p$ for every $p > 0$. The problem of finding the best $C$ in (8.38) is the same as that of maximizing $f * g * K_\lambda(0)$ over all nonnegative $f$ and $g$ with $\|f\|_p = \|g\|_q = 1$. By duality, still another way to state (8.38) is that

$$\|f * K_\lambda\|_{q'} \le C\|f\|_p$$

for all $f \in L^p$, where $\frac{1}{q} + \frac{1}{q'} = 1$, or that the linear map $f \to K_\lambda * f$, still denoted by $K_\lambda$, maps $L^p$ to $L^{q'}$ with operator norm $\|K_\lambda\|_{p,q'} \le C$. The function $f * K_\lambda$ is often called, for example in Stein (1970), the Riesz potential of $f$ of order $n - \lambda$.

We first explore the relation of (8.38) to another classical inequality for convolutions. If $p, q, r$ satisfy $p, q, r \ge 1$ and $\frac{1}{p} + \frac{1}{q} + \frac{1}{r} = 2$ then according to Young's inequality (Lieb and Loss, 1997, p. 90)

$$\int_{\mathbb{R}^{2n}} f(x)g(y)h(x-y)\,dx\,dy \le C\|f\|_p\|g\|_q\|h\|_r, \tag{8.39}$$

where $C$ is a constant depending on $p, q, r$, and $n$. Equivalently, the convolution, as a function from $L^p \times L^r$ to $L^{q'}$ satisfies the bound

$$\|f * h\|_{q'} \le C\|f\|_p\|h\|_r\,.$$

Since $K_\lambda$ just fails to belong to $L^{n/\lambda}$, the Hardy–Littlewood–Sobolev inequality does not follow from (8.39).

Conversely, let $p, q, r > 1$ and suppose that $h$ does not necessarily lie in $L^r$ but satisfies the weaker assumption that

$$\sup_{t>0}\left\{t^{-1/r}\mathcal{L}^n(h > t)\right\} < \infty.$$

One says in that case that $h$ belongs to the space *weak* $L^r$. Note for $0 < \lambda < 1$, the kernel $K_\lambda$ belongs to weak $L^{n/\lambda}$ by definition. Since

$$h^\# \le \sup_{t>0}\left\{t^{-1/r}\mathcal{L}^n(h > t)\}\, K_{n/r}\right\},$$

the Riesz–Sobolev inequality implies that

$$\|f * h\|_{q'} \le \sup_{t>0}\left\{t^{-1/r}\mathcal{L}^n(h > t)\right\}\, \|f^\# * K_{n/r}\|_{q'}\,.$$

The weak $L^r$-spaces may be endowed with norms $\|\cdot\|_{r,w}$ that are bounded above and below by constant multiples of $\sup_t t^{-1/r}\mathcal{L}^n(h > t)$. It follows from (8.38) that

$$\|f * h\|_{q'} \le C \sup_{t>0}\left\{t^{-1/r}\mathcal{L}^n(h > t)\right\}\, \|f^\#\|_p \le \tilde{C}\,\|h\|_{r,w}\,\|f\|_p\,.$$

Thus (8.39) holds with $\|h\|_{r,w}$ in place of $\|h\|_r$. See Lieb and Loss (1997, p. 98).

For most nonnegative functions $K$, it is not known how to maximize expressions of the form $f * g * K(0)$ when $f$ and $g$ vary over various $L^p$ and $L^q$ spaces, or even if extremals exist. But for (8.38), when $K = K_\lambda$, Lieb (1983) proved that nontrivial extremal functions do exist for all allowable values of $p, q$ and $\lambda$. That is, if we take the supremum of the left side of (8.38) over $f$

and $g$ with $\|f\|_p = 1$ and $\|g\|_q = 1$, then admissible functions $f$ and $g$ exist for which the supremum is achieved. For two special parameter configurations Lieb found explicit extremal functions and the best constants $C(\lambda, p, n)$. One such situation is when

$$p = q \in (1,2), \quad \lambda = 2n/p'$$

where $\frac{1}{p} + \frac{1}{p'} = 1$. The second situation is when

$$p \in (1,2), \quad q = 2, \quad \lambda = n\left(\frac{1}{2} + \frac{1}{p'}\right). \tag{8.40}$$

In both cases, $\lambda$ lies strictly between $n/2$ and $n$. We shall see at the end of this section that the second situation can be treated as a by-product of the first.

To attack the extremal problem when $p = q$, $\lambda = 2n/p'$, let $\mathcal{H}(f,g)$ be the left side of (8.38). The Fourier transform of $|x|^{-\lambda}$ is $B|x|^{\lambda-n}$, where $B = B(\lambda, n)$ is a positive constant (Stein, 1970, §V.1). Therefore $\mathcal{H}$ defines a positive definite quadratic form on $L^p$. By Parseval's formula and Schwarz's inequality, for nonnegative $f$ and $g$ we have

$$\begin{aligned} 0 \le \mathcal{H}(f,g) = B \int_{\mathbb{R}^n} \hat{f}(x)\overline{\hat{g}(x)}|x|^{\lambda-n}\,dx &\le \mathcal{H}(f,f)^{1/2}\mathcal{H}(g,g)^{1/2} \\ &\le \max\{\mathcal{H}(f,f), \mathcal{H}(g,g)\}. \end{aligned}$$

Write

$$\mathcal{H}(f) = \mathcal{H}(f,f) = \int_{\mathbb{R}^{2n}} f(x)f(y)|x-y|^{-\lambda}\,dx\,dy.$$

Then $\sup\{\mathcal{H}(f,g)\colon \|f\|_p = \|g\|_p = 1\} = \sup\{\mathcal{H}(f)\colon \|f\|_p = 1\}$, and the maximizers for $\mathcal{H}(f,g)$ over $\|f\|_p = \|g\|_p = 1$ are given by $(f,f)$ where $f$ maximizes $\mathcal{H}(f)$ over $\|f\|_p = 1$.

For $1 < p < 2$, define $h$ on $\mathbb{R}^n$ by

$$h(x) = C_{p,n}(1 + |x|^2)^{-n/p}, \tag{8.41}$$

where $C_{p,n}$ is chosen so that $\|h\|_p = 1$. It will emerge below that $C_{p,n} = 2^{n/p}\beta_n^{-1/p}$, where $\beta_n$ is the surface measure of the unit sphere in $\mathbb{R}^{n+1}$.

We can now state a version of Lieb's Theorem. Another version will be stated near the end of this section.

**Theorem 8.11** (Lieb's sharp HLS inequality) *For $1 < p < 2$, $\lambda = 2n/p'$, we have*

$$\sup\{\mathcal{H}(f)\colon f \in L^p(\mathbb{R}^n),\ \|f\|_p = 1\} = \mathcal{H}(h). \tag{8.42}$$

*Moreover, if* $\|f\|_p = 1$ *then* $\mathcal{H}(f) = \mathcal{H}(h)$ *if and only if*

$$f(x) = \delta^{-n/p}\, h\left(\frac{x-a}{\delta}\right)$$

*for some* $a \in \mathbb{R}^n$ *and* $\delta > 0$.

Thus, the maximum of (8.38) over $\|f\|_p = \|g\|_p = 1$ is achieved when $f = g = h$, and the only extremals have the form $h \circ S$ where $S$ is a composition of translations and dilations. The best constant $C$ for (8.38) when $p = q$ and $\lambda = \frac{2n}{p'}$ is (Lieb, 1983, p. 359; Lieb and Loss, 1997, p. 98)

$$C(\lambda, n) \equiv \mathcal{H}(h) = \pi^{\lambda/2} \frac{\Gamma(\frac{n-\lambda}{2})}{\Gamma(n - \frac{\lambda}{2})} \left\{ \frac{\Gamma(\frac{n}{2})}{\Gamma(n)} \right\}^{-1+\frac{\lambda}{n}}.$$

Another way to state the theorem is to say that the operator norm of the convolution operator with kernel $K_\lambda$ satisfies $\|K_\lambda\|_{p,p'} = C(\lambda, n)$.

By the Riesz–Sobolev inequality, $\mathcal{H}(f) \leq \mathcal{H}(f^\#)$. Thus, if a maximizer of $\mathcal{H}(f)$ over $\|f\|_p = 1$ exists then a symmetric decreasing maximizer also exists. Lieb's wonderful discovery was that there are additional operations that leave the HLS integral (8.38) unchanged. Under these operations, the set of functions obtained from $h$ by composition with translations and dilations is stable. The function $h$ turns out to be the unique fixed point of one of the operations followed by s.d.r. This insight suggests that Theorem 8.11 might be true, and gives clues about how to prove it.

The new operations arise from the fact that the HLS integral turns out to be *conformally invariant*. Let $T\colon \overline{\mathbb{R}^n} \to \mathbb{S}^n$ be the stereographic projection. Thus, if $x = (x_1, \ldots, x_n) \in \mathbb{R}^n$, $T$ is defined by $T(x) = s$, where $s = (s_1, \ldots, s_{n+1})$ is given by

$$s_i = \frac{2x_i}{1 + |x|^2}, \qquad 1 \leq i \leq n, \qquad s_{n+1} = \frac{1 - |x|^2}{1 + |x|^2}.$$

This is augmented by $T(\infty) = e_{n+1}$. The inverse map $T^{-1}(s) = x$ is given by

$$x_i = \frac{s_i}{1 + s_{n+1}}, \qquad 1 \leq i \leq n.$$

In §7.2 we saw how to transform integrals on $\mathbb{S}^n$ to integrals on a subset of $\mathbb{R}^n$ by passing to spherical coordinates. The map $T$ furnishes another way to transform integrals from $\mathbb{S}^n$ to $\mathbb{R}^n$. With $g_{ij}$ the metric coefficients as in §7.2, we have

$$g_{ij} = \frac{\partial T}{\partial x_i} \cdot \frac{\partial T}{\partial x_j} = \left(\frac{2}{1 + |x|^2}\right)^2 \delta_{ij},$$

and hence, writing $dx$ for $d\mathcal{L}^n$,

$$d\sigma_n(s) = J_T(x)\,dx, \quad dx = J_{T^{-1}}(s)\,d\sigma_n(s),$$

where

$$J_T(x) \equiv \left(\frac{2}{1+|x|^2}\right)^n, \quad J_{T^{-1}}(s) = (1+s_{n+1})^{-n}.$$

In particular,

$$\int_{\mathbb{R}^n} J_T(x)\,dx = \int_{\mathbb{S}^n} 1\,d\sigma_n = \beta_n\,.$$

Since the matrix $g_{ij}(x)$ is a multiple of the identity at every point $x \in \mathbb{R}^n$, the map $T$ preserves angles, hence is called a conformal map.

Define a linear operator $U = U_p$ which transforms functions $f$ on $\mathbb{R}^n$ to functions $Uf$ on $\mathbb{S}^n$ by

$$Uf(s) = J_{T^{-1}}^{1/p}(s) f(T^{-1}s) = (1+s_{n+1})^{-p/n} f(T^{-1}s).$$

Then $U$ defines an isometry from $L^p(\mathbb{R}^n, \mathcal{L}^n)$ to $L^p(\mathbb{S}^n, \sigma_n)$. Next observe that, with $h$ as in (8.41),

$$h(x) = \beta_n^{-1/p} J_T^{1/p}(x)\,,$$

and $J_{T^{-1}}(s)J_T(x) = 1$ when $s = Tx$. It follows that for $g$ on $\mathbb{R}^n$, the corresponding function $Ug$ on the sphere is constant if and only if $g$ is a constant multiple of $h$.

Here is the key ingredient in the proof of Theorem 8.11.

**Proposition 8.12** (Conformal invariance of the HLS integral) *For* $1 < p < 2$, $\lambda = 2n/p'$, $f, g \in L^p(\mathbb{R}^n)$, *and* $F = Uf$, $G = Ug$, *we have*

$$\int_{\mathbb{R}^n \times \mathbb{R}^n} f(x)g(y)|x-y|^{-\lambda}\,dxdy = \int_{\mathbb{S}^n \times \mathbb{S}^n} F(s)G(t)|s-t|^{-\lambda}\,d\sigma_n(s)\,d\sigma_n(t). \tag{8.43}$$

On the right, $|s-t|$ is the chordal distance from $s$ to $t$, that is, the Euclidean distance in $\mathbb{R}^{n+1}$. The identity can be verified by changing variables.

There is another sense in which the HLS integral is conformally invariant. Let $\gamma \in G(\mathbb{R}^n)$, the group of all Möbius transformations of $\overline{\mathbb{R}^n}$ onto $\overline{\mathbb{R}^n}$ introduced in §7.6, also known as the conformal group. Define an action $\gamma^*$ of $\gamma$ on functions $f$ by

$$\gamma^* f(x) = |J_{\gamma^{-1}}(x)|^{1/p} f(\gamma^{-1}x), \tag{8.44}$$

where $J$ is the Jacobian. Then $\|\gamma^* f\|_p = \|f\|_p$, and we have (Lieb and Loss, 1997, p. 106) the following companion to (8.43):

$$\int_{\mathbb{R}^n \times \mathbb{R}^n} f(x)g(y)|x-y|^{-\lambda}\, dxdy = \int_{\mathbb{R}^n \times \mathbb{R}^n} (\gamma^* f(x)(\gamma^* g(y))|x-y|^{-\lambda}\, dxdy.$$

The Jacobian of any Möbius transformation $\gamma \in G(\mathbb{R}^n)$ has the form

$$J_\gamma(x) = (2a)^n (a^2 + |x-v|^2)^{-n},$$

where $a > 0$ and $v \in \mathbb{R}^n$ are constants. The Jacobians of conformal transformations on $\mathbb{S}^n$ have the form

$$J_\gamma(s) = C_\zeta (1 - \zeta \cdot s)^{-n}, \tag{8.45}$$

where $\zeta \in \mathbb{B}^{n+1}$ and $C_\zeta = 2^n(1-|\zeta|^2)^{n/2}$.

Next, assume that $n \geq 2$. Let $O \in O(n+1)$ be the 90° rotation of $\mathbb{S}^n$ which carries $e_{n+1}$ to $e_n$, $e_n$ to $-e_{n+1}$, and fixes all the $e_i$ for $1 \leq i \leq n-1$. Define the operator $V$ which carries functions on $\mathbb{S}^n$ to functions on $\mathbb{S}^n$ by

$$VF(s) = F(O^{-1}s).$$

Let $W$ denote any of the maps $U$, $V$, $U^{-1}$, or $V^{-1}$. Then the linear map $W$ enjoys the following properties, which are easily checked:

**Fact 8.13** (Isometric property) $\|Wf\|_p = \|f\|_p$.

**Fact 8.14** (Order preserving property) *If $f \leq g$ a.e., then $Wf \leq Wg$ a.e.*

We can now prove Theorem 8.11 when $n \geq 2$. For $n = 1$, the proof is an exercise in Lieb and Loss (1997, chapter 4, Exercise 7).

*Proof of Theorem 8.11 when $n \geq 2$* Since $\mathcal{H}(f) \leq \mathcal{H}(f^\#)$ by the Riesz–Sobolev inequality, we may restrict the competition in (8.42) to symmetric decreasing $f$. Also, it will suffice to prove that $\mathcal{H}(f) \leq \mathcal{H}(h)$ for bounded $f$ with compact support.

Fix $f \in L^p(\mathbb{R}^n)$ with $f$ bounded, compactly supported, symmetric decreasing and with $\|f\|_p = 1$. Let $C$ be a constant such that $f \leq Ch$ a.e. on $\mathbb{R}^n$, and define

$$\mathcal{S}(f) = \{g \in L^p(\mathbb{R}^n) \colon \|g\|_p = 1,\ 0 \leq g \leq Ch \text{ a.e. on } \mathbb{R}^n,\\ g = g^\#,\ \mathcal{H}(g) \geq \mathcal{H}(f)\}.$$

Set

$$d(f,g) = \|f-g\|_p \quad \text{and} \quad d_0 = \inf\{d(g,h) \colon g \in \mathcal{S}(f)\}.$$

We want to show that $d_0 = 0$.

Take a sequence $\{g_k\}$ in $\mathcal{S}(f)$ such that $\lim_{k\to\infty} d(g_k, h) = d(g, h)$. Since the $g_k$ are symmetric decreasing, Lemma 1.44 provides us with a subsequence, also denoted $\{g_k\}$, and a symmetric decreasing function $g_0$ on $\mathbb{R}^n$ such that $\lim_{k\to\infty} g_k = g_0$ a.e. Since $g_k \le Ch$ and $h \in L^p(\mathbb{R}^n)$, the dominated convergence theorem shows that $g_k \to g_0$ in $L^p$. Moreover, $g_k(x)g_k(y)|x-y|^{-\lambda} \le C^2 h(x)h(y)|x-y|^{-\lambda}$, and, by the weak Young inequality (Lieb and Loss, 1997, Section 4.3, Eq. (7)), the function on the right is in $L^1(\mathbb{R}^{2n})$. The dominated convergence theorem shows that $\lim_{k\to\infty} \mathcal{H}(g_k) = \mathcal{H}(g_0)$. It follows that $g_0 \in \mathcal{S}(f)$, and that

$$d(g_0, h) = d_0.$$

Define

$$g_1 = U^{-1}VUg_0 \quad \text{and} \quad g_2 = g_1^{\#}.$$

The isometric and order preserving properties of $U, V$ and $U^{-1}$ and the conformal invariance of $\mathcal{H}$ imply that $d(g_1, h) = d(g_0, h) = d_0$ and that $g_2 \in \mathcal{S}(f)$.

Suppose that $g_1$ is not symmetric decreasing. Let

$$A = \{(x, y) \in \mathbb{R}^{2n} \colon g_1(x) < g_1(y),\ h(x) > h(y)\}.$$

Since $h$ is strictly decreasing on rays from the origin, the set $A$ equals the set $A_1$ in Lemma 1.38, with $g_1$ in place of $f$. Lemma 1.38 gives $\mathcal{L}^{2n}(A) > 0$. For $1 < p < \infty$, the function $\Psi$ on $\mathbb{R}^+ \times \mathbb{R}^+$ defined by $\Psi(x, y) = x^p + y^p - |x-y|^p$ belongs to the class $SAL_0$ defined in Chapter 2. See Fact 2.6 in §2.1. From Theorem 2.15(d) we deduce that $d(g_2, h) < d(g_1, h)$. But $d(g_1, h) = d_0$ and $g_2 \in \mathcal{S}(f)$. This contradicts the definition of $d_0$.

We conclude that $g_1 = U^{-1}VUg_0$ must be symmetric decreasing on $\mathbb{R}^n$. Since $T$ maps balls centered at the origin in $\mathbb{R}^n$ to spherical caps centered at $e_{n+1}$, it follows that $Ug_1$ must be symmetric with respect to the $e_{n+1}$-axis. Thus, there exists a function $\phi$ such that $Ug_1(s) = \phi(s_{n+1})$. On the other hand, $Ug_1 = VUg_0$ and $Ug_0$ is $e_{n+1}$-symmetric, so $VUg_0$ is $e_n$-symmetric, and thus there exists a function $\psi$ such that $VUg_0(s) = \psi(s_n)$. For $VUg_0$ to have both symmetries, it must be constant. Then $Ug_0$ is constant, which implies $g_0 = h$. Since $g_0 \in \mathcal{S}(f)$, we have $h \in \mathcal{S}(f)$, so that $\mathcal{H}(h) \ge \mathcal{H}(f)$. This proves that $d_0 = 0$ and $h$ is a maximizer of $\mathcal{H}$.

The "if" part of the uniqueness statement is clear. To prove the "only if" part, suppose that $\|f\|_p = 1$ and $\mathcal{H}(f) = \mathcal{H}(h)$. Then $\mathcal{H}(f) = \mathcal{H}(f^{\#})$, and it follows from Theorem 2.15(b), applied with $K(t) = t^{-\lambda}$, that $f(x) = f^{\#}(x+a)$ for some $a \in \mathbb{R}^n$. So to complete the description of extremal functions, we just have to verify that if $f$ is symmetric decreasing with $\|f\|_p = 1$ and $\mathcal{H}(f) = \mathcal{H}(h)$, then $f(x) = \delta^{-n/p} h(\delta^{-1}x)$ for some $\delta > 0$ .

Let $f$ be symmetric decreasing, with $\|f\|_p = 1$ and $\mathcal{H}(f) = \mathcal{H}(h)$. Consider the distance

$$d_0 = \inf_{\gamma \in G(\mathbb{R}^n)} d(\gamma^* f, h) = \inf_{\gamma \in G(\mathbb{R}^n)} d(f, \gamma^* h).$$

For every $\gamma \in G(\mathbb{R}^n)$, the function $\gamma^* h$ can be written as $\tau^*(\delta^* h)$, where $\tau$ is a translation and $\delta$ a dilation (see Lieb and Loss, 1997, Lemma 4.8). Since both $f$ and $\delta^* h$ are symmetric decreasing, their distance cannot be reduced by translation. Therefore, it suffices to minimize over dilations. Since $h$ is strictly positive, $d(f, h) < 2$. On the other hand, $d(f, \delta^* h)$ approaches 2 when the dilation factor $\delta$ approaches zero or $\infty$. Therefore the minimum is assumed, and $f = \delta^* h$ for some dilation $\delta$. Replacing $f$ with $(\delta^{-1})^* f$, we may assume that the infimum is assumed when $\delta$ is the identity, that is, $d_0 = d(f, h)$.

Let $g = U^{-1} V U f$ and $f_1 = g^*$. Since $f_1$ is a maximizer, we have $\mathcal{H}(g^*) = \mathcal{H}(f_1) = \mathcal{H}(g)$, and we deduce from Theorem 2.15(b) that $g$ is a translate of $f_1$. If $g$ is itself symmetric decreasing, then it follows as in the first part of the proof that $f = g = h$, and we are done. Otherwise, we have $d(f_1, h) < d(g, h)$ by Theorem 2.15(d). Since

$$d(g, h) = d(U^{-1} V U f, U^{-1} V U h) = d(f, h) = d_0,$$

this contradicts the definition of $d_0$. Thus, $g$ must be symmetric decreasing, and the proof is complete. □

For ease of reference, we give here another statement of Lieb's HLS inequality. It is assumed that $f$ and $g$ are nonnegative functions on $\mathbb{R}^n$, that $F = Uf$, $G = Ug$ where $U$ is the operator introduced before (8.43), that $1 < p < 2$, and that $\lambda = 2n/p'$. The $L^p$ norms for $f$ and $g$ are taken with respect to $\mathcal{L}^n$, and for $F$ and $G$ with respect to $\sigma_n$.

**Theorem 8.15**

$$\int_{\mathbb{R}^n \times \mathbb{R}^n} f(x) g(y) |x - y|^{-\lambda} \, dx \, dy \le C(\lambda, n) \|f\|_p \|g\|_p, \tag{8.46}$$

$$\int_{\mathbb{S}^n \times \mathbb{S}^n} F(s) G(t) |s - t|^{-\lambda} \, d\sigma_n(s) \, d\sigma_n(t) \le C(\lambda, n) \|F\|_p \|G\|_p, \tag{8.47}$$

*where*

$$C(\lambda, n) = \pi^{\lambda/2} \frac{\Gamma(\frac{n-\lambda}{2})}{\Gamma(n - \frac{\lambda}{2})} \left\{ \frac{\Gamma(\frac{n}{2})}{\Gamma(n)} \right\}^{-1+\frac{\lambda}{n}}. \tag{8.48}$$

Equality is achieved in (8.46) if and only if $f(x)$ and $g(x)$ are constant multiples of a function $(a^2+|x-v|^2)^{-n/p}$, where $a > 0$ and $v \in \mathbb{R}^n$. Equality in (8.47) is achieved if and only if $F$ and $G$ are constant multiples of $(1-\zeta\cdot s)^{-n/p}$ for some $\zeta \in \mathbb{B}^{n+1}$.

To conclude this section, let us see what happens when the parameters in the HLS inequality are given by (8.40):

$$1 < p < 2, \quad q = 2, \quad \lambda = n\left(\frac{1}{2}+\frac{1}{p'}\right) = n\left(\frac{3}{2}-\frac{1}{p}\right). \tag{8.49}$$

Then $\frac{n}{2} < \lambda < n$. Write again

$$\mathcal{H}(f,g) = \int_{\mathbb{R}^n\times\mathbb{R}^n} f(x)g(y)|x-y|^{-\lambda}\,dxdy \tag{8.50}$$

where now $f \in L^p$, $g \in L^2$. Then for fixed $f$,

$$\sup\{\mathcal{H}(f,g)\colon \|g\|_2 = 1\} = \|K_\lambda * f\|_2,$$

where $K_\lambda(x) = |x|^{-\lambda}$, so

$$\sup\{\mathcal{H}(f,g)\colon \|f\|_p = 1,\ \|g\|_2 = 1\} = \sup\{\|f * K_\lambda\|_2\colon \|f\|_p = 1\}.$$

Since $K_\lambda$ is even, we have that

$$\|f * K_\lambda\|_2^2 = \int_{\mathbb{R}^n} (f * K_\lambda * K_\lambda) f\,dx. \tag{8.51}$$

But $K_\lambda * K_\lambda = CK_{2\lambda-n}$ for a positive constant $C$ (see Stein, 1970, pp. 73–117; Lieb, 1983, p. 360). We conclude that maximizers $f$ for $\mathcal{H}(f,g)$ over $\|f\|_p = \|g\|_2 = 1$ are maximizers for the right-hand side of (8.51). By Theorem 8.11, the function $h$ in (8.41) is a maximizer. To achieve $\int_{\mathbb{R}^n}(h*K_\lambda)\,g\,dx = \|h*K_\lambda\|_2$ we should take $g = Ah * K_\lambda$ for a constant $A$. We have proved:

**Corollary 8.16** *With the situation of* (8.49), *and* $\mathcal{H}$ *given by* (8.50),

$$\sup\{\mathcal{H}(f,g)\colon \|f\|_p = 1,\ |g\|_2 = 1\} = \mathcal{H}(h, Ah * |x|^{-\lambda}), \tag{8.52}$$

*where* $h(x) = C_{p,n}(1+|x|^2)^{-n/p}$. *The constant* $C_{p,n}$ *is given in* (8.41) *and* $A$ *is chosen so that* $A\|h * K_\lambda\|_2 = 1$.

A formula for the best constant, that is, the right side of (8.52), appears as (3.4) on p. 359 of Lieb (1983). This best constant can also be described as the operator norm $\|K_\lambda\|_{p,2}$. On p. 368, Lieb notes that when $\lambda = n-1$ and $n \geq 3$ this result and duality produce the best constant in the Sobolev inequality $\|f\|_{2^*} \leq C\|\nabla f\|_2$ (see Theorem 4.23), where $2^* = \frac{2n}{n-2}$.

Apart from the parameter configurations treated in Theorem 8.11 and Corollary 8.16, the best constants for the HLS inequality (8.38) are not known.

## 8.8 Logarithmic Sobolev Inequalities

Beckner (1991, 1992, 1993) discovered that interesting consequences ensue when we let $\lambda$ approach zero or $n$ in Lieb's inequality (8.47). We first investigate what happens when $p \to 2$, that is, $\lambda \to n$. In this discussion, limits and derivatives at $p = 2$ are understood to be one-sided and taken from the left.

Spherical harmonics play a key role in Beckner's results. We will strive to convey the basic facts about spherical harmonics as they become relevant, and will sometimes make statements unsupported by proof or documentation. For more information see, for example, Stein and Weiss (1971) and Andrews et al. (1999, chapter 9).

Spherical harmonics of degree $k$ on $\mathbb{S}^n$ are restrictions to $\mathbb{S}^n$ of harmonic functions $u$ on $\mathbb{R}^{n+1}$, which are homogeneous polynomials of degree $k$. Thus, $u$ has the form

$$u(rx) = r^k Y(x), \quad 0 < r < \infty, \quad x \in \mathbb{S}^n,$$

and satisfies $\Delta u = 0$ in $\mathbb{R}^{n+1}$. The function $Y$ is then a spherical harmonic of degree $k$.

Let $\mathcal{H}^{k,n}$ denote the vector space of all spherical harmonics on $\mathbb{S}^n$ of degree $k$. In §7.4, we saw that in polar coordinates on $\mathbb{R}^{n+1}$, the Laplacian is given by

$$\Delta = \partial_{rr} + nr^{-1}\partial_r + r^{-2}\Delta_s,$$

where $\Delta_s$ is the Laplace operator on $\mathbb{S}^n$. If $Y \in \mathcal{H}^{k,n}$ it follows from $\Delta u = 0$ that

$$\Delta_s Y = -k(k+n-1)Y, \quad k \geq 0,$$

so that $\lambda_k \equiv k(k+n-1)$ is an eigenvalue for $\Delta_s$ and $Y$ is an eigenfunction for $\lambda_k$. It can be shown that $\{\lambda_k \colon k \geq 0\}$ comprises all eigenvalues of $\Delta_s$, and that each eigenfunction of $\lambda_k$ belongs to $\mathcal{H}^{k,n}$. Note that $\lambda_0 = 0$ and that the eigenspace $\mathcal{H}^{0,n}$ is the 1-dimensional space of constant functions. For $n = 1$ and $k \geq 1$ all $\mathcal{H}^{k,2}$ have dimension 2. The functions $\sin k\theta$ and $\cos k\theta$ furnish a basis for $\mathcal{H}^{k,2}$, as do $e^{ik\theta}$, $e^{-ik\theta}$ when we permit the functions to be complex valued. In general,

$$\dim \mathcal{H}^{k,n} = \binom{n+k}{k} - \binom{n+k-2}{k-2}, \quad k \geq 2,$$

while $\dim \mathcal{H}^{1,n} = n+1$ and $\dim \mathcal{H}^{0,n} = 1$.

Up to now, we have used the standard surface measure $\sigma_n$ on $\mathbb{S}^n$, for which $\sigma_n(\mathbb{S}^n) = \beta_n$. For the remaining part of the chapter, it will be more convenient to use the *normalized surface measure*

$$\nu_n = \frac{1}{\beta_n}\sigma_n.$$

Most of the time we will drop the subscript $n$. The measure $\nu$ is also called the *uniform probability measure* on $\mathbb{S}^n$. Often, we write $L^2 = L^2(\mathbb{S}^n, \nu)$. Unless otherwise specified, functions in $L^2$ are taken to be real-valued.

The operator $\Delta_s$ is self-adjoint on $L^2$. Since eigenspaces of $\Delta_s$ corresponding to different eigenvalues are orthogonal, we have $\mathcal{H}^{k,n} \perp \mathcal{H}^{m,n}$ when $m \neq k$. Finite linear combinations of spherical harmonics are dense in $L^2$. Thus, there exist orthonormal bases of $L^2$ which contain exactly $\dim \mathcal{H}^{k,n}$ functions from each $\mathcal{H}^{k,n}$. For $F \in L^2$, the *spherical harmonic expansion* of $F$ is

$$F = \sum_{k=0}^{\infty} Y_k,$$

where $Y_k$ is the orthogonal projection of $F$ onto the space $\mathcal{H}^{k,n}$, and the series converges in $L^2$. In particular, $Y_0$ is the constant function with value $\int_{\mathbb{S}^n} F\, d\nu$.

We are now ready to state Beckner's result.

**Theorem 8.17** (Beckner's logarithmic Sobolev inequality) *Let $F\colon \mathbb{S}^n \to \mathbb{R}^+$ be bounded and have spherical harmonic expansion $\sum_{k=0}^{\infty} Y_k$. Then*

$$\int_{\mathbb{S}^n} F^2 \log F\, d\nu \leq \|F\|_2^2 \log \|F\|_2 + \sum_{k=1}^{\infty} a_k(n) \int_{\mathbb{S}^n} Y_k^2\, d\nu, \tag{8.53}$$

*where*

$$a_k(n) = \sum_{j=0}^{k-1} \frac{n}{n+2j}. \tag{8.54}$$

For the proof, we need some more facts about spherical harmonics.

**Proposition 8.18** (Funk–Hecke formula) *Let $K\colon [-1, 1] \to \mathbb{R}$ be a measurable function such that $K(x \cdot e_1)$ is in $L^1(\mathbb{S}^n, \nu)$. There exist numbers $b_k$ depending on $K$ and $n$ such that*

$$\int_{\mathbb{S}^n} K(x \cdot y) Y_k(y)\, d\nu(y) = b_k Y_k(x), \quad Y_k \in \mathcal{H}^{k,n}, \quad k \geq 0.$$

See Andrews et al. (1999, Theorem 9.6.3). Once existence of the multipliers $b_k$ is known, they can be computed by evaluating the formula on a convenient choice of $Y_k$.

For $k \geq 0$, $\alpha > 0$ and $t \in [-1, 1]$, define functions $C_k^{\alpha}(t)$ by the generating relation

$$(1 - 2rt + r^2)^{-\alpha} = \sum_{k=0}^{\infty} C_k^{\alpha}(t) r^k, \quad 0 \leq r < 1.$$

When $\alpha = 0$, this relation may be replaced by

$$-\log(1 - 2rt + r^2) = 2\sum_{k=1}^{\infty} k^{-1} T_k(t) r^k, \quad 0 \le r < 1,$$

where the $T_k$ are the Chebyshev polynomials of the first kind. See Andrews et al. (1999, pp. 302–303). We will write $C_k^0(t) = T_k(t)$ for $k \ge 1$ and $C_0^0(t) = T_0(t) = 1$. The $C_k^\alpha$ turn out to be polynomials of degree $k$. We will call $C_k^\alpha$ the Gegenbauer polynomial of degree $k$ and type $\alpha$.

For fixed $n$, let $C_k = C_k^{(n-1)/2}$ denote the Gegenbauer polynomials of type $(n-1)/2$ on $\mathbb{S}^n$. We note that when $n = 2$, $C_k(t)$ is the $k$th Legendre polynomial. By Andrews et al. (1999, Theorem 9.6.3), the function

$$Z_k(x) = C_k(x \cdot e_1)$$

belongs to $\mathcal{H}^{k,n}$. It is called the *zonal harmonic* of degree $k$ with pole $e_1$. Evaluating the Funk–Hecke formula on the zonal harmonics with $x = 1$ yields the integral representation

$$b_k = C_k(1)^{-1} \frac{\beta_{n-1}}{\beta_n} \int_0^\pi K(\cos\theta) C_k(\cos\theta)\, (\sin\theta)^{n-1}\, d\theta\,, \tag{8.55}$$

see Andrews et al. (1999, Eq. (9.7.4)). Keep in mind that our calculations are for the uniform measure $\nu$ on $\mathbb{S}^n$, rather than the standard surface measure $\sigma_{n-1}$ on $\mathbb{S}^{n-1}$.

**Lemma 8.19** *Let $F \in L^2(\mathbb{S}^n)$ with spherical harmonic expansion $F = \sum_{k=0}^\infty Y_k$, and let $0 < \lambda < n$. Then*

$$\int_{\mathbb{S}^n \times \mathbb{S}^n} F(x)F(y)|x - y|^{-\lambda}\, d\nu(x) d\nu(y) = \sum_{k=0}^\infty b_k \int_{\mathbb{S}^n} Y_k^2(x)\, d\nu(x), \tag{8.56}$$

*where*

$$b_k = 2^{-\lambda} \frac{\Gamma(n)\Gamma(\frac{n-\lambda}{2})\Gamma(\frac{\lambda}{2} + k)}{\Gamma(\frac{n}{2})\Gamma(\frac{\lambda}{2})\Gamma(n + k - \frac{\lambda}{2})}, \quad k \ge 0. \tag{8.57}$$

*Proof* For $0 < \lambda < n$, let $K(t) = (2 - 2t)^{-\lambda/2}$. $K(x \cdot y) = |x - y|^{-\lambda}$ for $x, y \in \mathbb{S}^n$. By the Funk–Hecke formula, the identity

$$\int_{\mathbb{S}^n} |x - y|^{-\lambda} F(y)\, d\nu(y) = \sum_{k=0}^\infty b_k(\lambda, n) Y_k$$

holds with

$$b_k = C_k(1)^{-1} \frac{\beta_{n-1}}{\beta_n} \int_0^\pi (2 - 2\cos\theta)^{-\lambda/2} C_k(\cos\theta)(\sin\theta)^{n-1}\, d\theta,$$

see (8.55), and (8.56) follows by taking the inner product with $F$. To evaluate the $b_k$, we first insert the constants $C_k(1) = (n+k-2)!\,/(n-2)!$, and then change variables $t = -\cos\theta$ in the integral. Since $C_k(t)$ is even in $t$ when $k$ is even, and odd when $k$ is odd, we can apply formula (3) on p. 280 of Erdélyi et al. (1954). Collecting terms, we obtain the claim. □

*Proof of Theorem 8.17* We start from Lieb's inequality (8.47) on $\mathbb{S}^n$, with parameters $1 < p < 2$ and $\lambda = 2n/p'$. Taking into account the change of measure from $\sigma_n$ to $\nu$, the inequality becomes

$$C(\lambda,n)^{-1}\int_{\mathbb{S}^n\times\mathbb{S}^n} F(x)G(y)|x-y|^{-\lambda}\,d\nu(x)\,d\nu(y) \le \|F\|_p\|G\|_p, \tag{8.58}$$

where the $L^p$ norms are with respect to $\nu$, and the sharp constant has the value

$$C(\lambda,n) = 2^{-\lambda}\frac{\Gamma(n)\Gamma(\frac{n-\lambda}{2})}{\Gamma(\frac{n}{2})\Gamma(n-\frac{\lambda}{2})}\,.$$

In computing the constant, we have used the duplication formula

$$\Gamma(z)\Gamma(z+\frac{1}{2}) = 2^{1-2z}\pi^{\frac{1}{2}}\Gamma(2z)$$

with $z = n/2$ to write the last term in (8.48) as

$$\frac{\Gamma(\frac{n}{2})}{\Gamma(n)} = \frac{\beta_n}{2^n\pi^{\frac{n}{2}}}\,.$$

Take $F = G$ and use Lemma 8.19 to expand the left side of (8.58) in spherical harmonics,

$$\sum_{k=0}^{\infty} c_k(\lambda)\int_{\mathbb{S}^n} Y_k^2(x)\,d\nu(x) \le \|F\|_p^2\,. \tag{8.59}$$

Suppressing the dependence on $n$ in the notation, we compute for the constants $c_k(\lambda) = b_k(\lambda)/C(\lambda,n)$, where $b_k(\lambda)$ is given by (8.57), that is

$$c_k(\lambda) = \frac{\Gamma(\frac{\lambda}{2}+k)\Gamma(n-\frac{\lambda}{2})}{\Gamma(n+k-\frac{\lambda}{2})\Gamma(\frac{\lambda}{2})} = \prod_{j=0}^{k-1}\frac{\frac{1}{2}\lambda+j}{n-\frac{1}{2}\lambda+j} \tag{8.60}$$

for $k \ge 1$, and $c_0(\lambda) = 1$.

Let $A(p)$ be the function on the left side of (8.59), and let $B(p)$ be the function on the right. Since $\lambda \to n$ as $p \to 2$ and $c_k(n) = 1$ for each $k \ge 0$, we have that $A(2) = B(2) = 0$. Therefore we can differentiate the inequality at $p = 2$ to obtain

$$B'(2) \le A'(2). \tag{8.61}$$

We easily compute $B'(2) = \int_{\mathbb{S}^n} F^2 \log F\, d\nu - \|F\|_2^2 \log \|F\|_2$. By logarithmic differentiation,

$$\frac{dc_k}{d\lambda}(n)\frac{d\lambda}{dp}(2) = n\sum_{j=0}^{k-1} \frac{1}{n+2j} = a_k,$$

where $a_k = a_k(n)$ is given by (8.54). Thus

$$A'(2) = \sum_{k=0}^{\infty} a_k(n) \int_{sn} Y_k^2\, d\nu,$$

and by (8.61) the theorem is proved. □

Clearly, constant functions are extremals for Theorem 8.17. Furthermore, the inequality inherits the conformal invariance of the Hardy–Littlewood–Sobolev inequality from Proposition 8.12, that is, the transformations on the sphere corresponding to (8.44) with $p = 2$ leave both sides of (8.60) invariant. Therefore, the extremals include all functions of the form $F(x) = cJ_\gamma(x)^{1/2}$, where $J_\gamma$ is the Jacobian of a conformal transformation $\gamma$ on $\mathbb{S}^n$, and $c$ is a constant. As shown by Carlen and Loss (1992), these are the only extremals.

Let now $f$ be a bounded nonnegative function on $\mathbb{R}$. For $N \geq 1$, define $F_N$ on the sphere $\{x \in \mathbb{R}^{N+1} : |x| = \sqrt{N}\}$ by $F_N(x_1, \ldots, x_N) = f(x_1)$. Apply the logarithmic Sobolev inequality (8.53) to the function $F_N(\sqrt{N}x)$. By the Mehler–Poincaré formula in §7.8, as $N \to \infty$, the left-hand integral approaches $\int_{\mathbb{R}} f^2 \log f d\gamma$ and the first term on the right approaches $\|f\|_2^2 \log \|f\|_2$, where $\gamma$ is the 1-dimensional Gauss measure and the $L^2$ norms are taken with respect to $\gamma$. As explained by Beckner (1992, (6)), the terms $\int_{\mathbb{S}^n} Y_k^2$ on the right approach the squares of the coefficients in the Hermite expansion of $f$, and the $a_{k,N}$ approach $k$. This implies that the last term in (8.53) approaches $\int_{\mathbb{R}} |f'|^2 d\gamma$. Thus, we have

$$\int_{\mathbb{R}} f^2 \log f\, d\gamma \leq \|f\|_2^2 \log\|f\|_2 + \int_{\mathbb{R}} |f'|^2\, d\gamma\,. \tag{8.62}$$

This is the *logarithmic Sobolev inequality* of Gross (1975). The argument above shows that Gross's logarithmic Sobolev inequality is the infinite-dimensional limit of Beckner's logarithmic Sobolev inequality.

Using (8.62) along with the product structure of the $n$-dimensional Gauss measure $\gamma_n$, Gross (1975, p. 1074) also established the $n$-dimensional logarithmic Sobolev inequality:

$$\int_{\mathbb{R}^n} f^2 \log f\, d\gamma_n \leq \|f\|_2^2 \log\|f\|_2 + \int_{\mathbb{R}^n} |\nabla f|^2\, d\gamma_n, \tag{8.63}$$

for nonnegative bounded functions on $\mathbb{R}^n$, where the $L^2$ norms are taken with respect to $\gamma_n$.

## 8.9 Hypercontractivity

Let $H$ be the Ornstein–Uhlenbeck operator $\frac{1}{2}\Delta - x\cdot\nabla$ in $\mathbb{R}^n$, and let $\{e^{tH}\}$ denote the semigroup generated by $H$. Then for $1 \le p \le \infty$ and each $t > 0$, $e^{-tH}$ is a contraction on $L^p(\mathbb{R}^n, \gamma_n)$, that is, has operator norm $\le 1$. Since $e^{tH}$ takes constants onto themselves, the norm is exactly 1.

If $1 \le p < q \le \infty$ then for all small enough $t$, $e^{tH}$ has norm $> 1$ from $L^p$ to $L^q$. Motivated by considerations from quantum field theory, Nelson (1973) discovered a remarkable fact: If

$$e^{-2t} \le \frac{p-1}{q-1},$$

then $e^{tH}$ *is* a contraction from $L^p(\mathbb{R}^n, \gamma_n)$ into $L^q(\mathbb{R}^n, \gamma_n)$. Moreover, this result is sharp, since Nelson showed that if $e^{-2t} > \frac{p-1}{q-1}$, then $e^{tH}$ is no longer a contraction.

The phenomenon is now called hypercontractivity, the term having been coined in a paper by Simon and Høegh-Krohn (1972). Nelson's proof was complicated and probabilistic. Gross showed that Nelson's inequality follows readily from Gross's logarithmic Sobolev inequality. In fact, Gross's inequality and Nelson's inequality are equivalent. See, for example Gross (1975) or Ledoux (1994, p. 498).

Let us return now to the sphere $\mathbb{S}^n$ with its uniform probability measure $\nu$. Again following Beckner, we shall investigate an analogue of Nelson's theorem on $\mathbb{S}^n$. Take $F \in L^1(\mathbb{S}^n, \nu)$, with spherical harmonic decomposition $\sum_{k=0}^{\infty} Y_k$. Let $\mathbb{B}^{n+1}$ be the open unit ball in $\mathbb{R}^{n+1}$. Define $u$ on the closed ball by

$$u(rx) = \sum_{k=0}^{\infty} r^k Y_k(x), \quad r \in [0,1], \quad x \in \mathbb{S}^n.$$

Since each $r^k Y_k(x)$ is harmonic on $\mathbb{B}^{n+1}$, so is $u$. As $r \to 1$, $u(r\cdot) \to F$ in appropriate senses. Thus, $u$ solves the Dirichlet problem for $F$ in $\mathbb{B}^{n+1}$. There is also a Poisson integral representation

$$u(rx) = \int_{\mathbb{S}^n} F(y) \frac{1-r^2}{|rx-y|^{n+1}}\, d\nu(y), \quad r \in [0,1), \quad x \in \mathbb{S}^n.$$

Let $P_t$ denote the operator which carries $F$ to $u(e^{-t}\cdot)$. If $s, t > 0$ then $P_{s+t} = P_t \circ P_s$. Thus, the family $\{P_t\}$, $t \ge 0$ is a semigroup under composition, called the *Poisson semigroup* on $\mathbb{S}^n$. For $p \ge 1$, $P_t$ is a contraction from $L^p(\mathbb{S}^n, \nu)$ onto itself. That is, $\|P_t\|_{p,p} = 1$, where we use $\|T\|_{p,q}$ to denote the operator norm of a linear mapping $T$ from $L^p$ to $L^q$. The contractivity of $P_t$ on $L^p$ follows from subharmonicity of $|u|^p$, or from the Poisson integral representation. By considering the action of $P_t$ on constants, we see that $\|P_t\|_{p,q} \ge 1$ for every $p, q$.

Suppose now that $1 \leq p < q < \infty$. Then $\|P_t\|_{p,q} > 1$ if $t$ is close to 0, so that $P_t$ is in general not a contraction from $L^p(\mathbb{S}^n, \nu)$ to $L^q(\mathbb{S}^n, \nu)$. There is, however, a perfect analogue of Nelson's inequality:

**Theorem 8.20** (Hypercontractivity of the Poisson semigroup) *For* $1 \leq p < q \leq \infty$,

$$\|P_t\|_{p,q} = 1 \quad \textit{if and only if} \quad e^{-2t} \leq \frac{p-1}{q-1}.$$

*Proof* ($\Rightarrow$) For $\epsilon > 0$, consider the function $F(x) = 1 + \epsilon x \cdot e_1$ and its harmonic extension $u(rx) = 1 + \epsilon r x \cdot e_1$. By definition, $P_t F(x) = 1 + \epsilon e^{-t} x \cdot e_1$. Using a Taylor expansion to second order in $\epsilon$, and noting that all terms of odd order integrate to zero, we see that

$$\|F\|_p = 1 + \epsilon^2 \frac{p-1}{2} \int_{\mathbb{S}^n} |x \cdot e_1|^2 \, d\nu(x) + O(\epsilon)^4,$$

while

$$\|P_t F\|_q = 1 + \epsilon^2 e^{-2t} \frac{q-1}{2} \int_{\mathbb{S}^n} |x \cdot e_1|^2 \, d\nu(x) + O(\epsilon)^4.$$

We see that $P_t$ cannot be a contraction from $L^p$ to $L^q$ unless $e^{-2t}(q-1) \leq p-1$.

($\Leftarrow$) The deduction of hypercontractivity on the sphere from Beckner's logarithmic Sobolev inequality is nearly the same as the derivation of Nelson's inequality from Gross's inequality. The sphere case of the argument goes as follows. First, the semigroup property and contractivity of $P_t$ from $L^q$ into itself show that it suffices to prove $\|P_t\|_{p,q} = 1$ when $e^{-2t}(q-1) = p-1$. Since $|P_t F| \leq P_t |F|$, we need only consider nonnegative $F$ on $\mathbb{S}^n$. We may further assume that $F$ is bounded. Fix a nonnegative bounded $F$. Define functions $q(t,p)$ and $A(t,p) = A(t,p,F)$ by

$$q(t,p) = 1 + e^{2t}(p-1), \quad A(t,p) = \|P_t F\|_{q(t,p)}, \quad t > 0, \quad 1 \leq p < \infty.$$

We need to prove that $A(t,p) \leq A(0,p)$, for all $t$ and $p$. Take $p \in (1,\infty)$ and $t_0 > 0$. Set $p_0 = q(t_0,p)$ and $F_0 = P_{t_0} F$. By the semigroup property, we have for $t \geq t_0$ that $P_t F = P_{t-t_0} F_0$. Similarly, since $q(t,p)$ satisfies the differential equation $\partial_t q = 2(q-1)$ with initial condition $q(0) = p$, we have that $q(t,p) = q(t-t_0, p_0)$. It follows that

$$\frac{d}{dt} \|P_t F\|_{q(t,p_0)} \Big|_{t=t_0} = \frac{d}{dt} \|P_t(P_{t_0} F)\|_{q(t,p_0)} \Big|_{t=0},$$

that is, $A'(t_0, p_0) = A'(0, q(t_0, p_0))$. Therefore it suffices to prove that $A'(0,p) \leq 0$ for all $p \geq 1$.

To compute $A'(0,p)$, write

$$B(t,p) = \int_{\mathbb{S}^n} \exp\left\{q(t,p)\log u(e^{-t}x)\right\} d\nu(x),$$

where $u$ is the harmonic extension of $F$. Then $\log A(t,p) = \frac{1}{q(t,p)} \log B(t,p)$. Calculation gives $q(0,p) = p$, $q'(0,p) = 2(p-1)$, and

$$B'(0,p) = 2(p-1)\int_{\mathbb{S}^n} F^p \log F\, d\nu - p\int_{\mathbb{S}^n} F^{p-1} u_r\, d\nu,$$

where $u_r(x)$ is the radial derivative of $u(rx)$ at $r = 1$. More calculation gives

$$A'(0,p) = \frac{2(p-1)}{p}\left(\int_{\mathbb{S}^n} F^p \log F\, d\nu - \|F\|_p^p \log\|F\|_p\right) - \int_{\mathbb{S}^n} F^{p-1} u_r d\nu.$$

Thus, the inequality $A'(0,p) \le 0$ is equivalent to the inequality

$$\int_{\mathbb{S}^n} F^p \log F\, d\nu \le \|F\|_p^p \log\|F\|_p + \frac{p}{2(p-1)}\int_{\mathbb{S}^n} F^{p-1} u_r d\nu. \tag{8.64}$$

For $p = 2$, the integral in the last term on the right-hand side has spherical harmonic expansion

$$\int_{\mathbb{S}^n} F u_r\, d\nu = \sum_{k=1}^{\infty} \int_{\mathbb{S}^n} k Y_k^2\, d\nu.$$

From (8.54), we see that each $a_k(n) \le k$, and conclude from inequality (8.53) that (8.64) holds. Also, Green's formula and the identity $\Delta(u^2) = 2|\nabla u|^2$ for harmonic functions gives

$$\int_{\mathbb{S}^n} F u_r\, d\nu = \frac{1}{2}\int_{\mathbb{S}^n} \frac{\partial}{\partial r}(u^2)\, d\nu = \frac{1}{\beta_n}\int_{\mathbb{B}^{n+1}} |\nabla u|^2\, dx.$$

Thus, we can write (8.64) for $p = 2$ as

$$\int_{\mathbb{S}^n} F^2 \log F\, d\nu \le \|F\|_2^2 \log\|F\|_2 + \frac{1}{\beta_n}\int_{\mathbb{B}^{n+1}} |\nabla u|^2\, dx. \tag{8.65}$$

For general $p \in (1,\infty)$, set $G = F^{p/2}$, and let $v$ be the harmonic extension of $G$ to $\mathbb{B}^{n+1}$. By (8.65) applied to $G$,

$$\int_{\mathbb{S}^n} G^2 \log G\, d\nu \le \|G\|_2^2 \log\|G\|_2 + \frac{1}{\beta_n}\int_{\mathbb{B}^{n+1}} |\nabla v|^2\, dx,$$

which implies

$$\int_{\mathbb{S}^n} F^p \log F\, d\nu \le \|F\|_p^p \log\|F\|_p + \frac{2}{p\beta_n}\int_{\mathbb{B}^{n+1}} |\nabla v|^2\, dx. \tag{8.66}$$

Since $\nu$ is harmonic, Dirichlet's principle yields that

$$\int_{\mathbb{B}^{n+1}} |\nabla v|^2\,dx \le \int_{\mathbb{B}^{n+1}} |\nabla(u^{p/2})|^2\,dx = \frac{p^2}{4}\int_{\mathbb{B}^{n+1}} u^{p-2}|\nabla u|^2\,dx.$$

The harmonic function $u$ satisfies $\Delta u^p = p(p-1)u^{p-2}|\nabla u|^2$, so we obtain for the last term in (8.66) the bound

$$\begin{aligned}\frac{2}{p\beta_n}\int_{\mathbb{B}^{n+1}} |\nabla v|^2\,dx &\le \frac{1}{2(p-1)\beta_n}\int_{\mathbb{B}^{n+1}} \Delta u^p\,dx\\ &= \frac{p}{2(p-1)}\int_{\mathbb{S}^n} F^{p-1}u_r\,d\nu.\end{aligned}$$

This shows that (8.64) holds for $p$. The proof of Theorem 8.20 is complete. □

The proof of hypercontractivity just given did not require the full use of Beckner's inequality (8.53) but only of its consequence (8.65). That inequality becomes sharp in the limit $n \to \infty$, since then $a_k(n)$ approaches $k$. Conversely, hypercontractivity implies that $A(t,2) \le A(0,2)$ for every $t$, so that $A'(0,2) \le 0$, which is equivalent to (8.65).

## 8.10 Sharp Inequalities for Exponential Integrals

We saw in Section 8.9 that by letting $p \to 2$, that is, $\lambda \to n$, in Lieb's HLS inequality (8.56) on $\mathbb{S}^n$, one obtains Beckner's logarithmic Sobolev inequality (8.53). Beckner (1991, 1993) also studied the limit $p \to 1$, that is, $\lambda \to 0$. Here is the result. We remind the reader that for fixed $n \ge 1$, $\nu$ is the uniform probability measure on $\mathbb{S}^n$.

**Theorem 8.21** (Beckner) *For real-valued $F \in L^1(\mathbb{S}^n,\nu)$ with spherical harmonic expansion $\sum_{k=0}^{\infty} Y_k$, we have*

$$\log\int_{\mathbb{S}^n} e^F\,d\nu \le \int_{\mathbb{S}^n} F\,d\nu + \frac{1}{2n}\sum_{k=1}^{\infty} d_k(n)\int_{\mathbb{S}^n} |Y_k|^2\,d\nu, \tag{8.67}$$

*where*

$$d_k(n) = \frac{\Gamma(n+k)}{\Gamma(n)\Gamma(k)}. \tag{8.68}$$

The inequality is complementary to Jensen's inequality

$$\int_{\mathbb{S}^n} F\,d\nu \le \log\int_{\mathbb{S}^n} e^F\,d\nu\,.$$

For $n = 1$, it can be written as

$$\log\left\{\frac{1}{2\pi}\int_{-\pi}^{\pi} e^F\,d\theta\right\} \le \frac{1}{2\pi}\int_{-\pi}^{\pi} F\,d\theta + \frac{1}{4\pi}\int_{\mathbb{B}^2} |\nabla u|^2 dx, \tag{8.69}$$

where $u$ is the harmonic extension of $F$ to $\mathbb{B}^2$, as in §8.8. Inequality (8.69) is equivalent to the first Lebedev–Milin inequality (see Duren, 1983, p. 143), which plays an important role in the theory of univalent functions.

For $n = 2$ the constants are $d_k(2) = k(k+1) = \lambda_k$, where $\lambda_k$ is the $k$th nonzero eigenvalue of the spherical Laplacian $\Delta_s$ on $\mathbb{S}^2$. Using Green's formula on $\mathbb{S}^2$, (8.67) may be written as

$$\log\left\{\int_{\mathbb{S}^2} e^F\,dv\right\} \le \int_{\mathbb{S}^2} F\,dv + \frac{1}{4}\int_{\mathbb{S}^2} |\nabla_s F|^2 dv, \tag{8.70}$$

an inequality proved by Onofri (1982), which gives the sharp form of an inequality due to Moser and Trudinger.

Inequalities (8.69) and (8.70) play an important role in a study by Osgood, Phillips and Sarnak (1988) of maximization problems for determinants of the Laplacian on 2-manifolds. For $n \ge 4$, inequality (8.67) may be restated in terms of the conformal Laplacian on $\mathbb{S}^n$, also known as the *Paneitz operator.* Then one can obtain some higher dimensional versions of the Osgood–Phillips–Sarnak results. See Beckner (1991, 1993).

*Proof of Theorem 8.21* We start again by expanding the sharp Hardy–Littlewood–Sobolev inequality in spherical harmonics, see (8.59). By approximation, may assume that $F = \sum_{k=0}^{\infty} Y_k$ with only finitely many $Y_k$ nonzero. Replacing $F$ by $F + c$, we may assume also that $F \ge 0$. Take $1 < p < 2$ and nonnegative bounded $G$ on $\mathbb{S}^n$, which we write as $G = \sum_{k=0}^{\infty} Z_k$. Let $c_k = c_k(\lambda)$ be as in (8.60). Then

$$\begin{aligned}\int_{\mathbb{S}^n} FG\,dv &= \sum_{k=0}^{\infty}\int_{\mathbb{S}^n} Y_k Z_k\,dv \\ &\le \left(\sum_{k=0}^{\infty}\int_{\mathbb{S}^n} c_k^{-1} Y_k^2\,dv\right)^{1/2}\left(\sum_{k=0}^{\infty}\int_{\mathbb{S}^n} c_k Z_k^2\,dv\right)^{1/2} \\ &\le \left(\sum_{k=0}^{\infty}\int_{\mathbb{S}^n} c_k^{-1} Y_k^2\,dv\right)^{1/2} \|G\|_p,\end{aligned}$$

where the first inequality uses two applications of Schwarz's inequality and the last inequality is by (8.59). Taking the supremum over all $G \in L^p(\mathbb{S}^n, v)$ and writing

$$q = p' = \frac{2n}{\lambda},$$

we obtain a dual inequality to (8.59):

$$\|F\|_q^2 \le \sum_{k=0}^{\infty} c_k^{-1}(\lambda) \int_{\mathbb{S}^n} Y_k^2\, dv, \quad \lambda = \frac{2n}{q}, \quad 2 < q < \infty.$$

Replace $F$ by $1 + \frac{F}{q}$. The result may be written as

$$\int_{\mathbb{S}^n} \left(1 + \frac{1}{q}F\right)^q dv \le \left\{ \left(1 + \frac{Y_0}{q}\right)^2 + q^{-2} \sum_{k=1}^{\infty} c_k^{-1}(\lambda) \int_{\mathbb{S}^n} Y_k^2\, dv \right\}^{q/2}. \tag{8.71}$$

The left side of (8.71) approaches $\int_{\mathbb{S}^n} e^F\, dv$ as $\lambda \to 0$. By (8.60), we have for $0 < \lambda < n$ and $k \ge 1$,

$$c_k(\lambda)^{-1} = \prod_{j=0}^{k-1} \frac{n - \frac{1}{2}\lambda + j}{\frac{1}{2}\lambda + j} \le \frac{2}{\lambda} d_k$$

with $d_k$ given by (8.68); in fact $\lim_{\lambda \to 0} \lambda c_k(\lambda)^{-1} = 2d_k$. Therefore, the right side of (8.71) is bounded from above by

$$\left\{ 1 + q^{-1} \left( 2Y_0 + \frac{1}{n} \sum_{k=1}^{\infty} d_k \int_{\mathbb{S}^n} Y_k^2 \right) + q^{-2} Y_0^2 \right\}^{q/2}.$$

We insert this bound into (8.71), take logarithms, and send $q = 2n/\lambda \to \infty$. Using that $\log(1+x) = x + O(x^2)$ as $x \to 0$, we obtain

$$\log \int_{sn} e^F\, dv \le Y_0 + \frac{1}{2n} \sum_{k=1}^{\infty} d_k \int_{\mathbb{S}^n} Y_k^2.$$

Now $Y_0 = \int_{\mathbb{S}^n} F\, dv$, and the theorem is proved. □

Equality clearly holds in (8.67) when $F$ is constant. In the same way as for Beckner's logarithmic Sobolev inequality, there is a family of equality cases, generated by a conformal symmetry inherited from the Hardy–Littlewood–Sobolev inequality in the limit $\lambda \to 0$, see (8.44). Here we define the action of a conformal transformation $\gamma$ on a function $F\colon \mathbb{S}^n \to \mathbb{R}$ by

$$\gamma^* F(x) = F(\gamma^{-1} x) + \log |J_{\gamma^{-1}}(x)|.$$

This leaves both sides of (8.67) invariant. By (8.45),

$$\log |J_{\gamma^{-1}}(x)| = -n \log(1 - \zeta \cdot x) + \log C_\zeta$$

for some $\zeta \in \mathbb{B}^{n+1}$ and a suitable constant $C_\zeta$. Thus the functions $F(x) = -n \log(1 - \zeta \cdot x) + C$ are conformally equivalent to constants and achieve equality in (8.67). Carlen and Loss (1992, Theorem 5) proved that there are no other equality cases.

## 8.11 Notes and Comments

In Theorem 8.1, equality holds whenever $f(e^{i\theta}) = f^{\#}(e^{im\theta})$, $g(e^{i\theta}) = g^{\#}(e^{im\theta})$, and $h(e^{i\theta}) = h^{\#}(e^{im\theta})$, for some integer $m$. As far as I know, no conditions sufficient for strict inequality have been identified in the $\mathbb{S}^1$ case.

A proof of Theorem 8.3 appears also in Hardy et al. (1952), and a proof of Theorem 8.4 in Lieb and Loss (1997). Burchard (1996) provides a history of matters pertinent to the Riesz and Riesz–Sobolev theorems. Among other things, she notes Sobolev's proof was incomplete, because is uses a result of Lusternik on approximations of the s.d.r. by Steiner symmetrizations that has a gap. The first complete proof of Theorem 8.4 is apparently due to Brascamp, Lieb and Luttinger in Brascamp et al. (1974). The 1-dimensional case of Theorem 8.8, is due to Rogers (1957).

Concerning uniqueness for Theorems 8.3 and 8.4, if one of $f, g, h$ is already symmetric decreasing, then a theorem of Lieb (1976–1977) provides a strong uniqueness assertion, as does our Corollary 2.19. For the general three-function case of Theorems 8.3 and 8.4, the best results known are due to Burchard (1996), where it is proved, for example that if $f, g$, and $h$ are characteristic functions of sets, then, roughly speaking, either one of the sets essentially contains the vector sum of the other two, or else $A, B$, and $C$ are essentially affine transforms of a common ellipsoid whose centers satisfy a certain constraint.

The history of the Brunn–Minkowski inequality and some of its extensions is presented in the book by Schneider (1993, p. 314). For more about Brunn–Minkowski, see Federer (1969) and Burago and Zalgaller (1988), and a more recent survey by Gardner (2002). For related results in a somewhat different direction, see the paper of Brascamp and Lieb (1976). It inspired several new directions of research, some still actively pursued. See, for example Barthe (2003) and Bennett et al. (2008).

Christ (1984, Thm. 4.2) gave an extension of the Brascamp–Lieb–Luttinger theorem to more general integrands, and used them to prove some sharp estimates for $k$-plane transforms. Pfiefer (1990) rediscovered a special case of Christ's inequality and used it to prove sharp "random simplex theorems": If $k + 1$ points are randomly chosen from a bounded measurable set $E \subset \mathbb{R}^n$ with given measure, then the mean $k$-volume of the simplex spanned by the $k + 1$ points is maximal when $E$ is a ball. The case $k = n$ is due to Blaschke (1917, 1923). For discussion and some conjectures, see Baernstein and Loss (1997); for recent developments, see Paouris and Pivovarov (2012, 2017).

When $n \geq 2$, there is no known analogue of the general three-function Riesz–Sobolev inequality for symmetric decreasing rearrangement on $\mathbb{S}^n$.

The spherical version of Corollary 2.19 (Corollary 7.1) gives such a result when one of the functions is of the form $K(d(x,y))$ with decreasing $K$. Via polarization, extensions of this sphere theorem to integrands with many input functions have been studied by Draghici (2005), Morpurgo (2002), and Burchard and Schmuckenschläger (2001).

Inequalities for Dirichlet heat kernels and related Schrödinger operators, as in Theorem 8.9 provided the original motivation for the Brascamp–Lieb–Luttinger theorem. Lieb's sharp Hardy–Littlewood–Sobolev inequality in Theorem 8.11 is a rare instance when we can find the operator norm $\|T\|_{p,p'}$ of a convolution operator $Tf = K * f$. The proof of Lieb's sharp Hardy–Littlewood–Sobolev inequality given here essentially follows Lieb and Loss (1997). It is an open question whether polarization may lead to a simpler proof of the inequality.

The extremal functions and best constants in Young's inequality were found independently by Beckner (1975) and Brascamp and Lieb (1976). For uniqueness questions, see Lieb (1990) and Barthe (1998).

**Revisions to the chapter** The presentation in this chapter closely follows Baernstein's original manuscript, up to some reorganization of Sections 8.6–8.10. Additions by Almut Burchard include the conclusion of the proof of Theorem 8.9, discussion of the equality cases for Theorems 8.17 and 8.21, and the proof of the "only if" part of Theorem 8.20.

# 9

# The $\star$-Function

The "star function" associated to a function $u$ is a new function $u^\star$ obtained by indefinite integration of the rearrangement $u^\#$ of $u$, with respect to some specific symmetrization process. This chapter is devoted to star functions associated with s.d.r. on $\mathbb{R}^n$ and $\mathbb{S}^n$, and with Steiner and cap symmetrization. The star function first arose in complex analysis ($n = 2$) for circular or $(1, 2)$-cap symmetrization. That connection will be the theme of Chapter 11.

Our chief goal is to prove that if $u$ satisfies $-\Delta u = \mu$ in a weak (distributional) sense, where $\mu$ is a locally finite signed measure on the domain of $u$, then $-\Delta^\star u^\star \le \mu^\star$ on the domain of $u^\star$. Here $\Delta^\star$ is a certain differential operator associated to the Laplacian $\Delta$, and $\mu^\star$ is a measure related to $\mu$. Such results are called "subharmonicity properties." They will be used in Chapter 10 to prove comparison theorems for solutions of partial differential equations, and in Chapter 11 to prove sharp inequalities in complex analysis.

After defining the star function on a general measure space, in §9.1, an overview of the chapter is given at the end of §9.2. The theory is easiest to grasp, I think, in the case of $(n - 1, n)$-cap symmetrization, under the assumption that the domain of $u$ is a spherical shell. Accordingly, we do that case first.

The natural class of functions for which the subharmonicity results make sense appears to be the set of $u \in L^1_{loc}(\Omega)$ for which the distributional Laplacian $\Delta u$ is a locally finite measure on the open set $\Omega$. The collection of such $u$ will be denoted by $\mathcal{W}(\Omega)$. This class too is discussed in §9.2.

## 9.1 The $\star$-Function on General Measure Spaces

Let $(X, \mathcal{M}, \mu)$ be a nonatomic measure space with $A = \mu(X) < \infty$.

**Definition 9.1** For $f \in L^1(X)$, define $f^\star \colon [0,A] \to \mathbb{R}$ by

$$f^\star(x) = \sup\left\{\int_E f\,d\mu \colon E \in \mathcal{M},\ \mu(E) = x\right\}, \quad 0 \le x \le A,$$

where the sup is taken over all measurable sets $E \subset X$ with $\mu(E) = x$.

Recall that $f^* \colon [0,A] \to \mathbb{R}$ is the decreasing rearrangement of $f$. Our first result tells us that the sup in Definition 9.1 is actually achieved, and explains how the star function $f^\star$ is related to $f^*$.

**Proposition 9.2** *Let $f \in L^1(X)$. Then*

*(a) For each $x \in [0,A]$ there exists $E \in \mathcal{M}$ with $\mu(E) = x$ such that*

$$f^\star(x) = \int_E f\,d\mu.$$

*(b) $f^\star(x) = \int_0^x f^*(s)\,ds, \quad 0 \le x \le A.$*

*Proof* By Proposition 1.26, we have $f = f^* \circ T$, $\mu$-a.e. on $X$, where $T\colon (X,\mathcal{M},\mu) \to ([0,A],\mathcal{B},\mathcal{L})$ is measure preserving. If $x = 0$ or $x = A$, take $E$ to be the empty set or $X$, respectively. Suppose that $x \in (0,A)$. Let $t = f^*(x)$. Define $E = T^{-1}([0,x])$. Then $f \ge t$ a.e. on $E$ and $f \le t$ a.e. on $X \setminus E$. Let $F \in \mathcal{M}$ with $\mu(F) = x$. Then

$$\int_F f\,d\mu = \int_F \{(f-t)+t\}\,d\mu \le \int_X (f-t)^+\,d\mu + tx,$$

while

$$\int_E f\,d\mu = \int_X (f-t)^+\,d\mu + tx.$$

This proves (i). To get (ii), observe that since $T$ is measure preserving,

$$f^\star(x) = \int_E f\,d\mu = \int_E f^* \circ T\,d\mu = \int_0^x f^*(s)\,ds. \qquad \square$$

## 9.2 Preliminaries, and What Happens Next

### Locally Finite Measures

Let $X$ be a locally compact Hausdorff space and $\mathcal{B}_c(X)$ the set of all Borel sets $B \subset X$ such that $B \subset K$ for some compact $K \subset X$. We shall say that a set function $\mu\colon \mathcal{B}_c(X) \to \mathbb{R}$ is a *locally finite real-valued (regular Borel) measure* if for each compact $K \subset X$ the restriction of $\mu$ to $\mathcal{B}(K)$ belongs to the space

$M(K)$ of (finitely) real-valued (signed) regular Borel measures on $K$. The set of all locally finite real-valued measures on $X$ will be denoted by

$$M_{loc}(X).$$

For $\mu \in M(K)$, denote the total variation of $\mu$ on $K$ by $\|\mu\|_K$. Then $\|\cdot\|_K$ is a norm, under which $M(K)$ is a Banach space. By the Riesz Representation Theorem, $M(K)$ is the dual of the Banach space $C(K)$ of all continuous real-valued functions on $K$, equipped with the sup norm, with the natural pairing

$$(f, \mu) = \int_K f \, d\mu.$$

(See Folland (1999, Cor. 7.18).) If $f \in C_c(X)$ and $\mu \in M_{loc}(X)$ we define $(f, \mu)$ to be $\int_K f \, d\mu$ where $K$ is any compact $K \subset X$ containing the support of $f$. Under this pairing, $M_{loc}(X)$ is the dual of $C_c(X)$, when the two spaces are endowed with their usual topologies. See Rudin (1966), Schaefer (1971), or Köthe (1960).

If $\mu \in M_{loc}(X)$ is nonnegative, then $\mu$ is called a *Radon measure* on $X$. The set of all Radon measures on $X$ will be denoted by $M^+_{loc}(X)$. Clearly, if $\mu_1, \mu_2 \in M^+_{loc}(X)$, then $\mu_1 - \mu_2 \in M_{loc}(X)$. From the Jordan decomposition theorem (Folland, 1999, Lemma 7.15), it follows that the converse holds: for each $\mu \in M_{loc}(X)$, there exist $\mu_1, \mu_2 \in M^+_{loc}(X)$ such that $\mu = \mu_1 - \mu_2$.

### Weak Laplacians

Let $\Omega$ be an open set in $\mathbb{R}^n$, not necessarily bounded, and assume that $u$ belongs to the Sobelev space $W^{1,2}_0(\Omega)$. In §5.1 we defined $u$ to be a weak solution to the p.d.e. $\Delta u = -f$ if $f \in L^2(\Omega)$ and

$$\int_\Omega \nabla u \cdot \nabla g \, dx = \int_\Omega fg \, dx, \quad \forall g \in C^1_c(\Omega).$$

We define now another notion of weak solution. Let $u \in L^1_{loc}(\Omega)$. If there exists $\mu \in M_{loc}(\Omega)$ such that

$$\int_\Omega u \Delta g \, dx = -\int_\Omega g \, d\mu, \quad \forall g \in C^2_c(\Omega),$$

then we say that $u$ is a weak (or distributional) solution of

$$\Delta u = -\mu,$$

and that $u$ has weak (or distributional) Laplacian $-\mu$.

If $u \in C^2(\Omega)$ with pointwise Laplacian $\Delta u = -f$, then we call $u$ a classical solution to the equation, and the Gauss–Green theorem implies $u$ is also a weak solution. (More precisely, $\Delta u = -\mu$ where $\mu$ is the absolutely continuous measure with density $f$.)

If $n \geq 3$ then, as the reader is invited to show, the locally integrable function $u(x) = |x|^{2-n}$ has weak Laplacian $\Delta u = -(n-2)\beta_{n-1}\delta_0$ where $\beta_{n-1}$ is the surface area of the $(n-1)$-dimensional unit sphere; for $n = 2$ and $n = 1$ the corresponding results are $\Delta \log |x| = 2\pi\delta_0$ and $\Delta|x| = 2\delta_0$.

Define $\mathcal{W}(\Omega)$ to be the set of all $u \in L^1_{loc}(\Omega)$ such that $\Delta u = -\mu$ weakly for some $\mu \in M_{loc}(\Omega)$. Obviously $C^2(\Omega) \subset \mathcal{W}(\Omega)$, since $u \in C^2(\Omega)$ implies $\Delta u \in L^1_{loc}(\Omega)$, but some functions in $\mathcal{W}(\Omega)$ are not smooth, as we have seen with the Green function.

Using the duality between $C_c(\Omega)$ and $M_{loc}(\Omega)$ described in the previous subsection, one can show $u \in L^1_{loc}(\Omega)$ belongs to $\mathcal{W}(\Omega)$ if and only if for each open set $\Omega_1 \subset\subset \Omega$ there is a constant $C(\Omega_1)$ such that

$$\Big|\int_\Omega u\Delta g\, dx\Big| \leq C(\Omega_1) \sup_\Omega |g|,$$

for all $g \in C^2(\Omega)$ with support in $\Omega_1$.

### Subharmonic Functions

Let $\Omega$ be an open set. A function $u\colon \Omega \to \mathbb{R}\cup\{-\infty\}$ is said to be *subharmonic* in $\Omega$ if it is upper semicontinuous, not identically $-\infty$, and satisfies the sub-mean value property:

$$u(x) \leq \frac{1}{\beta_{n-1}} \int_{\mathbb{S}^{n-1}} u(x+ry)\, d\sigma_{n-1}(y)$$

for every $x \in \Omega$ lying at distance greater than $r$ to the boundary. Upper semicontinuity insures that subharmonic functions are locally bounded above, so that the mean value integral is well defined. It turns out always to be finite, and $u \in L^1_{loc}(\Omega)$. See Hayman and Kennedy (1976) for proofs of these and results stated below.

If $u \in C^2(\Omega)$ then $u$ is subharmonic in $\Omega$ if and only if $\Delta u \geq 0$ at every point of $\Omega$. For general subharmonic $u$, one can show that

$$(u, \Delta g) \equiv \int_\Omega u\, \Delta g\, dx \geq 0, \tag{9.1}$$

for every nonnegative $g \in C^2_c(\Omega)$, from which it follows that the distributional Laplacian $\Delta u$ of $u$ can be identified with a Radon measure in $M^+_{loc}(\Omega)$. In particular, $u \in \mathcal{W}(\Omega)$. This nonnegative measure is called the *Riesz measure* of $u$. If we write $\Delta u = -\mu$, then the Riesz measure of $u$ is $-\mu$, and $\mu \leq 0$ for subharmonic $u$.

For example, if $f$ is holomorphic and nonconstant in a plane domain $\Omega$, with zero set $\{z_j\}$, counting multiplicities, then $\log|f|$ is subharmonic in $\Omega$ with $\Delta(\log|f|) = 2\pi \sum_j \delta_{z_j}$.

Returning to (9.1), we note that, conversely, if $u$ is a function in $L^1_{loc}(\Omega)$ for which (9.1) holds for all nonnegative $g \in C^2_c(\Omega)$, then there exists a subharmonic function in $\Omega$ which agrees with $u$ ($\mathcal{L}^n$-a.e.).

A function $u = u_1 - u_2$ with $u_1, u_2$ subharmonic in $\Omega$ is said to be $\delta$-subharmonic in $\Omega$. At points where $u_1 = u_2 = -\infty$, their difference $u$ is not well defined. Thus $\delta$-subharmonic functions are defined only $\mathcal{L}^n$-a.e. If $u$ is $\delta$-subharmonic then its distributional Laplacian $\Delta u$ may be represented as

$$\Delta u = \mu_1 - \mu_2$$

with $\mu_1, \mu_2 \in M^+_{loc}(\Omega)$, so that $\Delta u$ may be identified with an element of $M_{loc}(\Omega)$, and hence $u$ being $\delta$-subharmonic in $\Omega$ implies $u \in \mathcal{W}(\Omega)$.

The converse is true too: Given $\nu \in M_{loc}(\Omega)$, there exists a $\delta$-subharmonic function $u$ in $\Omega$ such that $\Delta u = -\nu$. It follows that each $u \in \mathcal{W}(\Omega)$ is $\delta$-subharmonic, and hence that the set of all $\delta$-subharmonic functions in $\Omega$ is exactly the set $\mathcal{W}(\Omega)$. We use this fact later when proving Lemma 9.6.

**Preview of what happens next** For fixed $n \geq 2$, let now

$$A = A(R_1, R_2) = \{x \in \mathbb{R}^n \colon R_1 < |x| < R_2\},$$

where $0 \leq R_1 < R_2 \leq \infty$. Then $A$ is an open shell in $\mathbb{R}^n$, possibly unbounded and possibly with inner boundary the point $\{0\}$. Consider the rectangle

$$A^\star = \{(r,\theta) \in \mathbb{R}^2 \colon R_1 < r < R_2,\ 0 < \theta < \pi\},$$

which can be viewed as the upper half of an open annulus in $\mathbb{R}^2$ using polar coordinates $(r,\theta)$. For $u \in L^1_{loc}(A, \mathbb{R})$, define

$$u^\star \colon A^\star \to \mathbb{R}$$

by

$$u^\star(r,\theta) = \sup_E \int_E u(rs)\, r^{n-1}\, d\sigma_{n-1}(s),$$

where $\sigma_{n-1}$ is the canonical measure on $\mathbb{S}^{n-1}$ and the sup is taken over all Borel measurable $E \subset \mathbb{S}^{n-1}$ with $\sigma_{n-1}(E) = \sigma_{n-1}(\mathcal{K}(\theta))$. Here $\mathcal{K}(\theta)$ denotes the spherical cap on $\mathbb{S}^{n-1}$ with center $e_1$ and opening $\theta$. After a change of variable, the slice function $u^\star(r,\cdot)$ is the star function in the sense of §9.1 of $u$ restricted to the sphere of radius $r$.

Baernstein (1973, 1974) proved that if $n = 2$ and $u$ is subharmonic in an annulus, then $u^\star/r$ is subharmonic in the upper half annulus with polar coordinates $r$ and $\theta$ (see Corollary 9.10). Baernstein and Taylor (1976) proved that if $n \geq 3$ and $u$ is subharmonic in $A$, then $u^\star$ is a subsolution of another elliptic operator, to be called $\Delta^\star$, which is closely related to $\Delta$. When $n \geq 2$, one thus has a maximum principle and this leads to various applications, some of which are presented in Chapters 10 and 11.

In the rest of Chapter 9 we will generalize these "subharmonicity" results. They turn out to be consequences of inequalities involving $u^\#$ and $\mu^\#$, where $u \in \mathcal{W}(\Omega)$ with $\Delta u = -\mu$, and # denotes the appropriate symmetrization. The $u^\#$, $\mu^\#$ results will be called "pre-subharmonicity" inequalities.

In §9.5 we prove pre-subharmonicity inequalities for functions $u$ on shells $A$, that is, for $(n-1, n)$-cap symmetrization when $n \geq 2$. Then in §9.6 we present the corresponding subharmonicity inequalities. Section 9.7 deduces pre-subharmonicity and subharmonicity results for the sphere $\mathbb{S}^n$. Section 9.8 indicates the extension to $(k, n)$-cap symmetrization on shells when $1 \leq k \leq n-2$. In principle it would be possible to embed the $(n-1, n)$ proof on shells into the $(k, n)$ proof, but they will be separated for pedagogical reasons. Sections 9.9 and 9.10 treat s.d.r. on $\mathbb{R}^n$, and §9.11 extends to $(k, n)$-Steiner symmetrization.

## 9.3 A Measurability Lemma

Let $(X, \mathcal{M}, \mu)$ be a finite measure space, $(Z, \mathcal{N})$ a measurable space, and $f$ and $g$ real-valued functions on $X \times Z$ which are $\mathcal{M} \times \mathcal{N}$ measurable. Then the slice functions $f^z$ and $g^z$ are $\mathcal{M}$ measurable on $X$. Define

$$F(t, t', z) = \mu(f^z > t, g^z > t'), \quad (t, t') \in \mathbb{R}^2, \quad z \in Z.$$

**Lemma 9.3** *$F$ is $\mathcal{B}(\mathbb{R}^2) \times \mathcal{N}$ measurable.*

*Proof* Assume first that $f$ and $g$ are simple. Write

$$f = \sum_{i=1}^{m} t_i \chi_{A_i}, \quad g = \sum_{i=1}^{n} u_i \chi_{B_i},$$

where the $t_i$ and $u_i$ are real and the sets $A_i$, $B_i$ are in $\mathcal{M} \times \mathcal{N}$ with $A_1 \supset \cdots \supset A_m$, $B_1 \supset \cdots \supset B_n$. The possible values of $f$ and $g$ are $s_j \equiv \sum_{i=1}^{j} t_i$ and $v_k \equiv \sum_{i=1}^{k} u_i$ respectively, for $0 \leq j \leq m$, $0 \leq k \leq n$, where $s_0 = v_0 = 0$.

Set $I_i = [s_{i-1}, s_i)$, $J_j = [v_{j-1}, v_j)$. If $t \in I_i$ and $t' \in J_j$ then $(f > t, g > t') = A_i \cap B_j$. If $t \geq s_m$ or $t' \geq v_n$ then $(f > t, g > t')$ is empty. Passing to slices on each side, it follows that

$$F(t, t', z) = \sum_{j=1}^{n} \sum_{i=1}^{m} \mu(A_i^z \cap B_j^z) \chi_{I_i \times J_j}(t, t'), \quad (t, t') \in \mathbb{R}^2, \quad z \in Z.$$

The $\mu$-terms in the sum are $\mathcal{N}$ measurable functions of $z$ and the $\chi$-terms are Borel measurable functions on $\mathbb{R}^2$. Thus, $F$ is $\mathcal{B}(\mathbb{R}^2) \times \mathcal{N}$ measurable, and so Lemma 9.3 is proved for simple functions.

An easy argument shows that if measurable functions $f$ and $g$ are pointwise increasing limits of sequences $\{f_n\}$ and $\{g_n\}$ on $X \times Z$ then the corresponding functions $F_n$ converge pointwise to $F$ on $\mathbb{R}^2 \times Z$. Thus, the validity of Lemma 9.3 for simple functions implies its validity for nonnegative measurable functions. To pass to all pairs of measurable functions requires another short argument, which is left to the reader. □

With $f$ and $g$ as above, define

$$F_1(t, z) = \mu(f^z > t) \quad \text{and} \quad F_2(t, t', z) = \mu(f^z = t, g^z > t').$$

**Lemma 9.4** *$F_1$ is $\mathcal{B}(\mathbb{R}) \times \mathcal{N}$ measurable on $\mathbb{R} \times Z$ and $F_2$ is $\mathcal{B}(\mathbb{R}^2) \times \mathcal{N}$ measurable on $\mathbb{R}^2 \times Z$.*

*Proof* The measurability of $F_1$ follows as above from the formula

$$F_1(t, z) = \sum_{i=1}^{m} \mu(A_i^z)\chi_{I_i}(t), \quad t \in \mathbb{R}, \quad z \in Z.$$

The measurability of $F_2$ follows from measurability of $F$ and the relation

$$F_2(t, t', z) = \lim_{\epsilon \to 0} [F(t - \epsilon, t', z) - F(t + \epsilon, t', z)], \quad (t, t', z) \in \mathbb{R}^2 \times Z. \quad \square$$

## 9.4 Formulas for the Laplacian

A key step in the $\star$-function method is to change a differential inequality into an integral inequality. The next lemma lies at the heart of the matter: it expresses the Laplacian of a function as the difference between an average of the function over a small ball and the value of the function at the center of the ball.

Let $K: \mathbb{R}^n \to \mathbb{R}^+$ be a smooth nonnegative symmetric bump function on $\mathbb{R}^n$. That is, $K \geq 0$, $K \in C_c^\infty(\mathbb{R}^n)$, $K$ is symmetric decreasing on $\mathbb{R}^n$, the support of $K$ is contained in the open unit ball $\mathbb{B}^n \subset \mathbb{R}^n$, and $\int_{\mathbb{R}^n} K(x)\,dx = 1$. Set $K_\epsilon(x) = \epsilon^{-n} K(\epsilon^{-1} x)$ and denote convolution in $\mathbb{R}^n$ by $*$. If $\Omega$ is an open set in $\mathbb{R}^n$ and $f \in L^1_{loc}(\Omega)$, then $f * K_\epsilon$ is defined on the set $\Omega_\epsilon = \{x \in \Omega : d(x, \partial\Omega) > \epsilon\}$.

**Lemma 9.5** *For $u \in C^2(\Omega)$ and $K_\epsilon$ as above,*

$$\lim_{\epsilon \to 0} \epsilon^{-2}[K_\epsilon * u(x) - u(x)] = C_K \Delta u(x), \quad x \in \Omega, \tag{9.2}$$

*where*

$$C_K = \frac{1}{2n} \int_{\mathbb{B}^n} |x|^2 K(x)\, dx. \tag{9.3}$$

*Moreover, the convergence in* (9.2) *is uniform on compact subsets of* $\Omega$.

*Proof* To verify (9.2), we may assume by translation that $0 \in \Omega$ and that $x = 0$. For $r > 0$, set $K_0(r) = K(re_1)$. Then for small $\epsilon$,

$$K_\epsilon * u(0) = \epsilon^{-n} \int_0^\epsilon r^{n-1} K_0(r\epsilon^{-1})\, dr \int_{\mathbb{S}^{n-1}} u(ry)\, d\sigma_{n-1}(y).$$

If $u$ is harmonic in $\Omega$, the mean value property and the fact that $\int_{\mathbb{R}^n} K\, dx = 1$ imply $K_\epsilon * u(0) = u(0)$. If $u(x) = |x|^2$, then (9.3) implies $K_\epsilon * u(0) = 2C_K n\epsilon^2$. The functions $K_\epsilon * (x_i^2)$ have the same values at 0 for each $i = 1, \ldots, n$, by symmetry, and thus,

$$(K_\epsilon * (x_i^2))(0) = \frac{1}{n}(K_\epsilon * |x|^2)(0) = 2C_K\epsilon^2, \quad i = 1, \ldots, n.$$

Linear functions and products $x_i x_j$ with $i \neq j$ are harmonic. By Taylor's Theorem, a function $u \in C^2(\Omega)$ has the representation

$$u(x) = h(x) + \frac{1}{2}\sum_{i=1}^n \partial_{x_i x_i} u(0) x_i^2 + R(x), \tag{9.4}$$

where $h$ is harmonic with $h(0) = u(0)$ and $R(x) = o(|x|^2)$ as $x \to 0$. Thus $K_\epsilon * R(0) = o(\epsilon^2)$, and so (9.2) follows from (9.4).

To prove the uniformity statement, take a compact set $E \subset \Omega$ and an open set $E_1$ with $E \subset E_1 \subset\subset \Omega$. Write

$$u(x) = P(x, a) + R(x, a)$$

with $P$ the quadratic Taylor polynomial. The function $|x-a|^{-2}|u(x) - P(x, a)|$, extended to be zero on $\{(x, x) : x \in E\}$, is continuous on the compact set $E \times \overline{E_1}$, hence is uniformly continuous. It follows that

$$\lim_{x \to a} |x - a|^{-2} |u(x) - P(x, a)| = 0$$

uniformly for $a \in E$. The argument that gave (9.2) now yields the uniformity of convergence. □

For the next lemma, we continue to let $K$ and $\Omega$ be as above, but this time will allow $u$ to belong to the set $\mathcal{W}(\Omega)$ instead of being $C^2$. The weak Laplacian $\Delta u$ of $u$ is a locally finite measure on $\Omega$. We also require some additional notation. Define $G \colon \mathbb{R}^n \to [0, \infty]$ by

$$G(x) = \begin{cases} b_n(|x|^{2-n} - 1)^+, & n \geq 3, \\ b_2 \log^+(1/|x|), & n = 2, \\ b_1(1 - |x|)^+, & n = 1, \end{cases}$$

where, with $\beta_{n-1} = \sigma_{n-1}(\mathbb{S}^{n-1})$,

$$b_n = \begin{cases} 1/(n-2)\beta_{n-1}, & n \geq 3, \\ 1/2\pi, & n = 2, \\ 1/2, & n = 1. \end{cases}$$

Then $G$ is the Green's function of the unit ball in $\mathbb{R}^n$ with pole at 0, that is, $\Delta G = -\delta_0$ in $\mathbb{B}^n(0, 1)$ and $G(x) = 0$ for $|x| \geq 1$ (see Evans, 1998, pp. 22, 39).

Define a function $L$, depending on the bump function $K$, by

$$L(x) = \beta_{n-1} \int_0^\infty s K_0(s) G(x/s)\, ds, \qquad x \in \mathbb{R}^n. \tag{9.5}$$

Note that $G(x/s) = 0$ when $s < |x|$. Clearly $L$ is nonnegative, symmetric decreasing, and is supported in $\overline{\mathbb{B}^n(0, 1)}$. Also, if $n \geq 2$ then $x \to 0$, $L(x) \to \infty$ at the same rate as $G$. Furthermore,

$$\int_{\mathbb{R}^n} L(x)\, dx = \frac{1}{2n} \int_{\mathbb{R}^n} |x|^2 K(x)\, dx = C_K.$$

The second equality is from definition (9.3), and the first is left as an exercise for the reader.

**Lemma 9.6** *Let $u \in \mathcal{W}(\Omega)$ with $\Delta u = -\mu$ and $x \in \Omega$ with $d(x, \partial\Omega) > \epsilon$. Then*

$$\epsilon^{-2}[K_\epsilon * u(x) - u(x)] = -\int_\Omega L_\epsilon(x - y)\, d\mu(y). \tag{9.6}$$

More precisely, if $u$ is superharmonic in $\Omega$ then (9.6) holds for all the indicated $x$, and has the value $-\infty$ at points $x$ with $u(x) = \infty$. Similarly if $u$ is subharmonic, (9.6) again holds with both sides equalling $\infty$ at points $x$ with $u(x) = -\infty$. Since the set $(u = \infty)$ has $\mathcal{L}^n$-measure zero when $u$ is superharmonic, it follows that if $u$ is the difference of superharmonic functions then the two sides of (9.6) are well-defined and equal for almost every $x$ with $d(x, \Omega) > \epsilon$.

Also, recall that $L_\epsilon(x) = \epsilon^{-n} L(x/\epsilon)$.

*Proof of Lemma 9.6* Every $u \in \mathcal{W}(\Omega)$ is a difference of superharmonic functions (or subharmonic functions) by the observations in §9.2. Thus it suffices to prove (9.6) when $u$ is superharmonic, and $x = 0$. Since $\epsilon < d(0, \partial\Omega)$, the closure of $\mathbb{B}^n(0, \epsilon)$ is contained in $\Omega$. By the Riesz decomposition theorem (Hayman and Kennedy, 1976, Thm 3.9), for $0 < t < d(0, \partial\Omega)$ we have

$$u(\zeta) = H(\zeta) + P(\zeta), \quad \zeta \in \mathbb{B}^n(t), \tag{9.7}$$

where $H$ solves the Dirichlet problem in $\mathbb{B}^n(t)$ with boundary function $u$, and

$$P(\zeta) = \int_{\mathbb{B}^n(t)} G(\zeta, y, \mathbb{B}^n(t))\, d\mu(y)$$

is the Green's potential of $\mu$ over $\mathbb{B}^n(t)$. Note

$$G(\zeta, y, \mathbb{B}^n(t)) = t^{2-n} G(\zeta/t, y/t, \mathbb{B}^n(1)) = t^{2-n} G((\zeta - y)/t).$$

In (9.7), take the mean value over $|\zeta| = t$ and use that $P(\zeta) = 0$ when $|\zeta| = t$. The result is

$$\begin{aligned}\frac{1}{\beta_{n-1}} \int_{\mathbb{S}^{n-1}} u(tz)\, d\sigma_{n-1}(z) &= H(0)\\ &= u(0) - P(0)\\ &= u(0) - t^{2-n} \int_{\mathbb{B}^n(t)} G(y/t)\, d\mu(y). \end{aligned}\tag{9.8}$$

Now

$$\begin{aligned} K_\epsilon * u(0) &= \int_{\mathbb{R}^n} K_\epsilon(-y) u(y)\, dy = \epsilon^{-n} \int_{\mathbb{R}^n} K_0(|y|/\epsilon) u(y)\, dy\\ &= \epsilon^{-n} \int_0^\infty K_0(t/\epsilon) t^{n-1}\, dt \int_{\mathbb{S}^{n-1}} u(tz)\, d\sigma_{n-1}(z). \end{aligned}$$

With (9.8), and the change of variable $t = \epsilon s$, this implies

$$K_\epsilon * u(0) = \int_0^\infty K_0(s) s^{n-1} \Big[\beta_{n-1} u(0) - \beta_{n-1} \epsilon^{2-n} s^{2-n} \int_\Omega G(y/\epsilon s)\, d\mu(y)\Big]\, ds.$$

Since $\int_0^\infty K_0(s) s^{n-1} \beta_{n-1}\, ds = \int_{\mathbb{R}^n} K(y)\, dy = 1$, the last identity yields

$$\begin{aligned} K_\epsilon * u(0) - u(0) &= -\beta_{n-1} \epsilon^{2-n} \int_0^\infty K_0(s) s \int_\Omega G(y/\epsilon s)\, d\mu(y)\, ds\\ &= -\epsilon^{2-n} \int_\Omega \beta_{n-1} \int_0^\infty K_0(s) s G(y/\epsilon s)\, ds\, d\mu(y)\\ &= -\epsilon^{2-n} \int_\Omega L(y/\epsilon)\, d\mu(y). \end{aligned}$$

Fubini's Theorem may be used above since $\mu, G$ and $K_0$ are nonnegative. Division by $\epsilon^2$ completes the proof of Lemma 9.6. □

## 9.5 Pre-Subharmonicity on Shells

In this section, # will always denote $(n-1,n)$-cap symmetrization.

As in §9.2, let $A = \{x \in \mathbb{R}^n : R_1 < |x| < R_2\}$ be a shell in $\mathbb{R}^n$, where $n \geq 2$. Let $\mu$ be a locally finite measure on $A$, that is, $\mu \in M_{loc}(A)$. The Jordan and Lebesgue decompositions yield the representation

$$d\mu = f\, d\mathcal{L}^n + d\tau - d\eta, \tag{9.9}$$

where $f \in L^1_{loc}(A)$, and $\tau$ and $\eta$ are (nonnegative) Radon measures on $A$, singular with respect to each other and to $\mathcal{L}^n$.

For $x \in A$ write $x = ry$, where $y \in \mathbb{S}^{n-1}$ and $r \in (R_1, R_2)$. Define a map $P\colon A \to A$ by $P(x) = re_1$. Then $P(x)$ is the spherical projection of $x$ onto the positive $x_1$-axis in $\mathbb{R}^n$, and $PA \equiv P(A)$ is the open line segment in $\mathbb{R}^n$ from $R_1 e_1$ to $R_2 e_1$. Note that $PA \subset A$.

For the singular $\tau$, define $\tau^\# \in M_{loc}(PA)$ to be the pushforward measure on $PA$ induced by $P$ and $\tau$:

$$\tau^\#(E) = \tau(P^{-1}E)$$

for $E \in \mathcal{B}_c(PA)$. Extend $\tau^\#$ to be 0 on $\mathbb{R}^n \setminus PA$. Thus, $\tau^\#$ is the measure obtained from $\tau$ by spherically sweeping the mass to the positive $x_1$-axis. For example, if $\delta_a$ denotes a unit point mass at $a$, and if $\tau = \sum_{i=1}^N c_j \delta_{a_j}$ with $a_j \in A$ and $c_j > 0$, then $\tau^\# = \sum_{j=1}^N c_j \delta_{|a_j| e_1}$.

For $\mu \in M_{loc}(\Omega)$ having the decomposition (9.9), define the symmetrized measure $\mu^\# \in M_{loc}(A)$ to be

$$d\mu^\# = f^\# d\mathcal{L}^n + d\tau^\# - d\eta_\#,$$

where $\eta_\#$ is defined by $\eta_\#(E) = \eta^\#(-E)$. Thus, $\eta_\#$ is supported on the negative $x_1$-axis.

Let $u \in \mathcal{W}(A)$, so that as discussed in §9.2, the weak Laplacian $\Delta u$ can be identified with a locally finite measure on $A$. We say that $-\Delta u = \mu$ in the weak (or distributional) sense if

$$-\int_A u \Delta g \, d\mathcal{L}^n = \int_A g \, d\mu$$

for every $g \in C_c^2(A)$. More generally, given a continuous real-valued function $\phi(\cdot,\cdot)$ of two real variables such that the composition $\phi(r,u)$ is locally integrable, we say

$$-\Delta u = \phi(r,u) + \mu$$

in the weak sense if

$$-\int_A u\Delta g\, d\mathcal{L}^n = \int_A \phi(r,u) g\, d\mathcal{L}^n + \int_A g\, d\mu$$

for every $g \in C_c^2(A)$.

We can now state the **pre-subharmonicity theorem** for $(n-1,n)$-cap symmetrization. It says $-\Delta(u^{\#}) \le \phi(r,u^{\#}) + \mu^{\#}$ weakly when restricted to cap-symmetric, nonnegative test functions.

**Theorem 9.7** (Pre-subharmonicity for $(n-1,n)$-cap symmetrization) *Assume $u \in \mathcal{W}(A)$ satisfies*

$$-\Delta u = \phi(r,u) + \mu \tag{9.10}$$

*in the weak sense in $A$, where $\phi\colon (R_1,R_2)\times\mathbb{R} \to \mathbb{R}$ is continuous with $\phi(r,u)$ locally integrable on $A$ and $\mu \in M_{loc}(A)$.*

*If $g \in C_c^2(A)$ is nonnegative with $g = g^{\#}$ then*

$$-\int_A u^{\#}\Delta g\, d\mathcal{L}^n \le \int_A \phi(r,u^{\#}) g\, d\mathcal{L}^n + \int_A g\, d\mu^{\#}.$$

*Proof* The proof is broken into steps.

**Step 1:** Construction of a measure preserving $T$.

Since functions in $\mathcal{W}(A)$ are locally integrable, we have $u \in L^1_{loc}(A)$. Changing $u$ on a set of $\mathcal{L}^n$-measure zero, if necessary, we may assume that $u$ is Borel measurable on $A$. We write $A$ in spherical coordinates as $A = \mathbb{S}^{n-1}\times(R_1,R_2)$.

Define $\psi$ by

$$\psi(x) = x_1, \qquad x \in \mathbb{S}^{n-1},$$

and write

$$\sigma = \sigma_{n-1}.$$

Then $\psi$ has no flat spots with respect to $\sigma$, that is, $\sigma(\psi = t) = 0$ for all $t \in \mathbb{R}$. Recall that the slice functions $u^r$ are defined by $u^r(x) = u(x,r)$. For $(x,r) \in A$, define

$$T(x,r) = \sigma\big(u^r > u^r(x)\big) + \sigma\big(u^r = u^r(x),\ \psi > \psi(x)\big),$$

where sets of the form $(u^r > t)$ are understood to be subsets of $\mathbb{S}^{n-1}$, and extend $\psi$ to $A$ by setting $\psi(x,r) = \psi(x)$. Then

$$T = F_1\circ P_1 + F_2\circ P_2$$

where

$$F_1(t,r) = \sigma(u^r > t), \quad F_2(t,t',r) = \sigma(u^r = t, \psi^r > t'),$$

and

$$P_1(x,r) = (u^r(x), r), \quad P_2(x,r) = (u^r(x), \psi^r(x), r).$$

The functions $(x,r) \mapsto (u^r(x), r)$ and $(x,r) \mapsto (u^r(x), \psi^r(x), r)$ are Borel measurable mappings from $A$ into $\mathbb{R} \times (R_1, R_2)$ and $\mathbb{R} \times \mathbb{R} \times (R_1, R_2)$, respectively. From Lemma 9.4 it follows that $T$ is a Borel measurable function from $A$ into $[0, \beta_{n-1}]$.

By Proposition 1.26, for each fixed $r \in (R_1, R_2)$ the slice function $T^r$ is a measure preserving map from $(\mathbb{S}^{n-1}, \sigma)$ onto $[0, \beta_{n-1}]$. Moreover

$$u(x,r) = u^*(T(x,r), r), \qquad (x,r) \in A, \tag{9.11}$$

where $u^*(\cdot, r)$ is the decreasing rearrangement of $u(\cdot, r)$. This completes the proof of Step 1.

**Step 2:** Construction of $h$ with $h^\# = g$.

Let $g \in C_c^2(A)$ be nonnegative, and let $g^*(\cdot, r)$ denote the decreasing rearrangement of $g(\cdot, r)$. Let

$$T_1(x) = \sigma(\mathcal{K}(\theta)),$$

where $\cos\theta = x_1$. Then $T_1$ is the measure preserving map from $\mathbb{S}^{n-1}$ onto $[0, \beta_{n-1}]$ introduced in §7.1, for which $F^* \circ T_1$ equals $F^\#$ for functions $F$ on $\mathbb{S}^{n-1}$. Since the slice functions $g^r$ are symmetric decreasing on $\mathbb{S}^{n-1}$ by hypothesis, we have

$$g(x,r) = g^\#(x,r) = g^*(T_1(x), r), \qquad (x,r) \in A.$$

Using Chapter 7, we see that the continuity of $g$ on $A$ implies the continuity of $g^*$ on $[0, \beta_{n-1}] \times (R_1, R_2)$.

Define $h$ on $A$ by

$$h(x,r) = g^*(T(x,r), r). \tag{9.12}$$

The Borel measurability of $T$ and $g^*$ imply the Borel measurability of $h$ on $A$. Moreover, since $T(\cdot, r)$ is measure preserving from $(\mathbb{S}^{n-1}, \sigma)$ onto $[0, \beta_{n-1}]$, (9.12) implies that $h^\# = g$ on $A$. This equation also shows that $h$ is bounded and has compact support in $A$, since $g$ has these properties.

For fixed $r \in (R_1, R_2)$, it follows from (9.11), (9.12) and $g = g^\#$ that

$$\begin{aligned}
\int_{\mathbb{S}^{n-1}} u(x,r)h(x,r)\,d\sigma(x) &= \int_{[0,\beta_{n-1}]} u^*(t,r)g^*(t,r)\,dt \\
&= \int_{\mathbb{S}^{n-1}} u^*(T_1(x), r)g^*(T_1(x), r)\,d\sigma(x) \\
&= \int_{\mathbb{S}^{n-1}} u^\#(x,r)g(x,r)\,d\sigma(x),
\end{aligned}$$

and integration with respect to $r^{n-1}\,dr$ gives

$$\int_A uh\,d\mathcal{L}^n = \int_A u^\# g\,d\mathcal{L}^n. \tag{9.13}$$

Similarly,

$$\int_A \phi(r,u)h\,d\mathcal{L}^n = \int_A \phi(r,u^\#)g\,d\mathcal{L}^n. \tag{9.14}$$

This completes Step 2.

**Step 3:** Convolution inequalities.

In the remainder of the proof, $x$ is a point in $\mathbb{R}^n$ (earlier in the proof it was a point in $\mathbb{S}^{n-1}$). For functions $\xi$ and $\psi$ on $A$, write $(\xi,\psi) = \int_A \xi\psi\,d\mathcal{L}^n$ whenever the integral exists. Then (9.13) shows that

$$(u,h) = (u^\#, g).$$

Let $K$ be a smooth nonnegative symmetric bump function on $\mathbb{R}^n$, and let $\epsilon_0 = d(\operatorname{supp} g, \partial A)$. For $0 < \epsilon < \epsilon_0$, we may write

$$(u, K_\epsilon * h) = \int_{A\times A} \epsilon^{-n} K_0(\epsilon^{-1}|y-x|)u(x)h(y)\,dx\,dy, \tag{9.15}$$

where $K_0(r) = K(re_1)$ is decreasing and $*$ denotes convolution on $\mathbb{R}^n$. The integral in (9.15) satisfies the hypotheses of Theorem 7.6 in the ring-type set $A$. (Note the remarks after that theorem relax the nonnegativity requirement on $u$ and $h$ to just integrability.) Therefore the cap analogue of the Riesz-type inequality Corollary 2.20 is applicable, and we obtain from Step 2 that

$$(u, K_\epsilon * h) \le (u^\#, K_\epsilon * h^\#) = (u^\#, K_\epsilon * g). \tag{9.16}$$

This application of Riesz rearrangement is the key step in the proof, because, as we proceed to show, it leads to an integral inequality relating $\Delta(u^\#)$ and $\Delta u$, in the weak sense.

The truncation argument at the end of the proof of Theorem 8.1 shows that (9.16) still holds even if $u$ is not nonnegative. Thus, when $0 < \epsilon < \epsilon_0$ we have

$$(u^\#, K_\epsilon * g) \ge (u, K_\epsilon * h),$$

and hence

$$\begin{aligned}(u^\#, K_\epsilon * g - g) &= (u^\#, K_\epsilon * g) - (u^\#, g)\\ &\ge (u, K_\epsilon * h) - (u,h)\\ &= (u * K_\epsilon, h) - (u,h) = (K_\epsilon * u - u, h).\end{aligned} \tag{9.17}$$

(Concerning the second equality, for general functions $f_1, f_2, f_3$, the correct relation is $(f_1, f_2 * f_3) = (f_1 * \tilde{f}_2, f_3)$, where $\tilde{f}_2(x) = f_2(-x)$. In the case at hand, $f_2 = K_\epsilon$ satisfies $K_\epsilon = \tilde{K}_\epsilon$ and so equation (9.17) holds as stated.)

Suppose, for a moment, that $u \in C^2(A)$, and write the equation (9.10) as $-\Delta u = \phi(r,u) + f$ (meaning $f\, d\mathcal{L}^n = d\mu$). Divide (9.17) by $\epsilon^2$ and let $\epsilon \to 0$. Then it follows from Lemma 9.5, (9.14), and the Hardy–Littlewood inequality (the cap analogue of Corollary 2.16) that

$$\begin{aligned}(u^\#, \Delta g) &\geq (\Delta u, h) \\ &= -(\phi(r,u), h) - (f, h) \\ &\geq -(\phi(r,u^\#), g) - (f^\#, h^\#) = -\int_A \phi(r,u^\#) g\, d\mathcal{L}^n - \int_A g\, d\mu^\#.\end{aligned}$$

This completes the proof of Theorem 9.7 when $u \in C^2(A)$.

**Step 4:** General case: bounds for the $L$-integrals.

When $u$ is a general function in $\mathcal{W}(A)$ we must work harder. Lemma 9.6 and the equation $\Delta u = -\phi(r,u) - \mu$ show that if $0 < \epsilon < \epsilon_0$ then

$$\epsilon^{-2}[K_\epsilon * u(x) - u(x)] = -\int_A L_\epsilon(x-y)\phi(r,u(y))\, dy - \int_A L_\epsilon(x-y)\, d\mu(y)$$

whenever $d(x, \partial A) > \epsilon_0$; recall the kernel $L$ was defined in (9.5). Combined with (9.17), this gives for $0 < \epsilon < \epsilon_0$ that

$$\begin{aligned}&-\int_{A\times A} L_\epsilon(x-y)\, h(x)\phi(r,u(y))\, dy\, dx - \int_{A\times A} L_\epsilon(x-y)\, h(x)\, d\mu(y)\, dx \\ &\leq \epsilon^{-2}(u^\#, K_\epsilon * g - g).\end{aligned} \tag{9.18}$$

Recall $\mu$ decomposes into three parts: $d\mu = f\, d\mathcal{L}^n + d\tau - d\eta$. Thus the second integral in (9.18) decomposes into three additional integrals, which we estimate in turn.

*First integral.* Since $L$ has the same relevant properties as $K$, and since $f \in L^1_{loc}(A)$, the Riesz inequality is valid:

$$\int_{A\times A} L_\epsilon(x-y)h(x)f(y)\, dx\, dy \leq \int_{A\times A} L_\epsilon(x-y)g(x)f^\#(y)\, dx\, dy. \tag{9.19}$$

*Second integral.* For $R_1 < r < R_2$ set $g_0(r) = g(re_1)$. We claim that

$$\int_{A\times A} L_\epsilon(x-y)h(x)\, dx\, d\tau(y) \leq \int_{A\times A} L_\epsilon(x-y)g_0(|x|)\, dx\, d\tau^\#(y). \tag{9.20}$$

To see this, note that since $h^\# = g^\# = g$, we have $0 \le h(x) \le g_0(|x|)$ for all $x \in A$. Hence,

$$\begin{aligned}\int_{A\times A} L_\epsilon(x-y)h(x)\,dx\,d\tau(y) &\le \int_{A\times A} L_\epsilon(x-y)g_0(|x|)\,dx\,d\tau(y)\\ &= \int_{A\times A} L_\epsilon(x-|y|e_1)g_0(|x|)\,dx\,d\tau(y)\\ &= \int_{A\times A} L_\epsilon(x-|y|e_1)g_0(|x|)\,dx\,d\tau^\#(y),\end{aligned}$$

where in the second line we used rotational symmetry of $L$ and a change of variable, and in the third line we used that $\tau$ and $\tau^\#$ give the same integral for any radial function. After once more invoking rotational symmetry of $L$, we arrive at (9.20).

*Third integral.* Let $g_1(r) = g(-re_1)$. Then we have $0 \le g_1(|x|) \le h(x)$ for all $x \in A$. With simple modifications, the argument just given for $\tau$ works also for $\eta$, and produces the inequality

$$\int_{A\times A} L_\epsilon(x-y)h(x)\,dx\,d\eta(y) \ge \int_{A\times A} L_\epsilon(x-y)g_1(|x|)\,dx\,d\eta^\#(y), \tag{9.21}$$

and (9.18), (9.19), (9.20), and (9.21) yield

$$\begin{aligned}&-\int_A\int_A L_\epsilon(x-y)\,h(x)\,dx\,\phi(r,u(y))\,dy - \int_A\int_A L_\epsilon(x-y)g(x)\,dx f^\#(y)\,dy\\ &\quad-\int_A\int_A L_\epsilon(x-y)g_0(|x|)\,dx\,d\tau^\#(y) + \int_A\int_A L_\epsilon(x-y)g_1(|x|)\,dx\,d\eta^\#(y)\\ &\qquad\le \epsilon^{-2}\int_A u^\#(x)[K_\epsilon * g(x) - g(x)]\,dx.\end{aligned} \tag{9.22}$$

In §9.4, $L$ is a smooth symmetric decreasing nonnegative bump function in $\mathbb{R}^n$ with support in the unit ball and $\int_{\mathbb{R}^n} L\,dx = C_K$. From boundedness of $h$ and $g$, we have the following four limits:

$$\begin{aligned}\lim_{\epsilon\to 0}\int_A L_\epsilon(x-y)h(x)\,dx &= C_K h(y),\\ \lim_{\epsilon\to 0}\int_A L_\epsilon(x-y)g(x)\,dx &= C_K g(y),\\ \lim_{\epsilon\to 0}\int_A L_\epsilon(x-y)g_0(|x|)\,dx &= C_K g_0(|y|),\\ \lim_{\epsilon\to 0}\int_A L_\epsilon(x-y)g_1(|x|)\,dx &= C_K g_1(|y|).\end{aligned}$$

The first holds for almost every $y$, and further each term is bounded because $\left|\int_A L_\epsilon(x-y)h(x)\,dx\right|$ is dominated by $C_K$ times the maximum value of $h$. Since $g$ is continuous, the second, third, and fourth limits behave even better: they hold uniformly with respect to $y$ on compact subsets of $A$, and hence uniformly on $A$ (because $g$ has compact support). Also, by Lemma 9.5,

$$\lim_{\epsilon\to 0}\epsilon^{-2}[K_\epsilon * g(x) - g(x)] = C_K \Delta g(x)$$

uniformly for $x \in A$. Since $g$ and $h$ have compact support and $\phi(r,u)$, $u$, and $f$ are all locally integrable and $\eta$ and $\tau$ are in $M_{loc}(A)$, we may pass to limits under the integral signs in (9.22), and obtain

$$\begin{aligned} &-\int_A h(y)\phi(r,u(y))\,dy - \int_A g(y) f^{\#}(y)\,dy \\ &-\int_A g_0(|y|)\,d\tau^{\#}(y) + \int_A g_1(|y|)\,d\eta^{\#}(y) \le \int_A u^{\#}(x)\Delta g(x)\,dx. \end{aligned} \tag{9.23}$$

The first integral on the left equals $\int_A g\phi(r,u^{\#})\,d\mathcal{L}^n$ by (9.14). Since $\tau^{\#}$ is supported on the positive $x_1$-axis and $g_0(|y|) = g(|y|e_1)$ for $y \in A$, we have $\int_A g_0(|y|)\,d\tau^{\#}(y) = \int_A g(y)\,d\tau^{\#}(y)$. Similarly, $\int_A g_1(|y|)\,d\eta^{\#}(y) = \int_A g(y)\,d\eta_{\#}(y)$. Thus, the left side of (9.23) equals $-\int_A g\phi(r,u^{\#})\,d\mathcal{L}^n - \int_A g\,d\mu^{\#}$, and the proof of Theorem 9.7 is complete.

## 9.6 The ★-Function on Shells

As in §9.5, let $A = \{x \in \mathbb{R}^n \colon R_1 < |x| < R_2\}$ be a shell in $\mathbb{R}^n$, where $0 \le R_1 < R_2 \le \infty$ and $n \ge 2$, and continue to write # for $(n-1,n)$-cap symmetrization. As in §9.2, define the new domain $A^\star \subset \mathbb{R}^2$ to be the rectangle

$$A^\star = \{(r,\theta) \in \mathbb{R}^2 \colon R_1 < r < R_2,\ 0 < \theta < \pi\}.$$

For $u \in L^1_{loc}(A)$, define a new function $u^\star \colon A^\star \to \mathbb{R}$ by

$$u^\star(r,\theta) = \sup_E \int_E u(rx)\, r^{n-1}\, d\sigma_{n-1}(x), \tag{9.24}$$

with the sup taken over Borel measurable $E \subset \mathbb{S}^{n-1}$ having $\sigma_{n-1}(E) = \sigma_{n-1}(\mathcal{K}(\theta))$. Here $\mathcal{K}(\theta)$ is the spherical cap on $\mathbb{S}^{n-1}$ with center $e_1$ and opening $\theta$. Notice definition (9.24) includes a factor of $r^{n-1}$, whereas earlier work on the ★-function in complex analysis and for cap symmetrization in higher dimensions did not include such a factor. This change is explained in the notes at the end of the chapter. Corollary 9.10 will present the original ★-function subharmonicity result in the plane.

After a change of variable, the functions $u^\star(r,\cdot)$ defined in (9.24) become the ★-functions as defined in §9.1 for the slice functions $u(r,\cdot)$, as we now show. If

$$\widetilde{u}(r,t) = \sup_E \int_E u(rs)\, d\sigma_{n-1}(s),$$

where the sup is over Borel sets $E \subset \mathbb{S}^{n-1}$ having $\sigma(E) = t$, then

$$u^\star(r,\theta) = r^{n-1}\widetilde{u}\big(r, \sigma_{n-1}(\mathcal{K}(\theta))\big), \qquad (r,\theta) \in A^\star.$$

For $x \in \mathbb{S}^{n-1}$ let $T(x) = \sigma_{n-1}(\mathcal{K}(\theta))$, where $x\cdot e_1 = \cos\theta$. Then (see §7.1) $T$ is a measure preserving map of $\mathbb{S}^{n-1}$ onto $[0, \beta_{n-1}]$, and $u^\#(rx) = u^*(r, T(x))$, where the $*$ denotes decreasing rearrangement. Proposition 9.2(b) then yields

$$\widetilde{u}(r, \sigma_{n-1}(\mathcal{K}(\theta))) = \int_0^{\sigma_{n-1}(\mathcal{K}(\theta))} u^*(r,s)\, ds = \int_{\mathcal{K}(\theta)} u^\#(rs)\, d\sigma_{n-1}(s),$$

which, with the previous identity, gives

$$u^\star(r,\theta) = \int_{\mathcal{K}(\theta)} u^\#(rs)\, r^{n-1}\, d\sigma_{n-1}(s), \quad (r,\theta) \in A^\star.$$

**$J$-Operator**

Define the operator $J\colon L^1_{loc}(A) \to L^1_{loc}(A^\star)$ by

$$Ju(r,\theta) = \int_{\mathcal{K}(\theta)} u(rx)\, r^{n-1}\, d\sigma_{n-1}(x),$$

and note that

$$u^\star = Ju^\# \qquad \text{on } A^\star.$$

If $h\colon \mathbb{S}^n \to \mathbb{R}$ is nonnegative and measurable or integrable on $\mathbb{S}^n$, then the spherical coordinate formula (7.6) implies

$$\int_{\mathbb{S}^n} h\, d\sigma_n = \int_0^\pi \sin^{n-1}\varphi\, d\varphi \int_{\mathbb{S}^{n-1}} h(\cos\varphi, y\sin\varphi)\, d\sigma_{n-1}(y),$$

where, to avoid some parentheses, we write $y\sin\varphi$ instead of $(\sin\varphi)y$. Assuming $n \geq 3$, replacing $n$ by $n-1$, and replacing $h$ by $\chi_{\mathcal{K}(\theta)}h$, we obtain

$$\int_{\mathcal{K}(\theta)} h\, d\sigma_{n-1} = \int_0^\theta \sin^{n-2}\varphi\, d\varphi \int_{\mathbb{S}^{n-2}} h(\cos\varphi, y\sin\varphi)\, d\sigma_{n-2}(y). \tag{9.25}$$

For brevity we often express integrals as inner products. If $f$ and $g$ are functions on $A$ for which $fg \in L^1(A)$, we shall write

$$(f,g) = \int_A fg\, d\mathcal{L}^n.$$

If $F$ and $G$ are functions on $A^\star$ for which $FG \in L^1(A^\star)$, we shall write

$$(F,G) = \int_{A^\star} FG\,drd\theta \equiv \int_{R_1}^{R_2}\int_0^\pi F(r,\theta)G(r,\theta)\,d\theta dr$$

where the measure on $A^\star$ is simply $drd\theta$. It will be clear from the context whether $(\cdot,\cdot)$ refers to integration over $A$ or over $A^\star$.

**$J^t$-Operator**

Define the operator $J^t$ from $L^1_{loc}(A^\star) \to L^1_{loc}(A)$ by

$$J^tF(rx) = \int_\theta^\pi F(r,\varphi)\,d\varphi, \qquad r \in (R_1,R_2), \quad x \in \mathbb{S}^{n-1},$$

where $\theta \in (0,\pi)$ and $x \in \mathbb{S}^{n-1}$ are related by $x \cdot e_1 = \cos\theta$. Then the adjoint relation

$$\boxed{(Ju,G) = (u,J^tG)} \tag{9.26}$$

holds if, for example $u \in L^1_{loc}(A)$ and $G \in L^\infty_c(A^\star)$; this adjoint relation follows from (9.25) and Fubini's Theorem, when $n \geq 3$, since each side of (9.26) becomes

$$\int_{R_1}^{R_2}\int_0^\pi\int_0^\theta\int_{\mathbb{S}^{n-2}} G(r,\theta)\,u(r\cos\varphi, ry\sin\varphi)\,r^{n-1}\sin^{n-2}\varphi\,d\sigma_{n-2}(y)d\varphi d\theta dr.$$

The proof when $n = 2$ is even simpler, since the integrals in (9.26) each equal

$$\int_{R_1}^{R_2}\int_0^\pi\int_{-\theta}^\theta G(r,\theta)\,u(r\cos\varphi, r\sin\varphi)\,r\,d\varphi d\theta dr.$$

Thus, $J^t$ is indeed the adjoint operator of $J$, for all $n \geq 2$.

**$\Delta^\star$-Operator**

Let $n \geq 2$. By (7.10), if a function $u$ on $A$ depends only on $r$ and $\theta$, then its $n$-dimensional Laplacian is given in spherical coordinates by

$$\begin{aligned}\Delta u &= r^{1-n}\partial_r(r^{n-1}\partial_r u) + r^{-2}\sin^{2-n}\theta\ \partial_\theta(\sin^{n-2}\theta\ \partial_\theta u)\\ &= \partial_{rr}u + \frac{n-1}{r}\partial_r u + r^{-2}[\partial_{\theta\theta}u + (n-2)\cot\theta\ \partial_\theta u].\end{aligned}$$

For $n = 2$ the formula simplifies to

$$\Delta u = r^{-1}\partial_r(r\partial_r u) + r^{-2}\partial_{\theta\theta}u = \partial_{rr}u + r^{-1}\partial_r u + r^{-2}\partial_{\theta\theta}u.$$

Define operators $\Delta^\star$ and $\Delta^{\star t}$ acting on $C^2(A^\star)$ by

$$\Delta^\star F = \partial_r\big(r^{n-1}\partial_r(r^{1-n}F)\big) + r^{-2}\sin^{n-2}\theta\ \partial_\theta(\sin^{2-n}\theta\ \partial_\theta F)$$
$$= \partial_{rr}F - \frac{n-1}{r}\partial_r F + \frac{n-1}{r^2}F + r^{-2}[\partial_{\theta\theta}F - (n-2)\cot\theta\ \partial_\theta F], \tag{9.27}$$
$$\Delta^{\star t} F = r^{1-n}\partial_r(r^{n-1}\partial_r F) + r^{-2}\partial_\theta[\sin^{2-n}\theta\ \partial_\theta(\sin^{n-2}\theta\ F)]$$
$$= \partial_{rr}F + \frac{n-1}{r}\partial_r F + r^{-2}[\partial_{\theta\theta}F + (n-2)\ \cot\theta\ \partial_\theta F - (n-2)\csc^2\theta\ F].$$

For $n = 2$, this means

$$\Delta^\star F = \partial_r\big(r\partial_r(r^{-1}F)\big) + r^{-2}\partial_{\theta\theta}F = \partial_{rr}F - r^{-1}\partial_r F + r^{-2}F + r^{-2}\partial_{\theta\theta}F,$$
$$\Delta^{\star t} F = r^{-1}\partial_r(r\partial_r F) + r^{-2}\partial_{\theta\theta}F = \partial_{rr}F + r^{-1}\partial_r F + r^{-2}\partial_{\theta\theta}F. \tag{9.28}$$

In 2 dimensions $\Delta^{\star t}$ agrees with the polar coordinate form of $\Delta$. The operator $\Delta^\star$ does not equal $\Delta$, essentially because we chose the measure on $A^\star$ to be $drd\theta$ and not $r\,drd\theta$. The correct relation in 2 dimensions is

$$\Delta^\star(rF) = r\Delta F, \tag{9.29}$$

when the Laplacian is expressed in polar coordinates.

### Adjoint Relation and Main Identity

In all dimensions, integration by parts shows that if $F \in C^2(A^\star),\ G \in C_c^2(A^\star)$ then

$$\boxed{(\Delta^\star F, G) = (F, \Delta^{\star t}G)} \tag{9.30}$$

so that $\Delta^{\star t}$ is the adjoint of $\Delta^\star$ on $A^\star$, with respect to the measure $drd\theta$.

The theory of ★-functions is motivated by the following identity.

**Theorem 9.8** (Main identity for the ★-function) *For $u \in C^2(A)$, we have*

$$\boxed{\Delta^\star Ju = J\Delta u} \qquad \textit{on } A^\star. \tag{9.31}$$

*Proof of Main Identity when $n = 2$* We use complex notation:

$$Ju(re^{i\theta}) = \int_{-\theta}^{\theta} u(re^{i\varphi})\,r\,d\varphi,$$
$$\partial_\theta Ju(re^{i\theta}) = \big(u(re^{i\theta}) + u(re^{-i\theta})\big)r,$$
$$\partial_{\theta\theta} Ju(re^{i\theta}) = \big(\partial_\theta u(re^{i\theta}) - \partial_\theta u(re^{-i\theta})\big)r,$$
$$\Delta^\star Ju(re^{i\theta}) = (r\partial_{rr} + \partial_r)\int_{-\theta}^{\theta} u(re^{i\varphi})\,d\varphi + \big(\partial_\theta u(re^{i\theta}) - \partial_\theta u(re^{-i\theta})\big)r^{-1},$$
$$J\Delta u(re^{i\theta}) = \int_{-\theta}^{\theta} (\partial_{rr}u + r^{-1}\partial_r u + r^{-2}\partial_{\varphi\varphi}u(re^{i\varphi}))\,r\,d\varphi.$$

The right-hand sides of the third and fourth lines are equal. □

*Proof of Main Identity when $n \geq 3$* Following (7.8), write $\Delta = \Delta_r + r^{-2}\Delta_s$, where $\Delta_r$ and the $\Delta_s$ are the terms in $\Delta$ involving $\partial_r$ and $\partial_{\theta_j}$, respectively. Then, see (7.11), $\Delta_s$ is indeed the spherical Laplacian on $\mathbb{S}^{n-1}$. Similarly, write $\Delta^\star = \Delta_r^\star + r^{-2}\Delta_s^\star$. To prove (9.31), it suffices to verify it when $\Delta$ and $\Delta^\star$ are replaced by $\Delta_r$ and $\Delta_r^\star$, and by $\Delta_s$ and $\Delta_s^\star$. For $\Delta_r$ and $\Delta_r^\star$, the verification is easy: form $Ju$, then put the $r$-derivatives inside the integral.

To obtain (9.31) for $\Delta_s$ and $\Delta_s^\star$, we shall prove that $J(\Delta_s u)(r,\theta)$ and $\Delta_s^\star(Ju)(r,\theta)$ each equal

$$r^{n-1}\sin^{n-2}\theta \int_{\mathbb{S}^{n-2}} \partial_\theta[u(r\cos\theta, ry\sin\theta)]\, d\sigma_{n-2}(y) \tag{9.32}$$

when $0 < \theta < \pi, R_1 < r < R_2$. It will suffice to verify these identities when $r = 1$. Note that $\partial\mathcal{K}(\theta)$ is a Euclidean $(n-2)$-sphere in $\mathbb{R}^n$ with radius $\sin\theta$, and the Hausdorff measure $\mathcal{H}^{n-2}$ on $\partial\mathcal{K}(\theta)$ is uniform, with total mass $\beta_{n-2}\sin^{n-2}\theta$.

The definition of $J$ and the integration formula (9.25) imply that

$$\begin{aligned} Ju(1,\theta) &= \int_{\mathcal{K}(\theta)} u(x)\, d\sigma_{n-1}(x) \\ &= \int_0^\theta \sin^{n-2}\varphi\, d\varphi \int_{\mathbb{S}^{n-2}} u(\cos\varphi, y\sin\varphi)\, d\sigma_{n-2}(y), \end{aligned}$$

from which follows

$$\begin{aligned} \Delta_s^\star Ju(1,\theta) &= (\sin^{n-2}\theta)\partial_\theta \int_{\mathbb{S}^{n-2}} u(\cos\theta, y\sin\theta)\, d\sigma_{n-2}(y) \\ &= \sin^{n-2}\theta \int_{\mathbb{S}^{n-2}} \partial_\theta[u(\cos\theta, y\sin\theta)]\, d\sigma_{n-2}(y). \end{aligned}$$

Thus, $\Delta_s^\star Ju(1,\theta)$ equals the expression in (9.32) when $r = 1$.

On the other hand, the Gauss–Green Theorem for domains on $\mathbb{S}^{n-1}$, implies

$$J\Delta_s u(1,\theta) = \int_{\mathcal{K}(\theta)} \Delta_s u\, d\sigma_{n-1} = \int_{\partial\mathcal{K}(\theta)} \nabla_s u \cdot \nu\, d\mathcal{H}^{n-2}, \tag{9.33}$$

where $\nu$ is the outward pointing unit normal vector on $\partial\mathcal{K}(\theta)$. For $x \in \partial\mathcal{K}(\theta)$, write $x = (\cos\theta, y\sin\theta)$ with $y \in \mathbb{S}^{n-2}$. This provides a parametrization of $\partial\mathcal{K}(\theta)$, and the reader may show that

$$(\nabla_s u \cdot \nu)(x) = \partial_\theta[u(\cos\theta, y\sin\theta)].$$

The reader may also verify the change of variable formula

$$d\mathcal{H}^{n-2}(x) = \sin^{n-2}\theta\, d\sigma_{n-2}(y).$$

Substitution of these last two identities into (9.33) shows that $J\Delta_s u(1,\theta)$ also equals (9.32) when $r = 1$. Thus, the Main Identity is proved. □

We shall need an adjoint version of the Main Identity. Assume $u \in C^2(A)$ and $G \in C_c^2(A^\star)$. Then

$$\begin{aligned}(u, J^t \Delta^{\star t} G) &= (Ju, \Delta^{\star t} G) \\ &= (\Delta^\star Ju, G) = (J\Delta u, G) = (\Delta u, J^t G) = (u, \Delta J^t G).\end{aligned}$$

(The last equality follows from Green's formula, using the compact support of $G$.) Hence

$$J^t \Delta^{\star t} G = \Delta J^t G \tag{9.34}$$

for every $G \in C_c^2(A^\star)$.

A weak version of the Main Identity will also be needed. Take $v \in L^1_{loc}(A)$ and $G \in C_c^2(A^\star)$. Set $g = J^t G$. Then with the help of (9.34),

$$(Jv, \Delta^{\star t} G) = (v, J^t \Delta^{\star t} G) = (v, \Delta J^t G) = (v, \Delta g).$$

That is,

$$\int_{A^\star} Jv \, \Delta^{\star t} G \, dr d\theta = \int_A v \Delta g \, d\mathcal{L}^n \tag{9.35}$$

for all $v \in L^1_{loc}(A)$ and $G \in C_c^2(A^\star)$. Applying this formula to $v = u^\#$ with $u \in L^1_{loc}(A)$ gives

$$\int_{A^\star} u^\star \Delta^{\star t} G \, dr d\theta = \int_A u^\# \Delta g \, d\mathcal{L}^n. \tag{9.36}$$

**Locally Finite Measures**

Our next aim is to define a ★-operation for locally finite measures. As in §9.5, let $\mu \in M_{loc}(A)$. Then we have the decomposition $d\mu = f \, d\mathcal{L}^n + d\tau - d\eta$. We defined $\mu^\# \in M_{loc}(A)$ by the formula $d\mu^\# = f^\# \, d\mathcal{L}^n + d\tau^\# - d\eta_\#$, where $\#$ is $(n-1, n)$ cap symmetrization, $\tau^\#$ is a certain measure supported on the positive $x_1$-axis, and $\eta_\#$ is a certain measure supported on the negative $x_1$-axis.

Define now $\mu^\star \in M_{loc}(A^\star)$ by the formula

$$d\mu^\star = f^\star \, dr d\theta + d\tau^\star \tag{9.37}$$

where $f^\star$ is the star function of $f$ as defined in (9.24). To obtain $\tau^\star$, we view $\tau^\#$ as a measure on the line segment $PA$ (see §9.5), or equivalently on the interval $(R_1, R_2)$. Then define $\tau^\star$ to be the product measure $\tau^\# \times \mathcal{L}|_{(0,\pi)}$, where the subscript denotes restriction. Then $\tau^\star \in M_{loc}(A^\#)$ and Fubini's Theorem gives

$$\tau^\star(E) = \int_{(R_1, R_2)} \mathcal{L}(E^r) \, d\tau^\#(r), \qquad E \in \mathcal{B}_c(A^\star).$$

For example, if $E = (a,b) \times (\theta_1, \theta_2)$ is a rectangle in $A^\star$, then

$$\tau^\star(E) = (\theta_2 - \theta_1)\tau^\#\big((a,b)\big).$$

In the formation of $\mu^\star$ the mass $\eta_\#$ on the negative axis has been ignored, for reasons that will become clear below.

Take $G \in C_c(A^\star)$. Define $g = J^t G$. The definition of $J^t$ implies that $g = 0$ on the negative $x_1$ axis. Write $(G, \mu^\star) = \int_{A^\star} G\, d\mu^\star$. Then $(G, \mu^\star) = (G, f^\star) + (G, \tau^\star)$. Now,

$$(G, f^\star) = (G, Jf^\#) = (J^t G, f^\#) = (g, f^\#).$$

Also, if $G(r,\theta)$ is the characteristic function of a rectangle $(a,b) \times (\theta_1, \theta_2)$ in $A^\star$ then $g(re_1) = (\theta_2 - \theta_1)\chi_{(a,b)}(r)$ and we have

$$(G, \tau^\star) = (\theta_2 - \theta_1)\tau^\#\big((a,b)\big) = (g, \tau^\#).$$

By the theory of product measures, it follows that $(G, \tau^\star) = (g, \tau^\#)$ for all $G \in C_c(A^\star)$, where $g = J^t G$. From the last several identities, we conclude that $(G, \mu^\star) = (g, f^\#) + (g, \tau^\#)$. Furthermore, since $g$ vanishes on the part of $A$ along the negative $x_1$ axis, we have $(g, \eta_\#) = 0$. Subtracting this from the previous identity, we conclude that

$$\int_{A^\star} G\, d\mu^\star = \int_A g\, d\mu^\# \tag{9.38}$$

holds for all $G \in C_c(A^\star)$ and $g = J^t G$.

Let now $F \in L^1_{loc}(A^\star)$ and $\nu \in M_{loc}(A^\star)$. We say that the differential inequality

$$-\Delta^\star F \le \nu$$

holds in the weak sense, or in the sense of distributions, if

$$-\int_{A^\star} F \Delta^{\star t} G\, dr d\theta \le \int_{A^\star} G\, d\nu$$

for all nonnegative $G \in C_c^2(A^\star)$. For example, if $F \in C^2(A^\star), H \in L^1_{loc}(A^\star)$, and $-\Delta^\star F \le H$ holds at every point of $A^\star$, so that the inequality holds in the strong or classical sense, then the inequality also holds in the weak sense due to the adjoint property (9.30) for the differential operator.

Here, finally, is our "subharmonicity result."

**Theorem 9.9** (Subharmonicity property of the ★-function on shells) *If $u \in \mathcal{W}(A)$ satisfies*

$$-\Delta u = \phi(r, u) + \mu$$

*in the weak sense in A, where* $\phi\colon (R_1, R_2) \times \mathbb{R} \to \mathbb{R}$ *is continuous with* $\phi(r,u)$ *locally integrable on A and* $\mu \in M_{loc}(A)$*, then*

$$-\Delta^\star u^\star \leq J\phi(r, u^\#) + \mu^\star$$

*in the weak sense in the rectangle* $A^\star$.

When we write $J\phi(r, u^\#)$ in the theorem, we mean the $J$-operator applied to the function $\phi(r, u^\#) \in L^1_{loc}(A)$.

*Proof* Let $g = J^t G$ and combine (9.36), (9.26) and (9.38) with Theorem 9.7 (pre-subharmonicity). To apply Theorem 9.7, note that $g = g^\#$, and also $G \geq 0$ implies $g \geq 0$ in $A$. Also, the compact support of $G$ in $A^\star$ insures that $g$ is compactly supported in $A$, that $g$ is independent of $\theta$ when $\theta$ is near 0, and that $g$ equals zero when $\theta$ is close to $\pi$. These last facts and the hypothesis $G \in C_c^2(A^\star)$ make it easy to show $g \in C_c^2(A)$. □

The next corollaries remind us of the origins of the star function in the complex plane, and provide a foundation for Chapter 11.

**Corollary 9.10** (Subharmonicity of the ★-function in the plane) *If* $n = 2$ *and* $u \in L^1_{loc}(A)$ *satisfies*

$$\Delta u \geq 0$$

*in the weak sense in an annulus* $A \subset \mathbb{C}$*, then*

$$\Delta(u^\star/r) \geq 0$$

*in the weak sense in the upper half annulus. In other words, if u is subharmonic then so is* $u^\star/r$.

Remember that in the plane, the definition (9.24) of the ★-function gives $u^\star(r,\theta)/r = \sup_{|E|=2\theta} \int_E u(re^{i\psi})\, d\psi$.

*Proof* Let $\phi \equiv 0$. Define $\mu = -\Delta u$, so that $\mu \leq 0$ is a measure as explained in §9.2. Clearly $f \leq 0$ and $\tau = 0$ in the decomposition (9.9) of $\mu$, and so the definition (9.37) gives $\mu^\star \leq 0$. Now the corollary follows immediately from Theorem 9.9 and the relation (9.29) between $\Delta^\star$ and $\Delta$. □

**Corollary 9.11** (Log of a meromorphic function) *If* $f(z)$ *is meromorphic in a punctured disk* $\mathbb{D}(R) \setminus \{0\}$*, and has poles* $\{b_k\}$ *there listed with multiplicity, then*

$$\Delta\big((\log|f|)^\star/r\big) \geq -2\pi \sum_k \frac{s_k}{|b_k|}$$

*in the upper half disk, where* $s_k$ *is arclength measure on the circle of radius* $|b_k|$.

The conclusion of the corollary means that if $\eta$ is a nonnegative test function supported in the upper half disk $\mathbb{D}^+(R)$ then

$$\int_{\mathbb{D}^+(R)} \big((\log|f|)^\star/r\big)(\Delta\eta)\, r\,drd\theta \geq -2\pi \sum_k \int_0^\pi \eta(|b_k|e^{i\theta})\,d\theta. \qquad (9.39)$$

*Proof* Let

$$\mu = -\Delta(\log|f|) = 2\pi \sum_k \delta_{b_k} - 2\pi \sum_j \delta_{a_j}$$

where the $\{a_j\}$ are the zeros of $f$. The construction of $\mu^\star$ in (9.37) gives

$$\mu^\star = 2\pi \sum_k (\delta_{b_k})^\star = 2\pi \sum_k \mu_k$$

where $\mu_k$ is Lebesgue measure along the vertical line $r = |b_k|$ in the $r\theta$-plane. Hence Theorem 9.9 with $\phi \equiv 0$ says

$$\Delta^\star(\log|f|)^\star \geq -2\pi \sum_k \mu_k$$

in the weak sense in the rectangle $(0, R) \times (0, \pi)$. Applying this differential inequality to the test function $\eta(re^{i\theta})$, where $\eta(z)$ is nonnegative, smooth, and supported in the upper half disk, gives that

$$\int_0^\pi \int_0^R (\log|f|)^\star \Delta^{\star t}\big(\eta(re^{i\theta})\big)\,drd\theta \geq -2\pi \sum_k \int_0^\pi \eta(|b_k|e^{i\theta})\,d\theta.$$

In 2 dimensions $\Delta^{\star t} = \Delta$, by definition (9.28), and so multiplying and dividing by $r$ to get $r\,drd\theta$ in the last formula gives the conclusion (9.39) that we wanted. □

## 9.7 The ⋆-Function on the Sphere

In this section, unless otherwise indicated, # will denote symmetric decreasing rearrangement on $\mathbb{S}^n$, with $n \geq 1$. Our aim is to introduce the star function in this setting, and to deduce analogues of Theorems 9.7 and 9.9 by extending the functions from a sphere to a shell. We shall write $L^1(\mathbb{S}^n) = L^1(\mathbb{S}^n, \sigma_n)$, and

$$(f, g) = \int_{\mathbb{S}^n} fg\,d\sigma_n,$$

whenever the integral exists. We shall also write

$$(f, \mu) = \int_{\mathbb{S}^n} f\,d\mu$$

when $\mu \in M(\mathbb{S}^n)$, the set of real-valued (signed finite) regular Borel measures on $\mathbb{S}^n$, and $f$ is such that the integral exists. Recall $\Delta_s$ is the spherical Laplacian, defined in §7.2.

Denote by $\mathcal{W}(\mathbb{S}^n)$ the set of $u \in L^1(\mathbb{S}^n)$ for which the distributional spherical Laplacian $\Delta_s u$ can be represented by a (signed) measure. We say $-\Delta_s u = \mu$ in the weak sense on $\mathbb{S}^n$ (for given $\mu \in M(\mathbb{S}^n)$) if

$$-(u, \Delta_s g) = (g, \mu), \qquad \forall g \in C^2(\mathbb{S}^n).$$

More generally, if $\phi$ is continuous and $\phi(u)$ is integrable then we say

$$-\Delta_s u = \phi(u) + \mu$$

in the weak sense if

$$-(u, \Delta_s g) = (\phi(u), g) + (g, \mu), \qquad \forall g \in C^2(\mathbb{S}^n).$$

Construct the measure $\mu^\#$ as follows. Let $d\mu = f\, d\sigma_n + d\tau - d\eta$ be the Jordan–Lebesgue decomposition of $\mu$, where $f \in L^1(\mathbb{S}^n)$, and $\tau$ and $\eta$ are nonnegative Borel measures on $\mathbb{S}^n$ singular with respect to $\sigma_n$ and each other. Define $\mu^\# \in M(\mathbb{S}^n)$ by

$$d\mu^\# = f^\#\, d\sigma_n + d\tau^\# - d\eta_\#$$

where $\tau^\# = \tau(\mathbb{S}^n)\delta_{e_1}$ and $\eta^\# = \eta(\mathbb{S}^n)\delta_{-e_1}$.

The following pre-subharmonicity theorem says $-\Delta_s(u^\#) \le \phi(u^\#) + \mu^\#$ weakly when restricted to symmetric decreasing, nonnegative test functions on the sphere.

**Theorem 9.12** (Pre-subharmonicity theorem for s.d.r. on $\mathbb{S}^n$) *Assume $u \in \mathcal{W}(\mathbb{S}^n)$ satisfies $-\Delta_s u = \phi(u) + \mu$ in the weak sense on $\mathbb{S}^n$, where $\phi\colon \mathbb{R} \to \mathbb{R}$ is continuous with $\phi(u)$ integrable on $\mathbb{S}^n$ and $\mu \in M(\mathbb{S}^n)$.*

*If $g \in C^2(\mathbb{S}^n)$ is nonnegative with $g = g^\#$ then*

$$-\int_{\mathbb{S}^n} u^\# \Delta_s g\, d\sigma_n \le \int_{\mathbb{S}^n} \phi(u^\#) g\, d\sigma_n + \int_{\mathbb{S}^n} g\, d\mu^\#.$$

*Proof* We shall deduce Theorem 9.12 from the shell theorem, Theorem 9.7. Fix $R_1, R_2$ with $0 < R_1 < 1 < R_2 < \infty$. Write $A = \{x \in \mathbb{R}^{n+1} : R_1 < |x| < R_2\}$. We extend the functions and measures to $A$ as follows: for $x \in \mathbb{S}^n, r \in (R_1, R_2)$, let

$$\begin{aligned} &\widetilde{u}(rx) = u(x), \quad \widetilde{\phi}(r,z) = r^{-2}\phi(z), \\ &\widetilde{f}(rx) = r^{-2} f(x), \quad d\widetilde{\tau} = r^{n-2}\, dr\, d\tau(x), \quad d\widetilde{\eta} = r^{n-2}\, dr\, d\eta(x), \\ &\quad d\widetilde{\mu} = \widetilde{f}\, d\mathcal{L}^{n+1} + d\widetilde{\tau} - d\widetilde{\eta}. \end{aligned}$$

We claim that the extended function satisfies

$$-\Delta\widetilde{u} = \widetilde{\phi}(r,\widetilde{u}) + \widetilde{\mu} \tag{9.40}$$

in $A$, in the weak sense. Indeed, multiplying each term on the right by a test function $\widetilde{g} \in C_c^2(A)$ and integrating in spherical coordinates shows that

$$\int_A \widetilde{\phi}(r,\widetilde{u})\widetilde{g}\, d\mathcal{L}^{n+1} = \int_{R_1}^{R_2}\int_{\mathbb{S}^n} \phi(u)\widetilde{g}(rx)\, d\sigma_n(x)\, r^{n-2}\, dr,$$

$$\int_A \widetilde{g}\, d\widetilde{\mu} = \int_{R_1}^{R_2}\int_{\mathbb{S}^n} \widetilde{g}(rx)\, d\mu(x)\, r^{n-2}\, dr.$$

For the left side, we observe in spherical coordinates that

$$\begin{aligned}\int_A \widetilde{u}\Delta\widetilde{g}\, d\mathcal{L}^{n+1} &= \int_{R_1}^{R_2}\int_{\mathbb{S}^n} u(x)\Big[\frac{\partial}{\partial r}\Big(r^n\frac{\partial\widetilde{g}}{\partial r}\Big) + r^{n-2}\Delta_s\widetilde{g}(rx)\Big] d\sigma_n(x)\, dr,\\ &= \int_{R_1}^{R_2}\int_{\mathbb{S}^n} u(x)\Delta_s\widetilde{g}(rx)\, d\sigma_n(x)\, r^{n-2}\, dr,\end{aligned}$$

where we eliminated the first part of the integrand by integrating with respect to $r$ and using the compact support of $\widetilde{g}$. Since for each $r$ the function $x \mapsto \widetilde{g}(rx)$ is a $C^2$-smooth test function on the sphere, we may add the last three displayed equations and use the assumption $-\Delta_s u = \phi(u) + \mu$ on $\mathbb{S}^n$ to arrive at (9.40).

The shell pre-subharmonicity result, Theorem 9.7, now implies that if $\widetilde{g} \geq 0$ and $\widetilde{g} = \widetilde{g}^\#$ then

$$-\int_A \widetilde{u}^\# \Delta\widetilde{g}\, d\mathcal{L}^{n+1} \leq \int_A \widetilde{\phi}(r,\widetilde{u}^\#)\widetilde{g}\, d\mathcal{L}^{n+1} + \int_A \widetilde{g}\, d\widetilde{\mu}^\#,$$

where the #'s for all objects defined on $A$ denote $(n, n+1)$-cap symmetrization.

Consider a test function in separated form, $\widetilde{g}(rx) = \psi(r)g(x)$, where $\psi \in C_c^2(R_1,R_2)$ is a nonnegative function satisfying $\int_{R_1}^{R_2}\psi(r)r^{n-2}\, dr = 1$ and $g \in C^2(\mathbb{S}^n)$ is nonnegative with $g = g^\#$ on $\mathbb{S}^n$. Then $\widetilde{g} \geq 0$ and $\widetilde{g} = \widetilde{g}^\#$ on $A$. Substituting $\widetilde{g}$ into the preceding displayed formula gives (after calculations similar to those done already) that

$$-\int_{\mathbb{S}^n} u^\#\Delta_s g\, d\sigma_n \leq \int_{\mathbb{S}^n} \phi(u^\#)g\, d\sigma_n + \int_{\mathbb{S}^n} g\, d\mu^\#,$$

which proves the theorem. □

Next come the definitions of the ★-function and associated objects on the sphere:

$$Ju(\theta) = \int_{\mathcal{K}(\theta)} u \, d\sigma_n, \qquad \theta \in [0, \pi], \quad u \in L^1(\mathbb{S}^n),$$

$$J^t F(x) = \int_\theta^\pi F(\varphi) \, d\varphi, \qquad x \in \mathbb{S}^n, \quad x \cdot e_1 = \cos\theta, \quad F \in L^1[0, \pi],$$

$$u^\star(\theta) = Ju^\#(\theta) = \sup_E \int_E u \, d\sigma_n,$$

where the sup is taken over all Borel $E \subset \mathbb{S}^n$ with $\sigma_n(E) = \sigma_n(\mathcal{K}(\theta))$. The adjoint relation on the sphere (proved like (9.26) for shells) says

$$(Ju, G) = (u, J^t G) \tag{9.41}$$

for $u \in L^1(\mathbb{S}^n)$ and $G \in L_c^\infty(0, \pi)$.

Define differential operators

$$\Delta_s^\star F(\theta) = F''(\theta) - (n-1)(\cot\theta) F'(\theta),$$
$$\Delta_s^{\star t} F(\theta) = F''(\theta) + (n-1)(\cot\theta) F'(\theta) - (n-1)(\csc^2\theta) F(\theta),$$

for $F \in C^2(0, \pi)$. The operator $\Delta_s^{\star t}$ is the adjoint of $\Delta_s^\star$ with respect to $d\theta$ on $(0, \pi)$. The Weak Main Identity on $\mathbb{S}^n$ says

$$\int_0^\pi Jv \, \Delta_s^{\star t} G \, d\theta = \int_{\mathbb{S}^n} v \Delta_s g \, d\sigma_n \tag{9.42}$$

for all $v \in L^1(\mathbb{S}^n)$ and $G \in C_c^2(0, \pi)$, and choosing $v = u^\#$ with $u \in L^1(\mathbb{S}^n)$ implies

$$\int_0^\pi u^\star \, \Delta_s^{\star t} G \, d\theta = \int_{\mathbb{S}^n} u^\# \Delta_s g \, d\sigma_n. \tag{9.43}$$

Note that (9.42) is the $\mathbb{S}^n$ version of identity (9.35), and can be proved analogously or else deduced from that identity by homogeneous extension as used earlier in this section (the appropriate extension of $G$ is $\widetilde{G}(r, \varphi) = \psi(r) G(\varphi)$).

For $\mu \in M(\mathbb{S}^n)$ with Jordan–Lebesgue decomposition $d\mu = f \, d\sigma_n + d\tau - d\eta$, define $\mu^\star \in M\big((0, \pi)\big)$ by

$$d\mu^\star = f^\star \, d\theta + \tau(\mathbb{S}^n) \, d\theta.$$

Remember that all "starred" objects live on the interval $(\mathbb{S}^n)^\star = (0, \pi)$.

**Theorem 9.13** (Subharmonicity property of the ⋆-function on $\mathbb{S}^n$) *If $u \in \mathcal{W}(\mathbb{S}^n)$ satisfies*

$$-\Delta_s u = \phi(u) + \mu$$

*in the weak sense on $\mathbb{S}^n$, where $\phi\colon \mathbb{R} \to \mathbb{R}$ is continuous with $\phi(u)$ integrable on $\mathbb{S}^n$ and $\mu \in M(\mathbb{S}^n)$, then*

$$-\Delta_s^\star u^\star \le J\phi(u^\#) + \mu^\star$$

*in the weak sense on* $(0, \pi)$.

Spelled out, the conclusion of the theorem says that

$$-\int_0^\pi u^\star \Delta_s^{\star t} G\, d\theta \le \int_0^\pi J\phi(u^\#) G\, d\theta + \int_0^\pi G\, d\mu^\star \tag{9.44}$$

for every nonnegative $G \in C_c^2(0, \pi)$.

*Proof* Given a nonnegative $G \in C_c^2(0, \pi)$, set $g = J^t G$. Then $g = g^\#$ and $g$ is nonnegative, with $g \in C^2(\mathbb{S}^n)$ by construction (since the compact support of $G$ insures $g$ is constant on a small neighborhood of $e_1$, and on a small neighborhood of $-e_1$). Moreover, we have the identities

$$\int_0^\pi u^\star \Delta_s^{\star t} G\, d\theta = \int_{\mathbb{S}^n} u^\# \Delta_s g\, d\sigma_n,$$
$$\int_0^\pi J\phi(u^\#) G\, d\theta = \int_{\mathbb{S}^n} \phi(u^\#) g\, d\sigma_n,$$
$$\int_0^\pi G\, d\mu^\star = \int_{\mathbb{S}^n} g\, d\mu^\#,$$

by (9.43), (9.41), and an $\mathbb{S}^n$ version of (9.38) (which can be deduced from (9.38) by extension, as above). The desired inequality (9.44) now follows from the pre-subharmonicity result in Theorem 9.12.

Alternatively, one may deduce (9.44) by directly applying the shell subharmonicity result Theorem 9.9 to the homogeneous extension of $u$. □

## 9.8 The ★-Function for Cap Symmetrization on Ring-Type Domains

The setting here is that of Chapter 7. Thus, we fix $n \ge 3$ and $1 \le k \le n-2$. For $x \in \mathbb{R}^n$, write $x = (w, z)$, where $w \in \mathbb{R}^{k+1}$, $z \in \mathbb{R}^{m-1}$, and $k + m = n$. Let $X$ be an open subset of $\mathbb{R}^n$, and let $Z$ be the set of all $z \in \mathbb{R}^{m-1}$ such that the slice $X(z) \subset \mathbb{R}^{k+1}$ is nonempty. Recall $X$ is *ring-type* if, for each $z \in Z$, $X(z)$ equals a union of open shells in $\mathbb{R}^{k+1}$. That is, for each $z \in Z$ there is a nonempty open set $B(z) \subset \mathbb{R}^+$ such that

$$X = \{(ry, z) \in \mathbb{R}^n \colon\ r \in B(z), y \in \mathbb{S}^k, z \in Z\}.$$

**Example 9.14** In 3 dimensions, an open set that is rotationally symmetric about the $z$-axis is ring-type.

In this section, $X$ will always be ring-type.

If $u\colon X \to \mathbb{R}$ is $\mathcal{L}^n$-measurable, then for $z \in Z$ we can consider the slice function $u^z\colon X(z) \to \mathbb{R}$. The $(k,n)$-cap symmetrization $u^\#$ of $u$ is defined to be the function on $X$ obtained by performing a $(k, k+1)$ symmetrization of $u^z$ for each $z \in Z$. In this section, # denotes $(k,n)$-cap symmetrization, unless otherwise indicated.

Next, take $\mu \in M_{loc}(X)$, with Jordan–Lebesgue decomposition

$$d\mu = f\,d\mathcal{L}^n + d\tau - d\eta, \tag{9.45}$$

where $f \in L^1_{loc}(X)$, and $\tau$ and $\eta$ are (nonnegative) Radon measures on $X$ singular with respect to each other and to $\mathcal{L}^n$.

For $x = (w, z) \in \mathbb{R}^n$, write $x = (ry, z)$, where $y \in \mathbb{S}^k$, $r \geq 0$. Define a map $P\colon X \to X$ by $P(x) = (re_1, z)$, where $e_1$ denotes the first coordinate vector in $\mathbb{R}^{k+1}$. For each fixed $z$, $P$ is the spherical projection map of $\mathbb{R}^{k+1}$ onto the positive $x_1$-axis of $\mathbb{R}^{k+1}$. The image set $PX$ of $X$ by $P$ is

$$PX = \{(r, 0, \ldots, 0, z) \in \mathbb{R}^n\colon r \in B(z), z \in Z\},$$

where there are $k$ zeros following the $r$. Note that $PX \subset X$.

Define the Radon measure $\tau^\# \in M_{loc}(PX)$ to be the measure induced by $P$ and $\tau$. That is,

$$\tau^\#(E) = \tau(P^{-1}E)$$

for $E \in \mathcal{B}_c(PX)$. Extend the measure to $\mathbb{R}^n \setminus PX$ by defining $\tau^\#(E) = 0$ whenever $E \cap PX$ is empty. Thus, $\tau^\#$ is the measure obtained from $\tau$ by sweeping all the mass to the positive $x_1$-axis within each $z$-slice.

If $\mu \in M_{loc}(X)$ has the decomposition (9.45), define the symmetrized measure $\mu^\# \in M_{loc}(X)$ to be

$$d\mu^\# = f^\# d\mathcal{L}^n + d\tau^\# - d\eta_\#, \tag{9.46}$$

with $\eta_\#(E) \equiv \eta^\#(E')$, where $E' = \{(-w, z) \in X\colon (w, z) \in E\}$.

Write $\Delta$ for the Euclidean Laplacian on $\mathbb{R}^n$. Let $u \in \mathcal{W}(X)$. Given a continuous real-valued function $\phi$ of three variables such that $\phi(r, z, u)$ is locally integrable, we say

$$-\Delta u = \phi(r, z, u) + \mu$$

in the weak sense if

$$-\int_X u \Delta g\, d\mathcal{L}^n = \int_X \phi(r, z, u) g\, d\mathcal{L}^n + \int_X g\, d\mu$$

for every $g \in C_c^2(X)$.

**Theorem 9.15** (Pre-subharmonicity theorem for $(k,n)$-cap symmetrization) *Assume $u \in \mathcal{W}(X)$ satisfies $-\Delta u = \phi(r,z,u) + \mu$ in the weak sense in $X$, where $\phi\colon \mathbb{R}^+ \times Z \times \mathbb{R} \to \mathbb{R}$ is continuous with $\phi(r,z,u)$ locally integrable on $X$ and $\mu \in M_{loc}(X)$.*

*If $g \in C_c^2(X)$ is nonnegative with $g = g^\#$ then*

$$-\int_X u^\# \Delta g \, d\mathcal{L}^n \le \int_X \phi(r,z,u^\#) g \, d\mathcal{L}^n + \int_X g \, d\mu^\#.$$

*Proof* The proof of the $(n-1,n)$ case in Theorem 9.7 works here too. We just have to throw in some $z$-integrals. The main ingredient in the proof of Theorem 9.7 is a Riesz-type convolution inequality, and the reference given there for the $(n-1,n)$-inequality covers the $(k,n)$ case as well. □

Here is the resulting subharmonicity theorem, generalizing the $(n-1,n)$ case in Theorem 9.9:

**Theorem 9.16** (Subharmonicity property of the ★-function for $(k,n)$-cap symmetrization) *Let $u, \phi$ and $\mu$ be as in Theorem 9.15. Then $-\Delta^\star u^\star \le J\phi(r,z,u^\#) + \mu^\star$ in the weak sense on $X^\star$.*

*Proof* We shall define $X^\star, u^\star$, etc., in the remainder of the section, and will leave it to the reader to verify that the proof of Theorem 9.9 carries over to the $(k,n)$ case when $1 \le k \le n-2$. □

Recall that $X$ can be written as $X = \{(ry,z) \in \mathbb{R}^n : r \in B(z), y \in \mathbb{S}^k, z \in Z\}$, where $B(z)$ is a nonempty open subset of $\mathbb{R}^+$. Define

$$X^\star \equiv \{(r,\theta,z) : r \in B(z),\ \theta \in (0,\pi),\ z \in Z\} \subset \mathbb{R}^{m+1}.$$

Define the operator $J\colon L^1_{loc}(X) \to L^1_{loc}(X^\star)$ by

$$Ju(r,\theta,z) \equiv \int_{\mathcal{K}(\theta)} u(ry,z)\, r^k \, d\sigma_k(y),$$

where $\mathcal{K}(\theta)$ is the spherical cap on $\mathbb{S}^k$ with center $e_1$, opening $\theta$. The term $J\phi(r,z,u^\#)$ in the theorem means we apply the $J$-operator to the function $\phi(r,z,u^\#) \in L^1_{loc}(X)$.

If $f$ and $g$ are functions on $X$ for which $fg \in L^1(X, \mathcal{L}^n)$, write

$$(f,g) = \int_X fg \, d\mathcal{L}^n.$$

If $F$ and $G$ are functions on $X^\star$ for which $FG \in L^1(X^\star,\ drd\theta d\mathcal{L}^{m-1})$, write

$$(F,G) = \int_{X^\star} F(r,\theta,z) G(r,\theta,z) \, drd\theta d\mathcal{L}^{m-1}(z).$$

It will be clear from context whether $(\cdot,\cdot)$ refers to integration over $X$ or over $X^\star$.

Define a "transpose" operator

$$J^t : L^1_{loc}(X^\star,\, dr d\theta d\mathcal{L}^{m-1}) \to L^1_{loc}(X, \mathcal{L}^n)$$

by

$$J^t F(ry, z) = \int_\theta^\pi F(r, \varphi, z)\, d\varphi, \qquad r \in B(z), \qquad y \in \mathbb{S}^k,$$

where $\theta \in (0, \pi)$ and $y$ are related by $y \cdot e_1 = \cos\theta$. Then

$$(Jf, G) = (f, J^t G),$$

if, for example $f \in L^1_{loc}(X)$ and $G \in L^\infty_c(X^\star)$.

If a function $u$ on $X$ depends only on $r, \theta$ and $z$, then its $n$-dimensional (Euclidean) Laplacian is given by

$$\begin{aligned}\Delta u &= r^{-k}\partial_r(r^k \partial_r u) + r^{-2}(\sin^{1-k}\theta)\, \partial_\theta(\sin^{k-1}\theta\, \partial_\theta u) + \Delta_z u\\ &= \partial_{rr}u + \frac{k}{r}\partial_r u + r^{-2}[\partial_{\theta\theta}u + (k-1)\cot\theta\, \partial_\theta u] + \Delta_z u,\end{aligned}$$

where $\Delta_z u = \sum_{i=1}^{m-1} \partial_{z_i z_i} u$ is the Laplacian on $\mathbb{R}^{m-1}$. For $k = 1$ the formula simplifies to

$$\begin{aligned}\Delta u &= r^{-1}\partial_r(r\partial_r u) + r^{-2}\partial_{\theta\theta}u + \Delta_z u\\ &= \partial_{rr}u + r^{-1}\partial_r u + r^{-2}\partial_{\theta\theta}u + \Delta_z f.\end{aligned}$$

Define operators $\Delta^\star$ and $\Delta^{\star t}$ acting on $C^2(X^\star)$ by

$$\begin{aligned}\Delta^\star F &= \partial_r\big(r^k\partial_r(r^{-k}F)\big) + r^{-2}\sin^{k-1}\theta\, \partial_\theta(\sin^{1-k}\theta\, \partial_\theta F) + \Delta_z F\\ &= \partial_{rr}F - \frac{k}{r}\partial_r F + \frac{k}{r^2}F + r^{-2}[\partial_{\theta\theta}F - (k-1)\cot\theta\, \partial_\theta F] + \Delta_z F,\\ \Delta^{\star t} F &= r^{-k}\partial_r(r^k\partial_r F) + r^{-2}\partial_\theta[\sin^{1-k}\theta\, \partial_\theta(\sin^{k-1}\theta\, F)] + \Delta_z F\\ &= \partial_{rr}F + \frac{k}{r}\partial_r F + r^{-2}[\partial_{\theta\theta}F + (k-1)\cot\theta\, \partial_\theta F - (k-1)\csc^2\theta\, F]\\ &\quad + \Delta_z F.\end{aligned}$$

For $k = 1$, the expressions above reduce to

$$\begin{aligned}\Delta^\star F &= \partial_r\big(r\partial_r(r^{-1}F)\big) + r^{-2}\partial_{\theta\theta}F + \Delta_z F\\ &= \partial_{rr}F - \frac{1}{r}\partial_r F + \frac{1}{r^2}F + r^{-2}\partial_{\theta\theta}F + \Delta_z F,\\ \Delta^{\star t} F &= r^{-1}\partial_r(r\partial_r F) + r^{-2}\partial_{\theta\theta}F + \Delta_z F\\ &= \partial_{rr}F + \frac{1}{r}\partial_r F + r^{-2}\partial_{\theta\theta}F + \Delta_z F.\end{aligned}$$

In all dimensions, integration by parts and Fubini's Theorem show that if $F \in C^2(X^\star)$, $G \in C_c^2(X^\star)$ then

$$(\Delta^\star F, G) = (F, \Delta^{\star t} G),$$

so that $\Delta^{\star t}$ is the adjoint of $\Delta^\star$ on $X^\star$, with respect to the measure $drd\theta d\mathcal{L}^{m-1}$.

The Main Identity $\Delta^\star Ju = J\Delta u$ and its adjoint form $J^t \Delta^{\star t} G = \Delta J^t G$ and weak form $(Jv, \Delta^{\star t} G) = (v, \Delta g)$ (where $g = J^t G$) all follow easily from the corresponding formulas in §9.6 for cap symmetrization on shells.

For $u \in L^1_{loc}(X)$, define the function $u^\star$ on $X^\star$ by

$$u^\star(r, \theta, z) = Ju^\#(r, \theta, z) = \sup \int_E u(ry, z)\, r^k\, d\sigma_k(y),$$

where the sup is taken over all Borel measurable $E \subset \mathbb{S}^k$ with $\sigma_k(E) = \sigma_k(\mathcal{K}(\theta))$.

Suppose that $\mu \in M_{loc}(X)$ has Jordan–Lebesgue decomposition

$$d\mu = f\, d\mathcal{L}^n + d\tau - d\eta,$$

where $f \in L^1_{loc}(X)$, and $\tau$ and $\eta$ are (nonnegative) Radon measures on $X$ singular with respect to each other and to $\mathcal{L}^n$. The symmetrized measure $d\mu^\# = f^\#\, d\mathcal{L}^n + d\tau^\# - d\eta_\#$ was defined in (9.46).

We can view $\tau^\#$ as a measure on $PX$ and can view the set $X^\star$ as $PX \times (0, \pi)$. Define $\tau^\star$ to be the product measure $\tau^\# \times \mathcal{L}|_{(0,\pi)}$, where the subscript denotes restriction. Then $\tau^\star \in M_{loc}(X^\star)$, and Fubini's Theorem gives

$$\tau^\star(E) = \int_{PX} \mathcal{L}(E^{r,z})\, d\tau^\#(r, z), \quad E \in \mathcal{B}_c(X^\star),$$

with $E^{r,z} = \{\theta \in (0, \pi) : (r, \theta, z) \in E\}$. For example, if $E = (a, b) \times (\theta_1, \theta_2) \times E_1$ is a rectangular box in $X^\star$, where $E_1 \in \mathcal{B}(\mathbb{R}^{m-1})$, then

$$\tau^\star(E) = (\theta_2 - \theta_1) \int_{E_1} \int_{(a,b)} d\tau^\#(r, z).$$

Define

$$d\mu^\star = f^\star\, drd\theta d\mathcal{L}^{m-1} + d\tau^\star.$$

All objects relevant to the $\star$-function for $(k, n)$-cap symmetrization have now been defined, and we take Theorems 9.15 and 9.16 to be proved.

## 9.9 Pre-Subharmonicity Theorem for s.d.r. on Euclidean Domains

In this section # will denote symmetric decreasing rearrangement on a nonempty open set $\Omega \subset \mathbb{R}^n$. Thus $\Omega^\#$ equals $\mathbb{B}^n(0, R)$, the open ball centered at 0 having radius $R \in (0, \infty]$ determined by the requirement that $\Omega^\#$ and $\Omega$ have the same volume: $\alpha_n R^n = \mathcal{L}^n(\Omega^\#) = \mathcal{L}^n(\Omega)$.

A function $u \colon \Omega \to \mathbb{R}$ is said to satisfy **Condition B** if

$$\int_\Omega (u-t)^+ \, d\mathcal{L}^n < \infty \quad \forall\, t > \operatorname{ess\,inf}_\Omega u.$$

If $u$ satisfies Condition B then $u < \infty$ $\mathcal{L}^n$-a.e. and $u$ satisfies the finiteness condition $\mathcal{L}^n(u > t) < \infty$ for all $t > \operatorname{ess\,inf}_\Omega u$, so that $u^\#$ is well-defined and finite everywhere on $\Omega^\#$, except perhaps at $x = 0$. Furthermore,

$$\int_{\Omega^\#} (u^\# - t)^+ \, d\mathcal{L}^n < \infty \quad \forall\, t > \operatorname{ess\,inf}_{\Omega^\#} u^\#.$$

Hence $(u^\# - t)^+$ is integrable over each compact set in $\Omega^\#$, and so $u^\# \in L^1_{loc}(\Omega^\#)$. The converse holds too: if $u^\# \in L^1_{loc}(\Omega^\#)$ then $u$ satisfies Condition B.

Condition B holds in particular if $u$ is integrable and $\Omega$ has finite measure, since $(u-t)^+ \le |u| + |t|$.

Our aim in this section is to obtain "pre-subharmonicity inequalities" for $u^\#$ which, for appropriate $u$, run parallel to the cap results in Theorems 9.7 and 9.15. These inequalities will require more assumptions on $u$ and $\phi$ than were necessary for cap symmetrization on shells and ring-type domains, because the Dirichlet boundary condition on $\partial\Omega$ will necessitate a certain perturbation step in the proof.

Consider $\mu \in M_{loc}(\Omega)$, with Jordan–Lebesgue decomposition

$$d\mu = f \, d\mathcal{L}^n + d\tau - d\eta, \tag{9.47}$$

where $f \in L^1_{loc}(\Omega)$, and $\tau$ and $\eta$ are nonnegative locally finite measures on $\Omega$ which are singular with respect to $\mathcal{L}^n$ and to each other.

**Definition 9.17** Supposing in (9.47) that $f$ satisfies Condition B and $\tau(\Omega) < \infty$, we define $\mu^\# \in M_{loc}(\Omega^\#)$ by

$$d\mu^\# = f^\# \, d\mathcal{L}^n + d\tau^\#, \tag{9.48}$$

where $\tau^\# = \tau(\Omega)\delta_0$.

To verify that $\mu^\#$ belongs to $M_{loc}(\Omega^\#)$, recall $f$ satisfies Condition B if and only if $f^\# \in L^1_{loc}(\Omega^\#)$, while the point mass $\tau^\#$ belongs to $M_{loc}(\Omega^\#)$ since $\tau(\Omega) < \infty$.

The mass $\eta$ has been discarded in Definition (9.48). There exist alternative definitions of $\mu^\#$ that would take $\eta$ into account, but we will not pursue them.

If $\tau(\Omega) = \infty$ or if $f^\#$ does not satisfy Condition B then $\mu^\#$ is left undefined.

Recall from §9.2 that $\mathcal{W}(\Omega)$ is the set of all $u \in L^1_{loc}(\Omega)$ for which the distributional Laplacian $\Delta u$ is a locally finite measure. Given a continuous real-valued function $\phi$ on $\mathbb{R}$ such that $\phi(u)$ is locally integrable, and given a measure $\mu \in M_{loc}(\Omega)$, we say

$$-\Delta u = \phi(u) + \mu$$

in the weak (or distributional) sense if

$$-\int_{\Omega} u \Delta g \, d\mathcal{L}^n = \int_{\Omega} \phi(u) g \, d\mathcal{L}^n + \int_{\Omega} g \, d\mu$$

for every $g \in C_c^2(X)$.

Now we can state the pre-subharmonicity theorem for symmetric decreasing rearrangement on $\Omega$, which says $-\Delta(u^{\#}) \leq \phi(u^{\#}) + \mu^{\#}$ weakly with respect to symmetric decreasing test functions. One of the hypotheses involves the maximum of $|\phi|$ over nearby points, which is defined for $\kappa > 0$ by

$$M_{\kappa}\phi(\omega) = \max_{|\tilde{\omega}-\omega| \leq \kappa} |\phi(\tilde{\omega})|, \qquad \omega \in \mathbb{R}.$$

**Theorem 9.18** (Pre-subharmonicity theorem for s.d.r.) *Let $u \in \mathcal{W}(\Omega)$ satisfy*

$$u \geq 0 \quad in \quad \Omega$$

*and the Dirichlet boundary condition*

$$\lim_{x \to x_0, x \in \Omega} u(x) = 0, \quad \forall x_0 \in \partial\Omega. \tag{9.49}$$

*Suppose*

$$-\Delta u = \phi(u) + \mu$$

*in the weak sense in $\Omega$, where $\phi \colon \mathbb{R} \to \mathbb{R}$ is continuous, $\phi(u)$ is locally integrable, $f$ in decomposition* (9.47) *satisfies Condition B, and $\tau(\Omega) < \infty$. If either*

*(i) $u > 0$ in $\Omega$, or*
*(ii) $M_{\kappa}\phi\big(u(x)\big)$ is locally integrable on $\Omega$, for some $\kappa > 0$,*

*then $u$ satisfies Condition B on $\Omega$ and*

$$-\int_{\Omega^{\#}} u^{\#} \Delta g \, d\mathcal{L}^n \leq \int_{\Omega^{\#}} \phi(u^{\#}) g \, d\mathcal{L}^n + \int_{\Omega^{\#}} g \, d\mu^{\#} \tag{9.50}$$

*whenever $g \in C_c^2(\Omega^{\#})$ is nonnegative and symmetric decreasing in $\Omega^{\#}$.*

When $\Omega$ is unbounded, $\infty$ is regarded as a boundary point of $\Omega$ for the purposes of (9.49). For the conclusion (9.50), the measure $\mu^{\#}$ was defined in (9.48).

Condition (ii) in the theorem is satisfied if $|\phi(\omega)|$ is bounded or grows polynomially or exponentially as $\omega \to \infty$ (using here that $\phi(u)$ is assumed already to be locally integrable). Condition (ii) also holds if $\phi(\omega)$ is convex decreasing, since in that case $|\phi(\omega)| = O(\omega)$ as $\omega \to \infty$.

*Proof of Theorem 9.18* The Dirichlet boundary condition $u(x) \to 0$ as $x \to \partial\Omega$ implies that the set $(u > \gamma)$ has compact closure in $\Omega$, for each $\gamma > 0$. Since $u \in L^1_{loc}(\Omega)$ by the definition of $\mathcal{W}$, and since $\operatorname{ess\,inf} u = 0$, it follows that $u$ satisfies Condition B on $\Omega$. Hence $u^\#$ is locally integrable. Write

$$a = \mathcal{L}^n(g > 0), \quad b = \mathcal{L}^n(\Omega) = \mathcal{L}^n(\Omega^\#).$$

Then $0 \le a < b \le \infty$.

The theorem will be proved in six steps. First assume Condition (i), meaning $u$ is positive and so the set $(u > \gamma)$ expands to fill $\Omega$ as $\gamma \searrow 0$.

**Step 1**: (Rearranging $u$) Let $\psi\colon \Omega \to \mathbb{R}^+$ be a function with no flat spots. Define $T\colon \Omega \to [0, b]$ by

$$\begin{aligned} T(x) = {} & \mathcal{L}^n\big(\{y \in \Omega\colon u(y) > u(x)\}\big) \\ & + \mathcal{L}^n\big(\{y \in \Omega\colon u(y) = u(x) \text{ and } \psi(y) > \psi(x)\}\big). \end{aligned}$$

Notice $T(x) < b$ for all $x \in \Omega$, since the set $\{y \in \Omega\colon u(y) \ge u(x) > 0\}$ has compact closure in $\Omega$ (using here the positivity assumption on $u(x)$). A short argument using Proposition 1.26 implies that $T$ is a measure preserving map of $\Omega$ onto $[0, b)$, and that

$$u = u^* \circ T$$

holds $\mathcal{L}^n$-a.e. in $\Omega$. Here $*$ denotes decreasing rearrangement.

**Step 2**: (Rearranging $g$ to define a compactly supported $h$ in $\Omega$) Define $h\colon \Omega \to \mathbb{R}^+$ by

$$h = g^* \circ T.$$

Then $h$ and $g$ are equidistributed, so that $h^\# = g^\# = g$. Moreover, $h$ is bounded and nonnegative, since $g$ is.

Since $E = (u > \gamma)$ expands to fill $\Omega$ as $\gamma \searrow 0$, we may choose $\gamma > 0$ small enough that

$$\mathcal{L}^n(E) > a.$$

Take $x \in \Omega \setminus E$, so that $u(x) \le \gamma$ and hence $(u > u(x)) \supset E$ and so

$$T(x) \ge \mathcal{L}^n(u > u(x)) \ge \mathcal{L}^n(E) > a.$$

Since $g$ is symmetric decreasing, $g^*(t) = 0$ for all $t > a$. Thus,

$$h(x) = g^*(T(x)) = 0 \quad \forall\, x \in \Omega \setminus E,$$

and because $E \subset\subset \Omega$ we deduce that $h$ has compact support.

Since $u$ is locally integrable and $h$ is bounded with compact support,

$$\int_\Omega uh\, d\mathcal{L}^n = \int_\Omega (u^* \circ T)(g^* \circ T)\, d\mathcal{L}^n = \int_0^b u^* g^*\, d\mathcal{L} = \int_{\Omega^\#} u^\# g\, d\mathcal{L}^n \quad (9.51)$$

(using $g = g^\#$ in the last step), and all these integrals are finite. Similarly,

$$\int_\Omega \phi(u) h\, d\mathcal{L}^n = \int_{\Omega^\#} \phi(u^\#) g\, d\mathcal{L}^n. \quad (9.52)$$

**Step 3**: (Convolution inequalities) Let $K$ be a nonnegative symmetric decreasing smooth bump function supported in the unit ball of $\mathbb{R}^n$ with $\int_{\mathbb{R}^n} K(x)\, dx = 1$. Write

$$K_0(t) = K(te_1), \quad t \geq 0.$$

Then

$$\beta_{n-1} \int_0^\infty t^{n-1} K_0(t)\, dt = 1.$$

Set $u$ and $h$ equal to zero outside $\Omega$, and $u^\#$ and $g$ equal to zero outside $\Omega^\#$. For functions $f_1, f_2$ on $\mathbb{R}^n$, write

$$(f_1, f_2) = \int_{\mathbb{R}^n} f_1(x) f_2(x)\, d\mathcal{L}^n.$$

Then $(u, h) = (u^\#, g)$, by (9.51). We will show that

$$\begin{aligned}(u^\#, K_\epsilon * g - g) &= (u^\#, K_\epsilon * g) - (u^\#, g)\\ &\geq (u, K_\epsilon * h) - (u, h)\\ &= (u * K_\epsilon, h) - (u, h) = (K_\epsilon * u - u, h). \end{aligned} \quad (9.53)$$

The first and last equalities are obvious. The middle inequality follows from the Riesz rearrangement inequality on $\mathbb{R}^n$ (Theorem 8.4 or Corollary 2.19), since $g = h^\#$. As for the first equality in (9.53), for general functions $f_1, f_2, f_3$, the correct relation is $(f_1, f_2 * f_3) = (f_1 * \tilde{f}_2, f_3)$, where $\tilde{f}_2(x) = f_2(-x)$. In the case at hand we may take $f_2 = K_\epsilon$, so that the second equality holds as stated, since $K_\epsilon = \tilde{K}_\epsilon$.

This application of Riesz rearrangement is the central step of the proof, because inequality (9.53) and Lemma 9.6 will imply a suitable integral inequality between $\Delta(u^\#)$ and $\Delta u$, as we proceed to show.

If $g \equiv 0$ then the conclusion of Theorem 9.18 is true. Assume from now on that $g \not\equiv 0$. Assume $\epsilon > 0$ is so small that

$$\epsilon < \frac{1}{2}\operatorname{dist}(\operatorname{supp} g, \partial\Omega^{\#}) \quad \text{and} \quad \epsilon < \frac{1}{2}\operatorname{dist}(\operatorname{supp} h, \partial\Omega). \tag{9.54}$$

From (9.53) we have

$$(u^{\#}, K_\epsilon * g - g) \geq (h, K_\epsilon * u - u). \tag{9.55}$$

Next we find formulas for each side of (9.55).

By (9.54), if $x \in \operatorname{supp} h$ then $\operatorname{dist}(x, \partial\Omega) > \epsilon$. For such $x$, the formula

$$-\Delta u = \phi(u) + \mu \tag{9.56}$$

(which holds weakly in $\Omega$) and Lemma 9.6 together tell us that

$$\epsilon^{-2}[K_\epsilon * u(x) - u(x)] = -\int_\Omega L_\epsilon(x-y)\phi(u(y))\,dy - \int_\Omega L_\epsilon(x-y)\,d\mu(y), \tag{9.57}$$

where $L$ is a certain smooth nonnegative symmetric decreasing function on $\mathbb{R}^n$, supported in the unit ball and determined by $K$. Multiplying by $h(x)$ and integrating over $\Omega$ gives

$$\epsilon^{-2}(h, K_\epsilon * u - u) = -\int_\Omega (h * L_\epsilon)(y)\phi(u)\,dy - \int_\Omega (h * L_\epsilon)(y)\,d\mu(y). \tag{9.58}$$

If $x \in \operatorname{supp}(K_\epsilon * g - g)$ then $\operatorname{dist}(x, \partial\Omega^{\#}) > \epsilon$, by (9.54). For such $x$, Lemma 9.6 tells us

$$\epsilon^{-2}[K_\epsilon * g(x) - g(x)] = \int_{\Omega^{\#}} L_\epsilon(x-y)\Delta g(y)\,dy.$$

Multiplying by $u^{\#}(x)$ and integrating over $\Omega^{\#}$, we obtain

$$\epsilon^{-2}(u^{\#}, K_\epsilon * g - g) = \int_{\Omega^{\#}} u^{\#}(x)(L_\epsilon * \Delta g)(x)\,dx. \tag{9.59}$$

Substituting (9.58) and (9.59) into (9.55), we find

$$\int_{\Omega^{\#}} u^{\#}(x)(L_\epsilon * \Delta g)(x)\,dx \geq -\int_\Omega (h * L_\epsilon)(y)\phi(u)\,dy - \int_\Omega (h * L_\epsilon)(y)\,d\mu(y). \tag{9.60}$$

The remainder of the proof will evaluate the limit of these expressions as $\epsilon \to 0$.

**Step 4**: (Handling the $L$-integrals) Recall $\mu \in M_{loc}(\Omega)$ has Jordan–Lebesgue decomposition

$$d\mu = f\,d\mathcal{L}^n + d\tau - d\eta.$$

Since $h, L_\epsilon$ and $\eta$ are nonnegative, it follows that

$$\int_\Omega (h * L_\epsilon)(y)\, d\mu(y) \le \int_\Omega (h * L_\epsilon)(y) f(y)\, dy + \int_\Omega (h * L_\epsilon)(y)\, d\tau(y). \quad (9.61)$$

The first term on the right side of (9.61) is estimated in the next step of the proof, where we obtain that

$$\int_\Omega (h * L_\epsilon)(y) f(y)\, dy \le \int_{\Omega^\#} (g * L_\epsilon)(y) f^\#(y)\, dy. \quad (9.62)$$

Let us assume (9.62) for now.

Recall the constant $C_K = \int_{\mathbb{R}^n} L(x)\, d\mathcal{L}^n(x)$, as defined before the statement of Lemma 9.6 in §9.4. For the second term on the right of (9.61), one finds

$$\int_\Omega (h * L_\epsilon)(y)\, d\tau(y) \le C_K(\max g)\tau(\Omega) = C_K \int_{\Omega^\#} g\, d\tau^\#,$$

where the last step uses that $g = g^\#$ attains its maximum at the origin and the measure $\tau^\#$ is concentrated at the origin.

Then, from the estimates starting with (9.60), we have

$$\begin{aligned} &\int_{\Omega^\#} u^\#(x)(L_\epsilon * \Delta g)(x)\, dx \qquad (9.63) \\ &\ge -\int_\Omega (h * L_\epsilon)(y)\phi(u)\, dy - \int_{\Omega^\#} (g * L_\epsilon)(y) f^\#(y)\, dy - C_K \int_{\Omega^\#} g\, d\tau^\# \end{aligned}$$

when $\epsilon$ satisfies (9.54). From the uniform continuity and compact support of $\Delta g$, it easily follows that $\lim_{\epsilon\to 0} L_\epsilon * \Delta g = C_K \Delta g$ pointwise, and in fact uniformly. Thus,

$$\lim_{\epsilon\to 0} \int_{\Omega^\#} u^\#(x)(L_\epsilon * \Delta g)(x)\, dx = C_K \int_{\Omega^\#} u^\# \Delta g\, d\mathcal{L}^n. \quad (9.64)$$

By hypothesis, $f$ satisfies Condition B in $\Omega$. Thus, $f^\# \in L^1_{loc}(\Omega^\#)$. By the same argument as in (9.64), we have

$$\lim_{\epsilon\to 0} \int_{\Omega^\#} (g * L_\epsilon)(y) f^\#(y)\, dy = C_K \int_{\Omega^\#} g f^\#\, d\mathcal{L}^n.$$

Meanwhile, $\phi(u) \in L^1_{loc}(\Omega)$ and $h$ is bounded with compact support, and so by dominated convergence and (9.52),

$$\lim_{\epsilon\to 0} \int_\Omega (h*L_\epsilon)(y)\phi(u)\, dy = C_K \int_\Omega h\phi(u)\, d\mathcal{L}^n = C_K \int_{\Omega^\#} g\phi(u^\#)\, d\mathcal{L}^n. \quad (9.65)$$

Combining (9.63)–(9.65), it follows that

$$\begin{aligned}\int_{\Omega^\#} u^\# \Delta g \, d\mathcal{L}^n &\geq -\int_{\Omega^\#} g\phi(u^\#)\, d\mathcal{L}^n - \int_{\Omega^\#} gf^\# \, d\mathcal{L}^n - \int_{\Omega^\#} g\, d\tau^\# \\ &= -\int_{\Omega^\#} g\phi(u^\#)\, d\mathcal{L}^n - \int_{\Omega^\#} g\, d\mu^\#,\end{aligned} \tag{9.66}$$

as claimed in the theorem. It remains to prove (9.62).

**Step 5**: (Proof of (9.62)) We prove that if $\epsilon$ satisfies (9.54) then

$$\int_{\Omega\times\Omega} h(x)L_\epsilon(x-y)f(y)\,dy\,dx \leq \int_{\Omega^\#\times\Omega^\#} g(x)L_\epsilon(x-y)f^\#(y)\,dy\,dx.$$

Earlier, we extended $h$ to $\mathbb{R}^n$ by setting $h = 0$ outside $\Omega$. If also $f \geq 0$, then extend $f$ to $\mathbb{R}^n$ by setting $f = 0$ outside $\Omega$. Then the s.d.r. $f^\#$ on $\Omega^\#$, viewing $f$ as a function on $\Omega$, coincides with the restriction of $f^\#$ to $\Omega^\#$, viewing $f$ as a function on $\mathbb{R}^n$. The same holds for $h$, and we recall that $h^\# = g$.

Now we use again the Riesz rearrangement inequality for $\mathbb{R}^n$, in Theorem 8.4 or Corollary 2.19, which gives

$$\begin{aligned}\int_{\Omega\times\Omega} h(x)L_\epsilon(x-y)f(y)\,dy\,dx &= \int_{\mathbb{R}^n\times\mathbb{R}^n} h(x)L_\epsilon(x-y)f(y)\,dy\,dx \\ &\leq \int_{\mathbb{R}^n\times\mathbb{R}^n} h^\#(x)L_\epsilon(x-y)f^\#(y)\,dy\,dx \\ &= \int_{\Omega^\#\times\Omega^\#} g(x)L_\epsilon(x-y)f^\#(y)\,dy\,dx.\end{aligned}$$

This confirms (9.62) when $f \geq 0$.

Next we extend to general $f$, by a standard approximation process. For integers $k \geq 1$, define $f_k$ on $\Omega$ by $f_k = \max\{f, -k\}$. Then $f_k + k \geq 0$ on $\Omega$, so by the case just proved,

$$\int_{\Omega\times\Omega} h(x)L_\epsilon(x-y)\,[f_k(y)+k]\,dy\,dx \leq \int_{\Omega^\#\times\Omega^\#} g(x)L_\epsilon(x-y)\,[f_k^\#(y)+k]\,dy\,dx. \tag{9.67}$$

Here $f_k^\#$ is the s.d.r. of $f_k$ viewed as a function on $\Omega$.

Since $\epsilon$ satisfies (9.54), we have

$$\int_{\Omega\times\Omega} h(x)L_\epsilon(x-y)\,dy\,dx = C_K \int_\Omega h(x)\,dx,$$

$$\int_{\Omega^\#\times\Omega^\#} g(x)L_\epsilon(x-y)\,dy\,dx = C_K \int_{\Omega^\#} g(x)\,dx.$$

The integrals of $h$ and $g$ agree since $h^\# = g$, and so we conclude that

$$\int_{\Omega\times\Omega} h(x)L_\epsilon(x-y)\,dy\,dx = \int_{\Omega^\#\times\Omega^\#} g(x)L_\epsilon(x-y)\,dy\,dx,$$

which, with (9.67), implies

$$\int_{\Omega\times\Omega} h(x)L_\epsilon(x-y)f_k(y)\,dy\,dx \le \int_{\Omega^\#\times\Omega^\#} g(x)L_\epsilon(x-y)f_k^\#(y)\,dy\,dx. \quad (9.68)$$

It remains to let $k\to\infty$. Let

$$A(\epsilon)=\{(x,y)\in\Omega\times\Omega\colon x\in\operatorname{supp}h, |x-y|\le\epsilon\}.$$

Then $A(\epsilon)$ is a compact subset of $\Omega\times\Omega$, and

$$\int_{\Omega\times\Omega} h(x)L_\epsilon(x-y)f_k(y)\,dy\,dx = \int_{A(\epsilon)} h(x)L_\epsilon(x-y)f_k(y)\,dy\,dx. \quad (9.69)$$

The functions $h$ and $L_\epsilon$ are nonnegative and bounded, and $h$ has compact support in $\Omega$ by Step 2, while $f$ is locally integrable. It easily follows that $h(x)L_\epsilon(x-y)f^{\pm}(y)$ is in $L^1(A(\epsilon),d\mathcal{L}^{2n})$. As $k\to\infty$ the sequence $\{f_k\}$ decreases to $f$ pointwise in $\Omega$, so the sequence $h(x)L_\epsilon(x-y)f_k(y)$ decreases pointwise to $h(x)L_\epsilon(x-y)f(y)$ on $A(\epsilon)$. By the monotone convergence theorem applied to the sequence

$$\{h(x)L_\epsilon(x-y)(f^+(y)-f_k(y))\}_{k\ge1},$$

on $A(\epsilon)$, and using (9.69), we obtain

$$\lim_{k\to\infty}\int_{\Omega\times\Omega} h(x)L_\epsilon(x-y)f_k(y)\,dy\,dx = \int_{\Omega\times\Omega} h(x)L_\epsilon(x-y)f(y)\,dy\,dx.$$

A similar argument, using the fact that $f_k^\#\searrow f^\#$, shows that, when $\epsilon$ satisfies (9.54), we have

$$\lim_{k\to\infty}\int_{\Omega^\#\times\Omega^\#} g(x)L_\epsilon(x-y)f_k^\#(y)\,dy\,dx$$
$$=\int_{\Omega^\#\times\Omega^\#} g(x)L_\epsilon(x-y)f^\#(y)\,dy\,dx.$$

Recall that $f_k^\#$ is the s.d.r. of $f_k$ viewed as a function on $\Omega$.

Passing to the limit $k\to\infty$ in (9.68), we deduce that

$$\int_{\Omega\times\Omega} h(x)L_\epsilon(x-y)f(y)\,dy\,dx \le \int_{\Omega^\#\times\Omega^\#} g(x)L_\epsilon(x-y)f^\#(y)\,dy\,dx.$$

Inequality (9.62) is proved and with it Theorem 9.18, assuming Condition (i) holds (that $u$ is positive).

**Step 6**: Now assume Condition (ii) instead of Condition (i), so that $u\ge0$ and $M_{2\kappa}\phi(u)$ is locally integrable for some $\kappa>0$, and thus also locally integrable for each smaller value of $\kappa$. We explain the changes needed to Steps 1–5 of the proof above.

The set $(u > \gamma)$ need not expand to the whole domain as $\gamma \searrow 0$, since $u$ is no longer assumed positive and could even vanish on a set of positive measure. To get around this problem, we perturb $u$ as follows.

Fix a function $v \in C_c^2(\Omega)$ with $0 \leq v \leq 1$ such that $v$ has larger support than $g$, in the sense that

$$\mathcal{L}^n(v = 1) > \mathcal{L}^n(g > 0) = a.$$

This $v$ is chosen independently of $\kappa$.

Define the perturbation of $u$ to be

$$w = u + \kappa v,$$

so that $w \geq \kappa v \geq 0$. Choose $\gamma = \kappa/2$. Then

$$b_0 \equiv \mathcal{L}^n(w > \gamma) \geq \mathcal{L}^n(v = 1) > a.$$

Clearly $b_0 < \infty$, because the set $(w > \gamma)$ has compact closure in $\Omega$ due to the Dirichlet boundary condition.

Notice

$$-\Delta w = \phi(u) + \widetilde{\mu}$$

where $\widetilde{\mu} = \mu - \kappa \Delta v$. The preceding formula involves $\phi(u)$ rather than $\phi(w)$. To get a formula in terms of $w$, we observe $\phi(u) = \phi(w - \kappa v) \leq \phi_\kappa(w)$, where the continuous function $\phi_\kappa$ is defined by

$$\phi_\kappa(\omega) = \max_{|\tilde{\omega} - \omega| \leq \kappa} \phi(\tilde{\omega}), \qquad \omega \in \mathbb{R}.$$

(The difference between $\phi_\kappa$ and $M_\kappa \phi$ is that $M_\kappa \phi$ maximizes the absolute value $|\phi|$.) Thus

$$-\Delta w \leq \phi_\kappa(w) + \widetilde{\mu} \tag{9.70}$$

weakly in $\Omega$. Notice $\phi_\kappa(w) \in L^1_{loc}(\Omega)$ since $|\phi_\kappa(w)| \leq M_{2\kappa}\phi(u)$, which is locally integrable by Condition (ii).

*Changes to Step 1.* Now repeat Step 1 except with $u$ replaced by $w$ and $\Omega$ replaced by the set $\Omega_0 = (w > \gamma)$ (which we do not claim is open, just measurable and with compact closure) and $b$ replaced by $b_0 = \mathcal{L}^n(\Omega_0)$. Proposition 1.26 gives that $T$ is a measure preserving map from $\Omega_0$ onto $[0, b_0]$. The result of Step 1 is that $w = w^* \circ T$ almost everywhere in $\Omega_0$.

*Changes to Step 2.* In Step 2, replace $\Omega$ with $\Omega_0$ and replace $u$ with $w$. Argue as previously and define $h = g^* \circ T$ on $\Omega_0$ (noting that $T(\Omega_0) = [0, b_0]$ contains the interval $[0, a)$ on which $g^*$ is nonvanishing, since $b_0 > a$ by construction).

Extend $h$ to equal 0 on the complement of $\Omega_0$, so that $h$ is bounded and nonnegative with compact support in $\Omega$. The formulas at the end of Step 2 change to:

$$\int_\Omega wh\, d\mathcal{L}^n = \int_{\Omega_0} (w^* \circ T)(g^* \circ T)\, d\mathcal{L}^n = \int_0^{b_0} w^* g^*\, d\mathcal{L} = \int_{\Omega^\#} w^\# g\, d\mathcal{L}^n$$

and similarly

$$\int_\Omega \phi_\kappa(w) h\, d\mathcal{L}^n = \int_{\Omega^\#} \phi_\kappa(w^\#) g\, d\mathcal{L}^n,$$

where we recall that $h = 0$ off $\Omega_0$.

*Changes to Step 3.* Replace $u$ with $w$, and $\mu$ with $\widetilde{\mu}$. Instead of using (9.56) we call on (9.70), and so $\phi(u)$ is replaced by $\phi_\kappa(w)$.

*Changes to Step 4.* Replace $u$ with $w$, and $\phi(u)$ with $\phi_\kappa(w)$. The right side of (9.61) gains an additional term $\int_\Omega (h * L_\epsilon)(y)\big(-\kappa \Delta v(y)\big)\, dy$ coming from the term $-\kappa \Delta v$ in $\widetilde{\mu}$. Hence the conclusion (9.66) of Step 4 becomes

$$\begin{aligned} &\int_{\Omega^\#} w^\# \Delta g\, d\mathcal{L}^n \\ &\quad \ge -\int_{\Omega^\#} g \phi_\kappa(w^\#)\, d\mathcal{L}^n - \int_{\Omega^\#} g\, d\mu^\# - \kappa(\max g) \int_\Omega |\Delta v|\, d\mathcal{L}^n. \end{aligned} \tag{9.71}$$

*Changes to Step 5.* No change.

To complete the proof, we will let $\kappa \to 0$ in (9.71) and show $w$ converges suitably to $u$, and $\phi_\kappa(w^\#)$ converges to $\phi(u^\#)$, so that one obtains in the limit the desired conclusion (9.50) of the theorem. To justify the convergence, first observe

$$u \le w \le u + \kappa \qquad \Longrightarrow \qquad u^\# \le w^\# \le u^\# + \kappa,$$

so that $|w^\# - u^\#| \le \kappa \to 0$ at every point. Hence the term on the left of (9.71) converges as wanted, with $\int_{\Omega^\#} w^\# \Delta g\, d\mathcal{L}^n \to \int_{\Omega^\#} u^\# \Delta g\, d\mathcal{L}^n$ as $\kappa \to 0$. The third term on the right of (9.71) clearly tends to 0 as $\kappa \to 0$. The second term does not depend on $\kappa$. For the first term on the right of (9.71) we apply dominated convergence:

$$\phi_\kappa\big(w^\#(x)\big) = \phi_\kappa\big(u^\#(x) + O(\kappa)\big) \to \phi\big(u^\#(x)\big) \qquad \text{as } \kappa \to 0$$

by continuity of $\phi$, for each fixed $x \in \Omega^\#$, and the dominating function

$$\big|\phi_\kappa\big(w^\#\big)\big| \le M_{2\kappa}\phi\big(u^\#\big)$$

is integrable on the support of $g$ by Condition (ii). (This dominating function depends on the parameter $\kappa$, which is acceptable here since the value of $M_{2\kappa}\phi$ goes down as $\kappa$ gets smaller.) Thus the proof of Theorem 9.18 is complete. □

## 9.10 The ⋆-Function for s.d.r. on Euclidean Domains

In this section, # will continue to denote symmetric decreasing rearrangement of functions on $\Omega$, where $\Omega$ is an open set in $\mathbb{R}^n$. For $r \in (0, \infty]$, write

$$B(r) = \mathbb{B}^n(0, r),$$

and define $R$ by

$$B(R) = \Omega^{\#}.$$

Define also an interval $\Omega^\star \subset (0, \infty)$ by

$$\Omega^\star = (0, R).$$

Suppose that a function $u\colon \Omega \to \mathbb{R}$ satisfies Condition B. Define a new function $u^\star\colon \Omega^\star \to [-\infty, \infty]$ by

$$u^\star(r) = \sup \int_E u \, d\mathcal{L}^n = \int_{B(r)} u^{\#} \, d\mathcal{L}^n, \quad r \in \Omega^\star, \tag{9.72}$$

where the sup is taken over all Lebesgue measurable sets $E \subset \Omega$ with $\mathcal{L}^n(E) = \mathcal{L}^n(B(r))$. As explained in §9.9, Condition B insures that $u^\star(r)$ exists and is finite for each $r$, and that the second and third expressions are equal.

Next, define an operator $J$ which takes functions $v \in L^1_{loc}(\Omega^{\#})$ to functions $Jv$ on $\Omega^\star$ by

$$Jv(r) = \int_{B(r)} v \, d\mathcal{L}^n, \quad r \in \Omega^\star.$$

The function $Jv$ is defined everywhere in $\Omega^\star$, and is bounded on $[0, r]$ for each $r < R$.

If $u$ satisfies Condition B on $\Omega$, then we can take $v = u^{\#}$ in (9.72), and obtain

$$u^\star = Ju^{\#}, \quad \text{on} \quad \Omega^\star.$$

Define now an operator called $J^t$ which takes functions $G \in L^1(\Omega^\star)$ to functions $J^t G$ on $\Omega^{\#}$ by

$$J^t G(x) = \int_{|x|}^R G(s) \, ds, \quad x \in \Omega^{\#}.$$

If $G \in L^\infty_c(\Omega^\star)$, so that $G$ has compact support in $\Omega^\star$, and if $v \in L^1_{loc}(\Omega^{\#})$, then

$$\int_{\Omega^\star} Jv(r) G(r) \, dr = \int_{\Omega^\star} G(r) \, dr \int_{B(r)} v(x) \, dx = \int_{\Omega^{\#}} v(x) \, dx \int_{|x|}^R G(r) \, dr$$
$$= \int_{\Omega^{\#}} v(x) J^t G(x) \, dx,$$

and hence

$$\int_{\Omega^\star} Jv\, G\, d\mathcal{L} = \int_{\Omega^\#} vJ^t G\, d\mathcal{L}^n. \tag{9.73}$$

Thus, $J^t$ indeed is an adjoint operator, with respect to the appropriate measures.

Define differential operators $\Delta^\star$ and $\Delta^{\star t}$ operating on functions $G \in C^2(\Omega^\star)$ by

$$\Delta^\star G(r) = G''(r) - \frac{n-1}{r} G'(r), \tag{9.74}$$

$$\Delta^{\star t} G(r) = G''(r) + \Big(\frac{n-1}{r} G\Big)'(r). \tag{9.75}$$

If $F, G \in C^2(\Omega^\star)$ and at least one of these functions has compact support in $\Omega^\star$, then, using (9.74), (9.75) and integration by parts, we have an adjoint property for $\Delta^\star$ and $\Delta^{\star t}$:

$$\begin{aligned}\int_{\Omega^\star} [\Delta^\star F] G\, d\mathcal{L} &= \int_{\Omega^\star} [F''(r) - \frac{n-1}{r} F'(r)] G(r)\, dr \\ &= \int_{\Omega^\star} F(r)[G''(r) + \Big(\frac{n-1}{r} G\Big)'(r)]\, dr \\ &= \int_{\Omega^\star} F[\Delta^{\star t} G]\, d\mathcal{L}.\end{aligned} \tag{9.76}$$

We shall need some identities involving $\Delta^\star$ and $\Delta^{\star t}$. Firstly, take $v \in C^2(\Omega^\#)$. Define

$$I(r) = I(r, v) = \int_{\mathbb{S}^{n-1}} v(ry)\, d\sigma(y), \quad r \in \Omega^\star,$$

where we write $\sigma = \sigma_{n-1}$. From the definition of $J$ and Green's formula, we get

$$J\Delta v(r) = r^{n-1} \int_{\mathbb{S}^{n-1}} \frac{\partial v}{\partial r}(ry)\, d\sigma(y) = r^{n-1} I'(r).$$

Also, $Jv(r) = \int_0^r s^{n-1} I(s)\, ds$, so $(Jv)'(r) = r^{n-1} I(r)$, and

$$(\Delta^\star Jv)(r) = (r^{n-1} I(r))' - \frac{n-1}{r} r^{n-1} I(r) = r^{n-1} I'(r).$$

From these identities, we obtain the following Main Identity:

$$\boxed{J\Delta v = \Delta^\star Jv} \quad \text{on } \Omega^\star. \tag{9.77}$$

We shall need also an adjoint version of the Main Identity. For $G \in C_c^2(\Omega^\star)$, we have

$$J^t(\Delta^{\star t} G)(x) = \int_{|x|}^R [G''(r) + \Big(\frac{n-1}{r} G\Big)'](r)\, dr = -G'(|x|) - \frac{n-1}{|x|} G(|x|).$$

Also,

$$\Delta(J^t G)(x) = \Delta\left(\int_{|x|}^{R} G(s)\,ds\right)$$
$$= \left(\frac{d^2}{dr^2} + \frac{n-1}{r}\frac{d}{dr}\right)\int_r^R G(s)\,ds$$
$$= -G'(r) - \frac{n-1}{r}G(r) = -G'(|x|) - \frac{n-1}{|x|}G(|x|),$$

where $x \in \Omega^{\#}$ and we have written $r = |x|$. From the last two formulas, it follows that

$$J^t(\Delta^{\star t} G) = \Delta(J^t G) \quad \text{in} \quad \Omega^{\#}. \tag{9.78}$$

A weak version of the Main Identity will be helpful too. Take $v \in L^1_{loc}(\Omega^{\#})$ and $G \in C^2_c(\Omega^\star)$. Set $g = J^t G$. Then by the adjoint relation and (9.78),

$$(Jv, \Delta^{\star t} G) = (v, J^t \Delta^{\star t} G) = (v, \Delta J^t G) = (v, \Delta g).$$

That is,

$$\int_{\Omega^\star} Jv\, \Delta^{\star t} G\, d\mathcal{L} = \int_{\Omega^{\#}} v \Delta g\, d\mathcal{L}^n. \tag{9.79}$$

Applying this formula to $v = u^{\#}$ where $u \in L^1_{loc}(\Omega)$ satisfies Condition B gives that

$$\int_{\Omega^\star} u^\star \Delta^{\star t} G\, d\mathcal{L} = \int_{\Omega^{\#}} u^{\#} \Delta g\, d\mathcal{L}^n \tag{9.80}$$

for $g = J^t G$.

Let us return now to the setting of the pre-subharmonicity result, Theorem 9.18, where $d\mu = f\,d\mathcal{L}^n + d\tau - d\eta$, with $f$ satisfying Condition B in $\Omega$ and $\tau(\Omega) < \infty$. We defined $\mu^{\#} \in M_{loc}(\Omega^{\#})$ by the formula

$$d\mu^{\#} = f^{\#}\,d\mathcal{L}^n + d\tau^{\#}.$$

Now we define another measure, called $\mu^\star$, which belongs to $M_{loc}(\Omega^\star)$:

**Definition 9.19**

$$d\mu^\star = f^\star\,d\mathcal{L} + d\tau^\star, \quad \text{where} \quad d\tau^\star = \tau(\Omega)\,d\mathcal{L}.$$

If $G$ and $Gf^\star$ both belong to $L^1(\Omega^\star)$ then we have

$$\int_{\Omega^\star} G\,d\mu^\star = \int_{\Omega^\star} Gf^\star\,d\mathcal{L} + \int_{\Omega^\star} G\,d\tau^\star$$
$$= \int_{\Omega^\star} G(Jf^{\#})\,d\mathcal{L} + \tau(\Omega)\int_0^R G(r)\,dr$$

where in the second equality we used $f^\star = Jf^\#$. Hence

$$\begin{aligned}\int_{\Omega^\star} G\,d\mu^\star &= \int_{\Omega^\#} (J^tG)f^\#\,d\mathcal{L}^n + \tau(\Omega)J^tG(0)\\ &= \int_{\Omega^\#} (J^tG)f^\#\,d\mathcal{L}^n + \int_{\Omega^\#} J^tG\,d\tau^\# \\ &= \int_{\Omega^\#} J^tG\,d\mu^\#,\end{aligned} \tag{9.81}$$

where the first equality relies on (9.73) and the others follow from definitions.

Let now $F \in L^1_{loc}(\Omega^\star)$ and $\nu \in M_{loc}(\Omega^\star)$. We say that the differential inequality

$$-\Delta^\star F \le \nu$$

holds in the weak sense, or in the sense of distributions, if

$$-\int_{\Omega^\star} F\Delta^{\star t}G\,dr \le \int_{\Omega^\star} G\,d\nu$$

holds for all nonnegative $G \in C_c^2(\Omega^\star)$. If $F \in C^2(\Omega^\star), H \in L^1_{loc}(\Omega^\star)$, and $-\Delta^\star F \le H$ holds at every point of $\Omega^\star$, then we say that $-\Delta^\star F \le H$ holds in the strong or the classical sense, in which case (9.76) implies that the inequality also holds in the weak sense.

**Theorem 9.20** (Subharmonicity property of the s.d.r. on $\Omega \subset \mathbb{R}^n$) *Let $u \in \mathcal{W}(\Omega)$ satisfy*

$$u \ge 0 \quad in \quad \Omega$$

*and the Dirichlet boundary condition*

$$\lim_{x\to x_0,\, x\in\Omega} u(x) = 0, \quad \forall x_0 \in \partial\Omega.$$

*Suppose*

$$-\Delta u = \phi(u) + \mu$$

*in the weak sense in $\Omega$, where $\phi\colon \mathbb{R} \to \mathbb{R}$ is continuous, $\phi(u)$ is locally integrable, $f$ in decomposition (9.47) satisfies Condition B in $\Omega$, and $\tau(\Omega) < \infty$. If either*

*(i) $u > 0$ in $\Omega$, or*
*(ii) $M_\kappa\phi\big(u(x)\big)$ is locally integrable for some $\kappa > 0$,*

*then*

$$-\Delta^\star u^\star \le J\phi(u^\#) + \mu^\star$$

*in the weak sense in $\Omega^\star = (0, R)$.*

When $\Omega$ is unbounded, $\infty$ is regarded as belonging to $\partial\Omega$.

Recall from after Theorem 9.18 that Condition (ii) in the theorem is satisfied if $|\phi(\omega)|$ is bounded or grows polynomially or exponentially as $\omega \to \infty$, and in particular it holds if $\phi(\omega)$ is convex decreasing.

*Proof* Take nonnegative $G \in C_c^2(0, R)$. Take $0 < R_1 < R_2 < R$ such that $G = 0$ on $(0, R_1] \cup [R_2, R)$. Define $g$ on $\Omega^\# = B(R)$ by

$$g(x) = J^t G(x) = \int_{|x|}^{R} G(s)\, ds.$$

Then obviously $g \geq 0, g = g^\#$, and $g$ has compact support in $\Omega^\#$. Also, $g \in C^3(\Omega^\#)$ since $G \in C^2$ and $g(x)$ is constant for $|x| < R_1$. By (9.80), Theorem 9.18, (9.73), and (9.81) we have

$$\begin{aligned}\int_{\Omega^\star} u^\star \Delta^{\star t} G\, d\mathcal{L} &= \int_{\Omega^\#} u^\# \Delta g\, d\mathcal{L}^n \\ &\geq -\int_{\Omega^\#} g\phi(u^\#)\, d\mathcal{L}^n - \int_{\Omega^\#} g\, d\mu^\# \\ &= -\int_{\Omega^\star} GJ\phi(u^\#)\, d\mathcal{L} - \int_{\Omega^\star} G\, d\mu^\star.\end{aligned}$$

When using (9.81), recall that Condition B gives local integrability of $f^\#$ and hence local boundedness of $f^\star$, so that $Gf^\star \in L^1(\Omega^\star)$. □

## 9.11 The ★-Function for Steiner Symmetrization on Euclidean Domains

Throughout this section, # denotes $(k, n)$-Steiner symmetrization. Fix $n \geq 2$ and $1 \leq k \leq n - 1$. For $x \in \mathbb{R}^n$, write $x = (y, z)$, where $y \in \mathbb{R}^k$, $z \in \mathbb{R}^m$, and $k + m = n$. Let $\Omega$ be an open subset of $\mathbb{R}^n$, and let $Z$ be the open set of all $z \in \mathbb{R}^m$ such that the slice $\Omega(z) \subset \mathbb{R}^k$ is nonempty. Then $\Omega$ can be expressed in terms of its slices as

$$\Omega = \{(y, z) \in \mathbb{R}^n : y \in \Omega(z), z \in Z\}.$$

The $(k, n)$-Steiner symmetrization of $\Omega$ is obtained by replacing each slice with the ball in $\mathbb{R}^k$ of the same volume centered at the origin, and the $(k, n)$-Steiner symmetrization of a nonnegative function on $\Omega$ is obtained by performing the s.d.r. on each slice. These constructions are explained in detail at the beginning of Chapter 6.

**The Presubharmonicity Result**

Suppose $u\colon \Omega \to \mathbb{R}$ is $\mathcal{L}^n$-measurable and nonnegative. We say $u$ satisfies **Condition BS** if

$$\int_{\Omega(Z_0)} (u-t)^+ \, d\mathcal{L}^n < \infty$$

for all $t > 0$ and every compact set $Z_0 \subset Z$, where

$$\Omega(Z_0) = \{(y,z) \in \Omega : z \in Z_0\}.$$

If $u$ satisfies Condition BS, then on almost every slice the function $(u^z - t)^+$ is integrable for each $t > 0$ (recall $u^z \equiv u(\cdot, z)$). Hence the slice function $u^z$ satisfies finiteness condition (6.2); that is,

$$\mathcal{L}^k(u^z > t) < \infty \quad \text{for every } t > 0,$$

for $\mathcal{L}^m$-almost every $z \in Z$. Thus $u^\#$ is well-defined and $\mathcal{L}^n$-measurable by Chapter 6. Further, $u^\# \in L^1_{loc}(\Omega^\#)$ because for each compact set $E \subset \Omega^\#$, Condition BS guarantees

$$0 \le \int_E u^\# \, d\mathcal{L}^n \le \int_E (u^\# - t)^+ \, d\mathcal{L}^n + t\mathcal{L}^n(E) < \infty.$$

Now we develop the Steiner symmetrization of a measure.

**Definition 9.21** Consider $\mu \in M_{loc}(\Omega)$, with Jordan–Lebesgue decomposition

$$d\mu = f \, d\mathcal{L}^n + d\tau - d\eta. \tag{9.82}$$

Assume for some $c \in \mathbb{R}$ that $f + c$ is nonnegative and satisfies Condition BS, and define the Steiner symmetrization of $f$ to be

$$f^\# = (f+c)^\# - c \quad \text{on } \Omega^\#.$$

Assume $\tau(\Omega) < \infty$, write $P(y,z) = (0,z)$ for the projection of $\Omega$ onto $\mathbb{R}^m$, and define a Radon measure $\tau^\# \in M_{loc}(\Omega^\#)$ by

$$\tau^\#(E) = \tau(P^{-1}E)$$

for $E \in \mathcal{B}(\Omega^\#)$. (In other words, $\tau^\#$ is the measure obtained from $\tau$ by sweeping the mass to the origin on each $z$-slice.) Note $\tau^\#(\Omega^\#) = \tau(\Omega)$. The symmetrized measure $\mu^\# \in M_{loc}(\Omega^\#)$ is defined by

$$d\mu^\# = f^\# d\mathcal{L}^n + d\tau^\#.$$

(As in the s.d.r. case, the mass $\eta$ is discarded when passing to $\mu^\#$.)

Next we state the pre-subharmonicity theorem for $(k, n)$-Steiner symmetrization, which says $-\Delta(u^{\#}) \le \phi(z, u^{\#}) + \mu^{\#}$ weakly with respect to nonnegative, Steiner symmetric test functions. The statement of the theorem uses the maximum of $|\phi|$ over nearby points, defined by

$$(M_\kappa \phi)(z, \omega) = \max_{|\tilde{\omega}-\omega| \le \kappa} |\phi(z, \tilde{\omega})|, \qquad z \in Z, \quad \omega \in \mathbb{R}, \quad \kappa > 0.$$

**Theorem 9.22** (Pre-subharmonicity theorem for $(k, n)$-Steiner symmetrization) *Let $u \in \mathcal{W}(\Omega)$ satisfy*

$$u \ge 0 \quad \text{in} \quad \Omega$$

*and the Dirichlet boundary condition*

$$\lim_{x \to x_0,\, x \in \Omega} u(x) = 0 \tag{9.83}$$

*for all $x_0 = (y_0, z_0) \in \partial\Omega$ with $z_0 \in Z$. Suppose*

$$-\Delta u \le \phi(z, u) + \mu \tag{9.84}$$

*in the weak sense in $\Omega$, where $\phi \colon Z \times \mathbb{R} \to \mathbb{R}$ is continuous with $\phi\big(z, u(y, z)\big)$ locally integrable on $\Omega$. Suppose in decomposition (9.82) that $f + c$ is nonnegative and satisfies Condition BS, for some $c \in \mathbb{R}$, and that $\tau(\Omega) < \infty$. If either*

*(i) $u$ is lower semicontinuous and $u > 0$ in $\Omega$, or*
*(ii) $M_\kappa \phi\big(z, u(y, z)\big)$ is locally integrable on $\Omega$ for some $\kappa > 0$,*

*then $u$ satisfies Condition BS on $\Omega$ and*

$$-\int_{\Omega^{\#}} u^{\#} \Delta g \, d\mathcal{L}^n \le \int_{\Omega^{\#}} \phi(z, u^{\#}) g \, d\mathcal{L}^n + \int_{\Omega^{\#}} g \, d\mu^{\#}$$

*whenever $g \in C_c^2(\Omega^{\#})$ is nonnegative and Steiner symmetric ($g = g^{\#}$).*

Note the point at infinity ($y_0 = \infty$) is allowed in the Dirichlet boundary condition, when $\Omega$ is unbounded.

This theorem assumes the differential inequality $-\Delta u \le \phi(z, u) + \mu$, instead of a differential equation as assumed by other presubharmonicity theorems in this chapter. In the other theorems, one would gain no generality by using a differential inequality since any discrepancy in the inequality could be absorbed into the measure $\mu$ to arrive at an equality. In Theorem 9.22, though, the measure faces the additional constraint that its absolutely continuous part $f$ must be bounded below, so that $f + c$ is nonnegative for some constant $c$. Here the added generality of a differential inequality is useful, because if $f$ is not bounded below then we may replace $\mu$ in inequality (9.84) with the larger measure obtained by truncating $f$ from below at height $-c$.

*Proof* The set $(u > \gamma) \cap \Omega(Z_0)$ has compact closure in $\Omega$ for each $\gamma > 0$ and each compact set $Z_0 \subset Z$, thanks to the Dirichlet boundary condition (9.83), which says $u(x) \to 0$ as $x \to (y_0, z_0) \in \partial\Omega$ with $z_0 \in Z$. Since $u \in L^1_{loc}(\Omega)$ by assumption, it follows that $u$ satisfies Condition BS on $\Omega$.

Fix the test function $g$ in the theorem, write $A = (g > 0)$, and let

$$a(z) = \mathcal{L}^k(A(z)), \quad b(z) = \mathcal{L}^k\big(\Omega(z)\big),$$

for $z \in Z$. The functions $a(z)$ and $b(z)$ satisfy $0 \le a(z) < b(z) \le \infty$. Define a compact set

$$Z_0 = \{z \in Z\colon \text{the slice of } \Omega \text{ at height } z \text{ intersects } \operatorname{supp} g\}.$$

First assume Condition (i), saying $u$ is positive and lower semicontinuous. We adapt the s.d.r. case from Theorem 9.18. That proof had six steps, and we indicate below only the changes needed in each step. As we will see, some parts of the proof should be carried out slice-by-slice.

**Step 1**: The measure preserving map $T_z$ takes the slice $\Omega(z)$ onto $[0, b(z))$, for almost every $z$, with

$$u_z = (u_z)^* \circ T_z.$$

**Step 2**: Define $h : \Omega \to \mathbb{R}^+$ by

$$h_z = (g_z)^* \circ T_z$$

when $z \in Z_0$, and $h_z = 0$ otherwise. For joint measurability of $h(y, z)$, argue as in the proof of Theorem 9.7.

We claim $\gamma > 0$ can be chosen small enough that $E = (u > \gamma)$ satisfies

$$\mathcal{L}^k\big(E(z)\big) > a(z)$$

for all $z \in Z_0$. Indeed, one can construct a compact subset $C$ of $\Omega$ that has larger slices than the support of $g$, in the sense that $\mathcal{L}^k\big(C(z)\big) > a(z)$ for all $z \in Z_0$. This compact set must lie inside the set $E$ for some $\gamma$, since $E$ is open (by lower semicontinuity of $u$) and expands to fill $\Omega$ as $\gamma \to 0$ (by positivity of $u$).

The rest of Step 2 proceeds as before, with obvious adaptations. One sees $h$ has compact support, since it vanishes outside $(u > \gamma)$. The integrals in (9.51) and (9.52) should be taken over each slice.

**Step 3**: The differential inequality (9.84) replaces the equality (9.56), and so equalities (9.57) and (9.58) become inequalities "$\geq$". Also, $\phi(u)$ changes to

$\phi(z, u)$, and for Riesz rearrangement one uses the appropriate result for Steiner symmetrization, from Theorem 6.8.

**Step 4**: One must treat differently the second term on the right of (9.61), as follows. On each slice, $h$ is bounded by the largest value of $g$:

$$h(y, z) \leq g(0, z) = (g \circ P)(y, z),$$

where $P$ is the projection defined earlier. Hence $h \leq g \circ P$, and so after extending the measure $\tau$ to be zero outside $\Omega$ we have

$$\begin{aligned}\int_{\Omega} (h * L_\epsilon)(y)\, d\tau(y) &\leq \int_{\mathbb{R}^n} (g \circ P) * L_\epsilon \, d\tau \\ &\to C_K \int_{\mathbb{R}^n} (g \circ P)\, d\tau\end{aligned}$$

as $\epsilon \to 0$, using here that $(g \circ P) * L_\epsilon \to C_K (g \circ P)$ uniformly on $\mathbb{R}^n$ because $g$ is continuous with compact support. Thus

$$\limsup_{\epsilon \to 0} \int_{\Omega} (h * L_\epsilon)(y)\, d\tau(y) \leq C_K \int_{\Omega^\#} g \, d\tau^\#,$$

as needed for handling the second term on the right of (9.61).

**Step 5**: Recall the hypothesis that $f + c \geq 0$ satisfies Condition BS in $\Omega$, for some $c \in \mathbb{R}$. We write $f = (f + c) - c$ and apply the first part of the proof of Step 5 to $f + c$ (except using the Steiner version of Riesz rearrangement from Theorem 6.8). To handle the "$-c$" term, observe

$$\int_{\Omega \times \Omega} h(x) L_\epsilon(x - y)\, dy\, dx = \int_{\Omega^\# \times \Omega^\#} g(x) L_\epsilon(x - y)\, dy\, dx$$

by integrating out the $y$-variable and recalling $h$ and $g$ are equidistributed. Note the approximation part of the proof of Step 5 is no longer needed.

**Step 6**: Now assume Condition (ii) holds in the theorem. One follows the proof under Condition (i), in Steps 1–5 above, with a perturbation like in Step 6 of the proof of Theorem 9.18. The perturbing function $v \in C_c^2(\Omega)$ with $0 \leq v \leq 1$ is chosen to have larger support than $g$ at each height, in the sense that $\mathcal{L}^k(V(z)) > a(z)$ for each $z \in Z_0$, where $V = (v = 1)$. □

### The Subharmonicity Result

Next we develop a subharmonicity theorem for Steiner symmetrization, generalizing the s.d.r. case in Theorem 9.20. Denote by $R(z) \in (0, \infty]$ the radius of the ball $\Omega^\#(z) \subset \mathbb{R}^k$, and for each $z \in Z$ define an interval

$$\Omega^\star(z) = (0, R(z)).$$

Let

$$\Omega^\star = \{(r,z) \colon r \in \Omega^\star(z), z \in Z\} \subset \mathbb{R}^{m+1}.$$

Suppose that a nonnegative function $u \colon \Omega \to \mathbb{R}$ satisfies Condition BS. Define a new function $u^\star \colon \Omega^\star \to [0,\infty]$ by

$$u^\star(r,z) = \sup \int_E u(y,z)\, d\mathcal{L}^k(y) = \int_{\mathbb{B}^k(r)} u^\#(y,z)\, d\mathcal{L}^k(y), \tag{9.85}$$

for $r \in \Omega^\star(z), z \in Z$, where the sup is taken over all Lebesgue measurable sets $E \subset \Omega(z)$ with $\mathcal{L}^k(E) = \mathcal{L}^k(\mathbb{B}^k(r))$. (In other words, apply the ★-operator for the s.d.r. on each slice of the function.) By adapting the argument near the beginning of Section 9.9, Condition BS insures that $u^\#$ is locally integrable on slices and so $u^\star(r,z)$ exists and is finite for each $r$, for almost every $z$.

When $f + c$ is nonnegative and satisfies Condition BS for some $c \in \mathbb{R}$, the Steiner symmetrization of $f$ is defined by $f^\# = (f+c)^\# - c$ on $\Omega^\#$ and so $f^\star$ can be defined exactly as in (9.85), writing $f$ instead of $u$.

The operator $J \colon L^1_{loc}(\Omega^\#) \to L^1_{loc}(\Omega^\star)$ acts on each slice, with

$$(Jv)(r,z) = \int_{\mathbb{B}^k(r)} v(y,z)\, d\mathcal{L}^k(y), \quad r \in \Omega^\star(z),$$

where the definition makes sense for $\mathcal{L}^m$-almost every $z \in Z$. By (9.85),

$$u^\star = Ju^\# \quad \text{on} \quad \Omega^\star.$$

Next, given $G \in L^1(\Omega^\star)$ define

$$(J^t G)(y,z) = \int_{|y|}^{R(z)} G(r,z)\, dr, \quad y \in \Omega^\#(z).$$

If $G(\cdot, z) \in L^\infty_c(\Omega^\star(z))$ for each $z$, meaning $G$ is bounded with compact support on each slice of $\Omega^\star$, and if $v \in L^1_{loc}(\Omega^\#)$, then

$$\int_{\Omega^\star(z)} Jv(r,z)\, G(r,z)\, d\mathcal{L}(r) = \int_{\Omega^\#(z)} v(y,z) J^t G(y,z)\, d\mathcal{L}^k(y) \tag{9.86}$$

for almost every $z$, by arguing as for (9.73). Thus $J^t$ is an adjoint operator of $J$, with respect to the appropriate measures.

Define now differential operators $\Delta^\star$ and $\Delta^{\star t}$ operating on functions $G(r,z) \in C^2(\Omega^\star)$ by

$$\Delta^\star G = \partial_{rr} G - \frac{k-1}{r} \partial_r G + \Delta_z G, \tag{9.87}$$

$$\Delta^{\star t} G = \partial_{rr} G + \partial_r \Big(\frac{k-1}{r} G\Big) + \Delta_z G. \tag{9.88}$$

If $F, G \in C^2(\Omega^\star)$ and at least one of these functions has compact support in $\Omega^\star$, then using (9.87), (9.88) and integration by parts yields an adjoint property for $\Delta^\star$ and $\Delta^{\star t}$:

$$\int_{\Omega^\star} (\Delta^\star F) G \, d\mathcal{L}^{m+1} = \int_{\Omega^\star} F(\Delta^{\star t} G) \, d\mathcal{L}^{m+1}.$$

Further, for $v \in C^2(\Omega^\#)$ the Main Identity

$$\boxed{J\Delta v = \Delta^\star Jv} \quad \text{on } \Omega^\star$$

follows by the s.d.r. case in (9.77) except with an additional term involving $J\Delta_z v$, which equals $\Delta_z Jv$. Similarly one obtains an adjoint version of the Main Identity:

$$J^t(\Delta^{\star t} G) = \Delta(J^t G) \quad \text{in} \quad \Omega^\# \tag{9.89}$$

for $G \in C_c^2(\Omega^\star)$.

The weak version of the Main Identity says

$$\int_{\Omega^\star} Jv \, \Delta^{\star t} G \, d\mathcal{L}^{m+1} = \int_{\Omega^\#} v \Delta g \, d\mathcal{L}^n \tag{9.90}$$

where $v \in L^1_{loc}(\Omega^\#)$, $G \in C_c^2(\Omega^\star)$ and $g = J^t G$, as one proves by using the adjoint relation (9.86) and the adjoint main identity (9.89). If $u \geq 0$ satisfies Condition BS then putting $v = u^\#$ into (9.90) gives that

$$\int_{\Omega^\star} u^\star \Delta^{\star t} G \, d\mathcal{L}^{m+1} = \int_{\Omega^\#} u^\# \Delta g \, d\mathcal{L}^n. \tag{9.91}$$

Lastly, we define

$$d\mu^\star = f^\star \, d\mathcal{L}^{m+1} + d\tau^\star$$

where $d\tau^\star = d\mathcal{L}(r) d\tau^\#(z)$, recalling here that $\tau^\#$ can be regarded as a measure on $z \in Z$ because it concentrates where $y = 0$. If $G \in C_c(\Omega^\star)$ and $Gf^\star \in L^1(\Omega^\star)$ then

$$\int_{\Omega^\star} G \, d\mu^\star = \int_{\Omega^\#} J^t G \, d\mu^\#, \tag{9.92}$$

by adapting the proof of (9.81).

**Theorem 9.23** (Subharmonicity property of the ★-function for $(k, n)$-Steiner symmetrization) *Let $u \in \mathcal{W}(\Omega)$ satisfy*

$$u \geq 0 \quad \text{in} \quad \Omega$$

*and the Dirichlet boundary condition*

$$\lim_{x \to x_0, x \in \Omega} u(x) = 0$$

*for all* $x_0 = (y_0, z_0) \in \partial\Omega$ *with* $z_0 \in Z$. *Suppose*

$$-\Delta u \le \phi(z, u) + \mu$$

*in the weak sense in* $\Omega$, *where* $\phi\colon Z \times \mathbb{R} \to \mathbb{R}$ *is continuous with* $\phi\big(z, u(y, z)\big)$ *locally integrable on* $\Omega$. *Suppose in decomposition* (9.82) *that* $f + c$ *is nonnegative and satisfies Condition BS, for some* $c \in \mathbb{R}$, *and that* $\tau(\Omega) < \infty$. *If either*

(i) $u$ *is lower semicontinuous and* $u > 0$ *in* $\Omega$, *or*
(ii) $M_\kappa \phi\big(z, u(y, z)\big)$ *is locally integrable on* $\Omega$ *for some* $\kappa > 0$,

*then*

$$-\Delta^\star u^\star \le J\phi(z, u^\#) + \mu^\star$$

*in the weak sense in* $\Omega^\star$.

Remember in the boundary condition that $(\infty, z_0)$ is regarded as belonging to $\partial\Omega$ if the slice $\Omega(z_0)$ is unbounded.

When we write $J\phi(z, u^\#)$ in the theorem, we mean the $J$-operator applied to the function $\phi\big(z, u^\#(y, z)\big) \in L^1_{loc}(\Omega^\#)$.

Condition (ii) in the theorem holds if $\phi(\omega)$ is convex decreasing. This and other examples are explained after Theorem 9.18.

*Proof* Take a nonnegative $G \in C_c^2(\Omega^\star)$ and extend $G$ to equal 0 outside $\Omega^\star$. Choose $R_1 > 0$ such that $G(r, z) = 0$ whenever $r \in (0, R_1], z \in Z$. Define $g$ on $\Omega^\#$ by

$$g(y, z) = (J^t G)(y, z) = \int_{|y|}^{\infty} G(s, z)\, ds, \quad y \in \Omega^\#(z).$$

Then $g \ge 0, g = g^\#$, and $g$ has compact support in $\Omega^\#$. Also $g \in C^2(\Omega^\#)$, since $G \in C^2$ and $g(y, z)$ is independent of $y$ when $|y| \le R_1$.

By (9.91), Theorem 9.22, (9.86), and (9.92) we have

$$\begin{aligned} \int_{\Omega^\star} u^\star \Delta^{\star t} G\, d\mathcal{L}^{m+1} &= \int_{\Omega^\#} u^\# \Delta g\, d\mathcal{L}^n \\ &\ge -\int_{\Omega^\#} g\phi(z, u^\#)\, d\mathcal{L}^n - \int_{\Omega^\#} g\, d\mu^\# \\ &= -\int_{\Omega^\star} GJ\phi(z, u^\#)\, d\mathcal{L}^n - \int_{\Omega^\star} G\, d\mu^\star, \end{aligned}$$

which proves Theorem 9.23. (For the application of (9.92), note that Condition BS for $f + c$ implies local integrability of $f^\#$ and hence local integrability of $f^\star$, so that $Gf^\star \in L^1(\Omega^\star)$.) □

## 9.12 Notes and Comments

Some history of the ★-function can be found in Baernstein (1980, 1994, 2002) and in the references below. The subharmonicity property of the ★-function is due to Baernstein in 2 dimensions. The higher dimensional theory was developed by Baernstein and Taylor (1976).

The definition of the ★-function for cap-symmetrization in this chapter differs by a power of $r$ from the definition used in earlier works. More precisely, our definition of $u^\star$ for the $(n-1,n)$-cap case in §9.6 includes a factor of $r^{n-1}$, and in §9.8 for the $(k,n)$-cap case we include a factor of $r^k$. In other words, $u^\star$ involves integration with respect to surface area measure on the sphere of radius $r$, rather than surface measure on the unit sphere. This break with tradition is desirable for two reasons. First, it brings the definition into line with the Steiner case, where the ★-function is defined via integration with respect to surface area measure on lower dimensional slices. Second, and more significantly, the change permits a more natural definition of the measure $\tau^\star$ and of the $J$-operator acting on measures (as needed in §10.6), which would otherwise require artificial factors of $r^{1-n}$.

The ★-function is treated on shells and spheres in §9.5–9.7, but not on *subdomains* of those spaces. Subharmonicity theorems do hold on subdomains of spheres and shells, analogous to the results in §9.9–9.11 for subdomains of Euclidean space under s.d.r. and Steiner symmetrization. Such results are omitted here in the interests of simplicity. Chapter 11 applies circular symmetrization and the ★-function to subdomains of the complex plane. The references in that chapter provide more information.

For subharmonicity results under symmetrization on hyperbolic space, see the passing remark in (Baernstein, 1994, p. 63).

**Material added to the chapter.** Baernstein left no Notes or Comments for Chapter 9. This section was added by Richard Laugesen during revision of the manuscript. Corollaries 9.10 and 9.11 were added also, as was §9.11 on Steiner symmetrization, along with the change mentioned above to the definition of the ★-function.

Baernstein's manuscript proved the pre-subharmonicity Theorem 9.18 for s.d.r. with the help of a Green function perturbation on a certain intermediate domain. That perturbation has been changed to a bump function, in order to simplify the argument.

The semilinear term $\phi(u)$ in pre- and sub-harmonicity theorems throughout the chapter was added in revision, drawing on statements and proof sketches in (Baernstein, 1994, §5 and §6). The maximal function $M_\kappa\phi$ does not appear in that earlier work. We introduced it here in the s.d.r. and Steiner cases to handle the perturbation step of the pre-subharmonicity proof when $u$ is nonnegative but not necessarily positive everywhere (Condition (ii)). Perhaps other ways could be found around that technical difficulty.

Subharmonicity results first appeared in Baernstein (1973, 1974) as tools to solve extremal problems in complex analysis (see Chapter 11). In those early papers, the needed subharmonicity results appeared in the simpler setting of 2 dimensions, and were derived using various technical lemmas. Baernstein (1994) later developed a modern approach to subharmonicity properties. This new approach splits into two parts: first a pre-subharmonicity result whose proof makes clear use of symmetrization methods, specifically Riesz-type rearrangement inequalities, and second a straightforward deduction of subharmonicity from pre-subharmonicity, with the help of the main identity and adjoint relations that rely on integration by parts. The modern approach is followed consistently in this chapter – the details vary depending on the geometry at hand (cap symmetrization, s.d.r., Steiner, and so on), but the underlying method remains the same in each case.

# 10

# Comparison Principles for Semilinear Poisson PDEs

This chapter proves comparison principles of the following type: if $u$ solves the Poisson equation $-\Delta u = f$ on an open set $\Omega$ and $v$ solves the "symmetrized" equation $-\Delta v = f^\#$ on $\Omega^\#$, with $u$ and $v$ nonnegative and satisfying Dirichlet boundary conditions, then the $L^p$ norm of $u$ is smaller than that of $v$ for each $p \geq 1$. That is, the solution of Poisson's equation increases in the integral sense when the data in the equation is rearranged.

This comparison result is central to Nadirashvili's proof of Rayleigh's conjecture that the principal eigenvalue of a vibrating plate decreases under s.d.r. of the domain; see references in Section 5.3. Comparison theorems also provide practical bounds: under s.d.r. on a Euclidean domain, the solution $v$ depends only on the radial variable and thus can be written in terms of integrals of $f$. Hence the comparison theorem implies explicit a priori bounds on $u$ in terms of $f$ (see Talenti (1976b)).

The comparison principles are quite robust, and hold with the function $f$ replaced by a measure. For example, by using delta measures one may compare the $L^p$ norms of the Green functions solving the equations $-\Delta u = \delta_P$ and $-\Delta v = \delta_Q$, where $P$ is a point and $Q$ is its symmetrized point. Under s.d.r. on a Euclidean domain the symmetrized point is at the origin, whereas under $(n-1, n)$-cap symmetrization on a shell the symmetrized point lies on the $e_1$-axis at the same radius as $P$.

The results hold also when a nonlinearity $\phi(u)$ is added to the right side of Poisson's equation, with $\phi$ convex decreasing. Further, the nonnegativity assumption on $u$ can be dropped when working on shells and on the sphere. On shells, the Dirichlet boundary condition can be replaced in some circumstances by a Neumann condition.

This chapter restricts attention to the s.d.r., Steiner, and $(n-1, n)$-cap symmetrizations. The general $(k, n)$-cap symmetrization is not covered. Also, we consider only the Laplace operator. Comparison principles for more general elliptic operators and for the heat equation and parabolic operators have been treated in the literature. References to such results are provided in the end-of-chapter notes.

## 10.1 Majorization

We commence the chapter by connecting $\star$-functions to $L^p$ norms, through the concept of majorization. The $\star$-function on a finite measure space $(X, \mathcal{M}, \mu)$ was defined in Section 9.1, where we found

$$f^\star(x) = \int_0^x f^*(s)\,ds, \quad 0 \le x \le \mu(X),$$

with $f^*$ denoting the decreasing rearrangement of $f$.

**Proposition 10.1** (Majorization) *Assume $\mu(X) < \infty$ and let $f, g \in L^1(X)$. Then the following are equivalent.*

*(a) $\int_X \Phi(f)\,d\mu \le \int_X \Phi(g)\,d\mu$ for all increasing convex $\Phi\colon \mathbb{R} \to \mathbb{R}$.*
*(b) $\int_X (f-t)^+\,d\mu \le \int_X (g-t)^+\,d\mu$ for all $t \in \mathbb{R}$.*
*(c) $f^\star(x) \le g^\star(x)$ for all $x \in [0, A]$, where $A = \mu(X)$.*

When $f$ and $g$ obey the conditions of the Proposition, $g$ is said to *majorize* $f$, in the sense of Hardy–Littlewood–Pólya.

**Corollary 10.2** *Suppose $f, g \ge 0$ are integrable on a finite measure space. If $f^\star \le g^\star$ then the $L^p$-norm of $f$ is less than or equal to the norm of $g$:*

$$\|f\|_p \le \|g\|_p, \quad 1 \le p \le \infty. \tag{10.1}$$

The corollary follows from Proposition 10.1 by taking

$$\Phi(s) = \begin{cases} s^p & \text{when } s \ge 0, \\ 0 & \text{when } s < 0, \end{cases}$$

when $1 \le p < \infty$. For $p = \infty$, one simply passes to the limit in (10.1).

*Proof of Proposition 10.1* *(a)* $\Longrightarrow$ *(b)* is trivial since $\Phi(x) = (x-t)^+$ is increasing and convex. For *(b)* $\Longrightarrow$ *(c)*, take $x \in [0, A]$ and set $t = g^*(x)$. Since (b) holds for $f$ and $g$ on $X$, it also holds for $f^*$ and $g^*$ on $[0, A]$ (by equidistribution of $f$ and $f^*$ and of $g$ and $g^*$). Hence

$$\begin{aligned} f^\star(x) = \int_0^x [(f^*(s) - t) + t]\,ds &\le \int_0^x (f^*(s) - t)^+\,ds + tx \\ &\le \int_0^x (g^*(s) - t)^+\,ds + tx = g^\star(x). \end{aligned}$$

Suppose that (c) holds. Take $t \in \mathbb{R}$. If $t \geq \operatorname{ess\,sup} f$, then of course $(f-t)^+ = 0$ and (b) holds. If $t \leq \operatorname{ess\,inf} f$, then taking $x = A$ in (c), we get

$$\begin{aligned}\int_X (f-t)^+ d\mu &= \int_X (f-t)\, d\mu \\ &= f^\star(A) - tA \\ &\leq g^\star(A) - tA = \int_X (g-t)\, d\mu \leq \int_X (g-t)^+\, d\mu,\end{aligned}$$

and (b) holds. If $\operatorname{ess\,inf} f < t < \operatorname{ess\,sup} f$, take $x \in (0,A)$ such that $f^*(x) \leq t \leq f^*(x-)$. Then

$$\begin{aligned}\int_0^A (f^* - t)^+ ds &= \int_0^x (f^* - t)\, ds \\ &= f^\star(x) - tx \\ &\leq g^\star(x) - tx = \int_0^x (g^* - t)\, ds \leq \int_0^A (g^* - t)^+\, ds,\end{aligned}$$

so (b) holds in all cases.

It remains to prove that $(b) \implies (a)$. Let $\Phi$ be increasing and convex on $\mathbb{R}$. We may assume that $\Phi$ is nonconstant. Then there is a linear function $L(x) = ax + b$ with $a > 0$ such that $L \leq \Phi$ everywhere. Since $f \in L^1(X)$ and $\Phi(f) \geq L(f)$, it follows that $\int_X [\Phi(f)]^- \, d\mu < \infty$, so that the integral $\int_X \Phi(f)\, d\mu$ is well defined, possibly as $\infty$. Moreover, since $(b) \implies (c)$ and $\int_X f\, d\mu = f^\star(A)$, it follows that

$$\int_X L(f)\, d\mu = af^\star(A) + b \leq \int_X L(g)\, d\mu \tag{10.2}$$

for every linear $L$ on $\mathbb{R}$ with nonnegative slope.

By Zygmund (1968, p. 24), $\Phi$ is locally absolutely continuous on $\mathbb{R}$, hence is differentiable $\mathcal{L}$ a.e. At such points we denote its derivative by $\Phi'(x)$. The convexity of $\Phi$ implies that $\Phi'$ is increasing on its domain. The one-sided limits of $\Phi'$ exist at every $x \in \mathbb{R}$, and coincide with the corresponding one-sided derivatives of $\Phi$. We denote these values by $\Phi'(x-)$ and $\Phi'(x+)$.

Let us assume for the moment that $\Phi = 0$ on $(-\infty, 0]$. There is a positive Borel measure $\nu$ on $\mathbb{R}$ such that

$$\Phi'(x+) = \nu(-\infty, x], \quad x \in \mathbb{R}.$$

For $x > 0$, we have

$$\begin{aligned}\Phi(x) &= \int_{-\infty}^x \Phi'(t)\, dt = -\int_0^x \Phi'(t) \frac{d}{dt}(x-t)\, dt \\ &= \int_{(0,x]} (x-t)\, d\nu(t) + x\Phi'(0+) \\ &= \int_{(0,\infty)} (x-t)^+ d\nu(t) + (x-0)^+ \Phi'(0+).\end{aligned}$$

(For the integration by parts, see Folland (1999, Theorem 3.36).) The same formula holds trivially when $x \leq 0$.

Replacing $x$ by $f$ and integrating with respect to $\mu$, we see that the desired inequality

$$\int_X \Phi(f)\, d\mu \leq \int_X \Phi(g)\, d\mu \tag{10.3}$$

holds when $\Phi = 0$ on $(-\infty, 0]$.

Next, assume that $\Phi$ equals a linear function $L$ on $(-\infty, a]$ for some $a \geq 0$. Set $\Phi_1(x) = \Phi(x+a) - L(x+a)$. Then $\Phi_1 = 0$ on $(-\infty, 0]$ and $\Phi(x) = L(x) + \Phi_1(x-a)$. If $f$ and $g$ satisfy the assumptions of (b) then so do $f - a$ and $g - a$. From (10.2), and (10.3) applied to $\Phi_1$, we see that (10.3) holds when $\Phi$ is linear on some $(-\infty, a]$.

Finally, let $\Phi$ be an arbitrary convex increasing function on $\mathbb{R}$. Take a decreasing sequence $\{a_n\}$ with $a_n \to -\infty$. Let $\Phi_n$ be the function which equals $\Phi$ on $[a_n, \infty)$, and on $(-\infty, a_n]$ is the linear function $L_n$ with $L_n(a_n) = \Phi(a_n)$ and $L_n'(a_n) = \Phi'(a_n-)$. Then $\Phi_n$ is convex, and (10.3) holds for each $\Phi_n$. Also, $\Phi_n(x) \nearrow \Phi(x)$ as $n \nearrow \infty$ for each $x \in \mathbb{R}$. If $\int_X \Phi_1(g)\, d\mu = \infty$ then $\int_X \Phi(g)\, d\mu = \infty$, so (10.3) holds for $\Phi$. If $\int_X \Phi_1(g)\, d\mu$ is finite, then so is $\int_X \Phi_1(f)\, d\mu$. The monotone convergence theorem applied to $\Phi_n(f) - \Phi_1(f)$ shows that (10.3) holds in this case too. □

We will need a variant of Proposition 10.1 that does not require the convex function $\Phi$ to be increasing. We keep the hypothesis $A = \mu(X) < \infty$, and add a new hypothesis that the integrals of $f$ and $g$ agree.

**Proposition 10.3** (Majorization with equal integrals) *Let $f, g \in L^1(X)$, with $\int_X f\, d\mu = \int_X g\, d\mu$. Then the following are equivalent:*

*(a) $\int_X \Phi(f)\, d\mu \leq \int_X \Phi(g)\, d\mu$ for all convex $\Phi\colon \mathbb{R} \to \mathbb{R}$.*
*(b) $\int_X (f-t)^+\, d\mu \leq \int_X (g-t)^+\, d\mu$ for all $t \in \mathbb{R}$.*
*(c) $f^\star(x) \leq g^\star(x)$ for all $x \in [0, A]$.*

*Under these conditions, we have*

$$\begin{aligned} \|f\|_p &\leq \|g\|_p, \quad 1 \leq p \leq \infty, \\ \operatorname{ess\,sup} f &\leq \operatorname{ess\,sup} g, \\ \operatorname{ess\,inf} g &\leq \operatorname{ess\,inf} f, \\ \operatorname{osc} f &\leq \operatorname{osc} g. \end{aligned}$$

*Proof* The proof of Proposition 10.1 carries over without change, except for the implication $(b) \implies (a)$, which requires the additional observation that (10.2) now holds with equality for every $a \in \mathbb{R}$.

For the statement about $L^p$-norms, simply take $\Phi(s) = |s|^p$ when $1 \leq p < \infty$, and then let $p \to \infty$ to obtain the corresponding inequality for $L^\infty$-norms. For the inequalities on the essential supremum and infimum, use that

$$\operatorname{ess\,sup} f = \lim_{x \to 0} \frac{f^\star(x)}{x}, \qquad \operatorname{ess\,inf} f = \lim_{x \to A} \frac{f^\star(A) - f^\star(x)}{x},$$

and similarly for $g$, and note $f^\star(A) = \int_X f\,d\mu = \int_X g\,d\mu = g^\star(A)$. Then the oscillation of $f$ is clearly less than the oscillation of $g$, since $f$ has larger infimum and smaller supremum than $g$. □

So far in this section, we have assumed $f \in L^1(X)$ and $\mu(X) < \infty$. These hypotheses can be relaxed. Let $f\colon X \to \mathbb{R}$ be a $\mathcal{M}$-measurable function on $X$, where $A = \mu(X)$ can be either finite or not. We shall say that $f$ satisfies **Condition B** if:

$$\int_X (f - t)^+\,d\mu < \infty, \qquad \forall\, t > \operatorname{ess\,inf}_X f.$$

(Condition B appeared already in Euclidean space, in Sections 9.9 and 9.10.)

The integral on the left equals $\int_t^\infty \lambda(s)\,ds$, where $\lambda$ is the distribution function of $f$. It follows that Condition B implies $\lambda(t) < \infty$ for all $t > \operatorname{ess\,inf}_X f$. This is the finiteness condition needed to insure existence and finiteness of the decreasing rearrangement $f^*$ on $(0, A)$. The integral in Condition B does not change if $f$ is replaced by $f^*$, or by any other rearrangement of $f$. As the reader may easily show, $f$ satisfies Condition B if and only if $f^* \in L^1_{loc}[0, A)$.

From now on we shall write $\inf_X f$ instead of $\operatorname{ess\,inf}_X f$. If $\inf_X f < 0$ then Condition B implies $f^+ \in L^1(X)$, while if $\inf_X f \geq 0$ then $f^- = 0$. Thus, $\int_X f\,d\mu$ is well-defined, possibly as $\pm\infty$, and so is $\int_E f\,d\mu$ for $E \subset X$. Thus, when $f$ satisfies Condition B, we may again define the star function of $f$ by

$$f^\star(x) = \sup \int_E f\,d\mu, \quad x \in [0, A],$$

where the sup is taken over all $\mathcal{M}$-measurable sets $E \subset X$ with $\mu(X) = x$. If the sup is attained by a set $E$, we shall call $E$ a maximal set.

**Claim 10.4** *If $(X, \mathcal{M}, \mu)$ is a nonatomic measure space and $f$ satisfies Condition B on $X$, then $f^\star(x)$ is finite for all $0 \leq x < A$ and Proposition 10.1 is still true.*

*Proof* From the definition, $f^\star(0) = 0$, and every set of measure zero is a maximal set, so Claim 10.4 is true when $x = 0$. Take $x \in (0, A)$. Write $t = f^*(x)$. Let $x_1$ be the smallest value of $s$ such that $f^*(s) = t$. Then $0 \leq x_1 \leq x$. Let $E_1 = (f > t)$. Then $\mu(E_1) = x_1$. Also, $\mu(f = t) \geq x - x_1$. Since the space is nonatomic, Proposition 1.25 provides a subset $E_2$ of $(f = t)$ with

$\mu(E_2) = x - x_1$. Let $E = E_1 \cup E_2$. Then $\mu(E) = x$, and the argument in the proof of Proposition 10.1 shows that $E$ is a maximal set. Take some $t_1 > t$. Then $t_1 > \inf_X f$. On $E$, we have $t \le f \le (f - t_1)^+ + |t_1|$. Thus, by Condition B, $f \in L^1(E)$. In particular, $f^\star(x) = \int_E f\,d\mu$ is finite.

To obtain part (b) of Proposition 10.1, again take $x \in (0, A)$. Let $E$ be a maximal set, and let $f_0$ be the restriction of $f$ to $E$. Then $f^\star(x) = f_0^\star(x)$, since each equals $\int_E f\,d\mu$. Also, $f^*(s) = f_0^*(s)$ for all $s \in [0, x]$. Since $\mu(E) < \infty$, we can apply Proposition 10.1(b) with $X = E$ and $f = f_0$, thereby obtaining $f^\star(x) = f_0^\star(x) = \int_0^x f_0^*(s)\,ds = \int_0^x f^*(s)\,ds$. □

**Claim 10.5** *Suppose that X is as in Claim 10.4 and that f and g satisfy Condition B on X. Then Proposition 10.1 is still true, provided that in part (a) we require* $\int_X \Phi(f)\,d\mu \le \int_X \Phi(g)\,d\mu$ *hold only for all* nonnegative *increasing convex* $\Phi$.

*Proof* (*a*) $\implies$ (*b*) is again trivial. We noted above that if $f$ satisfies Condition B then $f^* \in L^1[0, x]$ for every $x < A$. Using this local integrability, one sees that the proof of (*b*) $\implies$ (*c*) still works with our new hypotheses.

To show that (*c*) $\implies$ (*a*), let $\{x_k\}, k \ge 1$, be an increasing sequence in $(0, A)$ with limit $A$. Let $f_k$ and $g_k$ be the restrictions of $f^*$ and $g^*$ to $[0, x_k]$. For each $k$, the hypotheses of the original Proposition 10.1 are satisfied by $f_k$ and $g_k$. Moreover, if $f$ and $g$ satisfy (c) for all $x \in [0, A)$, then $f_k^\star \le g_k^\star$ for all $s \in [0, x_k]$. From (*c*) $\implies$ (*a*) in the original Proposition 10.1, and $f^* = f_k$, $g^* = g_k$ on $[0, x_k]$, it follows that

$$\int_{[0,x_k]} \Phi(f^*(s))\,ds \le \int_{[0,x_k]} \Phi(g^*(s))\,ds.$$

for all increasing convex $\Phi$ on $\mathbb{R}$. If also $\Phi \ge 0$, then we can apply the monotone convergence theorem when $k \to \infty$ to obtain

$$\int_{[0,A]} \Phi(f^*(s))\,ds \le \int_{[0,A]} \Phi(g^*(s))\,ds,$$

which is the same as $\int_X \Phi(f)\,d\mu \le \int_X \Phi(g)\,d\mu$. □

As an exercise, we invite the reader to create a decreasing function $f$ on $X = [0, \infty)$ which satisfies Condition B, and an increasing convex function $\Phi$ on $\mathbb{R}$ for which $\int_X \Phi(f(s))^+\,ds = \int_X \Phi(f(s))^-\,ds = \infty$. Thus, $\int_X \Phi(f)\,d\mu$ need not be well-defined in our current setting. This is why we had to tighten condition (a) of Proposition 10.1. For similar reasons, we have refrained from formulating a Condition B version of Proposition 10.3. Of course, versions do exist, and one can devise them according to the special features of the functions $\Phi$ under study.

## 10.2 Weakly Convex and Weakly Subharmonic Functions

A continuous function $w\colon (a,b) \to \mathbb{R}$ is called convex in the weak (or distributional) sense if $w'' \geq 0$ in the weak sense, which means

$$\int_a^b v'' w \, d\mathcal{L} \geq 0$$

for each nonnegative function $v \in C_c^2(a,b)$. Strict convexity in the weak sense means the last inequality is strict for each $v$ (except when $v$ is the zero function).

Convexity in the weak sense implies convexity in the classical sense, as the next lemma shows. We rely on this fact later in the chapter.

**Lemma 10.6** *If a continuous function on an open interval is (strictly) convex in the weak sense, then it is (strictly) convex in the classical sense.*

*Proof* Suppose $w\colon (a,b) \to \mathbb{R}$ is continuous and convex in the weak sense. Let $K$ be a nonnegative, even, smooth function supported in $(-1,1)$ with integral 1. Then the mollification $w_\epsilon = w * K_\epsilon$ is smooth and has second derivative

$$w_\epsilon''(x) = \int_a^b w(y)(K_\epsilon)''(x-y)\, dy \geq 0,$$

by applying the weak convexity condition with test function $v = K_\epsilon$. Hence $w_\epsilon$ is classically convex and so satisfies the convexity condition

$$w_\epsilon\big(tx + (1-t)y\big) \leq tw_\epsilon(x) + (1-t)w_\epsilon(y)$$

whenever $a + \epsilon < x < y < b - \epsilon$ and $0 < t < 1$. Letting $\epsilon \to 0$ shows

$$w\big(tx + (1-t)y\big) \leq tw(x) + (1-t)w(y) \tag{10.4}$$

whenever $a < x < y < b$ and $0 < t < 1$, noting $w_\epsilon \to w$ pointwise by continuity of $w$. Thus $w$ is convex in the classical sense.

For the strictness statement in the lemma, we will prove the contrapositive. Suppose $w$ is convex in the weak sense (and so is classically convex as shown above) but that $w$ is not strictly convex. Then equality holds in the convexity condition (10.4) for some $x, y, t$. Hence the convex function

$$f(\tau) = w(\tau) - w(x)\frac{y-\tau}{y-x} - w(y)\frac{\tau - x}{y-x}, \qquad \tau \in [x,y],$$

satisfies $f(x) = f(y) = 0$ and also $f\big(tx + (1-t)y\big) = 0$. Therefore $f$ is constant by Fact 2.1, which means $w$ is linear on $[x,y]$, and so

$$\int_a^b u'' w \, d\mathcal{L} = 0$$

by integration by parts whenever $u \in C_c^2(x, y)$. Choosing a test function $u$ that is nonnegative and not the zero function, we conclude $w$ is not strictly convex in the weak sense. □

A strictly convex function cannot have an interior maximum point. A corresponding maximum principle holds in higher dimensions for functions that are strictly subharmonic in the weak (or distributional) sense, as the next lemma shows.

**Proposition 10.7** (Maximum principle) *Assume $w$ is continuous on an open set $\Omega$, the vector field $b$ is smooth, and the constant $c$ is positive. If*

$$\Delta w + b \cdot \nabla w \geq 2c$$

*in the weak (distributional) sense then $w$ has no interior maximum point.*

*Proof* Suppose to the contrary that $w$ does have an interior maximum, which after a translation we may take to occur at the origin. Define

$$v = w - \frac{c}{4n}|x|^2$$

so that $v$ has a strict maximum at the origin and

$$\Delta v + b \cdot \nabla v \geq c \tag{10.5}$$

on some sufficiently small ball $\mathbb{B}^n(0, r)$ centered at the origin.

Then

$$\max_{|x|=r} v(x) < v(0)$$

since $v$ has a strict maximum at the origin. Fix a bump function $K$ (nonnegative, smooth, supported in the unit ball, with integral 1), and recall $K_\epsilon(x) = \epsilon^{-n}K(x/\epsilon)$ for $\epsilon > 0$. Note $v * K_\epsilon \to v$ locally uniformly as $\epsilon \to 0$, due to the continuity of $v$, and so the preceding inequality continues to hold for the $\epsilon$-mollification of $v$, with

$$\max_{|x|=r}(v * K_\epsilon)(x) < (v * K_\epsilon)(0)$$

whenever $\epsilon$ is sufficiently small. Hence the smooth function $v * K_\epsilon$ achieves a local maximum at some point $x_\epsilon \in \mathbb{B}^n(0, r)$. The first derivatives must vanish and the second derivatives must be nonpositive at this maximum point, and so

$$\nabla(v * K_\epsilon)(x_\epsilon) = 0, \qquad \Delta(v * K_\epsilon)(x_\epsilon) \leq 0.$$

To make use of these facts, we apply the weak differential equality (10.5) to the test function $x \mapsto K_\epsilon(x_\epsilon - x)$, obtaining that

$$\int_\Omega v(x)(\Delta K_\epsilon)(x_\epsilon - x)\,dx + \int_\Omega v(x)\nabla \cdot [-b(x)K_\epsilon(x_\epsilon - x)]\,dx \geq c.$$

The first integral equals $\Delta(v * K_\epsilon)(x_\epsilon) \leq 0$. The second integral equals

$$\int_\Omega v(x)\nabla \cdot \big[\big(b(x_\epsilon) - b(x)\big)K_\epsilon(x_\epsilon - x)\big]\,dx - \int_\Omega v(x)b(x_\epsilon) \cdot \nabla\big[K_\epsilon(x_\epsilon - x)\big]\,dx.$$

That last integral is $-b(x_\epsilon) \cdot \nabla(v * K_\epsilon)(x_\epsilon) = 0$ and so can be discarded. Combining the above facts, one has

$$\int_\Omega v(x)\nabla \cdot \big[\big(b(x_\epsilon) - b(x)\big)K_\epsilon(x_\epsilon - x)\big]\,dx \geq c > 0. \tag{10.6}$$

We will show the left side of (10.6) approaches 0 as $\epsilon \to 0$, giving a contradiction and hence completing the proof.

Let $\epsilon \to 0$ through some sequence of values. After passing to a subsequence we may further assume $x_\epsilon \to x_0 \in \overline{\mathbb{B}^n(0,r)}$. Changing variable with $x \mapsto x_\epsilon - \epsilon x$ on the left side of (10.6) gives

$$\begin{aligned}
&-\int_{\mathbb{B}^n(0,1)} v(x_\epsilon - \epsilon x)\nabla \cdot \Big[\frac{b(x_\epsilon) - b(x_\epsilon - \epsilon x)}{\epsilon}K(x)\Big]\,dx \\
&\to -v(x_0)\int_{\mathbb{B}^n(0,1)} \nabla \cdot \big[Db(x_0)xK(x)\big]\,dx \qquad \text{as } \epsilon \to 0 \\
&= 0
\end{aligned}$$

by the divergence theorem, since the vector field $Db(x_0)xK(x)$ is supported in the unit ball. Thus the left side of (10.6) approaches 0, as we wanted to show. □

To deal with the ⋆-function in spherical shells later in this chapter, we need a weighted version of the maximum principle.

**Proposition 10.8** (Weighted maximum principle) *Assume $w(r,\theta)$ is continuous on an open subset of $\{(r,\theta) : r > 0, 0 < \theta < \pi\}$, and $n, c_1 > 0$ are constant. If*

$$r^{1-n}\partial_r\big(r^{n-1}\partial_r w\big) + r^{-2}\sin^{n-2}\theta\;\partial_\theta(\sin^{2-n}\theta\;\partial_\theta w) \geq c_1$$

*in the weak (distributional) sense then $w$ has no interior maximum point.*

*Proof* Suppose $w$ does have an interior maximum point. We will deduce a contradiction. Multiplying by $r^2$ shows that

$$r^{3-n}\partial_r\big(r^{n-1}\partial_r w\big) + \sin^{n-2}\theta\;\partial_\theta(\sin^{2-n}\theta\;\partial_\theta w) \geq c_2$$

in the weak sense on a neighborhood of the maximum, where $c_2 > 0$ is a new constant. Changing variable with $r = e^s$ implies

$$(\partial_{ss} + \partial_{\theta\theta})w + (n-2)\big(\partial_s - \cot\theta\,\partial_\theta\big)w \geq c_2$$

around the maximum point, which is impossible by Proposition 10.7. □

Next we present an elementary lemma, unrelated to convexity or subharmonicity, that is needed later in the chapter.

**Lemma 10.9** *Suppose $\eta\colon (0,1) \to \mathbb{R}$ is $C^2$-smooth with compact support and $g\colon (0,1) \to \mathbb{R}$ is integrable.*

*(a) If $g(t) = o(1)$ as $t \to 0$ then*

$$\int_0^\epsilon \big(\eta(t/\epsilon)\big)' g(t)\,dt \to 0 \qquad \textit{as } \epsilon \to 0.$$

*(b) If $g(t) = o(t)$ as $t \to 0$ then*

$$\int_0^\epsilon \big(\eta(t/\epsilon)\big)'' g(t)\,dt \to 0 \qquad \textit{as } \epsilon \to 0.$$

*(c) If $\limsup_{t\to 0} g(t) \leq 0$ and $\eta' \geq 0$ then*

$$\limsup_{\epsilon\to 0} \int_0^\epsilon \big(\eta(t/\epsilon)\big)' g(t)\,dt \leq 0.$$

*Proof* The first integral equals $\int_0^1 \eta'(t) g(\epsilon t)\,dt$, which approaches 0 as $\epsilon \to 0$ since $g(\epsilon t) \to 0$. The second integral equals $\int_0^1 \eta''(t) t g(\epsilon t)/(\epsilon t)\,dt$, which approaches 0 as $\epsilon \to 0$ since $g(\epsilon t) = o(\epsilon t)$. The third integral is less than or equal to the corresponding integral with $g^+$ instead of $g$, since $\eta' \geq 0$, and so applying part (a) to $g^+$ gives the desired result. □

## 10.3 Comparison Principles for s.d.r. on Euclidean Domains

The first comparison principles will be on Euclidean domains for the s.d.r. Before stating them, we need to extend the definition of the $J$-operator from functions to measures.

Write # for symmetric decreasing rearrangement on an open set $\Omega \subset \mathbb{R}^n, n \geq 1$, as treated in Section 9.10. Take $R \in (0, \infty]$ to be the radius of the ball $\Omega^\# = \mathbb{B}^n(R)$, and recall that

$$\Omega^\star = (0, R)$$

is an open interval, and the indefinite integral operator $J$ maps functions on $\Omega^\#$ to functions on $\Omega^\star$ by

$$(Ju)(r) = \int_{B(r)} u\, d\mathcal{L}^n, \quad r \in \Omega^\star.$$

Given a measure $\mu \in M_{loc}(\Omega^\#)$, define the measure $J\mu \in M_{loc}(\Omega^\star)$ by

$$d(J\mu) = \mu\big(B(r)\big)\, dr.$$

In particular, one has $\mu^\star = J(\mu^\#)$ because

$$\begin{aligned} d\mu^\star &= \big(f^\star(r) + \tau(\Omega)\big)\, dr \\ &= \mu^\#\big(B(r)\big)\, dr \\ &= d(J(\mu^\#)) \end{aligned}$$

by the definitions of $\mu^\#$ and $\mu^\star$ in Sections 9.9 and 9.10. Hence the hypothesis $\mu^\star \le J\nu$ in the next theorem holds with equality if $\nu = \mu^\#$.

The adjoint relation for measures says

$$(G, J\mu) = (J^t G, \mu) \tag{10.7}$$

for $\mu \in M_{loc}(\Omega^\#)$ and $G \in C_c(\Omega^\star)$, which is easy to verify since each side equals

$$\int_{\Omega^\#} \int_0^R \chi_{B(r)}(x) G(r)\, dr\, d\mu(x).$$

As a consequence, we see that the identity $\int_{\Omega^\star} G\, d\mu^\star = \int_{\Omega^\#} J^t G\, d\mu^\#$ in (9.81), which we used in proving Theorem 9.20, is just the adjoint relation (10.7) applied to $\mu^\#$ and combined with the formula $\mu^\star = J(\mu^\#)$.

### 10.3.1 Linear Equations

Our first comparison theorem treats linear Poisson equations. The conclusion is that the s.d.r. of $u$ is bounded by the average of $v$ over a sphere centered at the origin, written

$$\overline{v}(rx) = \frac{1}{\beta_{n-1}} \int_{\mathbb{S}^{n-1}} v(ry)\, d\sigma_{n-1}(y), \qquad |x| = 1,\ 0 \le r < R.$$

If $v$ is radial, then $\overline{v} = v$. Recall that $\lambda_1$ denotes the first eigenvalue of the Dirichlet Laplacian, which for the ball $\mathbb{B}^n(R)$ was computed in terms of Bessel zeros in Proposition 5.7.

**Theorem 10.10** (Linear Poisson) *Assume $u \in \mathcal{W}(\Omega), v \in \mathcal{W}(\Omega^\#)$ are weak solutions of the linear Poisson equations*

$$-\Delta u + cu = \mu, \qquad -\Delta v + cv = \nu,$$

*where $\mu \in M_{loc}(\Omega), \nu \in M_{loc}(\Omega^\#)$, and $c \in \mathbb{R}$ satisfies $c > -\lambda_1\big(\mathbb{B}^n(R)\big)$ (if $R < \infty$) or $c \geq 0$ (if $R = \infty$). Further assume $\mu$ decomposes as in (9.47) with $f$ satisfying Condition B and with $\tau(\Omega) < \infty$. Suppose $\mu^\star \leq J\nu$.*

*If $u$ is nonnegative and satisfies the Dirichlet boundary condition*

$$\lim_{x \to x_0, x \in \Omega} u(x) = 0$$

*for all $x_0 \in \partial\Omega$, and $\overline{v}$ satisfies the boundary condition $\liminf_{r \to R} \overline{v}(rx) \geq 0$, then $u^\star \leq Jv \leq v^\star$ on $\Omega^\star$. Hence the convex increasing integral means of $u$ are dominated by those of $v$:*

$$\int_\Omega \Phi(u)\, d\mathcal{L}^n \leq \int_{\Omega^\#} \Phi(v)\, d\mathcal{L}^n$$

*for every convex increasing $\Phi\colon \mathbb{R} \to \mathbb{R}^+$.*

*Furthermore, when $c = 0$ one has the stronger pointwise conclusion $u^\# \leq \overline{v}$ on $\Omega^\#$, and hence $\int_\Omega \Phi(u)\, d\mathcal{L}^n \leq \int_{\Omega^\#} \Phi(\overline{v})\, d\mathcal{L}^n$ for every increasing $\Phi\colon \mathbb{R} \to \mathbb{R}^+$.*

The notion of weak (distributional) solutions was defined in Chapter 9, the point being that all derivatives are moved onto the test function. The theorem is most often applied with $\nu = \mu^\#$, in which case the hypothesis $\mu^\star \leq J\nu$ holds with equality.

Remember that when an open set is unbounded, the point at infinity is regarded as a boundary point for the purposes of the Dirichlet boundary condition.

A comparison result on the gradient norm (rather than the norm of the function) is due to Talenti (1976b, Theorem 1(v)): if $-\Delta u = f$ and $-\Delta v = f^\#$ and $u$ and $v$ satisfy the Dirichlet boundary condition, then

$$\int_\Omega |\nabla u|^q\, d\mathcal{L}^n \leq \int_{\Omega^\#} |\nabla v|^q\, d\mathcal{L}^n, \qquad q \in (0, 2]. \tag{10.8}$$

His proof uses level set methods and differential inequalities.

*Proof of Theorem 10.10* The reader should concentrate on the case $c = 0$, in what follows. The proof when $c > 0$ has a similar "pointwise" flavor. When $c < 0$, integral arguments will be used and the proof is considerably more complicated.

The nonnegativity assumption on $u$ together with the Dirichlet boundary condition $u \to 0$ at $\partial\Omega$ implies $u^\# \to 0$ at $\partial\Omega^\#$. That is, $u^\#$ satisfies a Dirichlet boundary condition as $r \to R$. Also $\liminf_{r\to R} \overline{v}(rx) \geq 0$ by hypothesis. Hence

$$\limsup_{r\to R}(u^\# - \overline{v}) \leq 0. \tag{10.9}$$

We will use this inequality later in the proof.

We may apply the Subharmonicity Theorem 9.20 for $u$, since $u$ satisfies the Dirichlet boundary condition and $\phi(z) = -cz$ satisfies Condition (ii) in that theorem. Combining that result with the Weak Main Identity (9.79) for $v$ and with the adjoint relation (10.7), we have

$$-\Delta^\star u^\star + cu^\star \leq \mu^\star,$$
$$-\Delta^\star Jv + cJv = Jv,$$

in the weak sense, where the operator $\Delta^\star$ was defined in (9.74) as

$$\Delta^\star F(r) = r^{n-1}\big(r^{1-n}F'(r)\big)'. \tag{10.10}$$

Since $\mu^\star \leq Jv$ by hypothesis, we conclude that the absolutely continuous function

$$w = u^\star - Jv = J(u^\# - v)$$

satisfies

$$-\Delta^\star w + cw \leq 0 \tag{10.11}$$

weakly on the interval $(0, R)$. Changing variable with $\rho = r^n$ gives

$$n^2\rho^{2-2/n}\frac{d^2w}{d\rho^2} \geq cw \tag{10.12}$$

in the weak sense.

**Step (i)**: Suppose first that $c = 0$, so that (10.12) implies $d^2w/d\rho^2 \geq 0$ in the weak sense. Then $w$ is classically convex as a function of $\rho$, by Lemma 10.6. Convexity insures that the derivative $dw/d\rho$ is increasing. This derivative is defined a.e., and equals

$$\begin{aligned}\frac{dw}{d\rho} &= \frac{1}{nr^{n-1}}\frac{d}{dr}\int_{B(r)}(u^\# - \overline{v})\,d\mathcal{L}^n\\ &= \frac{1}{n}\beta_{n-1}(u^\# - \overline{v})(rx),\end{aligned}$$

as one sees by expressing the integral in spherical coordinates (noting $x$ can be any unit vector, since $u^\#$ and $\overline{v}$ are radial functions). Therefore $u^\# - \overline{v}$ is an increasing radial function. Its limiting value is less than or equal to zero by

(10.9), and so $u^{\#} - \overline{v} \le 0$ everywhere. Integrating over the ball $\mathbb{B}^n(r)$ implies $u^\star - Jv \le 0$, from which the desired inequality on integral means follows by Section 10.1.

**Step (ii)**: Suppose now that $c > 0$. Our goal is to show $w \le 0$. Suppose instead that $w$ is positive somewhere. Then the open set $\{\rho \colon w > 0\}$ is nonempty, and so we may write $(\rho_1, \rho_2) \subset (0, R^n)$ for a maximal interval (component) on which $w$ is positive. By (10.12) and positivity of $c$, we see $w$ is classically convex as a function of $\rho \in (\rho_1, \rho_2)$.

Observe that $w = 0$ at the left endpoint $\rho_1$, since either $\rho_1 = 0$, at which point $u^\star$ and $Jv$ and hence $w$ equal zero by definition, or else $\rho_1 > 0$, where $w$ must equal zero because otherwise the component $(\rho_1, \rho_2)$ could be extended to the left.

Convexity then implies $w$ is increasing as a function of $\rho \in (\rho_1, \rho_2)$. In particular, $w$ increases to a positive limiting value as $\rho \nearrow \rho_2$. Hence $\rho_2 = R^n$, because otherwise the component $(\rho_1, \rho_2)$ could be extended further to the right. Convexity also implies that the derivative of $w$ has a positive limiting value as $\rho \nearrow R^n$, which contradicts the fact in (10.9) that $\limsup_{r \to R}(u^{\#} - \overline{v}) \le 0$. This contradiction implies $w$ cannot be positive anywhere, and so $w \le 0$ as desired.

**Step (iii)**: Suppose lastly that $-\infty < c < \infty$ and $R < \infty$. We will prove the contrapositive of the theorem, by showing that if $w > 0$ somewhere then $c \le -\lambda_1\big(\mathbb{B}^n(R)\big)$. (Thus if $c > -\lambda_1\big(\mathbb{B}^n(R)\big)$, then $w \le 0$ and so $u^\star \le Jv$.) So we suppose $(r_1, r_2) \subset (0, R)$ is a maximal interval on which $w(r)$ is positive.

Write $F(x)$ for the first eigenfunction of the Laplacian on the ball $\mathbb{B}^n(R)$, so that

$$-\Delta F = \lambda_1\big(\mathbb{B}^n(R)\big)F$$

with $F$ being positive and radially decreasing on the ball and $F = 0$ on the boundary. Explicitly,

$$F(x) = (j_{p,1}r/R)^{-p} J_p(j_{p,1}r/R), \qquad x \in \mathbb{R}^n,$$

and $\lambda_1\big(\mathbb{B}^n(R)\big) = (j_{p,1}/R)^2$, where $r = |x|, p = \frac{n}{2} - 1$, and $j_{p,1}$ is the first positive root of the Bessel function $J_p$ (see Proposition 5.7). In dimension 1 these formulas remain valid and yield the usual trigonometric eigenfunction $F(x) = \sqrt{2/\pi}\cos(\pi r/2R)$ (coming from $n = 1, p = -1/2, J_{-1/2}(r) = \sqrt{2/\pi}\, r^{-1/2}\cos r$, and $j_{p,1} = \pi/2$).

We proceed in two cases.

Case 1: $r_2 = R$. In this case $w(r_2) \ge 0$, and $\limsup_{r \to r_2} w'(r) \le 0$ by (10.9). Define

$$f(x) = \chi_{[r_1, r_2]}(r)F(x),$$

so that $f \geq 0$. Then $r^{1-n}Jf(r)$ is smooth for $r \in [r_1, r_2]$ and equals zero at $r = r_1$; the only potential issue arises when $r_1 = 0$, and in that situation smoothness can be established by expanding $F$ in a power series to find $JF(r) = F(0)\alpha_n r^n + O(r^{n+1})$ and so on.

The weak form of the differential inequality (10.11) says that

$$-\int_0^R \big(G(r)r^{n-1})'r^{1-n}\big)'w(r)\,dr + \int_0^R G(r)cw(r)\,dr \leq 0$$

for all nonnegative functions $G \in C_c^2(0, R)$. We choose

$$G(r) = \psi_\epsilon(r)r^{1-n}Jf(r),$$

where $\epsilon > 0$ is small and $\psi_\epsilon(r)$ is a smooth cut-off function compactly supported in $(r_1, r_2)$ that increases from 0 to 1 for $r \in [r_1, r_1 + \epsilon]$, equals 1 for $r \in [r_1 + \epsilon, r_2 - \epsilon]$, and decreases from 1 to 0 for $r \in [r_2 - \epsilon, r_2]$. Specifically, we take $\psi_\epsilon(r) = \eta\big((r - r_1)/\epsilon\big)\eta\big((r_2 - r)/\epsilon\big)$ where $\eta(t)$ is a smooth function that equals zero for $t \leq 1/2$, increases from 0 to 1 for $1/2 \leq t \leq 1$, and equals 1 for $t \geq 1$.

After substituting our choice of $G$ into the weak form of the differential inequality, we arrive at:

$$\int_{r_1}^{r_2} \psi_\epsilon\big[-\beta_{n-1}f' + cr^{1-n}Jf\big]w\,dr \tag{10.13}$$

$$\leq \int_{r_1}^{r_1+\epsilon} \psi'_\epsilon\big[(r^{1-n}Jf)' + \beta_{n-1}f\big]w\,dr + \int_{r_1}^{r_1+\epsilon} \psi''_\epsilon r^{1-n}Jf\,w\,dr \tag{10.14}$$

$$+ \int_{r_2-\epsilon}^{r_2} \psi'_\epsilon\,\beta_{n-1}fw\,dr - \int_{r_2-\epsilon}^{r_2} \psi'_\epsilon r^{1-n}Jf\,w'\,dr, \tag{10.15}$$

where in the final line we carried out an integration by parts to eliminate $\psi''_\epsilon$. We consider each line separately. The divergence theorem implies

$$\beta_{n-1}r^{n-1}f'(r) = \beta_{n-1}r_1^{n-1}f'(r_1) + \int_{\{r_1<|x|<r\}} \Delta f\,dx \leq -\lambda_1 Jf(r)$$

since $f'(r_1) \leq 0$ and $\Delta F = -\lambda_1 F$. Hence the expression in (10.13) is greater than or equal to $(\lambda_1 + c)\int_{r_1}^{r_2} \psi_\epsilon r^{1-n}Jf\,w\,dr$, which converges as $\epsilon \to 0$ to $(\lambda_1 + c)\int_{r_1}^{r_2} r^{1-n}Jf\,w\,dr$. As $\epsilon \to 0$ the first term in (10.14) tends to 0 by Lemma 10.9(a) since $w(r_1) = 0$, while the second term in (10.14) tends to 0 by Lemma 10.9(b) since

$$r^{1-n}Jf(r)w(r) = O(r - r_1)o(1) = o(r - r_1)$$

as $r \searrow r_1$. The first term in (10.15) is $\leq 0$ since $\psi'_\epsilon \leq 0$ on that interval and $f$ and $w$ are positive. The second term in (10.15) has lim sup less than or equal to 0 as $\epsilon \to 0$ by Lemma 10.9(c), since $\limsup_{r\to r_2} r^{1-n}Jf(r)w'(r) \le 0$ by (10.9). Combining the preceding observations, we see that letting $\epsilon \to 0$ in (10.13)–(10.15) implies

$$(\lambda_1 + c)\int_{r_1}^{r_2} r^{1-n}Jf\, w\, dr \le 0.$$

Since $f$ and $w$ are positive on the interval $(r_1, r_2)$ we conclude $c \le -\lambda_1$, which completes the proof in Case 1.

Case 2: $r_2 < R$. In this case $w(r_2) = 0$. Define

$$f(x) = \chi_{[r_1,r_2]}(r)F(Lx)$$

where the constant $L > 1$ is chosen so that $Jf(r) > 0$ for $r \in (r_1, r_2)$ and $Jf(r_2) = 0$, as we now justify. If $L = R/r_2 > 1$, then $F(Lx) > 0$ whenever $|x| < r_2$, and so $Jf(r_2) > 0$. On the other hand, if $L = R'/r_2$ where $R' > R$ is the first positive root of $\partial F/\partial r = 0$ (that is, $R' = Rj'_{p,1}/j_{p,1}$ where $j'_{p,1}$ is the first positive root of $(r^{-p}J_p(r))' = 0$), then

$$\begin{aligned} Jf(r_2) &= \int_{\{r_1<|x|<r_2\}} F(Lx)\, dx \\ &= (r_2/R')^n \int_{\{R'r_1/r_2<|x|<R'\}} F(x)\, dx \\ &= -(\text{const.}) \int_{\{R'r_1/r_2<|x|<R'\}} \Delta F(x)\, dx \qquad \text{since } -\Delta F = \lambda_1 F, \\ &\le 0 \end{aligned}$$

by the divergence theorem, since $\partial F/\partial r < 0$ when $0 < r < R'$ and $\partial F/\partial r = 0$ when $r = R'$. Thus by the intermediate value theorem, for some choice of $L \in (R/r_2, R'/r_2]$ one has $Jf(r_2) = 0$. Notice the value of $Jf(r)$ decreases strictly when $L$ increases (with $r \in (r_1, r_2)$ fixed), because $F$ is strictly decreasing on $(0, R')$. It follows that $L$ is unique, and $Jf(r) > 0$ for $r \in (r_1, r_2)$.

Clearly $r^{1-n}Jf(r)$ is smooth for $r \in [r_1, r_2]$ and equals zero at $r = r_1$, just as in Case 1.

Substituting our choice of $G$ into the weak form of the differential inequality again gives (10.13)–(10.15). We handle (10.13) as in Case 1, using that $f'(r_1) \le 0, Jf \ge 0$ and $\Delta f = -L^2\lambda_1 f$ to arrive after letting $\epsilon \to 0$ at the expression $(L^2\lambda_1 + c)\int_{r_1}^{r_2} r^{1-n}Jf\, w\, dr$. The terms in (10.14) are shown to vanish in the limit as $\epsilon \to 0$, by arguing exactly as in Case 1. After integration by parts, the terms in (10.15) are equal to

$$\int_{r_2-\epsilon}^{r_2} \psi'_\epsilon\big[(r^{1-n}Jf)' + \beta_{n-1}f\big]w\, dr + \int_{r_2-\epsilon}^{r_2} \psi''_\epsilon r^{1-n}Jf\, w\, dr. \qquad (10.16)$$

As $\epsilon \to 0$, the first term in (10.16) tends to 0 by Lemma 10.9(a) since $w(r_2) = 0$, while the second term in (10.16) tends to 0 by Lemma 10.9(b) since

$$r^{1-n} Jf(r)w(r) = O(r - r_2)o(1) = o(r - r_2)$$

as $r \nearrow r_2$. Thus by letting $\epsilon \to 0$ in (10.13), (10.14) and (10.16), we find

$$(L^2\lambda_1 + c) \int_{r_1}^{r_2} r^{1-n} Jf \, w \, dr \leq 0.$$

Since $Jf$ and $w$ are positive on the interval $(r_1, r_2)$ we conclude $c \leq -L^2\lambda_1 < -\lambda_1$, finishing the proof of the contrapositive for Case 2. □

**Example 10.11** (Alternate proof of Rayleigh–Faber–Krahn theorem) The first eigenvalue of the Laplacian under Dirichlet boundary conditions was shown in Theorem 5.6 to decrease under symmetric decreasing rearrangement of the domain $\Omega \subset \mathbb{R}^n$. The proof involved rearranging the first eigenfunction to obtain a trial function on the ball $\Omega^\#$, using that the Dirichlet integral decreases under s.d.r. We develop now an alternate proof, in which a trial function on the ball is obtained by solving a rearranged eigenfunction equation. The Faber–Krahn theorem will follow because this trial function has larger $L^2$ norm than the original function, by the comparison theorem.

Let $u$ be the first eigenfunction of the Laplacian on a bounded open set $\Omega \subset \mathbb{R}^n$ that has smooth boundary, so that the eigenfunction equation

$$-\Delta u = \lambda_1(\Omega)u$$

holds and $u$ is continuous on $\overline{\Omega}$ and satisfies the Dirichlet boundary condition $u = 0$ on $\partial\Omega$. We may suppose $u \geq 0$, since the first eigenfunction does not change sign.

Construct a function $v$ satisfying the rearranged Poisson equation

$$-\Delta v = \lambda_1(\Omega)u^\#$$

on the ball $\Omega^\#$, with $v$ continuous on the closure of $\Omega^\#$ and $v = 0$ on $\partial\Omega^\#$. (The solution $v$ exists by the classical Theorem 4.3 in Gilbarg and Trudinger (1983), noting the data $u^\#$ is Lipschitz continuous by Theorem 3.6, because $u$ is too.) Then $v \geq 0$ by the minimum principle for superharmonic functions. Using $v$ as a trial function for the first eigenvalue on the ball, we see from Rayleigh's formula (5.9) that

$$\lambda_1(\Omega^\#) \leq \frac{\int_{\Omega^\#} |\nabla v|^2 \, dx}{\int_{\Omega^\#} v^2 \, dx} = \frac{-\int_{\Omega^\#} v \Delta v \, dx}{\int_{\Omega^\#} v^2 \, dx}$$

by Green's formula. Substituting $-\Delta v = \lambda_1(\Omega)u^\#$ and applying Cauchy–Schwarz gives

$$\lambda_1(\Omega^\#) \le \lambda_1(\Omega)\frac{\int_{\Omega^\#} vu^\#\,dx}{\int_{\Omega^\#} v^2\,dx} \le \lambda_1(\Omega)\left(\frac{\int_{\Omega^\#}(u^\#)^2\,dx}{\int_{\Omega^\#} v^2\,dx}\right)^{1/2}.$$

Further,

$$\int_{\Omega^\#}(u^\#)^2\,dx = \int_\Omega u^2\,dx \le \int_{\Omega^\#} v^2\,dx$$

by the comparison principle in Theorem 10.10 applied with $c = 0, \mu = \lambda_1(\Omega)u, \nu = \lambda_1(\Omega)u^\#$, so that $\mu^\star = J\nu$, and taking the convex increasing function to be $\Phi(t) = t^2$ when $t \ge 0$ and $\Phi(t) = 0$ when $t < 0$.

Combining the inequalities implies $\lambda_1(\Omega^\#) \le \lambda_1(\Omega)$, and so the first eigenvalue of $\Omega$ is greater than or equal to that of the ball $\Omega^\#$, which is the Faber–Krahn Theorem 5.6.

A similar but simpler argument proves Pólya's Theorem 5.17 that the torsional rigidity increases under s.d.r. of the domain: if $-\Delta u = 2$ in $\Omega$ and $-\Delta v = 2$ in $\Omega^\#$, with zero Dirichlet boundary conditions on each function, then $u$ and $v$ are nonnegative by the maximum principle, and so Theorem 10.10 guarantees that

$$T(\Omega) = 2\int_\Omega u\,dx \le 2\int_{\Omega^\#} v\,dx = T(\Omega^\#).$$

### 10.3.2 Semilinear Equations

Next we state a comparison theorem for semilinear equations. The mean value $\overline{v}$ of $v$ over the sphere of radius $r$ was defined before Theorem 10.10.

**Theorem 10.12** (Semilinear Poisson) *Assume $u \in \mathcal{W}(\Omega)$ and $v \in \mathcal{W}(\Omega^\#)$ are weak solutions of*

$$-\Delta u = \phi(u) + \mu, \qquad -\Delta v = \phi(v) + \nu,$$

*where $\phi$ is a decreasing convex function such that $\phi(u)$ and $\phi(v)$ are locally integrable on $\Omega$ and $\Omega^\#$ respectively, and where $\mu \in M_{loc}(\Omega), \nu \in M_{loc}(\Omega^\#)$. Further assume $\mu$ decomposes as in (9.47) with $f$ satisfying Condition B and with $\tau(\Omega) < \infty$. Suppose $\mu^\star \le J\nu$.*

*If $u$ is nonnegative and satisfies the Dirichlet boundary condition*

$$\lim_{x\to x_0,\, x\in\Omega} u(x) = 0$$

*for $x_0 \in \partial\Omega$, and $\overline{v}$ satisfies the boundary condition $\liminf_{r\to R}\overline{v}(rx) \geq 0$, then $u^\star \leq Jv \leq v^\star$ on $\Omega^\star$. Hence the convex increasing integral means of $u$ are dominated by those of $v$, with*

$$\int_{\Omega} \Phi(u)\, d\mathcal{L}^n \leq \int_{\Omega^\#} \Phi(v)\, d\mathcal{L}^n$$

*for every convex increasing $\Phi\colon \mathbb{R} \to \mathbb{R}^+$.*

The interesting case is when $\phi$ is nonlinear. When $\phi$ is linear, Theorem 10.10 gives a stronger conclusion because not only can linearly decreasing $\phi$ be handled ($c \geq 0$ in that theorem) but also certain linearly increasing $\phi$ (namely $-\lambda_1 < c < 0$).

**Example 10.13** Suppose $\alpha$ and $\beta$ are positive, bounded functions satisfying

$$\Delta \log \alpha = \alpha^2 \qquad \text{and} \qquad \Delta \log \beta = \beta^2$$

on $\Omega$ and $\Omega^\#$, respectively. Let $C > 0$ be an upper bound on $u$ and $v$. Then $u = \log C/\alpha$ and $v = \log C/\beta$ satisfy

$$-\Delta u = C^2 e^{-2u} \qquad \text{and} \qquad -\Delta v = C^2 e^{-2v}.$$

Note $\phi(z) = C^2 e^{-2z}$ is convex decreasing. Hence Theorem 10.12 can be applied, provided the Dirichlet boundary condition holds for $u$ and $\overline{v}$, which is true if $\alpha = C$ on $\partial\Omega$ and $\beta \leq C$ on $\partial\Omega^\#$. This example hints at applications to the hyperbolic metric in complex analysis; see §11.9.

*Proof of Theorem 10.12* Letting $w_0 = u^\star - Jv$, the goal is to prove $w_0 \leq 0$. Clearly $w_0 = J(u^\# - v)$ is absolutely continuous, with $w_0(0) = 0$ by definition. Arguing for a contradiction, we suppose instead that $w_0 > 0$ somewhere.

We will perturb $w_0$ to ensure strict inequalities later in the proof. Let

$$H(r) = r^n + r^{n-1/2},$$

and compute that

$$H > 0, \quad H' > 0, \quad \Delta^\star H = 0 - \frac{1}{2}(n - 1/2)r^{n-5/2} < 0$$

for all $r > 0$, where the operator $\Delta^\star$ is given in (10.10). Define

$$w = w_0 - \epsilon H = J(u^\# - v) - \epsilon H,$$

where $\epsilon > 0$ is chosen so small that $w > 0$ at some point. Notice $w(0) = 0$ by definition. Further, $w$ is decreasing when $r$ is close to $R$, as we now show. We have for almost every $r$ that

$$w'(r) = \beta_{n-1} r^{n-1}(u^\# - \overline{v})(rx) - \epsilon\big(nr^{n-1} + (n-1/2)r^{n-3/2}\big),$$

by expressing $J(u^\# - v)$ in spherical coordinates; note $x$ can be any unit vector, since $u^\#$ and $\overline{v}$ are radial functions. The Dirichlet boundary condition on $u$ implies $u^\#(rx) \to 0$ as $r \to R$, while the boundary condition on $\overline{v}$ says $\liminf_{r\to R} \overline{v}(rx) \geq 0$. Hence

$$\limsup_{r\to R} r^{1-n} w'(r) \leq -\epsilon\big(n + (n-1/2)R^{-1/2}\big) < 0$$

(which remains valid even if $R = \infty$), and so $w'(r) < 0$ for all $r$ close to $R$.

Therefore $w$ attains its maximum at some point $s \in (0, R)$. We will show no such maximum can occur.

The Subharmonicity Theorem 9.20 applies to $u$, since $u$ satisfies the Dirichlet boundary condition and the convex decreasing function $\phi$ satisfies Condition (ii) in that result. (The absolute value $|\phi(z)|$ is bounded by a linear function when $z$ is large, and so $M_\kappa \phi(z) = \max_{|\tilde{z}-z|\leq\kappa} |\phi(\tilde{z})|$ is also bounded by a linear function. Hence $M_\kappa\phi(u)$ is locally integrable because $u$ is locally integrable, and so Condition (ii) holds.) Combining that Subharmonicity Theorem with the Weak Main Identity (9.79) for $v$ and with the adjoint relations (9.73) and (10.7), we have

$$-\Delta^\star u^\star \leq J\phi(u^\#) + \mu^\star,$$
$$-\Delta^\star Jv = J\phi(v) + J\nu.$$

Subtracting and using the hypothesis $\mu^\star \leq J\nu$, we deduce

$$-\Delta^\star w \leq J\phi(u^\#) - J\phi(v) - \epsilon(n-1)r^{n-3}. \tag{10.17}$$

We must analyze the nonlinear terms on the right side of the differential inequality. Write $\phi'$ for the right-derivative of $\phi$, which exists at every point in $\mathbb{R}^+$ by convexity, and write $\phi'(u^\#(r))$ to mean $\phi'(u^\#(ry))$ with $|y| = 1$. We have

$$\begin{aligned}[J\phi(u^\#) - J\phi(v)](s) &= \int_{B(s)} \big(\phi(u^\#) - \phi(v)\big)\, d\mathcal{L}^n \\ &\leq \int_{B(s)} \phi'(u^\#)(u^\# - v)\, d\mathcal{L}^n\end{aligned}$$

by convexity of $\phi$. Using spherical coordinates on the ball $B(s)$,

$$\begin{aligned}[J\phi(u^\#) - J\phi(v)](s) &\leq \int_0^s \phi'(u^\#(r)) \frac{d}{dr}\big[u^\star(r) - Jv(r)\big]\, dr \\ &\leq \int_0^s \phi'(u^\#(r)) w'(r)\, dr,\end{aligned}$$

since $\phi' \leq 0$ and $H' > 0$. Write

$$\psi(r) = \phi'(u^\#(r)) \leq 0,$$

and observe $\psi$ is decreasing by convexity of $\phi$. The measure $d\psi$ is therefore nonpositive. Integration by parts yields

$$\begin{aligned}\int_0^s \phi'(u^\#(r))w'(r)\,dr &= -\int_0^s \psi(r)\frac{d}{dr}[w(s)-w(r)]\,dr \\ &= \psi(0)w(s) + \int_{(0,s)} (w(s)-w(r))\,d\psi(r) \\ &\leq 0\end{aligned}$$

since $\psi(0) \leq 0$ and $w(r) \leq w(s)$ for all $r \leq s$. Therefore, by above,

$$[J\phi(u^\#) - J\phi(v)](s) \leq 0. \tag{10.18}$$

From (10.18) we see the right side of (10.17) is a continuous function that is negative at $r = s$, and hence is negative on a neighborhood of $s$. Thus

$$-\Delta^\star w < 0$$

on a neighborhood of the maximum point $s$. On that neighborhood, we change variable with $\rho = r^n$ so that $-\Delta^\star w < 0$ becomes $d^2w/d\rho^2 > 0$ in the weak sense. Hence $w$ is strictly convex in the classical sense as a function of $\rho$, by Lemma 10.6, and so $w$ cannot attain its maximum at the interior point $s$ of the neighborhood. This contradiction completes the proof. □

We deduce a symmetry result from Theorem 10.12.

**Corollary 10.14** *Assume $v \in \mathcal{W}(\Omega^\#)$ is a nonnegative weak solution of the semilinear Poisson equation $-\Delta v = \phi(v) + \nu$, where $\phi$ is a decreasing convex function with $\phi(v)$ locally integrable on $\Omega^\#$ and the measure $\nu \in M_{loc}(\Omega^\#)$ decomposes as in (9.47) with $f$ satisfying Condition B and with $\tau(\Omega^\#) < \infty$.*

*If the measure is symmetric decreasing ($\nu = \nu^\#$) and $v$ satisfies the Dirichlet boundary condition on $\partial\Omega^\#$, then $v$ is symmetric decreasing: $v = v^\#$ a.e. on $\Omega^\#$.*

*Proof* Since $\nu^\star = J(\nu^\#) = J\nu$, and $v$ satisfies the hypotheses imposed on $u$ in Theorem 10.12, we conclude from the theorem that $v^\star \leq Jv \leq v^\star$, and so $Jv = Jv^\#$. It follows that $v = v^\#$ a.e. □

The statement of the corollary is particularly simple if the measure vanishes ($\nu \equiv 0$), since in that case it automatically has the desired decomposition and is symmetric decreasing.

## 10.4 Comparison Principle for Steiner Symmetrization on Euclidean Domains

Steiner comparison theorems follow as for s.d.r. in the previous section, with the additional complication of "outer" boundary points, meaning, boundary points that are (roughly speaking) parallel to the direction in which the Steiner symmetrization is performed.

Throughout this section, # denotes $(k, n)$-Steiner symmetrization, where $n \geq 2$ and $1 \leq k \leq n - 1$. We continue to take $\Omega$ to be an open subset of $\mathbb{R}^n$. As in Section 9.11, we write $y \in \mathbb{R}^k, z \in \mathbb{R}^m, k + m = n$, and let $Z$ be the open set of all $z \in \mathbb{R}^m$ such that the slice $\Omega(z) \subset \mathbb{R}^k$ is nonempty.

The open set $\Omega^\star \subset \mathbb{R}^{m+1}$ was defined in Section 9.11 by replacing each slice $\Omega(z)$ with an interval $\Omega^\star(z)$ with length equal to the radius of the ball $\Omega^\#(z)$. The $J$-operator acts on functions slice-by-slice, and acts on measures as follows. Given a measure $\mu \in M_{loc}(\Omega^\#)$, construct a measure $J\mu \in M_{loc}(\Omega^\star)$ by letting

$$(J\mu)(E) = \int_0^\infty \mu\big(B(r) \times E(r)\big)\, dr \qquad \text{for } E \in \mathcal{B}_c(\Omega^\star),$$

where $E(r) = \{z \in \mathbb{R}^m \colon (r, z) \in E\}$ is the $r$-slice of $E$. This definition is consistent with how the $J$ operator acts on functions, because if the measure is represented by a locally integrable function, say $d\mu = u(y, z)\, dydz$, then one can check $J\mu(E) = \int_E Ju(r, z)\, drdz$. Lastly, notice $J(\mu^\#) = \mu^\star$ because

$$\mu^\#\big(B(r) \times E(r)\big) = \int_{E(r)} f^\star(r, z)\, dz + \int_{E(r)} d\tau^\#(z)$$

by the definitions of $\mu^\#$ and $\mu^\star$ in Section 9.11.

The adjoint relation for measures says

$$(G, J\mu) = (J^t G, \mu) \tag{10.19}$$

for $\mu \in M_{loc}(\Omega^\#)$ and $G \in C_c(\Omega^\star)$. The proof is straightforward: each side equals

$$\int_{\Omega^\#} \int_{\Omega^\star(z)} \chi_{B(r)}(y) G(r, z)\, dr\, d\mu(x)$$

where $x = (y, z)$ and $|y| = r$. For example, if we apply the adjoint relation (10.19) to $\mu^\#$, and remember that $\mu^\star = J(\mu^\#)$, then we obtain the identity (9.92) used in proving Theorem 9.23.

The average of $v$ over a sphere centered at the origin in a slice is

$$\overline{v}(ry, z) = \frac{1}{\beta_{k-1}} \int_{\mathbb{S}^{k-1}} v(ry', z)\, d\sigma_{k-1}(y'), \qquad |y| = 1,\ 0 \leq r < R(z).$$

Write $J\phi(v)$ for $J\big(\phi(v(y, z))\big)$, and so on.

The next theorem gives a Steiner comparison principle for nonnegative subsolutions.

**Theorem 10.15** *Assume $u \in \mathcal{W}(\Omega)$ and $v \in \mathcal{W}(\Omega^\#)$ are weak solutions of*

$$-\Delta u \le \phi(u) + \mu, \qquad -\Delta v = \phi(v) + \nu,$$

*where $\phi$ is a decreasing convex function on $\mathbb{R}$ such that $\phi(u)$ and $\phi(v)$ are locally integrable on $\Omega$ and $\Omega^\#$ respectively, and $J(u^\#), J\phi(u^\#), Jv, J\phi(v)$ are continuous on $\Omega^\star$, and $\mu \in M_{loc}(\Omega)$, $\nu \in M_{loc}(\Omega^\#)$. Assume $\mu$ decomposes as in (9.82) with $f + c$ nonnegative and satisfying Condition BS, for some $c \in \mathbb{R}$, and that $\tau(\Omega) < \infty$ and $\mu^\star \le J\nu$.*

*Suppose further that*

$$\limsup_{(r,z)\to(r_0,z_0)} \int_{B(r)} [u^\#(y,z) - v(y,z)]\, d\mathcal{L}^k(y) \le 0 \qquad (10.20)$$

*for all $(r_0, z_0) \in \partial(\Omega^\star)$ with $z_0 \in \partial Z$, where the* lim sup *is taken over points $(r,z) \in \Omega^\star$; and if $Z$ is unbounded then also assume* (10.20) *holds when the* lim sup *is taken over all points $(r,z) \in \Omega^\star$ with $|z| \to \infty$.*

*If $u \ge 0$ and $u$ and $\overline{v}$ satisfy the Dirichlet boundary conditions*

$$\lim_{x\to x_0,\, x\in\Omega} u(x) = 0, \qquad \liminf_{x\to x_1,\, x\in\Omega^\#} \overline{v}(x) \ge 0,$$

*for all $x_0 = (y_0, z_0) \in \partial\Omega$ with $z_0 \in Z$ and all $x_1 = (y_1, z_1) \in \partial\Omega^\#$ with $z_1 \in Z$, then $u^\star \le Jv \le v^\star$ on $\Omega^\star$. Hence the convex increasing integral means of $u$ are dominated by those of $v$ on each slice, with*

$$\int_{\Omega(z)} \Phi\big(u(y,z)\big)\, d\mathcal{L}^k(y) \le \int_{\Omega^\#(z)} \Phi\big(v(y,z)\big)\, d\mathcal{L}^k(y)$$

*for every convex increasing $\Phi\colon \mathbb{R} \to \mathbb{R}^+$ and every $z \in Z$.*

The theorem implicitly assumes that the functions $u, \phi(u), v, \phi(v)$ are locally integrable on every slice, not just on almost every slice.

Assumption (10.20) can be regarded as an "outer boundary condition" since it is assumed to hold for $z_0 \in \partial Z$ (and for $|z| \to \infty$ if applicable). We assume $z_0$ is a finite point and $r_0 \in [0, \infty]$, in (10.20).

*Proof* Once again the task is to prove $u^\star \le Jv$, because $Jv \le v^\star$ by definition of $v^\star$ and the conclusion on convex integral means follows from Section 10.1.

Put $w_0 = u^\star - Jv = J(u^\# - v)$, so that $w_0$ is continuous on $\Omega^\star$ by hypothesis. We want to show $w_0 \le 0$. Suppose for the sake of deducing a contradiction that $w_0 > 0$ somewhere in $\Omega^\star$. Define a perturbing function

$$H(r) = r^k + r^{k-1/2},$$

so that $H(0) = 0$ and

$$H > 0, \quad H' > 0, \quad \Delta^\star H = 0 - \frac{1}{2}(k - 1/2)r^{k-5/2} < 0$$

for all $r > 0$, where the operator $\Delta^\star$ was defined in (9.87) by

$$\Delta^\star F = r^{k-1}\partial_r\big(r^{1-k}\partial_r F\big) + \Delta_z F. \tag{10.21}$$

Now let

$$w = w_0 - \epsilon H = u^\star - Jv - \epsilon H$$

where we fix $\epsilon > 0$ so small that $w$ is positive somewhere in $\Omega^\star$.

Take a supremizing sequence $(r_l, z_l) \in \Omega^\star$ for $w$, so that

$$0 < w(r_l, z_l) \nearrow \sup_{\Omega^\star} w \qquad \text{as } l \to \infty.$$

For each $l$, note $w(r, z_l)$ is a continuous function of $r$ that tends to 0 as $r \to 0$. Hence we may suppose

$$w(r, z_l) \le w(r_l, z_l) \qquad \text{for all } r \in (0, r_l), \tag{10.22}$$

simply by replacing $r_l$ (if necessary) with a smaller value at which $w(\cdot\,, z_l)$ achieves its maximum over $[0, r_l]$.

After passing to a subsequence, one of the following must hold:

(i) $z_l \to z_0 \in \partial Z$ or $|z_l| \to \infty$,
(ii) $z_l \to z_0 \in Z$ and $r_l \to 0$,
(iii) $z_l \to z_0 \in Z$ and $r_l \to r_0 \in \big(0, R(z_0)\big)$,
(iv) $z_l \to z_0 \in Z$ and $r_l \to r_0 \in \big[R(z_0), \infty\big]$.

We will rule out each of these possibilities, in order to arrive at the desired contradiction.

Case (i) is impossible by the outer boundary condition (10.20), which implies $\limsup w(r_l, z_l) \le 0$.

Case (ii) is impossible since $w = 0$ when $r = 0$, by definition.

Case (iii) will be ruled out using a maximum principle argument. Since $\phi$ is convex and decreasing, the absolute value $|\phi(\zeta)|$ is bounded by a linear function for $\zeta \ge 0$. Hence $M_\kappa\phi(\zeta) = \max_{|\widetilde{\zeta}-\zeta|\le\kappa} |\phi(\widetilde{\zeta})|$ is also bounded by a linear function for $\zeta \ge 0$, and so $M_\kappa\phi(u)$ is locally integrable on $\Omega$ because $u \ge 0$ is locally integrable. Thus Condition (ii) holds in the hypotheses of Subharmonicity Theorem 9.23. That theorem implies

$$-\Delta^\star u^\star \le J\phi(u^\#) + \mu^\star$$

in the weak sense on $\Omega^\star$. Also

$$-\Delta^\star Jv = J\phi(v) + Jv$$

in the weak sense, as one verifies by combining the Weak Main Identity (9.90) for $v$, the weak form of the equation $-\Delta v = \phi(v) + v$, the adjoint relation (9.86) applied with $\phi(v)$ instead of $v$, and the adjoint relation (10.19) for measures.

Subtracting the preceding formulas and using the hypothesis $\mu^\star \leq Jv$ implies that

$$-\Delta^\star w \leq J[\phi(u^\#) - \phi(v)] + \epsilon\Delta^\star H \tag{10.23}$$

in the weak sense on $\Omega^\star$. In Case (iii) the positive maximum of $w$ is attained at a point $(r_0, z_0)$ with $0 < r_0 < R(z_0), z_0 \in Z$. By arguing exactly as in the proof of Theorem 10.12 (which uses that $\phi$ is convex decreasing and $H' > 0$), we find

$$J[\phi(u^\#) - \phi(v)] \leq 0$$

at $(r_0, z_0)$, so that the right side of (10.23) is negative at that point, since $\Delta^\star H < 0$. The right side remains negative on some neighborhood $N_0$ of the point, by continuity, and so

$$\Delta^\star w > c_0$$

weakly on $N_0$, for some constant $c_0 > 0$. The maximum principle (Proposition 10.7) for the operator $\Delta^\star$ in (10.21) implies $w$ cannot have an interior maximum at the point $(r_0, z_0) \in N_0$. This contradiction eliminates Case (iii).

Lastly, suppose as in Case (iv) that $r_0 \in [R(z_0), \infty]$. We will show for all large indices $l$ that $w(r, z_l)$ is strictly decreasing as a function of $r \in (r_l - 2^{-l}, r_l)$, which contradicts (10.22) and thus completes the proof. Indeed, for such $r$-values we compute (with $y \in \mathbb{R}^k$, $|y| = 1$) that

$$r^{1-k}\frac{\partial w}{\partial r}(r, z_l) = \beta_{k-1}\big(u^\#(ry, z_l) - \overline{v}(ry, z_l)\big) - \epsilon\big(k + (k - 1/2)r^{-1/2}\big),$$

and this last expression is negative for all large $l$ in view of the Dirichlet boundary conditions on $u$ and $\overline{v}$ (noting that $(ry, z_l) \to (r_0 y, z_0) \in \partial\Omega^\#$ as $l \to \infty$). □

The stronger conclusion $u^\# \leq \overline{v}$ obtained in Theorem 10.10 for the s.d.r. when $c = 0$ need no longer hold under $(k, n)$-Steiner symmetrization, as one may show by an explicit example (Baernstein, 1994, p. 52).

A symmetry statement analogous to Corollary 10.14 can be proved. We leave it to the reader.

## 10.5 Comparison Principle on the Sphere

In this section, # will denote symmetric decreasing rearrangement on $\mathbb{S}^n$, with $n \geq 1$. The sphere has no boundary, and so there are no boundary conditions. Instead we will normalize the solutions by requiring the integral of $u$ over the sphere to be less than that of $v$.

Recall from Section 9.7 that

$$(Ju)(\theta) = \int_{\mathcal{K}(\theta)} u(x)\, d\sigma_n(x),$$

where $\mathcal{K}(\theta)$ denotes the open spherical cap on $\mathbb{S}^n$ with center $e_1$ and opening $\theta \in (0, \pi)$. Then $u^\star = Ju^\#$ on the interval $A^\star = (0, \pi)$. To extend the $J$ operator to measures $\mu \in M(\mathbb{S}^n)$, define $J\mu \in M(A^\star)$ by

$$d(J\mu) = \mu\big(\mathcal{K}(\theta)\big)\, d\theta.$$

This definition is consistent with the previous one for functions, since if the measure is represented by a locally integrable function, say $d\mu(x) = u(x)\, d\sigma_n(x)$, then the preceding definitions give $d(J\mu) = Ju(\theta)\, d\theta$. In addition,

$$\mu^\star = J(\mu^\#)$$

because

$$\begin{aligned} d\mu^\star &= \big(f^\star(\theta) + \tau(\mathbb{S}^n)\big)\, d\theta \\ &= \mu^\#\big(\mathcal{K}(\theta)\big)\, d\theta \\ &= d(J(\mu^\#)) \end{aligned}$$

by the definitions of $\mu^\star$ and $\mu^\#$ in Section 9.7. Note there that the open cap $\mathcal{K}(\theta)$ never contains the antipodal point $-e_1$ on the sphere at which the measure $\eta^\#$ is concentrated.

The adjoint relation for measures says

$$(G, J\mu) = (J^t G, \mu) \tag{10.24}$$

for $\mu \in M(\mathbb{S}^n)$ and $G \in C_c(0, \pi)$, which is easily shown because each side equals

$$\int_{\mathbb{S}^n} \int_0^\pi \chi_{\mathcal{K}(\theta)}(x) G(\theta)\, d\theta\, d\mu(x).$$

By the way, the identity $\int_0^\pi G\, d\mu^\star = \int_{\mathbb{S}^n} g\, d\mu^\#$ used in proving Theorem 9.13 can now be understood as just the adjoint relation (10.24) applied to $\mu^\#$ together with the formula $\mu^\star = J(\mu^\#)$.

We compare now the solution of a semilinear Poisson equation on the sphere with a solution to the symmetrized version of the same equation. Recall $\Delta_s$ is the spherical Laplacian, defined in Section 7.2.

**Theorem 10.16** *Assume $u, v \in \mathcal{W}(\mathbb{S}^n)$ are weak solutions of*

$$-\Delta_s u = \phi(u) + \mu, \qquad -\Delta_s v = \phi(v) + \nu,$$

*where $\phi\colon \mathbb{R} \to \mathbb{R}$ is a decreasing convex function with $\phi(u)$ and $\phi(v)$ integrable on $\mathbb{S}^n$, and where $\mu, \nu \in M(\mathbb{S}^n)$. Suppose $\mu^\star \leq J\nu$.*

*If*

$$\int_{\mathbb{S}^n} u\, d\sigma_n \leq \int_{\mathbb{S}^n} v\, d\sigma_n \tag{10.25}$$

*then $u^\star \leq Jv \leq v^\star$ on the interval $(0, \pi)$. Hence the convex increasing integral means of $u$ are dominated by those of $v$, with*

$$\int_{\mathbb{S}^n} \Phi(u)\, d\sigma_n \leq \int_{\mathbb{S}^n} \Phi(v)\, d\sigma_n \tag{10.26}$$

*for every convex increasing $\Phi\colon \mathbb{R} \to \mathbb{R}$. Further, if equality holds in* (10.25) *then* (10.26) *holds for all convex $\Phi$, and hence one obtains comparisons between $L^p$-norms and the oscillations of $f$ and $g$ as in Proposition 10.3.*

The hypothesis $\mu^\star \leq J\nu$ in the theorem holds with equality if the measure $\nu$ equals the s.d.r. of $\mu$, meaning $\mu^\# = \nu$.

Some normalizing condition such as (10.25) is needed in the theorem, since in the linear case ($\phi \equiv 0$) one can add arbitrary constants to $u$ and $v$ and they still solve their respective Poisson equations.

*Proof* Put $w_0 = u^\star - Jv = J(u^\# - v)$, so that $w_0$ is absolutely continuous on $[0, \pi]$ with $w_0(0) = 0$ by definition and $w_0(\pi) \leq 0$ by (10.25). The task is to prove $w_0 \leq 0$, so that $u^\star \leq Jv$. Suppose to the contrary that $w_0(\gamma_0) > 0$ for some $\gamma_0 \in (0, \pi)$. We will obtain a contradiction.

Let

$$H(\theta) = \int_0^\theta (\pi - \varphi) \sin^{n-1} \varphi\, d\varphi \tag{10.27}$$

so that $H$ is smooth. When $\theta \in (0, \pi)$, one has

$$H' > 0, \quad \Delta_s^\star H = -\sin^{n-1}\theta < 0,$$

where the operator $\Delta_s^\star$ was defined in Section 9.7 by

$$\Delta_s^\star F(\theta) = \sin^{n-1}\theta \big(\sin^{1-n}\theta F'(\theta)\big)'.$$

Define a perturbation of $w_0$ by

$$w = u^\star - Jv - \epsilon H,$$

where $\epsilon > 0$ is chosen so small that

$$w(\gamma_0) > \max\big(0, w(0), w(\pi)\big).$$

Then the maximum of $w$ on the interval $[0, \pi]$ is positive, and is attained at some point in the interior of the interval, not at an endpoint. Write $\gamma \in (0, \pi)$ for such a maximum point.

By the Subharmonicity Theorem 9.13 for $u$ and the Weak Main Identity (9.42) for $v$ (together with the adjoint relations (9.41) and (10.24)), we have

$$\begin{aligned} -\Delta_s^\star u^\star &\le J\phi(u^\#) + \mu^\star, \\ -\Delta_s^\star Jv &= J\phi(v) + Jv, \end{aligned}$$

in the weak sense. Subtracting the preceding formulas yields that

$$-\Delta_s^\star w \le J\phi(u^\#) + \mu^\star - J\phi(v) - Jv - \epsilon \sin^{n-1}\theta.$$

Since $\mu^\star \le Jv$, we deduce that in the weak sense,

$$-\Delta_s^\star w \le J[\phi(u^\#) - \phi(v)] - \epsilon \sin^{n-1}\theta. \tag{10.28}$$

Notice the right side of this inequality is a continuous function.

In the special case where $\phi \equiv 0$ we conclude

$$-\Delta_s^\star w < 0 \quad \text{in } (0, \pi),$$

and so by the maximum principle, $w$ cannot achieve a strict interior maximum, giving the desired contradiction. (For full details, see the change of variables argument at the end of the proof below.)

In the general case of convex decreasing $\phi$, one must work a little harder. Write $\phi'$ for the right-derivative of $\phi$, which exists by convexity, and write $u^\#(\theta)$ to mean $u^\#(x)$ where $x \cdot e_1 = \cos\theta$. We have

$$\begin{aligned} J[\phi(u^\#) - \phi(v)](\gamma) &= \int_{\mathcal{K}(\gamma)} \big(\phi(u^\#) - \phi(v)\big)\, d\sigma_n \\ &\le \int_{\mathcal{K}(\gamma)} \phi'(u^\#)(u^\# - v)\, d\sigma_n \\ &= \int_0^\gamma \phi'(u^\#(\theta)) \frac{d}{d\theta}\big[u^\star(\theta) - Jv(\theta)\big]\, d\theta \\ &\le \int_0^\gamma \phi'(u^\#(\theta)) w'(\theta)\, d\theta, \end{aligned}$$

where the first inequality follows from convexity of $\phi$, the next line follows by using "spherical coordinates" on $\mathcal{K}(\gamma)$, and the last inequality follows by definition of $w$ since $\phi' \le 0$ and $H' > 0$ on the interval $(0, \pi)$.

Write

$$\psi(\theta) = \phi'(u^{\#}(\theta)) \le 0,$$

and observe $\psi$ is decreasing by convexity of $\phi$. The measure $d\psi$ is therefore nonpositive. Integration by parts yields

$$\begin{aligned}
\int_0^{\gamma} \phi'(u^{\#}(\theta))w'(\theta)\, d\theta &= -\int_0^{\gamma} \psi(\theta)\frac{d}{d\theta}\big[w(\gamma) - w(\theta)\big]\, d\theta \\
&= \psi(0)w(\gamma) + \int_{(0,\gamma)} (w(\gamma) - w(\theta))\, d\psi(\theta) \\
&\le 0
\end{aligned}$$

since $\psi(0) \le 0$ and $w(\theta) \le w(\gamma)$ (by definition of $\gamma$ as the maximum point). Therefore, by above,

$$J[\phi(u^{\#}) - \phi(v)](\gamma) \le 0.$$

Hence the right side of (10.28) is negative when $\theta = \gamma$. It remains negative on some open neighborhood $N$ containing $\gamma$, by continuity. Thus

$$-\Delta_s^{\star} w < 0$$

weakly on $N$. Change variable on the neighborhood $N$ containing $\gamma$ by letting

$$\rho = \int_0^{\theta} \sin^{n-1}\varphi\, d\varphi,$$

so that $d\rho = \sin^{n-1}\theta\, d\theta$. The inequality $-\Delta_s^{\star} w < 0$ transforms to $d^2w/d\rho^2 > 0$, still in the weak sense, and so $w$ is strictly convex in the classical sense as a function of $\rho$ by Lemma 10.6. Hence $w$ cannot attain its maximum at an interior point of the neighborhood. But $w$ attains its maximum at the interior point $\theta = \gamma$, and so we have found a contradiction, as desired. □

When the function $v$ satisfies the hypotheses imposed on $u$ in Theorem 10.16, and the measure $\nu$ is symmetric decreasing ($\nu = \nu^{\#}$), we conclude from the theorem that $v^{\star} \le Jv \le v^{\star}$. Hence $Jv = Jv^{\#}$, from which it is easy to deduce that $v = v^{\#}$ a.e. Thus we immediately obtain the next symmetry result.

**Corollary 10.17** *Assume $v \in \mathcal{W}(\mathbb{S}^n)$ is a weak solution of the semilinear Poisson equation $-\Delta_s v = \phi(v) + \nu$, where $\phi\colon \mathbb{R} \to \mathbb{R}$ is a decreasing convex function with $\phi(v)$ integrable on $\mathbb{S}^n$, and the measure $\nu \in M(\mathbb{S}^n)$ satisfies $\nu = \nu^{\#}$. Then the function $v$ is symmetric decreasing, with $v = v^{\#}$ a.e. on $\mathbb{S}^n$.*

## 10.6 Comparison Principles on Shells

Annular shells yield interesting comparison results. Shells benefit from compactness of their cross-sections (spheres) while allowing either Dirichlet or Neumann conditions at the inner and outer boundaries. Suppose

$$A = \{x \in \mathbb{R}^n \colon R_1 < |x| < R_2\}$$

is an open shell in $\mathbb{R}^n, n \geq 2$, with $0 \leq R_1 < R_2 \leq \infty$. Write # for $(n-1,n)$-cap symmetrization, and recall from Section 9.6 the operator $J$ that acts by

$$(Ju)(r,\theta) = \int_{\mathcal{K}(\theta)} u(rx)\, r^{n-1}\, d\sigma_{n-1}(x),$$

where $\mathcal{K}(\theta)$ denotes the open spherical cap on $\mathbb{S}^{n-1}$ with center $e_1$ and opening $\theta$. Then $u^\star = Ju^\#$ on the rectangle

$$A^\star = \{(r,\theta) \in \mathbb{R}^2 \colon R_1 < r < R_2,\ 0 < \theta < \pi\}.$$

To define $J\colon M_{loc}(A) \to M_{loc}(A^\star)$ on measures, we let

$$(J\mu)(E) = \int_0^\pi \mu\big(E(\theta) \times \mathcal{K}(\theta)\big)\, d\theta \quad \text{for } E \in \mathcal{B}_c(A^\star)$$

where $E(\theta) = \{r \in (R_1,R_2) \colon (r,\theta) \in E\}$ is the $\theta$-slice of $E$ and we write

$$E(\theta) \times \mathcal{K}(\theta) = \{rx \colon r \in E(\theta), x \in \mathcal{K}(\theta)\}.$$

This definition of $J$ on measures is consistent with how $J$ acts on functions, because if the measure is represented by a locally integrable function, say $d\mu = u(rx)\, r^{n-1} d\sigma_{n-1}(x) dr$, then the definitions imply $J\mu(E) = \int_E Ju(r,\theta)\, dr d\theta$. Lastly, $J(\mu^\#) = \mu^\star$ because

$$\mu^\#\big(E(\theta) \times \mathcal{K}(\theta)\big) = \int_{E(\theta)} f^\star(r,\theta)\, dr + \tau^\#\big(E(\theta)\big)$$

and integrating over $\theta \in (0,\pi)$ yields $\mu^\star(E)$, by the definition of $\mu^\star$ in Section 9.6.

Once again we need an adjoint relation for measures. It says

$$(G, J\mu) = (J^t G, \mu) \tag{10.29}$$

for $\mu \in M(A)$ and $G \in C_c(A^\star)$. To prove the relation, simply notice each side equals

$$\int_A \int_0^\pi \chi_{\mathcal{K}(\theta)}(x) G(r,\theta)\, d\theta\, d\mu(y)$$

where $y = rx$. (Then since $\mu^\star = J(\mu^\#)$, we see the identity (9.38) used for Theorem 9.9 is simply the adjoint relation (10.29) applied to $\mu^\#$.)

### 10.6.1 Dirichlet Boundary Conditions

We start with a comparison theorem under $(n-1,n)$-cap symmetrization, with Dirichlet type boundary conditions at the inner and outer boundaries.

**Theorem 10.18** *Assume $u, v \in \mathcal{W}(A)$ are weak solutions of the semilinear Poisson equations*

$$-\Delta u = \phi(u) + \mu, \qquad -\Delta v = \phi(v) + \nu,$$

*where $\phi\colon \mathbb{R} \to \mathbb{R}$ is a decreasing convex function with $\phi(u)$ and $\phi(v)$ locally integrable on $A$, and where $\mu, \nu \in M_{loc}(A)$. Suppose $J(u^{\#}), J\phi(u^{\#})$ and $Jv, J\phi(v)$ are continuous on $(R_1, R_2) \times [0,\pi]$, and that $\mu^{\star} \leq J\nu$ on $(R_1, R_2) \times (0,\pi)$.*

*If $u^{\star} \leq Jv$ at $r = R_1$ and $R_2$, meaning*

$$\limsup_{0<\theta<\pi,\, r\to R_i} \int_{\mathcal{K}(\theta)} [u^{\#}(rx) - v(rx)]\, d\sigma_{n-1}(x) \leq 0, \qquad i = 1,2, \tag{10.30}$$

*then $u^{\star} \leq Jv \leq v^{\star}$ on $A^{\star}$. Hence the convex increasing integral means of $u$ are dominated by those of $v$, at each radius, with*

$$\int_{\mathbb{S}^{n-1}} \Phi(u(rx))\, d\sigma_{n-1}(x) \leq \int_{\mathbb{S}^{n-1}} \Phi(v(rx))\, d\sigma_{n-1}(x), \qquad R_1 < r < R_2,$$

*for every convex increasing $\Phi\colon \mathbb{R} \to \mathbb{R}$.*

The continuity assumptions on $J(u^{\#}), J\phi(u^{\#}), Jv, J\phi(v)$ in the theorem are valid if $u$ and $v$ are themselves continuous, but the assumptions cover more general situations too, such as where $u$ or $v$ has a mild singularity at some point in the shell. The continuity assumptions implicitly include the requirement that $u, \phi(u), v, \phi(v)$ are integrable on every sphere of radius $r \in (R_1, R_2)$ – not merely on almost every such sphere.

The boundary condition (10.30) holds if, for example, $u$ and $v$ are continuous and $u \leq v$ on the inner and outer boundaries of the shell. Thus it should be regarded as a Dirichlet condition at the shell boundary.

Balls can be handled in Theorem 10.18 by taking $R_1 = 0$, since a weak solution on the ball is automatically a weak solution on the shell with inner radius 0. The boundary condition (10.30) must still be verified at the inner boundary as $r \to 0$. (It does not hold automatically there, because (10.30) does not include a factor of $r^{n-1}$.) Of course, that boundary condition certainly holds if $u$ and $v$ are continuous at the origin and $u(0) \leq v(0)$.

Weak solutions on all of $\mathbb{R}^n$ can be handled similarly, by choosing $R_1 = 0$ and $R_2 = \infty$ in the theorem.

*Proof of Theorem 10.18* The task is to prove $u^\star \leq Jv$, since the other inequality $Jv \leq v^\star$ in the conclusion is trivial by definition of $v^\star$, and the conclusion on convex integral means follows from Section 10.1.

Put $w_0 = u^\star - Jv = J(u^\# - v)$, so that $w_0$ is continuous on $(R_1, R_2) \times [0, \pi]$ by hypothesis. We want to show $w_0 \leq 0$ in this rectangle. We will show (equivalently) that $r^{1-n} w_0 \leq 0$. The factor of $r^{1-n}$ will be important when applying the maximum principle, later in the proof.

Suppose for the sake of deducing a contradiction that $r^{1-n} w_0 > 0$ somewhere in the rectangle. On the lower side of the rectangle, where $\theta = 0$, we already have $w_0 = 0$ by definition of the $J$-operator. And at the left and right sides of the rectangle, $\limsup r^{1-n} w_0 \leq 0$ as $r \to R_1, R_2$, by the boundary condition (10.30).

Thus, by continuity, $r^{1-n} w_0$ has a positive maximum point somewhere in $(R_1, R_2) \times (0, \pi]$. The same is true on the smaller rectangle $(R_1', R_2') \times (0, \pi]$, provided we choose radii satisfying $R_1 < R_1' < R_2' < R_2$, with $R_1'$ close to $R_1$ and $R_2'$ close to $R_2$.

Define a perturbed function

$$w = w_0 - \epsilon Q = u^\star - Jv - \epsilon Q$$

where $\epsilon > 0$ and

$$Q(r, \theta) = r^{n-1} H(\theta) - r^{1/2}$$

and $H$ is as defined in (10.27) except with $n$ replaced by $n - 1$:

$$H(\theta) = \int_0^\theta (\pi - \varphi) \sin^{n-2} \varphi \, d\varphi.$$

(The term $-r^{1/2}$ in the definition of $Q$ is useful in Case 2 of the proof below.)

We may choose the perturbation parameter $\epsilon > 0$ small enough that $r^{1-n} w$ has a positive maximum point in $(R_1', R_2') \times (0, \pi]$, since $r^{1-n} w_0$ has such a maximum point and $r^{1-n} Q$ is bounded on $[R_1', R_2'] \times [0, \pi]$ (using that $R_1' > 0$). Denote such a maximum point for $r^{1-n} w$ by $(s, \gamma)$, so that $s \in (R_1', R_2')$ and $\gamma \in (0, \pi]$.

The operator $\Delta^\star$ was defined in (9.27) by

$$\Delta^\star F = \partial_r \big( r^{n-1} \partial_r (r^{1-n} F) \big) + r^{-2} \sin^{n-2} \theta \, \partial_\theta (\sin^{2-n} \theta \, \partial_\theta F). \tag{10.31}$$

Note for use below that

$$\begin{aligned} \Delta^\star Q &= -r^{n-3} \sin^{n-2} \theta - \frac{1}{2} \Big( n - \frac{3}{2} \Big) r^{-3/2} \\ &< 0 \end{aligned}$$

and $\partial_\theta Q = r^{n-1} H'(\theta) > 0$ for $\theta \in (0, \pi)$.

We will use a maximum principle argument to arrive at a contradiction. By the Subharmonicity Theorem 9.9 for $u$ and the Weak Main Identity (9.35) for $v$ (together with the adjoint relations (9.26) and (10.29)), we have

$$-\Delta^\star u^\star \leq J\phi(u^\#) + \mu^\star,$$
$$-\Delta^\star Jv = J\phi(v) + Jv,$$

in the weak sense. Subtracting the preceding formulas and using the hypothesis $\mu^\star \leq Jv$ implies that

$$-\Delta^\star w \leq J[\phi(u^\#) - \phi(v)] + \epsilon \Delta^\star Q \tag{10.32}$$

in the weak sense on $(R_1, R_2) \times (0, \pi)$. Observe the right side of this inequality is a continuous function on $(R_1, R_2) \times [0, \pi]$, by the continuity hypothesis in the theorem. To complete the proof we consider two cases.

Case 1. Suppose $\gamma < \pi$, so that inequality (10.32) is valid in the weak sense on each neighborhood of $(s, \gamma)$. By arguing exactly as in the proof of Theorem 10.16 (which uses that $\phi$ is convex decreasing and $\partial_\theta Q > 0$), we find

$$J[\phi(u^\#) - \phi(v)] \leq 0$$

at $(s, \gamma)$, so that the right side of (10.32) is negative at that point. The right side remains negative on some neighborhood $N$ of the point, by continuity, and so

$$\Delta^\star w > c$$

weakly on $N$, for some constant $c > 0$. Shrink $N$ if necessary, so that it lies in $(R_1', R_2') \times (0, \pi)$.

Then

$$\Delta^\star (r^{1-n} w) > c_1$$

weakly on $N$ for some new constant $c_1 > 0$, where we have defined a new operator by $\Delta^\star = r^{1-n} \Delta^\star r^{n-1}$; that is,

$$\Delta^\star F = r^{1-n} \partial_r (r^{n-1} \partial_r F) + r^{-2} \sin^{n-2}\theta \, \partial_\theta (\sin^{2-n}\theta \, \partial_\theta F).$$

The operator $\Delta^\star$ satisfies the maximum principle Proposition 10.8, and so $r^{1-n} w$ cannot have an interior maximum at the point $(s, \gamma) \in N$. This contradiction completes the proof in Case 1, where $\gamma < \pi$.

Incidentally, $\Delta^\star = \Delta$ when $n = 2$ and the Laplacian is expressed in polar coordinates.

Case 2. Suppose $\gamma = \pi$. The idea is similar to the first case except we work only along the top side of the rectangle, with just the $r$-variable, as follows. Notice

$$\begin{aligned} w(r,\pi) &= u^\star(r,\pi) - Jv(r,\pi) - \epsilon Q(r,\pi) \\ &= \int_{\mathbb{S}^{n-1}} [u(rx) - v(rx)]\, r^{n-1}\, d\sigma_{n-1}(x) - \epsilon Q(r,\pi). \end{aligned}$$

We will show that

$$-\Delta_r^\star w(r,\pi) < 0$$

weakly on a neighborhood of $r = s$, where $\Delta_r^\star$ is the purely radial part of the operator:

$$\Delta_r^\star F = \partial_r\big(r^{n-1}\partial_r(r^{1-n}F)\big).$$

Indeed, for any test function $G \in C_c^2(R_1,R_2)$, we have

$$\begin{aligned} &-\int_{R_1}^{R_2} w(r,\pi)\Delta_r^{\star t} G\, dr \\ &= -\int_{R_1}^{R_2} \big(u^\star - Jv\big)(r,\pi)\, r^{1-n}\partial_r\big(r^{n-1}\partial_r G\big)\, dr + \epsilon \int_{R_1}^{R_2} Q(r,\pi)\Delta_r^{\star t} G\, dr \\ &= -\int_A (u-v) r^{1-n}\partial_r\big(r^{n-1}\partial_r G\big)\, d\mathcal{L}^n + \epsilon \int_{R_1}^{R_2} \Delta_r^\star Q(r,\pi)\, G(r)\, dr \\ &= -\int_A (u-v)\Delta G\, d\mathcal{L}^n + \epsilon \int_{R_1}^{R_2} \Delta_r^\star Q(r,\pi)\, G(r)\, dr \\ &= \int_A \big(\phi(u) - \phi(v)\big) G\, d\mathcal{L}^n + \int_A G\,(d\mu - d\nu) + \epsilon \int_{R_1}^{R_2} \Delta_r^\star Q(r,\pi)\, G(r)\, dr, \end{aligned}$$

by the partial differential equations for $u$ and $v$. The first term equals $\int_{R_1}^{R_2} J[\phi(u^\#) - \phi(v)](r,\pi) G(r)\, dr$, by expressing the integral in spherical coordinates. The second term equals $\lim_{\delta\to 0} \frac{1}{\delta}\int_{(R_1,R_2)\times(\pi-\delta,\pi)} G\, dJ(\mu^\# - \nu)$, by definition of the $J$-functional on measures at the beginning of this section; then because $J(\mu^\#) = \mu^\star \le J\nu$ by hypothesis, we conclude the second term is $\le 0$ if $G$ is nonnegative. So we have shown

$$-\Delta_r^\star w(r,\pi) \le J[\phi(u^\#) - \phi(v)](r,\pi) + \epsilon \Delta_r^\star Q(r,\pi).$$

Again arguing as in the proof of Theorem 10.16, we find $J[\phi(u^\#) - \phi(v)] \le 0$ at $(s,\pi)$. Since $\Delta_r^\star Q(r,\pi) = -(n-3/2)/2r^{1/2} < 0$ by direct computation, we conclude from continuity that $-\Delta_r^\star w(r,\pi) < 0$ on a neighborhood $N \subset (R_1', R_2')$ of the point $r = s$, as we wanted to show.

Multiplying by $r^{n-1}$, we conclude that the continuous function $\widetilde{w}(r) = r^{1-n} w(r,\pi)$ satisfies

$$r^{n-1}\partial_r\big(r^{n-1}\partial_r \widetilde{w}\big) > 0$$

weakly on $N$, which means $\widetilde{w}$ is strictly convex in the weak sense as a function of $r^{2-n}$ (or $\log r$, when $n = 2$). Thus $\widetilde{w}$ is classically strictly convex by Lemma 10.6, and so it is impossible for $\widetilde{w}$ to have an interior maximum point at $s \in N$. This contradiction completes the proof in Case 2. □

As in previous sections, we deduce a symmetry result.

**Corollary 10.19** *Assume $v \in \mathcal{W}(A)$ is a weak solution of the semilinear Poisson equation $-\Delta v = \phi(v) + \nu$, where $\phi\colon \mathbb{R} \to \mathbb{R}$ is a decreasing convex function with $\phi(v)$ locally integrable on A, and the measure $\nu \in M_{loc}(A)$ is symmetric decreasing ($\nu = \nu^{\#}$). Suppose the functions $Jv, J\phi(v), J(v^{\#}), J\phi(v^{\#})$ are continuous on $(R_1, R_2) \times [0, \pi]$.*

*If the function $v$ is symmetric decreasing at $r = R_1$ and $R_2$, meaning*

$$\lim \int_{\mathcal{K}(\theta)} [v^{\#}(rx) - v(rx)]\, d\sigma_{n-1}(x) = 0$$

*when the limit is taken over points $(r, \theta) \in A^{\star}$ with $r \to R_1$ or $r \to R_2$, then $v$ is symmetric decreasing on the whole shell: $v = v^{\#}$ a.e. on A.*

The statement of the corollary simplifies considerably if $v$ is continuous on the closure of the shell and $v = v^{\#}$ on the inner and outer spheres.

**Remark** The continuity assumption on $J(u^{\#})$ and related functions in Theorem 10.18 can be weakened in certain circumstances by an approximation procedure. Suppose the theorem holds as stated for functions $u_k$ and $v_k$, under the continuity assumption. If $u_k \to u$ and $v_k \to v$ in a suitable sense as $k \to \infty$, then the $\star$-function and integral mean conclusions of the theorem would continue to hold for $u$ and $v$, for almost every $r \in (R_1, R_2)$.

The same observation applies to the earlier Theorem 10.15 about Steiner symmetrization, and Theorem 10.20 below for $(n-1, 1)$-cap symmetrization on the shell with Neumann boundary conditions.

### 10.6.2 Neumann Boundary Conditions

Next we develop a comparison theorem for the linear Poisson equation under a Neumann type boundary condition at the inner and outer shell boundaries. Suppose a locally integrable function $h$ has radial weak derivative $\partial h/\partial r$. We say $h$ satisfies radial differentiation through the integral if $h$ and $\partial h/\partial r$ are integrable on each sphere of radius $r$ and satisfy

$$\frac{\partial}{\partial r} \int_E h(rx)\, d\sigma_{n-1}(x) = \int_E \frac{\partial h}{\partial r}(rx)\, d\sigma_{n-1}(x), \qquad r \in (R_1, R_2), \qquad (10.33)$$

for every measurable set $E \subset \mathbb{S}^{n-1}$.

**Theorem 10.20** *Assume $u, v \in \mathcal{W}(A)$ are weak solutions of the linear Poisson equations*

$$-\Delta u + cu = \mu, \qquad -\Delta v + cv = \nu,$$

*where $\mu, \nu \in M_{loc}(A)$ and $c \geq 0$ is constant. Suppose the outer radius $R_2$ is finite, that $J(u^{\#})$ and $Jv$ are continuous on $(R_1, R_2) \times [0, \pi]$, that $u$ and $v$ satisfy radial differentiation through the integral in the sense of (10.33), and that $\mu^{\star} \leq Jv$ on $(R_1, R_2) \times (0, \pi)$. If $c = 0$ then further suppose for some $r \in (R_1, R_2)$ that*

$$\int_{\mathbb{S}^{n-1}} u(rx)\, d\sigma_{n-1}(x) \leq \int_{\mathbb{S}^{n-1}} v(rx)\, d\sigma_{n-1}(x). \tag{10.34}$$

*If $(\partial u/\partial n)^{\star} \leq J(\partial v/\partial n)$ at $r = R_1$ and $r = R_2$, by which we mean*

$$\limsup_{\substack{0 \leq \theta \leq \pi \\ r \to R_i}} \int_{\mathcal{K}(\theta)} \left[ \left( \frac{\partial u}{\partial n} \right)^{\#} (rx) - \frac{\partial v}{\partial n}(rx) \right] d\sigma_{n-1}(x) \leq 0, \qquad i = 1, 2, \tag{10.35}$$

*where $\partial/\partial n = (-1)^i \partial/\partial r$ is the outward normal derivative, then $u^{\star} \leq Jv \leq v^{\star}$ on $A^{\star}$ and hence the convex increasing integral means of $u$ are dominated by those of $v$:*

$$\int_{\mathbb{S}^{n-1}} \Phi(u(rx))\, d\sigma_{n-1}(x) \leq \int_{\mathbb{S}^{n-1}} \Phi(v(rx))\, d\sigma_{n-1}(x), \qquad R_1 < r < R_2,$$

*for every convex increasing $\Phi \colon \mathbb{R} \to \mathbb{R}$.*

The boundary condition (10.35) holds in particular if $u$ and $v$ are smooth and their normal derivatives vanish on the boundary of the shell.

The continuity assumptions on $J(u^{\#})$ and $Jv$ implicitly assume that $u$ and $v$ are integrable on each sphere of radius $r \in (R_1, R_2)$, and hence that $\mu^{\star}$ and $Jv$ are well defined on each sphere.

The conclusion of the theorem can be strengthened (via Proposition 10.3) to handle all convex $\Phi$ if $u$ and $v$ have the same integral over every concentric sphere; then one obtains also comparisons between $L^p$-norms and the oscillations of $u$ and $v$ on each sphere. The integrals of $u$ and $v$ will indeed agree on every concentric sphere in the case where $-\Delta u = f, -\Delta v = f^{\#}$ and $u$ and $v$ satisfy homogeneous Neumann boundary conditions and are normalized to have the same integral over the shell $A$; see Step 2 in the proof of (Langford, 2015b, Theorem 3.1).

*Proof of Theorem 10.20* We start by relating the boundary condition (10.35) to the radial derivative of the continuous function $w_0 = u^{\star} - Jv = J(u^{\#} - v)$. We will prove that

$$\liminf_{r\to R_1} \inf_{0\le\theta\le\pi} \frac{\partial(r^{1-n}w_0)}{\partial r}(r+,\theta) \ge 0, \tag{10.36}$$

$$\limsup_{r\to R_2} \sup_{0\le\theta\le\pi} \frac{\partial(r^{1-n}w_0)}{\partial r}(r-,\theta) \le 0, \tag{10.37}$$

where since we do not know these derivatives exist, in the first formula the derivative means the lim inf of difference quotients from the right and in the second formula the derivative means the lim sup of difference quotients from the left.

To prove the second formula (10.37), consider $R \in (R_1, R_2)$ and $\theta \in [0, \pi]$. Let $E \subset \mathbb{S}^{n-1}$ be a maximizing set for $u^\star(R,\theta)$, meaning that $u^\star(R,\theta) = \int_E u(Rx)\, R^{n-1} d\sigma_{n-1}(x)$ and $E$ has the same measure as $\mathcal{K}(\theta)$. By definition of $u^\star$ we have $r^{1-n}u^\star(r,\theta) \ge \int_E u(rx)\, d\sigma_{n-1}(x)$ for all $r$, with equality at $r = R$ by our choice of $E$. Taking the left derivative at $r = R$ gives:

$$\begin{aligned}
\left.\frac{\partial(r^{1-n}w_0)}{\partial r}\right|_{r=R-} &\le \frac{\partial}{\partial r}\left(\int_E u(rx)\, d\sigma_{n-1}(x) - \int_{\mathcal{K}(\theta)} v(rx)\, d\sigma_{n-1}(x)\right)\Bigg|_{r=R-} \\
&= \int_E \frac{\partial u}{\partial r}(Rx)\, d\sigma_{n-1}(x) - \int_{\mathcal{K}(\theta)} \frac{\partial v}{\partial r}(Rx)\, d\sigma_{n-1}(x) \\
&\le \int_{\mathcal{K}(\theta)} \left[\left(\frac{\partial u}{\partial r}\right)^{\#} - \frac{\partial v}{\partial r}\right](Rx)\, d\sigma_{n-1}(x).
\end{aligned}$$

The lim sup of this last expression as $R \to R_2$ (with $\theta$ allowed to vary) is $\le 0$ by (10.35) with $i = 2$, which proves (10.37). The proof is similar for (10.36).

Next we define a perturbing function. Fix a number $R_3 \in (R_1, R_2)$, and take a constant $C > 0$ such that

$$C \ge nr^2 - (n-1)R_3 r + 1$$

for all $r \in [R_1, R_2]$ (here we need finiteness of $R_2$). Let

$$Q(r,\theta) = r^{n-1}\big[(r - R_3)^2 + C\theta(\pi - \theta)\big].$$

For later reference, note $Q \ge 0$ and $\partial(r^{1-n}Q)/\partial r = 2(r - R_3)$. Define

$$w = w_0 - \epsilon Q = u^\star - Jv - \epsilon Q,$$

where $\epsilon > 0$. Subtracting $\epsilon Q$ from $w_0$ in (10.36) and (10.37) implies that

$$\liminf_{r\to R_1} \inf_{0\le\theta\le\pi} \frac{\partial(r^{1-n}w)}{\partial r}(r+,\theta) \ge -2\epsilon(R_1 - R_3) > 0, \tag{10.38}$$

$$\limsup_{r\to R_2} \sup_{0\le\theta\le\pi} \frac{\partial(r^{1-n}w)}{\partial r}(r-,\theta) \le -2\epsilon(R_2 - R_3) < 0. \tag{10.39}$$

The goal of the theorem is to prove $w_0 \le 0$, that is, $u^\star \le Jv$. Then the conclusion on convex integral means follows from Section 10.1, since $Jv \le v^\star$ by definition. Note $w_0$ is continuous on $(R_1, R_2) \times [0, \pi]$ by hypothesis.

Suppose for the sake of obtaining a contradiction that $w_0 > 0$ somewhere in the rectangle $A^\star = (R_1, R_2) \times (0, \pi)$. We will proceed in three steps: first show $r^{1-n}w$ attains its maximum on the top side $T = (R_1, R_2) \times \{\pi\}$ of the rectangle, then deduce the corresponding property for $w_0$, and finally obtain a contradiction.

**Step 1**: Show $r^{1-n}w$ attains its maximum on the top side $T$. Clearly $w > 0$ somewhere in the rectangle, whenever $\epsilon$ is sufficiently small, since we assume $w_0 > 0$ at some point in the rectangle. The boundary conditions (10.38) and (10.39) insure that $r^{1-n}w$ does not attain a maximum at the left or right sides of $A^\star$. It also does not attain its maximum on the bottom side where $\theta = 0$, because $Q \ge 0$ while both $u^\star$ and $Jv$ vanish at $\theta = 0$ by definition. Hence $r^{1-n}w$ attains its positive maximum value either in the interior of the rectangle $A^\star$ or else on the top side $T$. We proceed now to rule out the interior case.

One computes from the definition of $\Delta^\star$ in (10.31) that

$$\begin{aligned}\Delta^\star Q &= 2r^{n-3}\big[nr^2 - (n-1)R_3 r - C\big] - C(n-2)r^{n-3}(\pi - 2\theta)\cot\theta \\ &\le 2r^{n-3}\big[nr^2 - (n-1)R_3 r - C\big] \\ &\le -2r^{n-3} < 0\end{aligned}$$

on $A^\star$, by our earlier choice of $C$. The Subharmonicity Theorem 9.9 for $u$ and Weak Main Identity (9.35) for $v$ (using also the adjoint relations (9.26) and (10.29)) imply that

$$\begin{aligned}-\Delta^\star u^\star + cJ(u^\#) &\le \mu^\star, \\ -\Delta^\star Jv + cJv &= Jv,\end{aligned}$$

in the weak sense. Subtracting these two formulas and using the hypothesis $\mu^\star \le Jv$ gives

$$-\Delta^\star w + cw \le \epsilon(\Delta^\star Q - cQ) \le -2\epsilon r^{n-3} \tag{10.40}$$

in the weak sense on $A^\star$, where we used in the last inequality that $c \ge 0$ and $Q \ge 0$.

Suppose $r^{1-n}w$ attains its maximum at a point in the interior of the rectangle $A^\star$. Inequality (10.40) implies $\Delta^\star w > c_1$ on a neighborhood $N$ of the maximum point, for some constant $c_1 > 0$, since $w > 0$ at the maximum point and $c \ge 0$. Now we obtain a contradiction from the maximum principle Proposition 10.8 for the operator $\Delta^\star$, exactly as in the proof of Theorem 10.18. Hence $r^{1-n}w$ must attain its maximum on the top side $T$, which completes the proof of Step 1.

**Step 2**: Show $w_0$ is positive somewhere on the top side $T$. We have

$$\begin{aligned}
&\sup_{A^\star} r^{1-n} w_0 \\
&\quad \le \sup_{A^\star} r^{1-n} w + \epsilon \sup_{A^\star} r^{1-n} Q && \text{by definition of } w = w_0 - \epsilon Q \\
&\quad = \sup_{T} r^{1-n} w + \epsilon \sup_{A^\star} r^{1-n} Q && \text{by Step 1} \\
&\quad \le \sup_{T} r^{1-n} w_0 + \epsilon \sup_{A^\star} r^{1-n} Q && \text{since } w \le w_0 \text{ (recalling } Q \ge 0).
\end{aligned}$$

Letting $\epsilon \to 0$ shows $\sup_{A^\star} r^{1-n} w_0 \le \sup_T r^{1-n} w_0$, and since $w_0$ is positive somewhere in $A^\star$ we deduce $w_0$ is positive at some point of $T$.

**Step 3**: Deduce a contradiction. The idea is similar to Step 1 except we work only along the top side of the rectangle, with only the $r$-variable, as follows. Notice

$$\begin{aligned}
w_0(r,\pi) &= u^\star(r,\pi) - Jv(r,\pi) \\
&= \int_{\mathbb{S}^{n-1}} [u(rx) - v(rx)]\, r^{n-1}\, d\sigma_{n-1}(x).
\end{aligned}$$

We will show $\Delta_r^\star w_0(r,\pi) \ge c w_0(r,\pi)$ in the weak sense, where $\Delta_r^\star$ is the purely radial part of the operator:

$$\Delta_r^\star F = \partial_r\big(r^{n-1}\partial_r(r^{1-n}F)\big).$$

Indeed, for any test function $G \in C_c^2(R_1,R_2)$, we have

$$\begin{aligned}
\int_{R_1}^{R_2} w_0(r,\pi)\Delta_r^{\star t} G\, dr &= \int_{R_1}^{R_2} \big(u^\star - Jv\big)(r,\pi)\, r^{1-n}\partial_r\big(r^{n-1}\partial_r G\big)\, dr \\
&= \int_A (u - v)\Delta G\, d\mathcal{L}^n \\
&= c\int_A (u-v) G\, d\mathcal{L}^n - \int_A G\,(d\mu - d\nu),
\end{aligned}$$

by the partial differential equations for $u$ and $v$. Hence

$$\begin{aligned}
&\int_{R_1}^{R_2} w_0(r,\pi)\Delta_r^{\star t} G\, dr \\
&= \int_{R_1}^{R_2} w_0(r,\pi) G(r)\, dr - \lim_{\delta\to 0}\frac{1}{\delta}\int_{(R_1,R_2)\times(\pi-\delta,\pi)} G\, dJ(\mu^\# - \nu)
\end{aligned}$$

by spherical coordinates and using the definition of the $J$-functional on measures from the beginning of this section. Recalling the hypothesis

$J(\mu^{\#}) = \mu^{\star} \leq J\nu$, we see the third integral is $\leq 0$ if $G$ is nonnegative, and so we have shown

$$\Delta_r^{\star} w_0(r,\pi) \geq c w_0(r,\pi)$$

in the weak sense.

Now consider the open set $P \subset T$ on which $w_0(r,\pi)$ is positive; this set is nonempty by Step 2. On $P$ we have $\Delta_r^{\star} w_0(r,\pi) \geq 0$ if $c = 0$ or $\Delta_r^{\star} w_0(r,\pi) > 0$ if $c > 0$. Thus

$$r^{n-1}\partial_r\big(r^{n-1}\partial_r(r^{1-n}w_0)\big) \begin{cases} \geq 0 & \text{if } c = 0 \\ > 0 & \text{if } c > 0 \end{cases} \qquad \text{on } P,$$

so that $r^{1-n}w_0(r,\pi)$ is convex in the weak sense as a function of $r^{2-n}$ (or $\log r$, when $n = 2$), and indeed is strictly convex if $c > 0$. This convexity holds in the classical sense by Lemma 10.6.

Suppose $c > 0$. Then by strict convexity, $P$ extends either leftwards all the way to $r = R_1$ or rightwards all the way to $r = R_2$. In the leftwards case, the slope of $r^{1-n}w_0$ tends to a negative constant or $-\infty$ at $R_1$, which contradicts the boundary condition (10.36); in the rightwards case, the slope of $r^{1-n}w_0$ tends to a positive constant or $\infty$ at $R_2$, which contradicts (10.37). Either way we have a contradiction.

Suppose $c = 0$. Then $w_0(r,\pi) \leq 0$ for some $r \in (R_1, R_2)$ by hypothesis (10.34), and so $r^{1-n}w_0$ is not constant on $P$. Convexity of $r^{1-n}w_0(r,\pi)$ on $P$ then leads to a contradiction just like in the preceding paragraph. The proof of the theorem is complete. □

## 10.7 Notes and Comments

This chapter draws on material in Baernstein (1994) and elsewhere. The material on majorization in Section 10.1 was prepared by Albert Baernstein II. The rest of the chapter was written by Richard Laugesen, working in collaboration with Jeffrey Langford on the following sections: Section 10.3.1 for linear Poisson equations, Section 10.6 for shells, and these Notes and Comments.

Section 10.1: Majorization results for sums and integrals such as in Propositions 10.1 and 10.3 go back to Hardy, Littlewood and Pólya (1952). For more on majorization and its applications, see Marshall et al. (2011). An equality statement for infinite sums is due to Laugesen and Morpurgo (1998, Proposition 10).

Section 10.2: The maximum principles for weak subsolutions in Propositions 10.7 and 10.8 are special cases of a general elliptic result due to Littman (1959).

Sections 10.3–10.6: The history of comparison results for elliptic boundary value problems goes back to work of Szegö (1950, 1958) on the clamped and buckling plate problems. The best-known comparison result is due to Talenti (1976b). His methods are quite different from this chapter, being based ultimately on level set decompositions. Surveys of ensuing work in the field can be found in the papers of Baernstein (1994) and Talenti (2016). See also (Kesavan, 2006, chapter 3) and the remarks on the literature at the end of that chapter. We make no effort to produce a comprehensive history here. Instead, we mention below only a few works most relevant to the material in Chapter 10.

The comparison theorems for the Laplacian in this chapter extend to general second-order elliptic and parabolic operators. Such extensions go back to Talenti (1976b) in the elliptic case, and Alvino et al. (1991) for parabolic equations. See the remarks in (Baernstein, 1994, Sections 7 and 8) for the $\star$-function approach to extending Laplacian comparison results to these more general elliptic and parabolic operators.

The strong conclusion $u^{\#} \leq v$ in Theorem 10.10 for s.d.r. when $c = 0$ is due to Talenti (1976b). The proof of the theorem when $c < 0$ is adapted from (Alvino et al., 2002, Theorem 1.1). These authors assumed the right side of the equation to be a function, meaning $d\mu = f\, d\mathcal{L}^n$.

Talenti's gradient comparison in (10.8) seems to be special to s.d.r. It is not known to hold for other forms of symmetrization. Open problems for the gradient under symmetrization are mentioned by Baernstein (1994, §9).

The alternate proof of the Faber–Krahn Theorem in Example 10.11 is based on Kesavan's idea of comparing with the equation $-\Delta v = \lambda_1(\Omega)u^{\#}$ (see Kesavan 2006, Theorem 4.1.2). Due to the inclusion here of a Cauchy–Schwarz step, we require only the comparison of $L^2$ norms from Theorem 10.10, rather than Talenti's stronger pointwise comparison $u^{\#} \leq v$. Thus the alternate proof could be extended from s.d.r. to other kinds of rearrangement, such as Steiner symmetrization.

The linear case of Theorem 10.16 for comparison on the sphere appeared in work of Langford (2015a), assuming the right side is a function, $d\mu = f\, d\sigma_n$. Theorem 10.20 for cap symmetrization on a shell under Neumann boundary conditions builds on work of Langford (2015b).

The nonlinear term $\phi(u)$ in comparison theorems for partial differential equations was treated first (as far as we know) by Bandle (1980 §IV.2.4) for s.d.r. and by Weitsman (1986) for circular and cap symmetrization.

Interestingly, Bandle's method does not require convexity of $\phi$. Weitsman's technique for handling convex decreasing $\phi$ is used throughout this chapter.

Talenti (1979, §4.1) earlier stated a comparison result for quasilinear equations under s.d.r., but his symmetrized partial differential equation (satisfied by $v$) does not have the same form as the equation for $u$.

Drift term comparison theorems, such as for the Ornstein–Uhlenbeck operator $-\Delta + x \cdot \nabla$, have been developed by Borell (1985). See also González (2000); Hamel et al. (2011), and for applications to biLaplacian (plate) problems in Gauss space, see Chasman and Langford (2016).

Comparison theorems are proved on spheres and shells in Sections 10.5 and 10.6, but not on *subdomains* of those spaces. Comparison theorems do hold on subdomains of spheres and shells, analogous to the results in Sections 10.3 and 10.4 for subdomains of Euclidean space under s.d.r. and Steiner symmetrization. See for example the circular symmetrization comparison results in the next chapter: Theorems 11.22, 11.25–11.27, 11.33 and 11.34.

The $\star$-function method in this chapter makes it possible (and natural) to use a measure $\mu$ as data on the right side of the partial differential equation. The only result we know in the literature that gives a comparison theorem involving measures is a brief remark by Alvino et al. (1991, §II.4) for linear equations.

**Applications** Comparison results for s.d.r. are used in Nadirashvili's proof of Lord Rayleigh's conjecture for the vibrating clamped plate; see Chapter 5. This result and further applications are described in elegant detail in Kesavan (2006). A semilinear comparison result for circular symmetrization was used by Weitsman (1986) to obtain sharp estimates on the hyperbolic metric; see §11.9. Applications of the $\star$-function and circular symmetrization in complex analysis, in particular to Nevanlinna theory and geometric function theory, are presented in Chapter 11.

# 11

# The $\star$-Function in Complex Analysis

## 11.1 Introduction and Background

The $\star$-function already has been used in Chapters 9 and 10, with emphasis on applications to partial differential equations. However, it was originally developed in the 1970s to resolve several well-established problems in classical one-variable complex analysis. That context camouflaged its links to the general theory of symmetrization, with the relation first being made explicit in Baernstein (1974).

This chapter returns to $\star$-function's original focus, analytic and meromorphic functions defined in the plane or the unit disk. It may be read independent of the earlier chapters. Notation is as follows: $\mathbb{D}$ is the open unit disk, $\mathbb{S}$ is the unit circle, $\mathbb{D}(r)$ is the disk $\{|z| < r\}$ of radius $r$ and $\mathbb{S}(r)$ is its boundary $\{|z| = r\}$. (Other authors write $\mathbb{T}$ for the unit circle.) The upper half plane is $\mathbb{H} = \{z\colon 0 < \arg z < \pi\}$. If $E$ is a set in $\mathbb{R}$ or $\mathbb{S}$, then $\overline{E}$ is its closure and $|E|$ its Lebesgue measure.

The star function used in this chapter,

$$u^{\star}(re^{i\theta}) = \sup_{|E|=2\theta} \int_E u(re^{it})\, dt,$$

is the original one introduced by Baernstein (1973, 1974). This $u^{\star}$ differs by the factor $r$ from the star function $u^{\star}(r,\theta) = ru^{\star}(re^{i\theta})$ used in Chapter 9 for circular or $(1,2)$-cap symmetrization; see the discussion in the Notes at the end of Chapter 9. The essential fact that the operator $u \mapsto u^{\star}$ preserves subharmonicity appeared as Corollary 9.10, and the history and other references are discussed in Note 2 to this chapter.

The theory of the $\star$-function in the plane is also covered in the complex analysis monograph (Duren, 1983, chapter 7), which emphasizes applications to univalent functions, and in Hayman (1989 §9.1).

In this chapter, # refers to circular symmetrization, defined as $(1,2)$-cap symmetrization, §7.5, in which a function or domain is rearranged on concentric circular arcs symmetric about the positive real axis in the complex plane. To avoid ambiguity we employ the symbol $\star$ when our conventions conflict with those of the previous chapters. As an example, if $A$ is the annular region

$$A = \{re^{i\theta} : R_1 < r < R_2\}$$

considered in §9.6, then

$$\begin{aligned} A^{\#} &= \{re^{i\theta} \in \mathbb{C} \colon R_1 < r < R_2\}, \\ A^{\bigstar} &= \{(r,\theta) \in \mathbb{R}^2 \colon R_1 < r < R_2, 0 < \theta < \pi\}, \\ A^{\star} &= \{re^{i\theta} \in \mathbb{C} \colon R_1 < r < R_2, 0 < \theta < \pi\}. \end{aligned}$$

Thus a subrectangle of $A^{\bigstar}$ corresponds to the polar coordinate representation of an annular subset in $A^{\star}$.

The general pattern of much of this chapter is to study extremal problems in disks in which the competing functions $u$ are either subharmonic or $\delta$-subharmonic. It had been noticed that in many important cases, the presumed extremal function $v$ is harmonic with $v(re^{i\theta})$ symmetric decreasing on each circle $\mathbb{S}(r)$ (perhaps after a rotation). The $\star$-function was created to exploit this observation. Suppose $u$ is defined in $\mathbb{D}$ and $u^{\#}(re^{it})$ is its circular symmetrization. Let

$$u^{\star}(re^{i\theta}) = \int_{-\theta}^{\theta} u^{\#}(re^{it})\, dt, \quad V(re^{i\theta}) = \int_{-\theta}^{\theta} v(re^{it})\, dt, \tag{11.1}$$

for $0 \leq \theta \leq \pi$ and $r < 1$. The theory asserts that

$$u^{\star} - V$$

is subharmonic in $\mathbb{D}^{\star}$, so that $u^{\star} \leq V$ on $\mathbb{D}^{\star}$ if this is true on the boundary.

The $\star$-function and this type of maximum principle argument were developed in 1971, building on an insight of Edrei and Fuchs (1960) concerning meromorphic functions $f$ of order $\rho \leq 1$. (The order of a meromorphic function will be defined in §11.2.) It may be helpful to recall their insight here. If $f$ has order $\rho$ less than 1, then it has the elementary factorization

$$f(z) = Cz^q \prod_j (1 - z/a_j) \Big/ \prod_k (1 - z/b_k) \tag{11.2}$$

in terms of its zeros and poles; these simple products in general do not converge when $\rho \geq 1$, and the necessary additional factors introduced to ensure convergence make the associated factorization far more difficult to analyze. It may be shown directly when the factorization (11.2) is possible that functions with zeros on one ray and poles on the opposite ray have extremal behavior.

Edrei and Fuchs associated to a function $F$ with the representation (11.2) and having zeros/poles on the negative/positive axes the function

$$U(re^{i\theta}) = \int_{-\theta}^{\theta} \log|F(re^{it})|\,dt + 2\pi N(r,F),$$

which is harmonic in the upper half plane. Note that the $\theta$-dependence of $U$ is made by integrating on a set of measure $2\theta$ chosen to make $U$ as large as possible. (Here $N(r,F)$ is the (integrated) counting function of the number of poles in $\mathbb{D}(r)$, as defined in the next section, so that $N(r,F) = 0$ when $F$ is entire.) The ⋆-function adapts this process to all functions $f$ of order at most one, by replacing, for each $r$ and $\theta$, integration over a single arc of measure $2\theta$ by integration over a set $E(\theta) \subset \mathbb{S}$ of total measure $2\theta$, chosen so that the function

$$T^{\star}(re^{i\theta}) = \int_{E(\theta)} \log|f(re^{it})|\,dt + 2\pi N(r,f)$$

is as large as possible. The remarkable discovery is that $T^{\star}$ is subharmonic and dominated by $U$.

This chapter considers the theory's successes in one complex variable, including classical Nevanlinna theory (§11.2), univalent functions (§11.6), the harmonic conjugate function (§11.8), and properties of the hyperbolic metric (§11.9).

## 11.2 The Nevanlinna Characteristic $T$ and Its Extension $T^{\star}$

### Nevanlinna Theory

This section recalls the basics of classical Nevanlinna theory, ignoring most proofs and with no mention of the ⋆-function. It may be omitted by readers familiar with Nevanlinna's work.

There are many fine expositions of the subject (also known as value-distribution theory), named after its creator Rolf Nevanlinna (1970), and the main definitions require little advance preparation. The traditional texts are Hayman (1964) and Goldberg and Ostrovskii (2008) (a translation and updating of the 1970 Russian original). The latter reference also discusses extensions of the theory for functions defined in more general regions, in particular half planes.

If $f$ is meromorphic in a region $\Omega$, its Laplacian has a simple form. Of course, when $f$ is analytic at $z_0$ with $f(z_0) \neq 0$, $\log|f|$ is harmonic in a neighborhood of $z_0$, and so $\Delta(\log|f|) = 0$. In general, if $f$ has a zero of order $k$ at $z_0$, then

$$\Delta(\log|f|)(z) = 2\pi k\delta(z - z_0)$$

in a neighborhood of $z_0$, with $\delta$ the Dirac delta measure. If $z_0$ is a pole of order $k$, the only change is that $k$ on the right side is negative. (The converse is also true: a discrete measure $\mu$ which is a (signed) sum of masses of this type is the Laplacian of $\log|f|$ for some meromorphic function $f$, although $f$ is not unique.) By convention, a meromorphic function is nonconstant. Here we only consider functions meromorphic in disks $\mathbb{D}(R), R > 0$, or in $\mathbb{C}$.

Nevanlinna defines his characteristic function $T(r)$ as the sum

$$T(r) = T(r,f) = m(r,f) + N(r,f),$$

with

$$m(r,f) = \frac{1}{2\pi}\int_{-\pi}^{\pi} \log^+ |f(re^{i\theta})|\, d\theta,$$

$$n(r,f) = \text{number of poles of } f \text{ in the closed disk } (|z| \leq r), \text{ counted with multiplicity,}$$

$$N(r,f) = \int_0^r \frac{n(t,f) - n(0,f)}{t}\, dt + n(0,f)\log r,$$

where $\log^+ x \geq 0, \log x = \log^+ x - \log^- x$. Note that one of the summands defining $T$ is controlled by the moduli of poles of $f$ inside $\mathbb{D}(r)$ and the other by large values of $|f|$ on $\mathbb{S}(r)$. The function $T$ is always increasing, which is not obvious since $m(r,f)$ is not monotone.

Nevanlinna's first fundamental theorem shows that as far as the characteristic $T$ is concerned, each complex value enjoys equal standing with the poles: for every $a \in \mathbb{C}$,

$$T(r,f) = T\big(r, 1/(f - a)\big) + O(1) \qquad \text{as } r \to \infty. \tag{11.3}$$

When $a = 0$ this becomes a perceptive reinterpretation of Jensen's theorem, presented below in (11.8). Nevanlinna's insight was to rewrite Jensen's identity by apportioning the positive and negative contributions in the integral of $\log|f(re^{i\theta})|$ to $m(r,f)$ and $m(r, 1/f)$. This approach will be sketched after the Poisson–Jensen Lemma 11.1.

Since $f$ is nonconstant, $T$ is unbounded when $f$ is meromorphic in the plane, and so the $O(1)$ is an error term that can normally be ignored. It is common to use the abbreviated notations

$$T(r,a),\ m(r,a),\ N(r,a),\ n(r,a)$$

in place of

$$T(r,1/(f-a)),\ m(r,1/(f-a)),\ N(r,1/(f-a)),\ n(r,1/(f-a)).$$

By $N(r,\infty)$ we mean $N(r,f)$, the usual integrated pole-counting function, and similarly for the other quantities when $a=\infty$.

That $m(r,a)\geq 0$ yields the universal upper bound

$$N(r,a)\leq T(r,f)+O(1)$$

for all $a$, showing that $T$ controls the number of solutions to $f(z)=a$ in $\mathbb{D}(r)$.

A remarkably useful identity was discovered early in the development by Henri Cartan, suggesting that in the previous inequality $N(r,a)$ is usually the dominant term. Concretely stated, Cartan showed that

$$T(r,f)=\frac{1}{2\pi}\int_{-\pi}^{\pi}N(r,e^{i\theta})\,d\theta+\log^+|f(0)|,\qquad (0<r<R),\qquad (11.4)$$

which is valid regardless of the value of $f(0)$. An immediate and non-obvious consequence is that $T(r,f)$ is an increasing function of $r$, and is convex as a function of $\log r$. The charming proof of (11.4) warrants being reproduced at the end of this section; it will be used later, for example in Lemma 11.23 and §11.8.3.

Nevanlinna theory traditionally studies how properties of $f$ influence $m$ and $N$ under various hypotheses, especially how they depend on $a$. Most notably, Nevanlinna defines his defect (deficiency)

$$\delta(a)=\delta(a,f)=\liminf_{r\to\infty}\frac{m(r,a)}{T(r)}=1-\limsup_{r\to\infty}\frac{N(r,a)}{T(r)},\qquad (11.5)$$

the last equations a consequence of the first fundamental theorem. The most famous consequence of his calculus was his second fundamental theorem, which has as a simple corollary that

$$\sum_{a\in\widehat{\mathbb{C}}}\delta(a)\leq 2\qquad \text{(defect relation);}\qquad (11.6)$$

thus for most $a$ and most large $r$, $m(r,a)$ is negligible in comparison to $N(r,a)$.

Inequality (11.6) provided a new interpretation of E. Picard's famous theorem of 1876 that an entire function $f$ can omit at most one finite value (notice $f$ being entire already guarantees that $\delta(\infty)=1$). For example, $f(z)=e^z$ also omits $w=0$, which means that $\delta(0)=\delta(\infty)=1$, showing that (11.6) is best possible. Much of the appeal of Nevanlinna theory arose from recognizing that Picard's theorem, rather than closing a subject, opens a new one. It also provided a purely analytic proof of Picard's theorem, which seemed more direct than relying on the deep universal covering function (Proposition 11.31).

Nevanlinna's primary tool was his analysis of the Poisson–Jensen formula, which we present below in a slightly simpler formulation (Ahlfors 1978) using the functionals of Nevanlinna theory. Jensen's theorem is the special situation where $z_0 = 0$. Recall that the Poisson kernel for the disk $\mathbb{D}(R)$ is

$$P_R(r,\theta) = \frac{1}{2\pi}\frac{R^2 - r^2}{R^2 - 2Rr\cos\theta + r^2} \qquad (r < R, |\theta| \le \pi),$$

and the Poisson integral of an integrable function $g$ defined on $\mathbb{S}(R)$ is the harmonic function

$$G(z) = \int_{-\pi}^{\pi} P_R(r,\theta - t)g(t)\,dt \qquad (z = re^{i\theta} \in \mathbb{D}(r)).$$

**Lemma 11.1** (see Ahlfors 1978, p. 208) *Let $f$ be meromorphic in $\mathbb{D}(R)$ and continuous on $\mathbb{S}(R)$, with zeros $\{a_m\}$ and poles $\{b_n\}$. Let $z = re^{i\theta} \in \mathbb{D}(R)$. If $f(z) \neq 0, \infty$ then*

$$\begin{aligned}\log|f(z)| &= \int_{-\pi}^{\pi} P_R(r,\theta - t)\log|f(Re^{it})|\,dt \\ &\quad + \sum_j \log\left|\frac{R(z - a_j)}{R^2 - \overline{a_j}z}\right| - \sum_k \log\left|\frac{R(z - b_k)}{R^2 - \overline{b_k}z}\right|.\end{aligned} \tag{11.7}$$

*Proof* If $f$ has no zeros or poles the formula is immediate, since $\log|f|$ is then harmonic. In general, when there are zeros and poles, apply this argument to the function

$$\psi(z) = f(z)\prod_k \frac{R(z - b_k)}{R^2 - \overline{b_k}z} \Big/ \prod_j \frac{R(z - a_j)}{R^2 - \overline{a_j}z},$$

which has no zeros or poles. Note that the products, known as Blaschke factors, have absolute value 1 on $\mathbb{S}(R)$ and hence do not affect the Poisson integral in the formula. □

Jensen's formula is the special case $z = 0$ of the Poisson–Jensen formula (11.7): assuming $f(0) \neq 0, \infty$, the formula says

$$\begin{aligned}\log|f(0)| &= \frac{1}{2\pi}\int_{-\pi}^{\pi}\log|f(Re^{it})|\,dt - N(R,0) + N(r,\infty) \\ &= m(R,f) + N(R,f) - m(R,1/f) - N(R,1/f) \\ &= T(R,f) - T(R,1/f),\end{aligned} \tag{11.8}$$

since

$$m(R,f) = \frac{1}{2\pi}\int_0^{2\pi} \log^+ |f(Re^{it})|\, dt,$$
$$m(R,1/f) = \frac{1}{2\pi}\int_0^{2\pi} \log^- |f(Re^{it})|\, dt.$$

This is Nevanlinna's rewriting of Jensen's formula, to which we referred earlier.

Since $T$ is defined only on $\mathbb{R}$, its rate of increase is its most evident property. The standard measure of growth assigned to an increasing, unbounded function is its *order* $\rho$, defined in the case of $T(r)$ as

$$\rho = \limsup_{r\to\infty} \frac{\log T(r)}{\log r}. \tag{11.9}$$

Originally the order $\rho$ was used only when $f$ is entire, with $\log^+ M(r)$ in place of $T(r)$, where

$$M(r) = M(r,f) = \max_{|z|=r} |f(z)|;$$

$\rho$ is the same in either case, since an elementary consequence of the Poisson–Jensen formula is that

$$T(r) \le \log^+ M(r) \le \frac{r'+r}{r'-r} T(r') \qquad (r < r' < R).$$

The left inequality is obvious since $|f(re^{i\theta})| \le M(r)$, and to obtain the right inequality, take $R > r' = 2r$ when estimating the Poisson kernel in the Poisson–Jensen formula.

The most well-studied functions in classical Nevanlinna theory, which are also those most encountered in science, have $0 < \rho < \infty$ (and, in particular, have finite order). It was noticed later that the *lower order* $\mu$ is a more flexible notion, where

$$\mu = \liminf_{r\to\infty} \frac{\log T(r)}{\log r} \tag{11.10}$$

(also defined with $\log^+ M(r)$ in place of $T(r)$ for entire functions). The most familiar functions of applied mathematics and physics have $\mu = \rho$, but in general there are no restrictions other than $0 \le \mu \le \rho \le \infty$. By bringing the (lower) order into play, it is also possible to discover a rich theory that is not present for the full class of meromorphic functions. For example, if $\mu < \infty$, then equality can hold in (11.6) only when $\mu \ge 1$ and $2\mu (= 2\rho) \in \mathbb{N}$, see Drasin (1976). (For entire functions with $\mu < \infty$, equality is only possible when $\mu \in \mathbb{N}$, with $\exp z^k$ showing the result sharp, see Pfluger (1946).)

The order of $f$ is the order of $1/(f - a)$, for each complex number $a$, since Nevanlinna's first theorem (11.3) ensures that their $T$ functions differ by only a bounded amount. Similarly, the lower orders of $f$ and $1/(f - a)$ are the same.

To conclude this section, we prove the Cartan identity (11.4). The Jensen formula (Lemma 11.1 with $z = 0$) when applied to the function $a - z$ with $R = 1$ gives

$$\frac{1}{2\pi}\int_{-\pi}^{\pi} \log |a - e^{i\theta}|\, d\theta = \log^+ |a|, \tag{11.11}$$

as can be checked by considering $a = 0, 0 < |a| < 1$ and $1 < |a| < \infty$. Another appeal to Jensen's formula, using the meromorphic function $f(z) - e^{i\theta}$ and afterwards replacing $R$ with $r$, gives

$$\log |f(0) - e^{i\theta}| = \frac{1}{2\pi}\int_{-\pi}^{\pi} \log |f(re^{it}) - e^{i\theta}|\, dt + N(r, \infty) - N(r, e^{i\theta})$$

provided $f(0) \neq e^{i\theta}, \infty$. We may integrate the last equality with respect to $\theta$ and apply (11.11). Then the definition of $T$ yields that

$$\begin{aligned}\log^+ |f(0)| &= \frac{1}{2\pi}\int_{-\pi}^{\pi} \log^+ |f(re^{it})|\, dt + N(r, \infty) - \frac{1}{2\pi}\int_{-\pi}^{\pi} N(r, e^{i\theta})\, d\theta \\ &= T(r,f) - \frac{1}{2\pi}\int_{-\pi}^{\pi} N(r, e^{i\theta})\, d\theta,\end{aligned}$$

which is Cartan's formula (11.4).

### The Function $T^\star$

If $f$ is meromorphic in $\mathbb{D}(R)$, we define (for $z = re^{i\theta} \in \mathbb{D}^\star(R)$)

$$T^\star(re^{i\theta}, f) = T^\star(z) = m^\star(z, f) + 2\pi N(r, f),$$

where $N(r,f) = N(|z|, f)$ was already defined in §11.2 and

$$\begin{aligned} m^\star(z,f) &= m^\star(z) = m^\star(re^{i\theta}) \\ &= \sup_E \int_E \log |f(re^{it})|\, dt, \end{aligned}$$

the supremum taken over all measurable sets $E \subset [0, 2\pi)$ of measure $2\theta$.

The reader should be aware that the definition given here for $T^\star$ equals $2/\pi$ times the author's original definition in Baernstein (1973). The definition here agrees with Baernstein (1974) and later works where the $\star$-function does not involve a factor of $1/2\pi$. See Notes 1 and 2 at the end of this chapter.

The key property in the next proposition is the subharmonicity of $T^{\star}$. The circular symmetrization operator $g \mapsto g^{\#}$ in part (iii) is the same as in §9.6 with $n = 2$.

**Proposition 11.2** *If $f$ is meromorphic in $\mathbb{D}(R)$ then*

(i) *$T^{\star}$ is continuous on the closure except possibly at the origin and subharmonic in $\mathbb{D}^{\star}(R) = \{re^{i\theta} : 0 < r < R, 0 < \theta < \pi\}$;*
(ii) *$T(r) = (1/2\pi) \sup_{0<\theta<\pi} T^{\star}(re^{i\theta})$ and*

$$T^{\star}(r) = 2\pi N(r, \infty), \qquad T^{\star}(-r) = 2\pi N(r, 0) + 2\pi \log |c|$$

*where $c$ is the leading coefficient in the series for $f$ at the origin, meaning $f(z) = cz^{\ell}(1 + O(z))$ for some integer $\ell$;*
(iii) *$T^{\star}(re^{i\theta}) = \int_{-\theta}^{\theta} (\log |f|)^{\#}(re^{it})\, dt + 2\pi N(r, \infty)$.*

*Proof* Corollary 9.11 says

$$\Delta(m^{\star}) \geq -2\pi \sum_k \frac{s_k}{|b_k|}$$

in the half disk $\mathbb{D}^{\star}(R)$, where the $\{b_k\}$ are the poles of $f$ in the punctured disk $\mathbb{D} \setminus \{0\}$, and $s_k$ is arclength measure on the circle of radius $|b_k|$. Meanwhile

$$\Delta N(r, \infty) = \sum_k \Delta \log^{+} \frac{r}{|b_k|} = \sum_k \frac{s_k}{|b_k|},$$

where the distributional Laplacian of $\log^{+} r/|b_k|$ was found as follows. Suppose $b \neq 0$ and $\eta(z)$ is smooth with compact support in $\mathbb{D}^{\star}(R)$, or indeed in the plane with the origin removed. By applying Green's formula on the region $(|z| > |b|)$,

$$\int (\Delta\eta) \left(\log^{+} \frac{r}{|b|}\right) r\, dr d\theta = \int_{(|z|=|b|)} \eta(z) \frac{\partial}{\partial r} \left(\log \frac{r}{|b|}\right) ds(z) = \int \eta \frac{ds}{|b|}.$$

Combining the above formulas gives $\Delta\big(m^{\star} + 2\pi N(r, \infty)\big) \geq 0$ in the half disk, which is subharmonicity of $T^{\star}$.

Parts (ii) and (iii) of the proposition follow readily from the definitions. The last claim in part (ii) is the special case $z = 0$ of the Poisson–Jensen Lemma 11.1, applied to $g(z) = r^{\ell} f(z)/z^{\ell}$ on the circle $\{|z| = r\}$, with $\ell$ chosen so that $g(0) \neq 0, \infty$, ensuring $|g| = |f|$ on $\{|z| = r\}$. This shows that the definition of $N(r,f)$ in §11.2 properly incorporates the case that $f(0) = \infty$. □

## 11.3 Pólya Peaks and the Local Indicator of $T^{\star}$

The definitions of the order and lower order in (11.9) and (11.10) refer to a meromorphic function $f$ defined in the entire plane. We need a concrete reformulation for computations. An effective approach has been to focus on $f$ in annuli centered at the sequence $\{r_n\}$ of Pólya peaks (of order $\lambda$) of $T(r,f)$. Such peaks may be associated to any increasing real function $S(r)$.

**Definition 11.3** Let $S(r)$ be an increasing function for $r > 0$. The function $S$ has Pólya peaks of order $\lambda$ ($0 < \lambda < \infty$) if there is a sequence of intervals $I(n) = \{r\colon |\log(r/r_n)| \le k_n\}$ on which

$$S(r) \le (1+\varepsilon_n)\left(\frac{r}{r_n}\right)^{\lambda} S(r_n),$$

where $r_n, k_n \to \infty$ and $\varepsilon_n \to 0$.

This definition was originally introduced only for $\lambda = \rho$ and $S(r) = T(r)$, but systematically exploited later primarily by Edrei. The notion of Pólya peaks makes sense for any $\lambda > 0$, although it should be clear from the definition that for a given $\lambda$, the function $S$ need not have Pólya peaks of that order, and when it does none of the sequences $(r_n)$, $(k_n)$, $(\varepsilon_n)$ are unique. The expression on the right side is a scaled copy of the simple function $r^{\lambda}$, with $r$ in a sequence of intervals $I(n)$ whose (logarithmic) lengths are unbounded as $n \to \infty$.

Theorem 11.4 below ensures that if the meromorphic function $f$ has order $\rho$ or lower order $\mu$, then there are peaks for $S(r) = T(r,f)$ of order $\lambda$ or $\mu$ whenever either of these lies in $(0,\infty)$.

**Theorem 11.4** (Existence theorem for Pólya peaks; Edrei, 1965) *Let $S(r)$ be an increasing function for which either the lower order $\mu$ or order $\rho$ are finite and nonzero. Then $S$ has Pólya peaks of order $\lambda$ on an interval $I$ with $[\mu,\rho] \subset I$, and $I$ is closed relative to $(0,\infty)$.*

Theorem 11.4 has been especially useful when considering properties of extremal functions, but we do not consider this here. Note that functions such as $S(r) = e^r$ or $S(r) = \log r$ do not have peaks of any order.

Pólya peaks also have a natural connection to $T^{\star}$, using the indicator function $h(\theta)$, which we develop here. The (Phragmén–Lindelöf) indicator was originally developed to study entire functions of finite order, with Cartwright (1956) and Levin (1980) among the excellent classical sources. The analytic property on which it is based is that if $f$ is entire, then the function

$$u(z) = \log|f(z)|$$

is subharmonic. This will reduce many properties of entire functions to elementary real analysis. The notion of "localizing" the indicator to Pólya peak intervals is due to Edrei (1970). The subharmonicity of $T^{\star}$ (in Proposition 11.2) enables the theory of the indicator to be extended to meromorphic $f$. This allows many applications to be obtained by a systematic procedure, a principle advanced by Rossi and Weitsman (1983).

Introducing the indicator may appear as a detour from traditional complex analysis, and does require some preparation. However, it makes many arguments more transparent and allows simpler formulations of many results. For example, contrast the statement of the spread relation in Proposition 11.7 below with the more cumbersome formulation in Theorem II in the Foreword. The procedure that produces the indicator is described shortly, and may be applied in more general situations to any subharmonic function defined in a sector. We consider only the indicator associated to $T^{\star}$ at a subsequence of Pólya peaks of $T(r)$.

As motivation, observe that the simplest functions in the $(r, \theta)$-plane have variables separate, and Pólya peaks already isolate the first variable. Thus, if we consider the test function

$$G(z) = r^{\lambda} g(\theta)$$

and assume $G$ is $C^2$, the Laplace operator has the simple form

$$\Delta(g) = r^{\lambda-2} s(\theta),$$

where $s$ depends purely on the factor $g$:

$$s(\theta) = g''(\theta) + \lambda^2 g(\theta). \tag{11.12}$$

In terms of the elliptic (non-homogeneous) equation

$$H''(\theta) + \lambda^2 H(\theta) = 0, \tag{11.13}$$

the indicator $h(\theta)$ is a subsolution to (11.13), and since the general solution of (11.13) is $H(\theta) = A \cos \lambda\theta$, we call $h$ subtrigonometric ($\lambda$-subtrigonometric). The geometric interpretation of this notion is discussed after Lemma 11.6.

The function $h$ arises using the following procedure. (Here only the indicator associated to the subharmonic function $T^{\star}(z)$ is considered, but the process may be applied to entire functions, since $\log|f(z)|$ is subharmonic, or to functions analytic in an angle.) Let $(r_n)$ be a sequence of Pólya peaks (of order $\lambda$) associated to $T(r)$. This yields an obvious family of comparison functions $\{V_n(r)\}$, by

$$V_n(r) = 2\pi T(r_n) \left(\frac{r}{r_n}\right)^{\lambda} \qquad (|\log(r/r_n)| \le k_n)$$

(note that in general the union of the domains of the $\{V_n\}$ is not an interval). For each $k > 1$ let $I(n,k)$ be the interval $|\log(r/r_n)| \le k$. We then define

$$h_k(\theta) = \limsup_{r\to\infty,\ r\in\cup_n I(n,k)} \frac{T^{\star}(re^{i\theta})}{V_n(r)}, \qquad h(\theta) = \lim_{k\to\infty} h_k(\theta),$$

for $0 \le \theta \le \pi$. This yields $h$ as the indicator of $T^{\star}$ (relative to the sequence $\{V_n(r)\}$). Note that the factor $2\pi$ was introduced in the definition of $V_n(r)$ to ensure that (iv) hold in the next lemma.

The main properties of the indicator are listed in the next lemma.

**Lemma 11.5** *The indicator $h$ of $T^{\star}$ relative to a sequence $(r_n)$ of Pólya peaks (of order $\lambda$) of $T^{\star}$ satisfies the following properties:*

(i) $h(\theta)$ is continuous on $0 \le \theta \le \pi$;
(ii) *given $\varepsilon > 0$ and $k > 1$, there exists $n_0 = n_0(k, \varepsilon)$ with*

$$T^{\star}(re^{i\theta}) \le (h(\theta) + \varepsilon)V_n(r) \qquad (\theta \in [0, \pi],\ r \in I(n,k), n \ge n_0);$$

(iii) *as a distribution,*

$$s(\theta) = h''(\theta) + \lambda^2 h(\theta)$$

*is a positive measure (and thus these derivatives of $h$ may be integrated);*
(iv) $\|h\|_{\infty} = 1$;
(v) $h(0) \le 1 - \delta(\infty),\ \ h(\pi) \le 1 - \delta(0)$;
(vi) *the (left or right) derivative of $h$ satisfies*

$$h'(\theta) \le 2 \limsup_{r\to\infty, r\in\cup I(n)} \frac{(\log|f|)^{\#}(re^{i\theta})}{V_n(r)};$$

*in addition, if $f$ is entire, then we have at the Pólya peaks*

$$h'(0) \ge 2 \limsup_{n\to\infty} \frac{\log M(r_n)}{V_n(r_n)}. \tag{11.14}$$

*Proof* Much of this lemma follows from the general theory of the indicator, supplemented by (ii) from Proposition 11.2; detailed arguments (which rely on the Pólya peaks inequalities) are in the references already cited. The first two statements of the lemma are immediate from general properties of the indicator, and while (iii) may be recovered from Levin (1980, p. 57), it also follows directly from the connection between $\Delta$ and $s$ in (11.12). It follows from (iii) that there is a representative with $h \in W^{1,\infty}$ of period $\pi$; in fact, $h$ is $C^1$ aside from $h'$ having at most countably many (positive) jump discontinuities.

Assertion (iv) follows at once from (ii) of Proposition 11.2, while (v) is also a consequence of (ii); there is only inequality because the lim inf which defines the Nevanlinna deficiency requires data for all large $r$, not just those $r \in \cup I(n)$.

We sketch the proof of the first inequality in (vi), postponing until §11.4.1 the verification of (11.14), since it uses notions which are extraneous at this point.

By definition, $h = \lim h_k$ and $h'$ is a limit of difference quotients, which we take to be the right derivative. Thus, given $\varepsilon > 0$ we may choose $\psi > \theta$ so that

$$h'(\theta) \le \frac{h(\psi) - h(\theta)}{\psi - \theta} + \varepsilon.$$

We chase definitions of the terms in the numerator. First if $K$ is so large that $h(\psi) \le h_k(\psi) + \varepsilon(\psi - \theta)$ when $k > K$, we may choose a subsequence $r_m$ of the peaks and $s_m \in [e^{-k} r_m, e^k r_m]$ with

$$h_k(\psi) \le \frac{T^{\star}(s_m e^{i\psi})}{V_m(s_m)} + \varepsilon_m,$$

and $\varepsilon_m \to 0$ $(m \to \infty)$. Also, for any $k$ and all $m$,

$$h(\theta) \ge h_k(\theta) \ge \frac{T^{\star}(s_m e^{i\theta})}{V_m(s_m)} - \eta_m \qquad (\eta_m \to 0,\ n \to \infty),$$

and (iii) of Proposition 11.2 gives for $\phi \in (-\pi, \pi)$ that

$$\frac{\partial T^{\star}}{\partial \phi}(re^{i\phi}) = 2(\log|f|)^{\#}(re^{i\phi}).$$

We combine these with the mean value theorem (with suitable $\phi \in (\theta, \psi)$) to write the initial inequality for $h'$ as

$$\begin{aligned} h'(\theta) &\le \frac{2(\log|f|)^{\#}(s_m e^{i\phi})}{V(s_m)} + \frac{\varepsilon_m + \eta_m}{\psi - \theta} + \varepsilon \\ &\le \frac{2(\log|f|)^{\#}(s_m e^{i\theta})}{V(s_m)} + \frac{\varepsilon_m + \eta_m}{\psi - \theta} + \varepsilon, \end{aligned}$$

where we could replace $\phi$ with the smaller value $\theta$ in the last inequality since $(\log|f(se^{i})|)^{\#}$ decreases in $t$. Take lim sup as $m \to \infty$ and then let $\varepsilon \to 0$. □

That $h$ is subtrigonometric is the main conclusion of the next lemma, and provides the key ingredient for the proofs which follow: the graph of a subtrigonometric function possesses an elegant geometric interpretation which parallels the characteristic property of real convex functions. That the process is being applied to $T^{\star}$ explains why its domain is the interval $0 \le \theta \le \pi$.

**Lemma 11.6** *Suppose $T^{\star}$ has Pólya peaks of order $\lambda > 0$, and the indicator $h$ is produced by the procedure described in the preamble to Lemma 11.5.*

*If $0 \le \theta_1 < \theta_2 < \theta_3 \le \pi$ with $\theta_3 - \theta_1 < \pi/\lambda$, then*

$$\begin{vmatrix} h(\theta_1) & \cos\lambda\theta_1 & \sin\lambda\theta_1 \\ h(\theta_2) & \cos\lambda\theta_2 & \sin\lambda\theta_2 \\ h(\theta_3) & \cos\lambda\theta_3 & \sin\lambda\theta_3 \end{vmatrix} \ge 0,$$

*where equality holds if $h$ is a trigonometric function:*

$$H(\theta) = a\cos\lambda\theta + b\sin\lambda\theta.$$

*This relation has a limiting form if we take, for example, $\theta_2 = \theta_1 + \delta$ and let $\delta \to 0$:*

$$h(\theta_1)\lambda\cos\lambda(\theta - \theta_1) + h'(\theta_1)\sin\lambda(\theta - \theta_1) \le h(\theta)$$

*provided $\theta < \theta_1 + \pi/\lambda$.*

The determinant inequality is an attractive way to display a complicated relation that may be exploited in several ways. For example, think of $\theta_2 = \theta$ as a variable, and solve for $h(\theta_2)$ The other side is the trigonometric function of order $\lambda$ with $h(\theta_1)$ and $h(\theta_3)$ given, and so controls $h(\theta_2)$. The restriction that the intervals considered have length at most $\pi/\lambda$ arises from the Sturm–Liouville theory applied to (11.13). The inequality is often rephrased as stating that $h$ is trigonometrically convex, or subtrigonometric.

In practice, the lemma may be summarized by a comparison principle: *if $\alpha < \beta, \alpha < \theta$ with $\max(\beta - \alpha, \theta - \alpha) < \pi/\lambda$, and $H$ is a sinusoid (of order $\lambda$) having*

$$h(\alpha) \le H(\alpha), \quad h(\beta) \le H(\beta),$$

*then*

$$\begin{aligned} h(\theta) &\le H(\theta) \text{ if } \theta \in (\alpha, \beta); \\ h(\theta) &\ge H(\theta) \text{ if } \theta \in (\beta, \alpha + \pi/\lambda). \end{aligned} \tag{11.15}$$

An obvious corollary of this is that if $h$ and $H$ agree at $\alpha$ or $\beta$, then

$$h'(\alpha) \le H'(\alpha) \text{ or } h'(\beta) \ge H'(\beta)$$

as appropriate.

The limiting case of (11.15), $\beta \to \alpha$, is: *if $H$ is sinusoidal (of order $\lambda$) and*

$$H(\alpha) = h(\alpha), \; H'(\alpha) = h'(\alpha), \tag{11.16}$$

*then*

$$H(\theta) \le h(\theta) \qquad (|\theta - \alpha| < \pi/\lambda). \tag{11.17}$$

The proof of Lemma 11.6 uses the Pólya peak inequalities and the maximum principle. For example, if

$$\Theta: \ = \theta_3 - \theta_1 < \pi/\lambda,$$

let $H(\theta) = A \cos \lambda(\theta - \tau)$, with $A, \tau$ chosen so $H(\theta_1) = h(\theta_1), H(\theta_3) = h(\theta_3)$. For fixed $\varepsilon > 0$ and $\widehat{\theta} = (\theta_1 + \theta_2)/2$, let

$$V(z) = V(re^{i\theta}) = r^\lambda \left[h(\theta) - (H(\theta) + 2\varepsilon \cos \lambda(\theta - \widehat{\theta}))\right]$$

in each region

$$\Omega = \Omega_n = \{z\colon r_n/M \leq |z| \leq Mr_n,\ \theta_1 \leq \theta \leq \theta_3\},$$

where the $r_n$ are the Pólya peaks of $T^\star$ and $M$ is fixed but, according to Definition 11.3, may be taken arbitrarily large.

Note that $V$ is subharmonic because $r^\lambda h(\theta)$ is subharmonic (using Proposition 11.5(iii)) while $r^\lambda H(\theta)$ and $r^\lambda \cos \lambda(\theta - \widehat{\theta})$ are harmonic. Hence by the maximum principle, $V$ is bounded on each $\Omega$ by the harmonic extension (Poisson integral) of its values on $\partial\Omega$. Each $\Omega$ is a wedge-like region, whose boundary consists of two long radial segments and arcs of two circles. The Pólya peak property ensure that when $n$ is large, relative to points on $\{|z| = r_n\}$, $\Omega$ will appear to be a near-angular sector.

First, since $h = H$ at $\theta_1, \theta_2$ the definition of $\widehat{\theta}$ ensures that on the radial segments, if $M$ is fixed (but large)

$$V(re^{i\theta}) \leq -2\varepsilon r^\lambda \cos \lambda\Theta/2 \qquad (r_n/M \leq r \leq Mr_n, \theta = \theta_1, \theta_2)$$

for some fixed $\varepsilon$ which depends only on $\theta$. On the portion of $\partial\Omega$ composed of the two circular arcs, there is less precise information; however, since $H$ and $h$ are bounded,

$$V(re^{i\theta}) \leq Kr^\lambda, \qquad (r = r_n/M, Mr_n; \theta_1 \leq \theta \leq \theta_2),$$

for some constant $K$.

To deduce from these data that

$$V(r_n e^{i\theta}) \leq 0 \qquad (\theta_1 \leq \theta \leq \theta_2)$$

(thus pointwise on the "central arc" of $\Omega$) is where the key condition $\Theta < \pi/\lambda$ is essential. According to Lemma 9.12 of Hayman (1989)

$$V(r_n e^{i\theta}) \leq r_n^\lambda(\epsilon(M) - 2\varepsilon \cos \lambda\Theta/2) \qquad (\theta_1 \leq \theta \leq \theta_2), \tag{11.18}$$

where $\epsilon(M)$ incorporates the boundary values on the two circular arcs and $\epsilon(M) \to 0$ as $M \to \infty$. Since (11.18) is valid for any $\varepsilon > 0$, we deduce that

$$h(\theta) \leq H(\theta) \qquad (\theta_1 \leq \theta \leq \theta_2),$$

and if equality holds for any $\theta \in (\theta_1, \theta_2)$, the maximum principles yields that $h = H$ on the entire interval.

Note that a sinusoid (of order $\lambda$) has the general form

$$H(\theta) = A\cos\lambda(\theta - \theta_0),$$

but the additional properties (iv) and (vi) usually ensure that $A = 1$, as we will see below, for example in one case of Proposition 11.10. This ambiguity is not present in the limiting situation (11.16).

## 11.4 Applications of $T^{\star}$ to Nevanlinna Theory

We follow the analysis of Rossi and Weitsman (1983) to obtain several key results as consequences of a general method. The definitions in §11.2 of the deficiency $\delta(a) = \delta(a,f)$ and of $m(r,a)$ ensure that when $\delta(a) > 0$, the quantity $|f(re^{i\theta}) - a|$ is small on a significant $\theta$-set for large $r$ (with an analogous interpretation when $a = \infty$). The first application of the $\star$-function makes these heuristics precise (for sharpness, see Baernstein, 1973).

**Proposition 11.7** (Spread relation; Baernstein, 1973) *Let $f$ be meromorphic in the plane with $\delta = \delta(\infty) > 0$, and suppose that $S(r) = T(r,f)$ has Pólya peaks of order $\lambda$, $0 < \lambda < \infty$. Then*

$$\liminf_{n\to\infty} \left|\{\theta:\ \log|f(r_n e^{i\theta})| > 0\}\right| \geq \min\left\{2\pi, \frac{4}{\lambda}\sin^{-1}\sqrt{\frac{\delta}{2}}\right\}. \tag{11.19}$$

*Proof* Choose $\{\beta_n\}$ with $(1/2\pi)T^{\star}(r_n e^{i\beta_n}) = T(r_n)$. According to the definition of $T^{\star}$,

$$\left|\{\theta:\ \log|f(r_n e^{i\theta})| > 0\right| = 2\beta_n,$$

so that (taking a subsequence which we relabel as $\{\beta_n\}$) we need only show

$$\beta = \liminf \beta_n \geq \min\{\pi, (2/\lambda)\sin^{-1}\sqrt{\delta/2}\}. \tag{11.20}$$

If $\beta = \pi$, there is nothing to prove, so we suppose $\beta < \pi$ and also that $\delta > 0$. We then claim that the indicator $h$ attains its maximum at the interior point $\theta = \beta$. Since $\delta > 0$, assertion (v) yields that $h(0) \leq 1 - \delta$. On the other hand, since $h$ is continuous, $|h| \leq 1$, and $\beta_n \to \beta$, we have from conclusion (ii) that

$$T(r_n) = \frac{1}{2\pi}T^{\star}(r_n e^{i\beta_n}) \leq (T(r_n) + \varepsilon)h(\beta),$$

so letting $n \to \infty$ and then $\varepsilon \to 0$ gives that $h(\beta) = 1$. Properties (iii) and (iv) then show that $h'(\beta)$ must exist, and hence $h'(\beta) = 0$.

The sinusoid $H(\theta) = \cos\lambda(\theta - \beta)$ has the same value and derivative as $h$ at $\beta$. The comparison principle (11.17) now implies that

$$\cos\beta\lambda = H(0) \leq h(0) \leq 1 - \delta.$$

Thus, from the half-angle formula,

$$\sin^2\frac{\beta\lambda}{2} = \frac{1-\cos\beta\lambda}{2} \geq \frac{\delta}{2}, \tag{11.21}$$

which finishes the proof. □

The spread relation was anticipated much earlier by Teichmüller (1939), but proposed independently by Edrei, incorporating the additional feature that it is realized at the Pólya peaks.

Recall that the definition of deficiency (11.5) involves division by $T(r)$ when computing the lim inf. Thus the significance of the hypothesis $\delta(\infty) > 0$ is greater than might appear at first glance; see for example the phrasing of Theorem II in the Foreword. Baernstein (1973) defines $\delta(\infty)$ by considering a function $\Lambda(r) \to \infty$ with $\Lambda(r) = o(T(r))$ and the lim inf of measures of sets

$$\{\theta : |f(re^{i\theta})| > e^{\Lambda(r)}\} \qquad (r \to \infty).$$

This means that if $a \neq \infty$, and we replace $f$ by $1/(f - a)$ and obtain an analogous relation for $w = a$, these $\theta$-sets associate to distinct $a$ will be asymptotically disjoint.

Inequality (11.19) gives a lower bound for the **spread** of the value $w = \infty$ (or $w = a$) in terms of $\lambda$ and the value $\delta(\infty)$. ($\delta(a)$) The spread is the number $\beta$ defined by the left side of inequality (11.20).

For completeness, we sketch Edrei's striking application of the spread relation, which in turn sharpens (11.6) when we are given that $f$ has (lower) order $\mu < 1$:

**Corollary 11.8** (Edrei, 1973) *If $f$ has lower order $\mu \in (0, 1)$ then*

$$\sum_{a\in\widehat{\mathbb{C}}} \delta(a,f) \leq \begin{cases} 1 & 0 < \mu \leq 1/2, \\ 2 - \sin\pi\mu & 1/2 \leq \mu < 1. \end{cases}$$

*Proof* We may assume $f$ has more than one deficient value, since otherwise the corollary follows immediately from $\delta(a) \leq 1$. List the deficiencies as $\delta_1, \delta_2, \delta_3, \ldots > 0$. Each $\delta_j$ corresponds to a value $w = a_j$ with $\delta(a_j,f) = \delta_j$, and we associate to each $a_j$ its spread $\beta_j$. Recall from Section 11.2 that $1/(f - a)$ has the same lower order $\mu$ as $f$, for each complex value $a$.

Applying the spread relation from Proposition 11.7 to $1/(f - a_j)$ (with $\lambda = \mu$ the lower order) ensures that $\beta_j > 0$ for each $j$. Informally, the spread means that when $n$ is large, $|f - a_j|$ is exponentially small on a set of angular measure about $2\beta_j$ on the circle $S(r_n)$. We have mentioned that these sets are (asymptotically ) disjoint, and so

$$\sum_j \beta_j \leq \pi .$$

Since $\beta_j > 0$ and the sum has at least two terms ($f$ has at least two deficient values), we conclude $\beta_j < \pi$ for each $j$. Thus by (11.21),

$$\sin^{-1}\sqrt{\delta_j/2} \leq \beta_j\mu/2.$$

Upon summing over $j$, the right side is at most $\pi\mu/2$. Corollary 11.8 now follows from an elementary extremal problem, in the next lemma. □

**Lemma 11.9** *Let $\mu \in (0, 1]$ and $m \geq 1$. If $0 \leq \delta_1, \ldots, \delta_m \leq 1$ with*

$$\sum_{j=1}^{m} \sin^{-1}\sqrt{\delta_j/2} \leq \pi\mu/2$$

*then*

$$\sum_{j=1}^{m} \delta_j \leq \begin{cases} 1 & 0 < \mu \leq 1/2, \\ 2 - \sin\pi\mu & 1/2 \leq \mu \leq 1. \end{cases}$$

*Proof* First, let $f(s) = \sin^2 s + \sin^2(C - s)$ where $0 \leq C \leq \pi/2$. Trigonometric identities show that $f(s) = 1 - \cos C \cos(2s - C)$, and so $f(s)$ is increasing when $C/2 \leq s \leq C$. Letting $t = C - s$, we deduce that if $0 \leq t \leq s \leq \pi/4$ then

$$\sin^2 s + \sin^2 t \leq \begin{cases} \sin^2(s+t) & s+t \leq \pi/4, \\ \sin^2(\pi/4) + \sin^2(s+t-\pi/4) & s+t \geq \pi/4. \end{cases} \tag{11.22}$$

This inequality will be used repeatedly in the proof.

Write $\tau_j = \sin^{-1}\sqrt{\delta_j/2}$. Then $\tau_j \leq \pi/4$ for each $j$, and $\sum_j \tau_j \leq \pi\mu/2$. When $m = 1$ the lemma follows directly from $\delta_1 \leq 1$. When $m = 2$,

$$\delta_1 + \delta_2 = 2\sin^2\tau_1 + 2\sin^2\tau_2 \leq \begin{cases} 2\sin^2(\tau_1+\tau_2) & \tau_1+\tau_2 \leq \pi/4, \\ 1 + 2\sin^2(\tau_1+\tau_2-\pi/4) & \tau_1+\tau_2 \geq \pi/4, \end{cases}$$

by inequality (11.22). Hence

$$\delta_1 + \delta_2 \leq \begin{cases} 1 & \tau_1+\tau_2 \leq \pi/4, \\ 2 - \sin(\pi\mu) & \tau_1+\tau_2 \geq \pi/4, \end{cases}$$

by using $\tau_1 + \tau_2 \le \pi\mu/2$ and $1 + 2\sin^2(\theta - \pi/4) = 2 - \sin(2\theta)$. Note if $0 < \mu \le 1/2$ then $\tau_1 + \tau_2 \le \pi/4$, while for any $\mu$ one has $1 \le 2 - \sin(\pi\mu)$. Thus the lemma is proved when $m = 2$.

For $m \ge 3$ we proceed by induction. From inequality (11.22) we see (depending on whether $\tau_1 + \tau_2$ is smaller or larger than $\pi/4$) that the sum $\delta_1 + \delta_2$ increases when we replace the pair $(\tau_1, \tau_2)$ with $(\tau_1 + \tau_2, 0)$ or with $(\pi/4, \tau_1 + \tau_2 - \pi/4)$. In the first case, one $\tau$-value has become 0, meaning only $m - 1$ values of $\delta$ need be summed, so that the desired inequality follows from the induction hypothesis.

The point of the second case is that we may assume $\tau_1 = \pi/4$. Then $\tau_2 + \tau_3 \le \pi/4$ (recalling $\sum_j \tau_j \le \pi\mu/2 \le \pi/2$), and so inequality (11.22) implies that the sum $\delta_2 + \delta_3$ increases when the pair $(\tau_2, \tau_3)$ is replaced with $(\tau_2 + \tau_3, 0)$. Having reduced the number of $\delta$-values to be summed, we again call on the induction hypothesis to complete the proof of Lemma 11.9. □

When Nevanlinna introduced his theory, he made several conjectures which turned out to be overly optimistic (Fuchs, 1967, §4). Our next result gives a short proof of what appears to be the first of these conjectures to be confirmed, albeit decades later.

**Proposition 11.10** (Paley conjecture, Govorov theorem; Govorov, 1969; Petrenko, 1969) *Let $f$ be entire with Pólya peaks $r_n$ for $T(r)$ of order $\lambda, 0 < \lambda < \infty$. Then*

$$\liminf_{n\to\infty} \frac{T(r_n)}{\log M(r_n)} \ge \begin{cases} (\sin\pi\lambda)/\pi\lambda, & 0 < \lambda \le \frac{1}{2}, \\ 1/\pi\lambda, & \lambda > \frac{1}{2}. \end{cases}$$

*Proof* Since $f$ is entire, $h(0) = 0$, and $\|h\|_\infty = 1$. First, take $\mu \ge 1/2$. The sinusoid function $H(\theta) = \sin\lambda\theta$ agrees with $h$ at 0 and dominates $h$ at $\theta = \pi/(2\lambda)$. Thus $H \ge h$ and so $h'(0) \le H'(0) \le \lambda$.

When $\lambda < 1/2$, we consider $H(\theta) = \sin\lambda\theta \cdot \csc\pi\lambda$, which agrees with $h$ at 0 and dominates $h$ at $\theta = \pi$. Thus, (11.14) of Lemma 11.5 yields

$$\limsup_n \frac{\log M(r_n)}{\pi V(r_n)} = \limsup_n \frac{\log M(r_n)}{\pi T(r_n)} \le h'(0) \le H'(0) = \lambda\csc\pi\lambda,$$

which completes the proof. □

The next result is from Edrei and Fuchs (1960):

**Proposition 11.11** (Ellipse theorem) *Let $f$ be meromorphic with peaks of order $\lambda \in (0, 1)$ and set*

$$u = 1 - \delta(0), \qquad v = 1 - \delta(\infty).$$

*Then in addition to the obvious restrictions* $0 \le u, v \le 1$, *we have*

$$\sin^2 \pi\lambda \le u^2 + v^2 - 2uv\cos\pi\lambda.$$

The region of possible values $(u, v)$ is illustrated in chapter 5 of Goldberg and Ostrovskii (2008). It is the portion of the first quadrant inside the square $Q\colon -1 \le u, v \le 1$ which lies outside the ellipse

$$E\colon\ u^2 + v^2 - 2uv\cos\pi\lambda = \sin^2\pi\lambda.$$

This ellipse is symmetric about the lines $u = \pm v$, and meets $\partial Q$ at the points $(1, \cos\pi\lambda), (\cos\pi\lambda, 1)$. Thus when $\lambda < 1/2$, the major axis of $E$ is a subset of the line $u = v$, while when $\lambda > 1/2$ the major axis is contained in the line $u = -v$.

*Proof* Let $2\beta$ be the spread for $f$ (i.e., for $w = \infty$). We only consider the possibility $0 < \beta < \pi$ here. Since $h$ is subtrigonometric and attains its maximum at $\beta$, we know that $h(\beta) = 1,\ h'(\beta) = 0$.

Property (ii) of $T^\star$ (in Proposition 11.2) gives $h(\pi) \le 1 - \delta(0) = u,\ h(0) \le 1 - \delta(\infty) = v$. The subtrigonometric property of $h$ will complement these to

$$h(\pi) \ge \cos\lambda(\pi - \beta) \qquad h(0) \ge \cos\lambda\beta. \tag{11.23}$$

Certainly, $h$ is dominated by $H(\theta) = \cos\lambda(\theta - \beta)$ when $\theta = \beta$. Thus, by (11.17), $H$ also dominates $h$ at $\theta = 0, \pi$, and (11.23) follows.

Trigonometry thus yields that

$$\begin{aligned}\sin\lambda\pi &\le u\sin\lambda\beta + v\sin\lambda(\pi - \beta)\\ &= (u - v\cos\lambda\beta)\sin\lambda\beta + v\sin\lambda\pi\cos\lambda\beta.\end{aligned}$$

When this inequality is squared, the proposition follows from Schwarz's inequality. □

Let us now return to the situation of the spread theorem, where (11.19) provides precise information on the size of the set

$$\{\theta\colon\ \log|f(re^{i\theta})| > 0\}$$

where $|f| > 1$, at least when $r = r_n$, a Pólya peak of order $\lambda$ of $T(r)$. This raises the possibility of replacing 0 in the definition of the set by $\alpha T(r)$; here it is necessary to include the factor $T(r)$, which is gratuitous when $\alpha = 0$. The situation $\alpha > 0$ is covered by the next result, but, as discussed in the preamble to Proposition 11.13, the case $\alpha < 0$ is less likely to have a simple answer.

**Proposition 11.12** (Anderson and Baernstein, 1978) *Suppose $f$ is meromorphic and has Pólya peaks of order $\lambda \in (0, \infty)$ with deficiency $\delta(\infty) = \delta \in (0, 1]$, and let $\alpha \in (0, \delta)$. Let $\sigma_0 > 0$ be the least positive solution to the equation*

$$\cos\lambda\sigma_0 = 1 - \delta,$$

*and for $x \in [0, \sigma_0]$ define $\varphi(x)$ by*

$$\varphi(x) = \frac{\pi\lambda(\cos\lambda x - (1-\delta))}{\sin\lambda x + \lambda(\pi - x)\cos\lambda x}.$$

*Then*

$$\limsup_{r\to\infty} \left|\{\theta : \log|f(re^{i\theta})| > \alpha T(r,f)\}\right| \geq \min(2\beta, 2\pi), \tag{11.24}$$

*where $r \in I(n)$ and $\beta$ is the smallest positive solution to $\alpha = \varphi(\beta)$.*

The left side of (11.24) can be thought of as (twice) the "$\alpha$-spread" of $f$. When $\alpha = 0$, this becomes Proposition 11.7.

*Proof* We only consider the case $\beta < \pi$. Let $J(r,\alpha)$ be the interval where $(\log|f|)^{\#}(re^{i\theta}) > \alpha V(r)$, and define $\beta$ by

$$\beta = \frac{1}{2}\limsup\nolimits_{r\to\infty,\ r\in I(n)} |J(r,\alpha)|,$$

where, as in the construction of the indicator $h$, $I(n)$ is an interval about $r_n$ whose logarithmic length suitably tends to $\infty$. Since we are using decreasing rearrangements, this means that for $\beta' > \beta$, $|\beta' - \beta|$ small,

$$(\log|f|)^{\#}(re^{i\theta}) \leq \alpha V(r)$$

when $\theta$ is outside $[0, \beta']$ and $r > r_0, r \in \cup I(n)$.

Again, we rely on properties of the indicator $h$. Here, the (indeterminate) point $\beta'$ provides first-order data, which with care will allow use of (11.17). Since $T^{\star}$ uses the decreasing rearrangement of $\log|f(re^{i\theta})|$, we also have that $h'(\theta) \leq \alpha/\pi$ for $\beta' \leq \theta \leq \pi$. But since $\alpha > 0$, $\|h\|_\infty = 1 = h(\gamma)$ for some $\beta' \leq \gamma \leq \pi$. Thus

$$1 - h(\beta') \leq (\gamma - \beta')\frac{\alpha}{\pi} \leq (\pi - \beta')\frac{\alpha}{\pi}.$$

We want to apply the string of inequalities below at $\theta = 0$, i.e., for $\theta - \beta' > -\pi/\lambda$:

$$\begin{aligned} h(\theta) &\geq h(\beta')\cos\lambda(\theta - \beta') + h'(\beta')\frac{\sin\lambda(\theta-\beta')}{\lambda} \\ &\geq \left[1 - (\pi - \beta')\frac{\alpha}{\pi}\right]\cos\lambda(\theta - \beta') + \frac{\alpha}{\pi}\frac{\sin\lambda(\theta - \beta')}{\lambda}. \end{aligned}$$

According to (11.17), the first inequality is valid since the right side is the $\lambda$-trigonometric function which agrees with $h$ to first order at $\beta'$. The second line is a $\lambda$-trigonometric function $H$ with

$$H(\beta') \geq h(\beta'),\ \ H'(\beta') \geq h'(\beta').$$

Since $h(0) \le 1 - \delta$ (via (v) of Lemma 11.5),

$$\alpha \ge \frac{\pi\lambda(\cos\lambda\beta' - (1-\delta))}{\sin\lambda\beta' + \lambda(\pi - \beta')\cos\lambda\beta'},$$

and the proposition follows on letting $\beta' \to \beta$. □

We close with three results. The first is not sharp and it applies under only restricted conditions; unlike the situation in Proposition 11.12, when $\alpha < 0$ one cannot extend $h$ outside an interval of length $2\beta$ in any natural way other than in very special situations.

**Proposition 11.13** (Essén et al. (1993)) *Let $\sigma_0$ be as in Proposition 11.12, but now consider the function*

$$\psi(x) = \pi\lambda\sin\lambda(\sigma_0 - x) \qquad (x \in [\sigma_0, \sigma_0 + \pi/2\lambda],\ x \le \pi).$$

*With $f$, $\lambda$, and $\delta$ as in the previous proposition, let $\alpha \in [-\pi\lambda, 0)$. Then* (11.24) *holds with $\beta$ the smallest positive solution to $\alpha = \psi(\beta)$.*

(Note that $\psi(\sigma_0) = 0$ and that $\psi$ strictly decreases in the relevant range. Also, this proposition together with Propositions 11.7 and 11.12 consider the spreads for all $\alpha$, with this being the least satisfactory.)

*Proof* Recall that $h(0) \le 1 - \delta = \cos\lambda\sigma_0$. Since $\beta < \sigma_0 + \pi/2\lambda < \pi/\lambda$, (11.17) may be used to link $h(0)$ with $h(\beta)$. Thus

$$h(\beta)\cos\lambda\beta - h'(\beta)\frac{\sin\lambda\beta}{\lambda} \le h(0) \le \cos\lambda\sigma_0.$$

Choose $\sigma$ to be the least solution to $h(\theta) = 1$. If $\sigma = \pi$ the left side of (11.24) is $2\pi$ for any $\alpha < 0$, in which case the proposition is obvious. According to the spread theorem, $\sigma_0 \le \sigma < \pi$. Thus, the scheme for (11.17) controls $h$ at $\beta$ and $\sigma$, and so

$$\cos\lambda(\beta - \sigma_0) < \cos\lambda(\beta - \sigma) \le h(\beta),$$

which we interpolate in the first estimate:

$$\begin{aligned} h'(\beta) &\ge \lambda\frac{h(\beta)\cos\lambda\beta - \cos\lambda\sigma_0}{\sin\lambda\beta} \\ &\ge \lambda\frac{\cos\lambda(\beta-\sigma_0)\cos\lambda\beta - \cos\lambda\sigma_0}{\sin\lambda\beta} \\ &\ge \lambda\frac{\cos\lambda\sigma_0(\cos^2\lambda\beta - 1) + \sin\lambda\beta\sin\lambda\sigma_0\cos\lambda\beta}{\sin\lambda\beta} \\ &= \lambda(-\cos\lambda\sigma_0\sin\lambda\beta + \sin\lambda\sigma_0\cos\lambda\beta) = \lambda\sin\lambda(\sigma_0 - \beta). \end{aligned} \tag{11.25}$$

The result follows from the first assertion of (vi) of Lemma 11.5. □

The next result was conjectured by Anderson and Baernstein (1978), and proved in Essén et al. (1993); the derivation here relies on the indicator.

**Corollary 11.14** *In the previous proposition, suppose in addition that* $\lambda < 1/2$ *and* $\delta < 1 - \cos \pi\lambda$. *Then*

$$\limsup_{r\to\infty} \left[ \inf_{\theta} \frac{\log |f(re^{i\theta})|}{T(r)} \right] \geq -\pi\lambda \left( (1-\delta)\sin \pi\lambda - \sqrt{\delta(2-\delta)}\cos \pi\lambda \right).$$

In the previous proposition, take $\pi$ in place of $\beta$ in (11.25); this is permitted since $\lambda < 1/2$. Since $\cos \lambda\sigma_0 = 1 - \delta$, the analysis then yields that

$$\begin{aligned} h'(\pi) &\geq \lambda \frac{\cos \pi\lambda \cos \lambda(\pi - \sigma_0) - (1-\delta)}{\sin \pi\lambda} \\ &= -\lambda((1-\delta)\sin \pi\lambda - \cos \pi\lambda \sin \lambda\sigma_0) \\ &= -\lambda \left( (1-\delta)\sin \pi\lambda - \cos \pi\lambda \sqrt{1-(1-\delta)^2} \right) \\ &= -\pi\lambda \left( (1-\delta)\sin \pi\lambda - \sqrt{\delta(2-\delta)}\cos \pi\lambda \right). \end{aligned}$$

An appeal to (vi) in Lemma 11.5 again completes the argument.

The final result of this section is from Fuchs (1974).

**Corollary 11.15** *In the previous proposition suppose in addition that* $1/2 < \lambda < 1$ *and* $1 - \sin \pi\lambda < \delta$. *Then*

$$\limsup_{r\to\infty} \left[ \inf_{\theta} \frac{\log |f(re^{i\theta})|}{T(r)} \right] \geq \pi\lambda \sin \lambda(\sigma_0 - \pi).$$

Replace $\beta$ by $\pi$ in (11.25) and once more use (vi) from Lemma 11.5.

### 11.4.1 Proof of (11.14) in Lemma 11.5

Here is the promised proof of the final claim (11.14) of Lemma 11.5. Note that this inequality is realized precisely at the peaks of $T^{\star}$. The argument has two steps. The first is the more formal, choosing $k > 1$ and manipulating definitions. The second, Lemma 11.16, is a reformulation of Fuchs' ingenious application of the Poisson–Jensen formula, which relies also on an elegant argument of Cartan.

The definition of derivative will yield a function $\Psi(r), 0 < \Psi(r) \to 0$, such that for each fixed $k$,

$$h'(0) \geq 2 \limsup_{r\to\infty} \frac{(\log |f|)^{\#}(re^{i\Psi(r)})}{V_n(r)} - o(1) \quad (r \in \cup I(n,k)), \tag{11.26}$$

where $J_n$ is the interval $|\log(r/r_n)| \leq k$. The Lemma sharpens (11.26) to hold at $r = r_n$ with $\Psi(r) = 0$. This precision depends on an estimate from Fuchs (1963), which we reformulate and prove as (11.27). It is important in many other situations, and we include the elegant and general counting argument of Cartan in Lemma 11.17, which is a key component.

In order to obtain (11.26), given $\varepsilon > 0$, the definition of derivative produces $\theta_0 > 0$ with

$$h'(0) \geq \frac{h(\theta)}{\theta} - \varepsilon \qquad (0 < \theta < \theta_0)$$

(recall that $h(0) = 0$). Item (ii) of Lemma 11.5 shows that

$$h(\theta) \geq T^\star(re^{i\theta})/V_n(r) - \varepsilon_n$$

uniformly in $\theta$ as $r \to \infty$, $r \in I(n,k)$ for any fixed $k$. Thus if $\theta_n \to 0$ so that $\varepsilon_n/\theta_n \equiv \varepsilon'_n \to 0$, we have from (iii) of Proposition 11.2 and the mean-value theorem a function $\Psi(r)$, $\Psi(r) \leq \theta_n$, with

$$h'(0) \geq 2\frac{(\log|f|)^\#(re^{i\Psi(r)})}{V_n(r)} - \varepsilon'_n - \varepsilon_n = 2\frac{(\log|f|)^\#(re^{i\Psi(r)})}{V_n(r)} - o(1),$$

for $r \in I(n,k)$ and large enough $n$. This yields (11.26).

The advance of (11.14) over (11.26) is to replace $(\log|f|)^\#(re^{i\Psi(r)})$ by $\log M(r_n,f) = (\log|f|)^\#(r_n)$, with error of the same nature. Although $(\log|f|)^\#$ is a decreasing rearrangement in $\theta$, we need to find a thick set of $r \in \cup I(n,k)$ where its decrease is controlled.

**Lemma 11.16** *Let $h$ be meromorphic in the plane with characteristic $T(r) = T(r,h)$, and $(r_n)$ a sequence of its peaks of order $0 < \lambda < \infty$. Given $1 < \sigma \leq 2$, there is $K = K(\lambda,\sigma)$ such that if $J$ is a $\theta$-interval of length $\delta < 1/2$, then*

$$r\int_J \frac{|h'(re^{i\theta})|}{|h(re^{i\theta})|}\,d\theta \leq \left(K\delta\log\frac{1}{\delta}\right)T(r,h) \tag{11.27}$$

*for large $n$ and*

$$r \in [r_n, \sigma r_n] \setminus E_n,$$

*where the exceptional set $E_n$ has Lebesgue measure $|E_n| \leq (\sigma - 1)r_n/2$.*

Let us accept this Lemma for the moment and use it to complete the proof of (11.14); the lemma itself is verified in the next subsection.

Given $\eta > 0$, choose $0 < \delta < 1/2$ with $K\delta\log(1/\delta) < \eta$. Let $1 < \sigma \leq 2$. For each $r$ choose $\alpha = \alpha(r)$ with $|f(re^{i\alpha})| = M(r)$. If $|\Psi(r)| < \delta$ then there is an angle $\phi = \phi(r)$ with $|\phi| < \delta$ such that $(\log|f|)^{\#}(re^{i\Psi(r)}) = \log|f(re^{i(\alpha+\phi)})|$. So

$$\begin{aligned}\log M(r) &= \log|f(re^{i\alpha})| - \log|f(re^{i(\alpha+\phi)})| + (\log|f|)^{\#}(re^{i\Psi(r)}) \\ &\leq r\int_{\alpha}^{\alpha+\phi}\left|\frac{f'}{f}\right| d\theta + (\log|f|)^{\#}(re^{i\Psi(r)}) \\ &\leq \eta T(r) + (\log|f|)^{\#}(re^{i\Psi(r)}), \qquad r \in [r_n, \sigma r_n] \setminus E_n,\end{aligned}$$

by Lemma 11.16 with $f$ in place of $h$. If this is incorporated in (11.26), we obtain that

$$h'(0) \geq 2\frac{\log M(r)}{V_n(r)} - o(1) - 2\eta\frac{T(r)}{V_n(r)} \quad (r \notin E_n). \tag{11.28}$$

Now take $s = s_n \in [r_n, \sigma r_n]$, $s \notin E_n$. Then the last estimate also holds at $r = s$, so that since $M(r)$ increases and $r_n$ is a Pólya peak of $T$, (11.28) becomes

$$\begin{aligned}h'(0) &\geq 2\frac{\log M(s) - \eta T(s)}{V_n(s)} - o(1) \\ &\geq 2\frac{\log M(r_n) - \eta(s/r_n)^{\lambda}T(r_n)}{V_n(s)} - o(1).\end{aligned}$$

The ratio $s/r_n$ may be taken arbitrarily close to 1 by choosing $\sigma - 1$ appropriately small, and since $\eta > 0$ is arbitrary, this completes the proof of (11.14).

### Proof of the Fuchs Lemma 11.16

The starting point is differentiation of the Poisson–Jensen formula, Lemma 11.1. Thus for $R > 1$ and $|z| < R$,

$$\begin{aligned}\frac{h'}{h}(z) = \frac{1}{2\pi}\int_{-\pi}^{\pi}\frac{2Re^{it}}{(Re^{it} - z)^2}\log|h(Re^{it})|\,dt + \sum_{|\alpha_\mu|<R}\frac{R^2 - |\alpha_\mu|^2}{(R^2 - \overline{\alpha}_\mu z)(z - \alpha_\mu)} \\ - \sum_{|\beta_\nu|<R}\frac{R^2 - |\beta_\nu|^2}{(R^2 - \overline{\beta}_\nu z)(z - \beta_\nu)} \qquad (|z| < R)\end{aligned}$$

with the $\alpha, \beta$ being the zeros and poles of $h$. Note that

$$\begin{aligned}\frac{1}{2\pi}\int_{-\pi}^{\pi}\big|\log|h(Re^{it})|\big|\,dt &= m(R,0) + m(R,\infty) \\ &\leq T(R,0) + T(R,\infty) \leq 2T(R) + O(1).\end{aligned}$$

Thus if $|z| < r < R$ and the $\{c_k\}$ are the zeros and poles of $h$ in $B(R)$, routine estimates and the fact that $T(R) \to \infty$ yield for all large $r$ that

$$\begin{aligned} r\left|\frac{h'}{h}\right| &\leq 5T(R)\frac{Rr}{(R-r)^2} + r\sum_c \frac{R^2-|c|^2}{R(R-|c|)}\frac{1}{|z-c|} \\ &\leq 5T(R)\frac{Rr}{(R-r)^2} + 2r\sum_c \frac{1}{|z-c|}. \end{aligned}$$

Hence if $r < R$, and $J$ is any interval on $\mathbb{S}(r)$ of $\theta$-length $\delta < 1/2$, then the quantity to be estimated in Lemma 11.16 is

$$r\int_J \left|\frac{h'}{h}\right| d\theta \leq 5\delta T(R)\frac{Rr}{(R-r)^2} + 2R\sum_k \int_J \frac{1}{|re^{i\theta}-c_k|}\, d\theta. \tag{11.29}$$

The choice(s) of $r$ will depend on an elegant geometric insight which lies behind a famous lemma of H. Cartan on the minimum modulus of a polynomial. The connection is shown in Remark 11.18 at the end of this section. The disks in Cartan's proof correspond to intervals on $[0,\infty)$ in our presentation – the points $c_k$ in the next lemma are complex numbers, but the conclusion is about the nonnegative numbers $|c_k|$.

**Lemma 11.17** *Let $c_1,\ldots,c_m$ be complex numbers, and $\varepsilon > 0$. Then there is an exceptional set $F$ of intervals having total measure $|F| \leq 4\varepsilon$ such that when $r \notin F$ and $j = 1,\ldots,m$, at most $(j-1)$ of the numbers $|c_k|$ lie at distance $\leq j\varepsilon/m$ from $r$.*

*Proof* The argument is amazingly simple, yet subtle.

Choose $\lambda'$ and $x'$ so that the closed interval

$$I = \{x\colon |x-x'| \leq \lambda'\varepsilon/m\}$$

contains exactly $\lambda'$ of the points $|c_k|$ and $\lambda'$ is the largest integer with this property; here is why this can be done. Begin with a given $c_k$ and consider intervals $|x-|c_k|| \leq \varepsilon/m$, $|x-|c_k|| \leq 2\varepsilon/m,\ldots$, terminating the process at $\lambda_0\varepsilon/m$ when there are exactly $\lambda_0$ of the points $|c_k|$ in the associated interval and no larger multiple will have this property. Note that the first interval $|x-|c_k|| \leq \varepsilon/m$ contains at least 1 point, and if it contains more than 1 point then the interval $|x-|c_k|| \leq 2\varepsilon/m$ contains at least 2 points, and so on, until the process terminates. Thus pairs $(x',\lambda')$ do exist, and we may take $\lambda_1$ and a closed interval $I_1$ about $x_1$ with $\lambda_1$ maximal. Note that $\lambda_1 \leq m$.

Now repeat the process with the remaining $m-\lambda_1$ of the $|c_k|$, and determine the largest $\lambda_2$ and closed interval $I_2$, with $I_2$ having length $2\lambda_2\varepsilon/m$, such that exactly $\lambda_2$ of the remaining $|c_k|$ lie in the interval $I_2$. We have $\lambda_2 \leq \lambda_1$ by maximality of $\lambda_1$. Continuing in this manner we obtain intervals $I_1, \ldots, I_p$ that exhaust the full collection of $|c_k|$.

The intervals must be disjoint. For if $I_1$ intersected $I_\ell$ for some $\ell \geq 2$ then the union $I_1 \cup I_\ell$ would lie in an interval of length $2(\lambda_1 + \lambda_\ell)\varepsilon/m$ containing at least $\lambda_1 + \lambda_\ell$ points, which would contradict the maximality of $\lambda_1$. Thus $I_1$ cannot intersect any intervals $I_\ell$ chosen later in the construction. Similarly $I_2$ intersects no later interval, and so on.

Consider the system of intervals $\Gamma : I'_1, \ldots, I'_p$ with the same midpoints but twice the lengths. The $I'_1, \ldots, I'_p$ have lengths $4\lambda_1\varepsilon/m, \ldots, 4\lambda_p\varepsilon/m$, and so cover a set of measure at most

$$4(\lambda_1 + \cdots + \lambda_p)\frac{\varepsilon}{m} = 4\varepsilon.$$

This condition defines the exceptional set $F = \cup_\ell I'_\ell$, with $|F| \leq 4\varepsilon$.

Suppose $r$ lies outside $\Gamma$, that is, $r \notin F$, and let $I(r)$ be the interval centered at $r$ of length $2\lambda\varepsilon/m$, where $\lambda$ is an arbitrary positive integer. This interval cannot intersect any of the intervals $I_\ell$ having length greater than or equal to $2\lambda\varepsilon/m$, because the intervals in $\Gamma$ have twice the length of the original $\{I_\ell\}$. The remaining $I_\ell$ (if any) all have length smaller than $2\lambda\varepsilon/m$, and so each one contains at most $\lambda - 1$ of the points $|c_k|$. Hence $I(r)$ can contain at most $\lambda - 1$ of the points $|c_k|$, or else we would have chosen a larger interval at some stage of the construction. Thus the lemma is proved. □

With Lemma 11.17 established, we return to the complex plane and next estimate the number $m$ of zeros and poles of the function $h$ of Lemma 11.16, in the disk of radius $R$. The definitions in §11.2 and Nevanlinna's first theorem show that

$$N(r, a) \leq T(r) + O(1),$$

for each $a$, and so

$$\begin{aligned} m &= n(R, 0) + n(R, \infty) \\ &\leq \frac{1}{\log 2} \int_R^{2R} \frac{n(t, 0) + n(t, \infty)}{t}\, dt \\ &\leq \frac{1}{\log 2}(N(2R, 0) + N(2R, \infty)) \\ &\leq \frac{2}{\log 2}(T(2R) + C), \end{aligned} \tag{11.30}$$

with $C$ a generic constant.

Now we can confront the sum in (11.29). Given $r$, partition the collection of zeros and poles $c$ into two classes: (*I*) those with $\big|r - |c|\big| < \delta r/2$ and (*II*) the others.

First consider the sum corresponding to (*I*). It is clear that

$$r \int_J \frac{d\theta}{|re^{i\theta} - c|}$$

is maximized when $J$ has its midpoint at $\theta = \arg c$. Let $\gamma = \big|r - |c|\big|/r < \delta/2$. Then

$$\begin{aligned} r \int_J \frac{d\theta}{|re^{i\theta} - c|} &\le 2r \left[ \int_0^\gamma + \int_\gamma^{\delta/2} \right] \frac{d\theta}{\big|re^{i\theta} - |c|\big|} \\ &\le \frac{2\gamma r}{|r - |c||} + 2r \int_\gamma^{\delta/2} \frac{d\theta}{\big|\Im m(re^{i\theta} - |c|)\big|} \\ &\le 2 + 2r \int_\gamma^{\delta/2} \frac{d\theta}{r \sin\theta} \le 2 + \pi \int_\gamma^{\delta/2} \frac{d\theta}{\theta} \\ &\le 2 + \pi \log \frac{\delta r}{2\,|r - |c||} \qquad (c \in (I)). \end{aligned}$$

Let $\kappa = (\sigma - 1)/12$ and $\varepsilon = \kappa R/4$. If $r$ is outside the exceptional set $F$ in Lemma 11.17 , then after arranging the points $c$ in terms of increasing distance from $r$, the $j$th element has

$$\big|r - |c|\big| > \frac{j\kappa R}{4m}. \tag{11.31}$$

Hence class (*I*) contains at most $M$ elements, where $M$ is the largest integer with

$$\frac{M\kappa R}{4m} \le \frac{\delta r}{2}, \qquad \text{that is,} \quad M \le \frac{2\delta rm}{\kappa R} = M_0. \tag{11.32}$$

Thus for $r \notin F$, we have

$$\begin{aligned} \sum_{c \in (I)} \left( 2 + \pi \log \frac{\delta r}{2\big|r - |c|\big|} \right) &\le 2M_0 + \pi \sum_1^M \log \frac{2\delta rm}{\kappa Rj} \\ = 2M_0 + \pi \sum_{j=1}^M \log \frac{M_0}{j} &< 2M_0 + \pi \sum_{j=1}^M \int_{j-1}^j \log \frac{M_0}{x}\, dx \\ < 2M_0 + \pi \int_0^{M_0} \log \frac{M_0}{x}\, dx &= (2 + \pi) M_0 < 6M_0. \end{aligned} \tag{11.33}$$

Next we treat group (*II*). For the first $M$ elements of group (*II*) we bound $\big|r - |c|\big|$ from below by $\delta r/2$, and for the remaining elements (if any) we use the lower bound in (11.31). Hence when (11.33) is included,

$$\begin{aligned} r\sum_c \int_J \frac{d\theta}{|re^{i\theta}-c|} &< 6M_0 + r\delta\left(M\frac{2}{\delta r} + \sum_{j=M+1}^{m} \frac{4m}{j\kappa R}\right) \\ &\le 8M_0 + \frac{4r\delta m}{\kappa R}\sum_{j=M+1}^{m}\frac{1}{j} \\ &\le 8M_0 + 2M_0\left(1 + \int_{M_0}^{m}\frac{dx}{x}\right) \\ &\le 10M_0 + 2M_0 \log^+ \frac{m}{M_0}. \end{aligned}$$

By this estimate and (11.29),

$$r\int_J \left|\frac{h'}{h}\right| d\theta \le 5\delta T(R)\frac{Rr}{(R-r)^2} + 2\left(10M_0 + 2M_0\log^+\frac{m}{M_0}\right).$$

We can now wrap up the proof of Lemma 11.16. With $R = 6r_n$, where $r_n$ is a peak of $T$, consider $r \in [r_n, 2r_n]$. Write $L = L(\lambda)$ for a constant depending only on $\lambda$. Then (11.30) and (11.32) show that

$$m \le LT(r_n), \qquad M_0 \le L\frac{\delta}{\kappa}m \lesssim L\frac{\delta}{\kappa}T(r_n),$$

which simplifies our estimate to

$$r\int_J\left|\frac{h'}{h}\right| d\theta \le L\delta T(r_n) + L\frac{\delta}{\kappa}\left(1 + \log^+\frac{\kappa}{\delta}\right)T(r_n).$$

Noting that $T(r_n) \le T(r)$ because $T$ is increasing, we obtain Lemma 11.16 with $K = K(\lambda, \sigma)$. The exceptional set has size $|E_n \cap [r_n, \sigma r_n]| \le 4\varepsilon = \kappa R = (\sigma - 1)r_n/2$.

**Remark 11.18** Our proof of Lemma 11.17 is based on the account in Levin (1980, p. 19), but phrased in terms of intervals on the real line rather than circles (disks) in the plane. Cartan's approach *immediately* yields his estimate on the minimum modulus of a (monic) polynomial: *given $H > 0$ and complex numbers $c_1, c_2, \ldots, c_m$, there is a system of finitely many closed disks in the plane whose sum of radii is $2H$ and*

$$|z - c_1| \cdot |z - c_2| \cdots |z - c_m| > \left(\frac{H}{e}\right)^m$$

*when $z$ is outside the system.* Indeed, take $z$ disjoint from the disks $C_1', \ldots, C_p'$ that are analogous to the doubled intervals $I_1', \ldots, I_p'$ in the proof of Lemma 11.17 (with $\varepsilon = H$), and arrange the points $c_k$ in order of increasing distance from $z$. Then for each $k$ the argument gives

$$|z - c_k| > k\frac{H}{m},$$

so that

$$|z - c_1|\,|z - c_2| \cdots |z - c_m| > m!\left(\frac{H}{m}\right)^m > \left(\frac{H}{e}\right)^m$$

where the final step relies on the Stirling-type estimate $m! > (m/e)^m$, proved by

$$\log m! = \sum_{j=2}^{m} \log j > \int_1^m \log x\, dx = \log(m/e)^m + 1.$$

A remarkable special case is that given any monic polynomial, we may find a collection of disks containing its zeros and with radii summing to $2e$ such that outside the disks, the polynomial has magnitude greater than 1, independent of the degree of the polynomial.

## 11.5 Interlude: Subordination and Lehto's Theorem

This section introduces two important tools from geometric complex analysis. The simplicity with which they may be formulated and proved conceals their remarkably sophisticated consequences, which have been mined for decades. The first, subordination, is a standard method to produce extrema in classes of mappings, by identifying extremal mappings as those whose range is "maximal" in the comparison class. It is used below in our proof of Theorem 11.26, and a comparison with the elaborate argument in §11.8.2 shows the ⋆-function may be interpreted as an extension of subordination. Lehto's theorem, our second tool, reveals a relation between the counting function $N$ and subharmonicity and is also relevant to the proof of Corollary 11.24.

**(a) Subordination.** In principle, subordination is an immediate consequence of the Schwarz lemma, which in turn is the maximum principle applied to $\omega(z)/z$.

**Lemma 11.19** (Schwarz) *If $\omega$ is an analytic map of $\mathbb{D}$ to itself with $\omega(0) = 0$, then*

$$|\omega'(0)| \le 1, \qquad |\omega(z)| \le |z| \quad (z \in \mathbb{D}).$$

*Equality is possible if and only if $\omega(z) = cz$ for some $c$ with $|c| = 1$.*

The principle of subordination (Littlewood 1944) is nearly a century old, and reveals one reason that conformal mappings (univalent functions) play an extremal role in geometrical complex analysis. We suppose that $f, g$ are meromorphic in $\mathbb{D}$. If

$$f(z) = g(\omega(z))$$

where $\omega$ satisfies the Schwarz lemma, then we write $f \prec g$ and say $f$ is *subordinate* to $g$. In most situations, $f$ and $g$ are analytic, but the definition allows that the ranges be a Riemann surface. In the most common situations, $g(\mathbb{D}) \subset \mathbb{C}$ and $g$ is univalent, in which case $\omega = g^{-1} \circ f$.

**Lemma 11.20** (Subordination) *Let $f, g$ be analytic in $\mathbb{D}$ with $f(\mathbb{D}) \subset g(\mathbb{D})$, $f(0) = g(0)$, and suppose $g$ is univalent. Then $f \prec g$,*

$$|f'(0)| \leq |g'(0)|$$

*and*

$$M(r,f) \leq M(r,g) \qquad (0 < r < 1).$$

*Proof* Apply the Schwarz lemma to the analytic function $\omega = g^{-1} \circ f$: $f \prec g$. Thus $|(g^{-1} \circ f)'(0)| \leq 1$ and $|(g^{-1} \circ f)(z)| \leq |z|$. The first inequality with the chain rule shows that $|f'(0)| \leq |g'(0)|$, and the second shows the $f$-image of $B(0, r)$ is contained in $g(B(0, r))$ if $0 < r < 1$. In particular,

$$M(r,f) = \max_{|z|=r} |f(z)| \leq \max_{|z|\leq r} |g(z)| = \max_{|z|=r} |g(z)| = M(r, g).$$

□

**(b) Lehto's theorem on $N(r, \zeta)$.** Lehto (1953) observed an important property of Nevanlinna's function $N(r, \zeta)$, which uses an insight similar to Cartan's (as shown in the proof given in Hayman (1989)). It is valid only for holomorphic functions.

**Theorem 11.21** *Let $r < R$ and $f$ be holomorphic in $B(R)$. Then $N(r, \zeta)$ is subharmonic in the $\zeta$-plane except at $\zeta = f(0)$. Near $\zeta = f(0)$ the function*

$$N(r, \zeta) + \log|f(0) - \zeta|$$

*is subharmonic.*

*Proof* This is immediate from Jensen's theorem applied to $f(z) - \zeta$:

$$N(r, \zeta) + \log|f(0) - \zeta| = \frac{1}{2\pi} \int_{-\pi}^{\pi} \log|f(re^{i\theta}) - \zeta|\, d\theta,$$

and the expression on the right side is obviously subharmonic. That expression will also be relevant later, and so merits a definition:

$$\mathcal{I}(r,\zeta,f) = \frac{1}{2\pi}\int_{-\pi}^{\pi} \log|f(re^{i\theta}) - \zeta|\,d\theta. \tag{11.34}$$

□

**Remark.** The function $N(r,\zeta)$ alone is not subharmonic. One can view the term $\log|f(0) - \zeta|$ as a correction to make $N$ subharmonic, and that is what will be important in §11.7.

## 11.6 The ⋆-Function and Univalent Functions

Recall that a region is an open, connected set. An analytic function $f\colon X \to Y$, where $X$ and $Y$ are regions in $\mathbb{C}$, is *univalent* if $f$ is 1–1; thus $X$ and $Y$ are conformally equivalent. The situation most analyzed is when $f$ belongs to the class $S$ of "schlicht" functions, which means $X$ is the unit disk $\mathbb{D}$ and $f(0) = 0$, $f'(0) = 1$. Two members of $S$ which are extremal for many problems are the identity function $f(z) = z$ and the Koebe function

$$k(z) = \sum_{n\geq 1} nz^n = \frac{z}{(1-z)^2}.$$

Intuitively, the identity function has the smallest possible image among functions in $S$, while $k$ has the largest image: it maps $\mathbb{D}$ conformally to the whole plane with the interval $(-\infty, -1/4]$ removed.

De Branges (1985) solved the most famous problem about $S$ by proving Bieberbach's conjecture on the coefficients of schlicht functions:

$$\text{if } f \in S \text{ then } |a_n| \leq n, \text{ equality only for } k \text{ and its rotations.} \tag{11.35}$$

Earlier, the ⋆-function was used by Baernstein (1974) to show the Koebe function is maximal for a different collection of extremal problems. Namely, $k$ maximizes $L^p$-means of univalent functions.

**Theorem 11.22** (Baernstein 1974, Theorems 1 and 2) *Let $\Phi\colon \mathbb{R} \to \mathbb{R}$ be convex. If $f \in S$ then*

$$\int_{-\pi}^{\pi} \Phi(\log|f(re^{i\theta})|)\,d\theta \leq \int_{-\pi}^{\pi} \Phi(\log|k(re^{i\theta})|)\,d\theta.$$

*If equality holds for some strictly convex $\Phi$ and some $r$, $0 < r < 1$, then $f(z) = e^{i\alpha}k(e^{-i\alpha}z)$ for some real $\alpha$.*

Recall that a convex function $\Phi\colon \mathbb{R} \to \mathbb{R}$ is *strictly convex* if it is convex and not linear on any subinterval.

The special cases $\Phi(x) = e^{px}$ ($p \in \mathbb{R}$) and $\Phi(x) = x^+$ in the theorem imply that

$$\begin{gathered}\int_{-\pi}^{\pi} |f(re^{i\theta})|^p \, d\theta \le \int_{-\pi}^{\pi} |k(re^{i\theta})|^p \, d\theta, \\ \int_{-\pi}^{\pi} \log^+ |f(re^{i\theta})| \, d\theta \le \int_{-\pi}^{\pi} \log^+ |k(re^{i\theta})| \, d\theta, \\ M(r,f) \le M(r,k).\end{gathered} \tag{11.36}$$

The $L^p$ result for the special case $p = 1$ along with Cauchy's theorem gives the coefficient estimate $|a_n| \le (e/2)n$, which is weaker than proved in de Branges' result (11.35).

Although Theorem 11.22 seems unable to yield sharp coefficient estimates for the full class $S$, it has been applied with $p < 0$ in Baernstein and Schober (1980) to give a short proof of Loewner's sharp estimates for the coefficients of the inverses to functions in $S$; note that if $f \in S$, then

$$g(w) = f^{-1}(w) = w + b_2 w^2 + \cdots$$

is also univalent in a neighborhood of the origin.

Moreover, Theorem 11.22 also holds for a class of functions more general than $S$, the normalized *weakly univalent* functions (Hayman 1989, 1994): $f$ is weakly univalent if $f(\mathbb{D})$ contains $\{|z| = r\}$ for all $r < R_0 \in (0, \infty)$ but not any larger circle. Whether the refined estimate (11.35) is valid for weakly univalent functions remains open.

Recently, Ronen Peretz Peretz (2017) showed that estimates for individual coefficients do not follow in general from Theorem 11.22.

The estimates in Theorem 11.22 are related to the geometry of the range of $f$ through the next lemma, which relies on the Cartan identity (11.4) with simplifications due to univalence, since

$$N(r, \infty) \equiv 0, \qquad N(r, \rho a) = \log^+ |r/f^{-1}(\rho a)|.$$

The proof also uses Green function $G(z, a, \Omega)$; this function was introduced in §5.6, but has a special role in the 2-dimensional theory, see for example Hayman and Kennedy (1976), Tsuji (1975); it is the value at $z$ of the properly normalized (positive) harmonic function in $\Omega \setminus \{a\}$ having boundary values zero on $\partial\Omega$ and a logarithmic pole at $a$.

**Lemma 11.23** *Let $f \in S$ and write $\Omega$ for the range of $f$. If $u$ is the Green function of $\Omega$ with pole at 0, then for $0 < r < 1$ and $0 < \rho < \infty$,*

$$\int_{-\pi}^{\pi} \log^+ \frac{|f(re^{i\theta})|}{\rho} \, d\theta = \int_{-\pi}^{\pi} (2\pi u(\rho e^{it}) + \log r)^+ \, dt.$$

*Proof* Cartan's identity (11.4) applied to $f/\rho$ allows expressing the left side as

$$\int_{-\pi}^{\pi} N(r, \rho e^{it}, f)\, dt + 2\pi \log^{+} |f(0)|/\rho, \tag{11.37}$$

but here $f(0) = 0$ because $f \in S$. Since $f$ is univalent, the counting function $N$ becomes

$$N(r, \zeta) = \log^{+}(r/|f^{-1}(\zeta)|)$$

and so the lemma follows upon substituting into (11.37), since the Green function for $\Omega$ is

$$u(\zeta) = \frac{1}{2\pi} \log \frac{1}{|f^{-1}(\zeta)|}.$$

□

*Proof of Theorem 11.22.* Both $\log|f|$ and $\log|k|$ are harmonic away from the origin, and they have the same mean value on each circle $S(r)$:

$$\frac{1}{2\pi}\int_{-\pi}^{\pi} \log|f(re^{i\theta})|\, d\theta = \frac{1}{2\pi}\int_{-\pi}^{\pi} \log|k(re^{i\theta})|\, d\theta = \log r$$

due to the normalizations $f(z)/z = 1 + O(z)$ and $k(z)/z = 1 + O(z)$. Thus the equivalences in Proposition 10.3 will yield Theorem 11.22 as a consequence of the inequality $(\log|f|)^{\star} \le (\log|k|)^{\star}$ on the upper half of $\mathbb{D}$ or, with $\rho = e^{t}$ in the proposition,

$$\int_{-\pi}^{\pi} \log^{+} \frac{|f(re^{i\theta})|}{\rho}\, d\theta \le \int_{-\pi}^{\pi} \log^{+} \frac{|k(re^{i\theta})|}{\rho}\, d\theta;$$

here $0 < r < 1, 0 < \rho < \infty$.

Let

$$\Omega = f(\mathbb{D}),\ \Omega_1 = k(\mathbb{D}),\ u(\zeta) = G(\zeta, 0, \Omega),\ v(\zeta) = G(\zeta, 0, \Omega_1),$$

where $G$ is the appropriate Green function extended to be zero outside $\Omega$ or $\Omega_1$. In view of Lemma 11.23 we want to show for each $0 < \rho < \infty$ and $t = -\frac{1}{2\pi}\log r > 0$ that

$$\int_{-\pi}^{\pi} (u(\rho e^{i\theta}) - t)^{+}\, d\theta \le \int_{-\pi}^{\pi} (v(\rho e^{i\theta}) - t)^{+}\, d\theta.$$

(Since $u$ and $v$ are nonnegative, this last inequality certainly holds for $t \le 0$ once it is proved for $t = 0$.) With $\mathbb{H}$ the upper half plane, the inequality in Theorem 11.22 thus reduces to an inequality for the ⋆-function applied to Green functions: we want

$$u^{\star} \le v^{\star} \text{ in } \mathbb{H}. \tag{11.38}$$

The essential, and by now familiar, guiding principle is that $u^\star$ is subharmonic in the upper half plane while $v^\star$ is harmonic there. To obtain (11.38) we employ the maximum principle. For $\varepsilon > 0$ set

$$Q(\rho e^{i\theta}) = u^\star(\rho e^{i\theta}) - v^\star(\rho e^{i\theta}) - \varepsilon\theta \qquad (\rho > 0,\ 0 < \theta < \pi). \tag{11.39}$$

Then $Q$ is subharmonic in $\mathbb{H}$ and continuous on $\overline{\mathbb{H}} \setminus \{0\}$. We investigate its boundary values.

If $d$ is the radius of the largest disk contained in $\Omega = f(\mathbb{D})$, then $d \geq 1/4$ (Koebe's theorem), and so

$$u(\zeta) = -\frac{1}{2\pi} \log|\zeta| + u_1(\zeta), \tag{11.40}$$

with $u_1$ harmonic in $|\zeta| < d$, and the normalization $f(z) = z + O(z^2)$ ensures that $u_1(0) = 0$. This gives that

$$u^\star(\rho e^{i\theta}) = -\frac{\theta}{\pi} \log\rho + o(1),$$

uniformly in $\theta$ as $\rho \to 0$, and the same reasoning applies to $v$. Hence

$$\limsup_{\zeta\to 0} Q(\zeta) \leq 0 \qquad (\zeta \to 0).$$

Since $u$ and $v$ tend to 0 at $\infty$, the same is true of their star functions, and so (11.39) now yields

$$\limsup_{\zeta\to\infty} Q(\zeta) \leq 0.$$

It is immediate from the definition of the ★-function that on the positive real axis,

$$Q(\rho) = 0 \quad (\rho > 0).$$

Thus we are left to consider $Q$ on the negative axis. With $u_1$ from (11.40), we have $u_1$ harmonic near 0 with $u_1(0) = 0$, and so

$$\begin{aligned} u^\star(\rho e^{i\pi}) &= \int_{-\pi}^{\pi} u(\rho e^{i\theta})\, d\theta = -\log\rho + \int_{-\pi}^{\pi} u_1(\rho e^{i\theta})\, d\theta \\ &= -\log\rho \qquad (0 < \rho < d). \end{aligned}$$

The relation analogous to (11.40) for $v$ is

$$v_1(\zeta) = v(\zeta) + \frac{1}{2\pi}\log|\zeta|,$$

but now $v_1$ is subharmonic in the entire plane and harmonic in a neighborhood of 0 (specifically, for $|\zeta| < 1/4$), with $v_1(0) = 0$. Hence

$$v^\star(\rho e^{i\pi}) = \int_{-\pi}^{\pi} v(\rho e^{i\theta})\, d\theta = -\log\rho + \int_{-\pi}^{\pi} v_1(\rho e^{i\theta})\, d\theta$$
$$\geq -\log\rho \qquad (0 < \rho < \infty).$$

When this is reconciled with the behavior of $u^\star$ near the negative axis, we find that

$$Q(\rho e^{i\pi}) \leq 0 \quad (0 < \rho < d).$$

To complete the analysis of $Q$ on $\partial\mathbb{H}$, we show the maximum of $Q$ cannot occur in the interval $(\infty, -d]$ on the negative real axis. Let $\zeta = \rho_0 e^{i\pi}$ with $\rho_0 \in [d, \infty)$. The symmetric decreasing rearrangement of the function $\theta \mapsto u(\rho_0 e^{i\theta})$ is continuous, since $u$ is continuous on the circle. Hence by differentiating (11.1),

$$\frac{\partial u^\star}{\partial\theta}(\rho_0 e^{i\theta}) = 2u^\#(\rho_0 e^{i\theta}) \qquad (0 \leq \theta \leq \pi),$$

where at $\theta = \pi$ the derivative is taken from below. Since $\rho_0 \geq d$, we have $u^\#(\rho_0 e^{i\pi}) = \inf_\varphi u(\rho_0 e^{i\varphi}) = 0$, which gives $(\partial u^\star)/(\partial\theta) = 0$ at $\theta = \pi$, and similarly $(\partial v^\star)/(\partial\theta) = 0$ at $\theta = \pi$. An appeal to the definition of $Q$ in (11.39) now yields

$$\frac{\partial Q}{\partial\theta}(\rho_0 e^{i\pi}) = -\varepsilon < 0.$$

Therefore the maximum of $Q$ does not occur at $\rho_0 e^{i\pi}$.

Since the maximum occurs somewhere on the boundary, and we have shown $Q \leq 0$ at all other boundary points, we conclude $Q \leq 0$ in $\mathbb{H}$. Since $\varepsilon$ is arbitrary positive, on letting $\varepsilon \to 0$ we find that $u^\star \leq v^\star$, and the theorem is proved. □

For the equality case, an argument that does not use the ☆-function, see Baernstein (1974).

## 11.7 Complements to the Univalent Integral Means Theorem

Theorem 11.22 on integral means has several applications: many are covered by Baernstein (1974) and Hayman (1989), while applications to moments of equilibrium measures in potential theory are developed by Laugesen (1993) and Baernstein et al. (2011). In this section we concentrate on applications to Green functions, holomorphic functions, and harmonic measures.

A weaker form of Lemma 11.23 holds when $f$ is not univalent:

**Corollary 11.24** *Let $f: \mathbb{D} \to \Omega$ be holomorphic, and suppose $\Omega$ has a Green function. If $u$ is the Green function for $\Omega$ with pole at $f(0)$, then*

$$\int_{-\pi}^{\pi} \log^+ \frac{|f(re^{i\theta})|}{\rho}\, d\theta \le \int_{-\pi}^{\pi} (2\pi u(\rho e^{it}) + \log r)^+ \, dt + 2\pi \log^+ \frac{|f(0)|}{\rho}$$

*whenever $0 < r < 1$ and $0 < \rho < \infty$.*

*Proof* We adapt the proof of Lemma 11.23. The function

$$2\pi u(f(z)) + \log |z|$$

is superharmonic in $\mathbb{D}$ (including at $z = 0$) and its boundary values are nonnegative on $\partial\mathbb{D}$. Thus if $r \in (0, 1)$ the set $\Delta_r = \{\zeta : 2\pi u(\zeta) > -\log r\}$ contains the image of $B(0, r)$, and hence $N(r, \zeta) = 0$ for $\zeta \notin \Delta_r$. But Theorem 11.21 implies that

$$N(r, \zeta) - \log r - 2\pi u(\zeta)$$

is subharmonic in $\Delta_r$ with non-positive boundary values, so that

$$N(r, \zeta) \le 2\pi u(\zeta) + \log r \qquad (\zeta \in \Delta_r).$$

This last inequality holds for all $\zeta \in \mathbb{C}$ when we apply the operation $[\cdot]^+$ to the expression on the right. We use this to estimate $N(\rho e^{it})$ in the term (11.37). □

The next result will be used later in our discussion of corollaries to Theorem 11.34 in §11.9.

**Theorem 11.25** (Baernstein 1974, Theorem 5) *Suppose that $\Omega$ and $\Omega^\#$ have Green functions. If $\Phi: \mathbb{R} \to \mathbb{R}$ is convex increasing then*

$$\int_{-\pi}^{\pi} \Phi(G(re^{i\theta}, a, \Omega))\, d\theta \le \int_{-\pi}^{\pi} \Phi(G(re^{i\theta}, |a|, \Omega^\#)\, d\theta$$

*for each $a \in \Omega$, $r \in (0, \infty)$.*

A by-product of the proof is worth stating explicitly:

the Green function of the circularly symmetric domain $\Omega^\#$ with pole at $a > 0$ is symmetric decreasing on each circle $|z| = r$.

The case in which $\Omega^\#$ is simply connected and $a = 0$ was settled in Jenkins (1955) using a different formulation. That (11.38) might be improved to

$$u^\# \le v \text{ in } \Omega^\#$$

was raised in Hayman (1967, Problem 5.17), but shown to be false by Pruss (1996).

Theorem 11.25 implies in particular that the maximum of the Green function is attained on the symmetrized domain:

$$\sup_{|z|=r} G(z, a, \Omega) \leq \sup_{|z|=r} G(z, |a|, \Omega^{\#}) = G(r, |a|, \Omega^{\#})$$

(recalling that the inequality between $L^{\infty}$ norms in Corollary 10.2 follows from the inequality between integral means). As shown in Hayman (1989, Theorem 9.4), this implies the classical result of Pólya and Szegö that symmetrization increases mapping radius. For simply connected $\Omega$ and $a = 0$, the Green function inequality was first obtained in Krzyż (1959).

Similar procedures yield other comparison theorems. For example, there is a sharpening of Theorem 11.25:

**Theorem 11.26** (Baernstein 1974, Theorem 6) *Let $\Omega$ be a domain in $\mathbb{C}$ with $\Omega^{\#}$ simply connected and not the full plane. Suppose $f$ is holomorphic in $\mathbb{D}$ with $f(\mathbb{D}) \subset \Omega$. Let $F$ be the conformal map of $\mathbb{D}$ onto $\Omega^{\#}$ normalized by $F(0) = |f(0)|$. If $\Phi\colon \mathbb{R} \to \mathbb{R}$ is increasing and convex then*

$$\int_{-\pi}^{\pi} \Phi(\log |f(re^{i\theta})|)\, d\theta \leq \int_{-\pi}^{\pi} \Phi(\log |F(re^{i\theta})|)\, d\theta.$$

Note that nothing is asserted when $\Omega^{\#}$ is not simply connected.

*Proof* Here is a sketch of a proof. First we suppose that $\Omega$ is simply-connected. If $f$ is the conformal map of $\mathbb{D}$ to $\Omega$, the argument in Theorem 11.22 shows that Theorem 11.26 is equivalent to Theorem 11.25. When $f$ is not univalent, it is subordinate to the conformal $g$ mapping $\Omega$ to $\mathbb{D}$ with $g(0) = f(0)$ (subordination was discussed in §11.5). The integral means in Theorem 11.26 increase when $f$ is replaced by $g$ (see for example, Hayman 1989, p. 76). Thus the theorem holds when $\Omega$ is simply connected.

For general $\Omega$, recall that we have assumed that $\Omega^{\#}$ is simply connected. Lehto's Theorem 11.21 shows that $w \mapsto N(r, w, f)$ is subharmonic other than having a logarithmic pole at $w = f(0)$, and so an analysis similar to that of Theorem 11.25 yields that

$$N^{\star}(r, w, f) \leq \left(2\pi G(w, F(0), \Omega^{\#}) + \log r\right)^{\star}, \qquad (w \in \mathbb{H}^{\star}, 0 < r < 1).$$

Theorem 11.26 follows from this and Lemma 11.23 along with the argument from the proof of Theorem 11.22; now let $r \to 1$ using

$$u(w) = N(r, w, f), \quad v(w) = G(w, F(0), \Omega^{\#}) \quad (0 < r < 1).$$

□

Other integral means comparisons involve meromorphic univalent functions, functions in annuli, or functions satisfying various constraints. In addition to Baernstein (1974), a partial list of papers includes Godula and Nowak (1987); Kirwan and Schober (1976); Laugesen (1993); Solynin (1996); Yang (1994). Leung (1979) used ☆-functions to solve extremal problems for integral means of derivatives of starlike and other special classes of univalent functions.

There are also applications to harmonic measure (see also §5.6). Here is a model case. Let $\Omega$ be a region in $\mathbb{C}$ for which the Dirichlet problem is solvable, and $E$ a Borel set in $\mathbb{C} \setminus \Omega$. Then $\omega(z, E, \Omega)$, the harmonic measure of $E$ at $z$ (with respect to $\Omega$) may be thought of as the value of the value at $z$ of the harmonic function in $\Omega$ with boundary values 1 on $E \cap \partial\Omega$ and 0 on $\partial\Omega \setminus E$; $\omega$ may always be extended to the complement of $\Omega$ by taking $\omega(z, E, \Omega) = 0$ outside $\Omega$; standard references are Tsuji (1975) and Hayman (1989). Theorem 11.22 in this setting gives the following application, where for specificity we suppose that $\Omega \subset \mathbb{D}$, not necessarily simply connected, and $E = \partial\mathbb{D}$ (thus $\omega \equiv 0$ when $\Omega \subset\subset \mathbb{D}$).

**Theorem 11.27** (Baernstein 1974, Theorem 6) *Let* $\Omega \subset \mathbb{D}$, $\Phi\colon \mathbb{R} \to \mathbb{R}$ *an increasing convex function and* $0 < r < 1$. *Then*

$$\int_{-\pi}^{\pi} \Phi(\omega(e^{i\theta}, \partial\mathbb{D}, \Omega))\, d\theta \le \int_{-\pi}^{\pi} \Phi(\omega(re^{i\theta}, \partial\mathbb{D}, \Omega^{\#}))\, d\theta.$$

*Proof* We give only an outline. Let $u$ and $v$ be the harmonic measure of $\Omega$ and $\Omega^{\#}$ respectively. The argument for Theorem 11.22 adapts to show that $u^{\star} \le v^{\star}$ in $\mathbb{D}^{+}$, which is then equivalent to Theorem 11.27. □

It can be shown also that the extremal function $v$ is symmetric decreasing on each circle $\{|z| = r\}$. Thus $u^{\star} \le v^{\star}$ gives a result of Haliste (1965):

$$\sup_{|z|=r} \omega(z, \partial\mathbb{D}, \Omega) \le \sup_{|z|=r} \omega(z, \partial\mathbb{D}, \Omega^{\#}) = \omega(r, \partial\mathbb{D}, \Omega^{\#}). \tag{11.41}$$

In turn, (11.41) and the maximum principle yield what is known as *Beurling's projection theorem*, a fundamental result in 2-dimensional potential theory. Let $A$ be the set of $r \in (0, 1]$ such that $\Omega$ does not contain the full circle $|z| = r$, and let $\Omega^{\#\#} = \mathbb{D} \setminus \{-r\colon r \in A\}$. Of course $\Omega^{\#} \subset \Omega^{\#\#}$ and this leads to

$$\sup_{|z|=r} \omega(z, \mathbb{D}, \Omega) \le \sup_{|z|=r} \omega(z, \mathbb{D}, \Omega^{\#\#}) \quad (0 < r < 1).$$

Although the discussions in this chapter have been using circular symmetrization, one can derive parallel results using Steiner symmetrization (Sections 9.11 and 10.4). Steiner symmetrization of domains increases integral means of Green functions and harmonic measures. For an informal discussion with references, see Baernstein (2002, §7.2).

## 11.8 The Conjugate Function

If $u$ is harmonic and real in the disk $\mathbb{D}$, there is a harmonic function $v$, the function *conjugate* to $u$, so that $f = u + iv$ is analytic in $\mathbb{D}$; a normalization such as $v(0) = 0$ makes $v$ unique. In this section we discuss a relation between $v$ and $u$ in terms of their boundary values.

Because of its connections to Fourier analysis (Zygmund 1968), there is a rich history relating how a function $u$ harmonic in $\mathbb{D}$ may be recovered from its boundary values on $\mathbb{S}$, where the boundary value $u(re^{i\theta})$ always is defined as the radial limit:

$$u(e^{i\theta}) = \lim_{r\to 1} u(re^{i\theta});$$

this theory has been successfully developed to have major influence in PDEs, geometric measure theory, and other disciplines in 2 and more dimensions.

There are subtleties, since there exist non-zero harmonic functions with radial boundary values 0 everywhere – for example the function

$$\partial P(r,\theta)/\partial\theta = -\sum_{n\geq 1} nr^n \sin n\theta = -\Im m\left(\frac{1+z}{1-z}\right)^2$$

where $P(r,\theta)$ is the Poisson kernel with $r = 1$; see §11.2. For this reason, pairing $u$ to its values on $\mathbb{S}$ requires additional information. For example, if $u \in L^p(\mathbb{S})$ with $p > 1$ then for $z \in \mathbb{D}$, $u(z)$ is the Poisson integral of its boundary values (cf. Hoffman 1962, chapters 3 and 4). However, it is possible that the means

$$\int_{-\pi}^{\pi} |u(re^{i\theta})|\, d\theta \qquad (0 < r < 1) \tag{11.42}$$

are uniformly bounded without $u(e^{i\theta})$ being an integrable function, the most compelling example being when $u$ is the Poisson kernel.

In its place, when (11.42) is uniformly bounded, there is (see Tsuji 1975, chapter IV) a signed real-valued measure $\mu$ on $\mathbb{S}$ with $\|\mu\| < \infty$ ($\|\mu\|$ is its total variation) with

$$f(z) = u + iv = \int_{-\pi}^{\pi} \frac{e^{i\varphi} + z}{e^{i\varphi} - z}\, d\mu(\varphi), \tag{11.43}$$

and $f$ has nontangential limits a.e. on $\mathbb{S}$. Moreover

$$\|\mu\| = \lim_{r\to 1} \frac{1}{2\pi} \int_{-\pi}^{\pi} |\Re e f(re^{i\theta})| \, d\theta \qquad (11.44)$$
$$v(e^{i\theta}) = \Im m f(e^{i\theta}) \qquad \text{(a.e. } \theta).$$

Thus in this section we consider harmonic functions $u$ which are real parts of functions $f$ represented as (11.43), and remark that when $f$ is in a Hardy space $H^p$ for some $p > 1$, we may take $d\mu = (1/2\pi)u(e^{i\theta})\,d\theta$; this is thoroughly discussed in Ahlfors (1978, pp. 167–168).

If $u \in L^p(\mathbb{S})$ with $1 < p < \infty$, a theorem of Marcel Riesz (1928) asserts that $v \in L^p(\mathbb{S})$ and there is a constant $C_p$ with

$$\|v\|_p \le C_p \|u\|_p, \qquad 1 < p < \infty. \qquad (11.45)$$

Riesz's theorem fails when $p = 1$, since there are integrable functions whose conjugates are not integrable. The replacement for (11.45) was discovered (in fact a few years earlier) by Kolmogorov, a "weak" 1–1 inequality on the distribution function:

$$\big|\{|v| \ge t\}\big| \le \frac{C}{t}\|u\|_1, \qquad t > 0, \qquad (11.46)$$

and from this directly follows that

$$\|v\|_p \le C_p \|u\|_1, \qquad 0 < p < 1, \qquad (11.47)$$

complementing (11.45)

The best constants for (11.45), (11.46), and (11.47) were discovered decades later; the first in Pichorides (1972) and the other two in Davis (1974), Davis (1976) respectively. Davis's results attracted additional attention since they relied on probability (Brownian motion).

Inequalities (11.46) and (11.47) were rederived in Baernstein (1978), the proofs also yielding the best constants. We will prove Theorem 11.28, which is from that paper, after which (11.47) (with the best constant $C_p$ displayed in (11.50)) will follow at once from assertion (i) of that theorem. Theorem 11.29 breaks new ground since it involves means of real and imaginary parts of analytic functions, and Proposition 11.30 shows the relevance of Nevanlinna's integrated counting function $N$ to the subject. We do not include the proof of (11.46), since the arguments in Baernstein (1978) do not use the ⋆-function.

### 11.8.1 A Relation to Conformal Mappings

Davis identifies the best constant $C$ in (11.46) as $\Theta_1$, with

$$\Theta_1^{-1} = \frac{2}{\pi^2} \int_0^\pi \left| \log \cot \frac{\theta}{2} \right| d\theta.$$

That this is sharp in (11.46) when $t = 1$ is clear since

$$g_1(z) = \frac{2}{\pi} \log \frac{1+z}{1-z}, \tag{11.48}$$

maps $\mathbb{D}$ conformally to the strip $\{|\Im m\, g_1(z)| < 1\}$.

When $p < 1$ the sharp constant $\Theta_p$ in (11.47) is associated with the conformal map

$$g_0(z) = \frac{2z}{1-z^2}, \tag{11.49}$$

which maps $\mathbb{D}$ to the doubly-split plane

$$\mathsf{Q}_0 = \mathbb{C} \setminus \{it : t \in (-\infty, -1] \cup [1, \infty)\}$$

(note that these domains do not depend on $p \in (0, 1)$). Now we have

$$\Theta_p^p = \frac{1}{2\pi} \int_0^{2\pi} |g_0(e^{it})|^p = \frac{1}{2\pi} \int_{-\pi}^{\pi} |\csc t|^p \, dt, \tag{11.50}$$

which is finite when $p < 1$, but not when $p = 1$. The function $g_0$ has the representation (11.43), where the support of its associated measure $\nu_0$ is two points: $\nu_0(\{1\}) = 1/2 = -\nu_0(\{-1\})$. In general, when using (11.43), $\nu$ is used for the extremal measure, and the corresponding mappings and domains are decorated with subscripts.

The identification of $\Theta_p$ in (11.50) as the sharp constant for $C_p$ in (11.47) is a limiting case of the next theorem, which in turn is a consequence of our main result, which is stated as Theorem 11.29 in the next section. We also discuss there why it is a more natural vehicle.

Here we present our intermediate result and then show how $\Theta_1$ follows.

**Theorem 11.28** *Let $\mu$ be a real (signed, Borel) measure on $\mathbb{S}$, let $f$ be defined by* (11.43)*, and assume that $\|\mu\| \le 1, |\mu(\mathbb{S})| = b, 0 < b < 1$. Let $g = g_b$ be associated to the specific measure $\mu = \nu_b$ of* (11.51) *below.*

*Then for $0 < r < 1$:*

(i) $\int_{-\pi}^{\pi} |f(re^{i\theta})|^p \, d\theta \le \int_{-\pi}^{\pi} |g(re^{i\theta})|^p \, d\theta \quad (0 < p \le 2)$,
(ii) $\int_{-\pi}^{\pi} |\Im m f(re^{i\theta})|^p \, d\theta \le \int_{-\pi}^{\pi} |\Im m\, g(re^{i\theta})|^p \, d\theta \quad (1 \le p \le 2)$,
(iii) $\int_{-\pi}^{\pi} |\Re e f(re^{i\theta})|^p \, d\theta \le \int_{-\pi}^{\pi} |\Re e\, g(re^{i\theta})|^p \, d\theta \quad (1 \le p < \infty)$.

*Remark.* The limiting case $\|\mu\| = 1 = |\mu(\mathbb{S})|$ is not covered by this theorem, but follows directly. For that case, $\mu$ must be of one sign, say positive. The extremal measure has unit charge at $z = 1$, and corresponds to $g_1$ of (11.48), so that $f$ is subordinate to $g_1$. (Subordination was discussed in §11.5.)

### 11.8.2 The Constant $\Theta_1$; Preparation for Theorem 11.29

Theorem 11.28 will be obtained in §11.8.5 as a consequence of Theorem 11.29. Assuming Theorem 11.29 proved, we show that $\Theta_p$ being the best constant in (11.47) follows from (i) in Theorem 11.28. Normalize so in (11.43) $\|\mu\| = 1$. According to (11.44), $|v(e^{i\theta})| \le |f(e^{i\theta})|$, so if $r \to 1$ in (i),

$$\int_{-\pi}^{\pi} |g_b(e^{i\theta})|^p \, d\theta \le \int_{-\pi}^{\pi} |g_0(e^{i\theta})|^p \, d\theta \qquad (0 < p < 1, 0 < b < 1)$$

where the $\{g_b\}$ were introduced in §11.8.1, and $g_0$ is from (11.49).

To compute the right integral, we consider intervals on $\mathbb{S}$ on which $\cos\theta$ is positive or negative, and find that

$$\begin{aligned}\int_0^{\pi} |g_b(e^{i\theta})|^p \, d\theta &= \int_{-\pi}^{\pi} \left|\frac{1 + b\cos\theta}{\sin\theta}\right|^p d\theta \\ &= \int_0^{\pi} \frac{(1 + b\cos\theta)^p + (1 - b\cos\theta)^p}{|\sin\theta|^p} \, d\theta.\end{aligned}$$

But when $p < 1$, $x^p$ is a concave function of $x$, and so the integral is largest when $b = 0$. Thus $\Theta_p$ is the sharp bound when $p < 1$.

Theorem 11.29, which is stated below in this section, implies Theorem 11.28 but is awkward to formulate. It is a continuum of results, each requiring a bounded simply connected (Steiner symmetric) domain $Q$ and scalar $b$, $0 \le b < 1$; collectively, these are the domains $\mathcal{S}(Q, b)$, with $f\colon \mathbb{D} \to Q$ conformally and $f(0) = b$. Note that the hypothesis depends only on the real parts.

Theorem 11.28, in particular identifying the constants in (11.47), uses domains $\mathsf{Q}_b$ which are limiting cases of the domains $\mathcal{S}(Q, b)$. The "universal" domain $\mathsf{Q}_0$ has already been introduced when discussing the function $g_0$ from (11.49); it is $\mathbb{C}$ with the slits $\{it\colon t \in (-\infty, -1] \cup [1, \infty)\}$ deleted. When $0 < b < 1$, the extremal $g_b$ has the representation (11.43) with measure $\nu = \nu_b$, where

$$\nu_b(\{1\}) = (1 + b)/2, \quad \nu_b(\{-1\}) = -(1 - b)/2, \tag{11.51}$$

and zero otherwise. In particular, $g_b(0) = b$ and $g_b$ maps $\mathbb{D}$ conformally to

$$\mathsf{Q}_b = \mathbb{C} \setminus \{it\colon t \in (-\infty, -(1 - b^2)^{1/2}] \cup [(1 - b^2)^{1/2}, \infty)\}.$$

The more elaborate Theorem 11.29, as well as Theorem 11.28 when $b \neq 0$, thus will provide information not in Davis (1976). It is based on Baernstein (1978, Theorem 4). The theorem has the flavor of symmetrization, but, as we have remarked, Theorem 11.29 maximizes integral means over families of mappings into $Q$. Associated to $Q$ is $B > 0$ (in nontrivial cases) with

$$Q \cap \{\Re e\, w = 0\} = \{i[-B, B] : t \in \mathbb{R}\}.$$

Next, for $0 < \beta \leq B$, let $Q_\beta$ be the domain obtained from $Q$ by deleting the two intervals $[-iB, -i\beta]$ and $[i\beta, iB]$. Then, for each $b \in Q \cap \mathbb{R}$, we have a unique conformal mapping $h_{\beta,B} : \mathbb{D} \to Q_\beta$ subject to

$$h_{\beta,B}(0) = b, \quad h'_{\beta,B} > 0,$$

and this defines the collection of mappings

$$\mathcal{S}(Q, b) = \{h_{\beta,b} : 0 < \beta \leq B\}.$$

Our main result shows that for each choice of $b = f(0) \in \mathbb{R}$, the $p$-norms of mappings $f : \mathbb{D} \to Q$ are extremal for the associated conformal $h \in \mathcal{S}(Q, b)$.

**Theorem 11.29** *Let $f$ be analytic in $\mathbb{D}$ with $f(0)$ real and $f(D) \subset Q$, where $Q$ is as above. Suppose, in addition, that $h \in \mathcal{S}(Q, f(0))$ and*

$$\int_{-\pi}^{\pi} |\Re e f(e^{i\theta})|\, d\theta \leq \int_{-\pi}^{\pi} |\Re e\, h(e^{i\theta})|\, d\theta. \tag{11.52}$$

*Then for $0 < r < 1$:*

(i) $\int_{-\pi}^{\pi} |f(re^{i\theta})|^p\, d\theta \leq \int_{-\pi}^{\pi} |h(re^{i\theta})|^p\, d\theta \qquad (0 < p \leq 2)$,
(ii) $\int_{-\pi}^{\pi} |\Im m f(re^{i\theta})|^p\, d\theta \leq \int_{-\pi}^{\pi} |\Im m\, h(re^{i\theta})|^p\, d\theta \qquad (1 \leq p \leq 2)$,
(iii) $\int_{-\pi}^{\pi} |\Re e f(re^{i\theta})|^p\, d\theta \leq \int_{-\pi}^{\pi} |\Re e\, h(re^{i\theta})|^p\, d\theta \qquad (1 \leq p < \infty)$.

By subordination (see (Baernstein 1989, p. 841) this holds, for example, for $h = h_{\beta,b}$). In §11.8.5 we note that it is also valid for certain functions $g_b$ in the class discussed in 11.51.

**Remarks** If $h \in \mathcal{S}(Q, f(0))$ then $h(0) = f(0)$, so that the functionals $N$ and $\mathcal{I}$ ($\mathcal{I}$ being from (11.34)) may be used interchangeably. Inequalities (11.52) for all $p \geq 2$ are also valid when $h$ is the conformal map $h_{B,b}$, but it is not known whether they hold for all $h \in \mathcal{S}(Q, f(0))$.

### 11.8.3 Proof of Theorem 11.29

Since the assertions of the theorem are quantitative inequalities for all $r$, we may assume $f$ is analytic on $\mathbb{S} = \partial\mathbb{D}$. The formulation relies on image domains and real parts of functions, so we translate the picture to the $w$-plane, using an exponential change of variables and the counting function $N$.

The versatility of the function $[\cdot]^+$ as a building block for convex functions is important here; recall its essential role in Nevanlinna theory (§11.2). In particular, if $x$ is real and $p > 0$ then

$$\begin{aligned} e^{px} &= p^2 \int_{-\infty}^{\infty} (x-t)^+ e^{pt}\, dt, \\ |x|^p &= p(p-1) \int_0^{\infty} (|x|-t)^+ t^{p-2}\, dt \qquad (t \ge 0), \\ (|x|-t)^+ &= (x-t)^+ + (x+t)^+ - (x+t) \qquad (t \ge 0). \end{aligned} \tag{11.53}$$

With $s > 0$, Cartan's identity (11.4) when applied to $s^{-1}f(z)$ yields

$$\int_{-\pi}^{\pi} \left[\log|f(re^{i\theta})| - \log s\right]^+ d\theta = \int_{-\pi}^{\pi} N(r, se^{i\varphi})\, d\varphi + 2\pi \log^+ \left|\frac{f(0)}{s}\right|,$$

so, if $x = \log|f(re^{i\theta})|$, $t = \log s$, the first identity of (11.53) gives that

$$\begin{aligned} &\int_{-\pi}^{\pi} |f(re^{i\theta})|^p\, d\theta \\ &\quad = p^2 \int_{-\infty}^{\infty}\int_{-\infty}^{\infty} N(r,w)|w|^{p-2}\, dudv + 2\pi|f(0)|^p \qquad (p > 0). \end{aligned} \tag{11.54}$$

The corresponding formulas for $|\Re e f|$ and $|\Im m f|$ in place of $|f|$ rely on the second and third identities of (11.53). We start with

$$N(r, e^{t+i\varphi}, e^f) = \sum_{-\infty}^{\infty} N(t + i(\varphi + 2\pi k), f).$$

On integrating with respect to $\varphi$ and applying Cartan's formula to $e^f$, we obtain

$$\int_0^{2\pi} [\Re e f(re^{i\theta}) - t]^+ d\theta = \int_{-\infty}^{\infty} N(r, t+iv) dv + 2\pi[\Re e f(0) - t]^+.$$

Thus the third identity in (11.53) gives that

$$\begin{aligned} &\int_0^{2\pi} |\Re e f(re^{i\theta})| - t)^+ \\ &\quad = \int_{-\infty}^{\infty} [N(r, t+iv) + N(r, -t+iv)]\, dv + 2\pi(|\Re e f(0)| - t)^+. \end{aligned} \tag{11.55}$$

The second identity leads to an analogue of (11.54):

$$\begin{aligned}\int_0^{2\pi} &|\Re e f(re^{i\theta})|^p \, d\theta \\ &= p(p-1)\int_{-\infty}^{\infty}\int_{-\infty}^{\infty} N(r,w)|u|^{p-2}\, dudv + 2\pi\, |\Re e f(0)|^p\end{aligned} \tag{11.56}$$

when $p > 1$, and the same analysis applies to $|\Im m F|$. Note that the case $t = 0$ in (11.55) gives that

$$\int_0^{2\pi} |\Re e f(re^{i\theta})| d\theta = 2\int_{-\infty}^{\infty} N(r, iv) dv + 2\pi\, |\Re e f(0)|. \tag{11.57}$$

We next convert the hypotheses of Theorem 11.29 to ones amenable to our methods. We prove:

**Proposition 11.30** *Let $f$ and $h$ be as in Theorem 11.29. Then if $w = u + iv$ with $v > 0$ we have*

$$\int_{-v}^{v} N(r, u + is, f)\, ds \le \int_{-v}^{v} N(r, u + is, h)\, ds. \qquad (0 < r < 1).$$

Before proving Proposition 11.30 in the next section, let us assume it is established and obtain Theorem 11.29. For assertion (i), note that if $p \in (0, 2)$ then

$$|w|^{p-2} = |u + iv|^{p-2}$$

is a decreasing function of $v$ for each fixed $u$. The proposition with integration by parts then yields

$$\int_{-\infty}^{\infty} N(r, u + iv, f)|w|^{p-2}\, dv \le \int_{-\infty}^{\infty} N(r, u + iv, h)|w|^{p-2}\, dv \quad (0 < p \le 2)$$

for every $u$. Since $f(0) = h(0)$, (i) then is immediate from (11.54). Assertion (iii) follows from (11.56) with $v = \infty$ in Proposition 11.30. Next, let $1 < p \le 2$ and note that (11.56) is valid using the argument for (iii) if $\Re e f$ is replaced by $\Im m f$ and $u$ by $v$. Finally, to obtain (ii) when $p = 1$, if $v \to 0$ in Proposition 11.30 we have that $N(r, u, f) \le N(r, u, h)$, and using (11.57) apply parallel arguments to $\Im m f$.

### 11.8.4 Key Steps Toward Proposition 11.30

As suggested by the formulation of Proposition 11.30, the function $N$ (and so Lehto's theorem) is relevant here. We observe that

$$N(1, w, h) = \begin{cases} \log |h^{-1}(w)|^{-1} & (w \in \mathbb{D}) \\ 0 & (w \notin \mathbb{D}) \end{cases}$$

is $2\pi$ times the Green function of $h(\mathbb{D})$ with pole at $w = h(0)$ and that when $r < 1$

$$N(r, w, h) = [N(1, w, h) + \log r]^{+} \qquad (0 < r < 1).$$

The main issue in obtaining Proposition 11.30 is for $r = 1$; the case $r < 1$ needs only few extra lines in Baernstein (1978, p. 848). The novel issue in the formulation of Theorem 11.29 is the appearance of the real and imaginary parts of $f$ and $h$. For example, if $r \in (0, 1]$, then $N(r, u + iv, h)$ is an even function of $v$ which decreases as $v$ increases, and its analysis calls for a Steiner ⋆-function, based on vertical segments rather than circular arcs. The machinery to do this has been developed earlier, in §9.11. However, complex analysis offers another method, which was the approach used in Baernstein (1978), and which will be followed here.

The exponential mapping reduces Steiner ⋆-function issues to the standard situation of symmetrization on circular arcs. Thus, $w = e^z$ will transform $\Omega = h(\mathbb{D})$ to a circularly symmetric region which, in general, will be neither simply connected nor a homeomorphic image in $\mathbb{C}$, but if $a$ is sufficiently small, the image of $h(\mathbb{D})$ by $e^{az}$ avoids these issues. Further, Steiner symmetrization of $N(1, u + iv, h)$ $(N(1, r, u + iv, f))$ with respect to $v$ corresponds to circular symmetrization on arcs centered on the real axis. Details of this procedure are outlined in Baernstein (1978, §5). Here we simplify notation by taking $a = 1$ and suppose that these issues have been resolved.

The conditions in §11.8.2 assert that the dominant region $Q$ is bounded and Steiner symmetric as is each domain $h(\mathbb{D})$.

Associated to a function $f$ analytic in $\{|z| \le r\}$ is another function in the upper half plane:

$$N^{\star}(r, w, f) = N^{\star}(r, u + iv, f) = \sup_{E} \int_{E} N(r, u + is, f)\, ds, \qquad (v > 0),$$

with the sup now over all subsets of $\mathbb{R}$ with $|E| = 2v$. Since $N$ is nonnegative with compact support, $N^{\star}$ is well defined.

Proposition 11.30 thus becomes an immediate consequence of the stronger inequality (again manifesting the centrality of the ⋆-function )

$$N^{\star}(r, u + iv, f) \le N^{\star}(r, u + iv, h),$$

and the proof will follow from arguments already encountered in the proof of Theorem 11.22, now with no need to consider behavior at $\infty$.

This procedure depends on the remark following Theorem 11.21 that although $N(r, w, f)$ is not subharmonic, it differs from being subharmonic by a fixed addend which depends only on $f(0)$, while $N(r, w, h)$ differs from a harmonic factor by the same addend. Thus we need only show that the subharmonic function

$$N^{\star}(r, w, f) - N^{\star}(r, w, h)$$

is nonpositive.

This calls for the maximum principle. Whenever $\varepsilon > 0$, the function

$$P(w) = N^{\star}(1, w, f) - N^{\star}(1, w, h) - \varepsilon v$$

is subharmonic in the upper half of the bounded region $h(\mathbb{D})^{+}$ and continuous on its closure. Let $M (= M(\varepsilon))$ be the maximum of $P$; Proposition 11.30 will follow once we show that $M = 0$ independent of $\varepsilon$.

Since $\Omega = h(\mathbb{D})^{+} \subset Q$, only points $w_0 = u_0 + iv_0 \in \partial\Omega$ (so that $v_0 \geq 0$) are locations of possible extrema (maximum principle), and the hypotheses $f(0) = h(0)$ and (11.52) will be important. When $v_0 = 0$, all three constituent terms in $P$ are zero, so $P \equiv 0$, and the strategy becomes to show that $v_0 > 0$ is not possible.

If the vertical line $\{u = u_0\}$ does not intersect $\overline{\Omega}$, both terms $N^{\star}$ vanish, and since $v_0 > 0$ we see $P < 0$. Otherwise, $\Omega \cap \{\Re w = u_0\}$ is a segment $[-v_0, v_0]$ with $v_0 = v_0(u_0)$. Details for the case $v_0 > 0, u_0 \neq 0$ are in Baernstein (1978, p. 847), and depend on the fact that

$$N^{\star}(1, u_0 + iv, f) = \int_{-v}^{v} A(t)\, dt \qquad (0 < v < \infty),$$

and A$(t)$ is continuous except when $u_0 = b$ and $v = 0$; however, we have already disposed of the possibility $v = 0$ as location of a maximum.

We present the special case $u_0 = 0$, so that $w_0 = iv_0$ with $v_0 > 0$; this shows the importance of our hypothesis (11.52). Let the intersection of $h(\mathbb{D})$ with the imaginary axis be the segment $[-v_0, v_0]$. An appeal to (11.57) (and remembering $f(0)$ and $h(0)$ are real) shows that

$$N^{\star}(1, iv_0, h) = \int_{-\infty}^{\infty} N(1, iv, h)\, dv = \frac{1}{2}\int_{-\pi}^{\pi} |\Re e\, h(e^{i\theta})|\, d\theta - \pi\,|h(0)|,$$

and since $N \geq 0$,

$$N^{\star}(1, iv_0, f) \leq \int_{-\infty}^{\infty} N(1, iv, f)\, dv = \frac{1}{2}\int_{-\pi}^{\pi} |\Re e f(e^{i\theta})|\, d\theta - \pi\,|f(0)|;$$

we recall (11.52) and see that $P(w_0) \le -\varepsilon v_0 < 0$: there is no maximum at $w_0$. This means that $P(w) \le 0$ on $h(\mathbb{D})^+$, and hence on its closure.

This settles the proposition when $r = 1$, and the reader is referred to p. 848 of Baernstein (1978) for the reduction when $r < 1$.

### 11.8.5 Proof of Theorem 11.28

After this preparation, Theorem 11.28 is not difficult to prove. Let $f$ be a possible extremal. Assume that $f(0) = \mu(\mathbb{S}) = b$ where $0 \le b < 1$ (the limiting case $b = 1$ was considered after the statement of Theorem 11.28). For the moment, assume that when $R < 1$,

$$\frac{1}{2\pi}\int_{-\pi}^{\pi} |\Re e\, f(Re^{i\theta})|\, d\theta < 1; \tag{11.58}$$

this issue will be resolved at the end of this section.

Assuming (11.58), we recall (11.44) and take $\rho \in (0, 1)$ with

$$\frac{1}{2\pi}\int_{-\pi}^{\pi} |\Re e\, f(Re^{i\theta})|\, d\theta \le \frac{1}{2\pi}\int_{-\pi}^{\pi} |\Re e\, g_b(\rho e^{i\theta})|\, d\theta.$$

Choose $\sigma$ sufficiently large that $M(R,f) < \sigma$, $M(\rho, g_b) < \sigma$. Recall from §11.8.1 that the range of $g_b$ is the plane slit from $i\lambda$ to $i\infty$ and $-i\infty$ to $-i\lambda$, with $\lambda = \lambda(b)$, and since by hypothesis $h(0) = b$, we have $h \in \mathcal{S}(|w| < \sigma, f(0))$, using the notation introduced before the statement of Theorem 11.29. This means that $g_b$ is subordinate to $h$, and so the right side of the last inequality does not decrease when $g_b$ is replaced by $h$. Thus if $r \in (0, 1)$ and $0 < p \le 2$,

$$\int_{-\pi}^{\pi} |f(Re^{i\theta})|^p\, d\theta \le \int_{-\pi}^{\pi} |h(re^{i\theta})|^p\, d\theta \le \int_{-\pi}^{\pi} |g_b(re^{i\theta})|^p\, d\theta;$$

the left inequality is a consequence of Theorem 11.29, and the right inequality follows from subordination. Assertion (i) follows on letting $R \to 1$, and the other assertions are obtained in the same manner.

To conclude the proof, we note that $f$ is not constant, and we need only justify (11.58). In the case that $\|\mu\| < 1$, this is immediate, so suppose $\|\mu\| = 1$. In that case, the integral in (11.58) would be identically 1 for all $z$ in the annulus $A = A(R, 1)$, and so $|\Re e f|$ would be harmonic in $A$. However, a harmonic function cannot have an interior minimum and so $\Re e f$ would have constant sign. This negates the assumption $b < 1$ since in this case we would have

$$1 = \lim_{r\to 1}\left|\frac{1}{2\pi}\int_{-\pi}^{\pi} \Re e\, f(re^{i\theta})\, d\theta\right| = b.$$

Hence (11.58) is valid.

## 11.9 Symmetrization and the Hyperbolic Metric

This section relies upon parts of Chapter 10 and is in many ways complementary to Sections 11.6 and 11.8. We no longer restrict ourselves to simply connected domains, although simply connected domains appear in an extremal role (not necessarily as domains in $\mathbb{C}$). Our treatment has overlap with Hayman (1989, chapter 9), but Hayman's approach is somewhat idiosyncratic, since it avoids Riemann surfaces. A good source for the preliminary material is Ahlfors (1973), and only the barest outline is given here. The key fact is there is metric, *the hyperbolic metric*, associated to any plane domain whose complement contains at least two points (alternatively, a domain whose complement with respect to the Riemann sphere has at least three points) and this metric has many monotonicity properties. While one can define the metric directly, much as in Hayman, it is best motivated by recalling its background.

These insights have been influential in algebraic topology, and play an important role in the original proof of Picard's theorem (recall §11.2). Picard's original proof was astonishingly short, and relied on Proposition 11.31. The fundamental group and covering transformations are considered basic tools of topology.

### The Universal Cover and Hyperbolic Metric

Let $\Omega \subset \mathbb{C}$ be an arbitrary domain whose complement contains at least two points. Fix $z_0 \in \Omega$. A standard topological construction associates a *fundamental group* $\Gamma$ to $\Omega$ with base point $z_0$. We consider all closed curves $\gamma\colon [0,1] \to \Omega$, normalized by $\gamma(0) = \gamma(1) = z_0$. Reparametrization identifies $\gamma$ with its image in $\Omega$. Then $\Gamma$ may be endowed with a group structure, with $-\gamma(t) = \gamma(1-t)$ and the addition $\gamma_1 + \gamma_2$ being defined as

$$(\gamma_1 + \gamma_2)(t) = \begin{cases} \gamma_1(2t) & 0 \le t \le 1/2, \\ \gamma_2(2t-1) & 1/2 \le t \le 1. \end{cases}$$

The identity (neutral) element is the curve $e(t) \equiv z_0$, $0 \le t \le 1$. The standard notation for the fundamental group is $\Gamma = \pi_1(\Omega)$.

Although in principle the definition depends on the choice of $z_0$, replacing $z_0$ by a point $z_1 \ne z_0$ converts $\Gamma$ to an isomorphic group.

Being a (non-commutative) group, $\Gamma$ has a family of subgroups, which range from $\Gamma$ itself down to the trivial group $\{e\}$. A topological construction based on liftings of curves in $\gamma$ associates to any subgroup $\Gamma_1$ of $\Gamma$ a *covering surface* $\Omega_1$ which is locally homeomorphic to $\Omega$ (in fact this local homeomorphism endows $\Omega_1$ with a complex structure so that $\Omega_1$ is a Riemann surface). Thus, the projection

$$\Pi_1 \colon \Omega_1 \to \Omega \tag{11.59}$$

is open, conformal, and made unique by the choices

$$\Pi_1(z_1) = z_0, \qquad \Pi_1'(z_1) = e^{i\alpha}, \alpha \in [0, 2\pi),$$

where $z_1 \in \Omega_1$ and $\alpha$ may be freely chosen. The derivative in the second equation is with respect to the induced conformal structure of $\Omega_1$ (and so is not canonical). If $z_1$ is replaced by $z_1^*$, with induced map $\Pi_1^*$, there is a conformal self map $\varphi_1$ of $\Omega_1$ with $\varphi_1(z_1^*) = z_1$ and

$$\Pi_1^* = \Pi_1 \circ \varphi_1.$$

The map $\varphi_1$ thus is a *cover transformation* of $\Omega_1$.

In particular, when $\Gamma_1 = \{e\}$ is the subgroup consisting of the identity element, $\Omega_1$ is simply connected, and is known as the *universal cover* of $\Omega$, denoted $\widetilde{\Omega}$. The universal cover is a purely topological phenomenon, but when the complement of $\Omega$ has at least two points, $\widetilde{\Omega}$ has the following remarkable analytic property, a consequence of the uniformization theorem:

**Proposition 11.31** (Ahlfors 1973, chapter 10) *Let $\Omega \subset \mathbb{C}$ be a domain whose complement contains at least two points and let $\widetilde{\Omega}$ be its universal cover. Then there is a complex-analytic homeomorphism*

$$F \colon \widetilde{\Omega} \to \mathbb{D},$$

*which is unique under the normalizations*

$$F(\widetilde{z_0}) = w_0 \qquad \arg(F'(\widetilde{z_0})) = \alpha \in [0, 2\pi),$$

*where $\widetilde{z_0}, w_0$ may be chosen arbitrarily.*

A corollary to this proposition endows these domains $\Omega$ with a remarkable metric. Let $F$ be the map produced by the proposition and $\Pi^* \colon \widetilde{\Omega} \to \Omega$ the specific projection in (11.59) with $\Omega_1 = \widetilde{\Omega}$. Since these are open mappings and $\mathbb{D}$ is simply connected, the composite

$$\phi = \Pi^* \circ F^{-1} \colon \mathbb{D} \to \Omega, \quad \phi(w_0) = z_0, \ \arg \phi'(w_0) = \alpha \in [0, 2\pi), \tag{11.60}$$

is (single valued and) holomorphic (monodromy theorem). If we change $z_0$ and $\alpha$ in (11.60), the resulting map $\phi^*$ may be factored

$$\phi^* = \phi \circ h,$$

with $h$ a conformal self-map of $\mathbb{D}$ (hence a Möbius transformation).

**Remark** In the study of Riemann surfaces, a surface conformally equivalent to $\mathbb{D}$ is called hyperbolic. This justifies calling a plane domain $\Omega$ having at least two points in its complement hyperbolic.

The mapping $\phi$ of (11.60) allows the hyperbolic metric (Poincaré metric) $\rho$ for $\Omega$ to be defined. Let $z_0 \in \Omega$ and $\phi\colon \mathbb{D} \to \Omega$ with $\phi(w_0) = z_0$, and set

$$\rho(z_0) = \rho(z_0, \Omega) = \frac{2}{(1 - |w_0|^2)|\phi'(w_0)|},$$

which is uniquely determined (the choice of $\arg \phi'(w_0)$ is irrelevant; the factor 2 gives $\rho$ curvature $-1$). Also $\rho$ is $C^\infty$ in $\Omega$,

$$\Delta \log \rho(z) = \rho^2 \qquad (z \in \Omega), \tag{11.61}$$

and $\rho$ is *complete*:

$$\lim_{z \to \zeta} \rho(z) = \infty \qquad (\zeta \in \partial\Omega \cap \mathbb{C}). \tag{11.62}$$

The book of Hayman (1989) is worth consulting for a more detailed picture. In particular, $\rho$ is the maximal solution of (11.61) in $\Omega$.

This metric plays an important role in metric geometry. For example,

**Proposition 11.32** (Principle of the hyperbolic metric; Nevanlinna 1970, §III.3; Hayman 1989, §9.4.1) *Let $z_0 \in \Omega_1$, $\zeta_0 \in \Omega_2$ and $f$ be holomorphic in a neighborhood of $z_0$ with $f(z_0) = \zeta_0$ and $f(\Omega_1) \subset \Omega_2$. Then*

$$\rho(\zeta_0, \Omega_2)|f'(z_0)| \le \rho(z_0, \Omega_1),$$

*and equality holds only when $\Omega_1 = \Omega_2$ and $f$ is a conformal homeomorphism.*

Hyperbolic distance in $\Omega$ has an intrinsic definition,

$$d(z, z') = \inf_\gamma \int_\gamma \rho(\zeta)\, |d\zeta|,$$

the infimum taken over all rectifiable curves $\gamma : [0, 1] \to \Omega$ with $\gamma(0) = z, \gamma(1) = z'$. The integral is the hyperbolic length of $\gamma$. When $\Omega = \mathbb{D}$ this specifies to

$$d(z, z') = \inf_\gamma \int_\gamma \frac{2}{1 - |z|^2}\, |dz|.$$

Proposition 11.32 yields a monotonicity property of the metric. For example, if $\Omega_1 \subset \Omega_2$, we take $f$ the identity map and see that

$$\rho(z_0, \Omega_1) \ge \rho(z_0, \Omega_2),$$

with equality possible only when the domains coincide.

### Symmetrization and the Universal Covering Map

The next theorem, from Weitsman (1986), extends part of Theorem 11.22 in §11.6 to comparisons involving the Poincaré metric:

**Theorem 11.33** *Let $\Omega \subset \mathbb{C}$ be a hyperbolic domain and $\Omega^\#$ its circular symmetrization. Let*

$$u_\Omega = -\log \rho_\Omega, \qquad \Omega(r) = \{\theta \in [-\pi, \pi) : re^{i\theta} \in \Omega\}.$$

*If $\Phi : \mathbb{R} \to \mathbb{R}$ is an increasing convex function and $\Omega(r)$ is nonempty, it follows that*

$$\int_{\Omega(r)} \Phi(u_\Omega(re^{i\theta}))\, d\theta \leq \int_{\Omega^\#(r)} \Phi(u_{\Omega^\#}(re^{i\theta}))\, d\theta.$$

*In other words,*

$$(u_\Omega)^\star(z) \leq (u_{\Omega^\#})^\star(z) \qquad (z \in \Omega^\star \equiv \Omega^\# \cap \mathbb{H}).$$

Theorem 11.33 is a consequence of a more general result about circular symmetrization in partial differential equations, due to Weitsman (1986). The result is close to Theorem 10.18 in this book for cap symmetrization (see also Theorem 10.12 and Example 10.13 for s.d.r.). For completeness, we state Weitsman's theorem. Notice that the functions $u$ and $v$ need not be $\infty$ at the boundary; compare with (11.62).

**Theorem 11.34** (Weitsman 1986, Theorem 1) *Let $B, C$ be real constants ($B \leq \infty, C < \infty, C \leq B$), $\Omega$ be a bounded region in $\mathbb{C}$, and $0 < g \in C^2(\mathbb{R})$ be strictly increasing and convex. If $u \in C^2(\Omega)$, $v \in C^2(\Omega^\#)$ satisfy*

$$\begin{cases} \Delta u = g(u) \\ \lim_{z \to \zeta} u(z) = B \ \ (\forall \zeta \in \partial\Omega); \end{cases}$$

$$\begin{cases} \Delta v \geq g(v) \\ \limsup_{z \to \zeta} v(z) \leq C \ \ (\forall \zeta \in \partial\Omega^\#), \end{cases}$$

*then*

$$v(r) \leq u(z) \qquad (\forall z \in \Omega, |z| = r).$$

The next corollary reveals the power of this theorem, including a remarkable extremal behavior of universal covering mappings.

**Corollary 11.35** *Let $\Omega$ and $\Omega^\#$ be as in Theorem 11.34.*

(i) *If $r \in (0, \infty)$ and $\Omega(r)$ is nonempty then*

$$\min_{\theta \in \Omega(r)} \rho_\Omega(re^{i\theta}) \geq \min_{\theta \in \Omega^\#(r)} \rho_{\Omega^\#}(re^{i\theta}) = \rho_{\Omega^\#}(r).$$

(ii) *Let $f$ be holomorphic in $\mathbb{D}$ with $f(\mathbb{D}) \subset \Omega$, and let $\psi$ be the universal cover map of $\mathbb{D}$ to $\Omega^\#$ normalized so that $\psi(0) = |f(0)|$. Then*

$$|f'(0)| \le |\psi'(0)| \text{ and } M(r,f) \le M(r,\psi), \qquad 0 < r < 1.$$

*Remark.* There is a natural supplement to (i) whose verification requires a separate argument (also based on the ⋆-function), which is given in Weitsman (1979), but omitted here: If $\Omega$ is a *symmetric* region about the positive real axis, then the hyperbolic metric is monotone about the axis. In particular

$$\theta \mapsto \rho_{\Omega^\#}(re^{i\theta}) \tag{11.63}$$

is symmetric increasing for $\theta \in \Omega^\#(r)$.

Note that when $\Omega^\#$ is simply connected, so that the universal cover map $\psi$ is univalent, assertion (ii) coincides with the final assertion in (11.36), and thus one part of Theorem 11.22 extends to a far more general situation. The complications in establishing generalizations of the other parts arise from the difficulty of handling regions of arbitrary connectivity, as well as the issue of boundary behavior. The analogues of the other assertions in (11.36) thus remain open. The chapter concludes with proofs of the two assertions in the corollary.

*Proof of claim* (i) We exhaust $\Omega$ by an increasing union of bounded regions $\Omega_n$. With $\rho_n$ the hyperbolic metric of $\Omega_n$, again $\rho_n \to \rho$ normally. Thus in a fixed $\Omega_n$, $\rho_n \equiv \infty$ on $\partial\Omega_n$ ($\rho_n$ is complete) while $\rho_\Omega$ is finite there. This means we may apply Weitsman's Theorem 11.34 for each $n$ and let $n \to \infty$ with $u = \log \rho_n$, $v = \log \rho_{\Omega^\#}$, and $B = \infty$, $C = \sup_{\partial\Omega_n^\#} v$.

*Proof of claim* (ii) Normalize so that $f(0) = \psi(0) = R > 0$. From the classical subordination principle (according to §11.5 (a)), we may suppose that $f$ is the universal cover map onto $\Omega$.

An elementary reduction reduces to the situation that $\psi$ is real when $z$ is real, as we now explain. Both $\psi(z)$ and $\overline{\psi(\bar{z})}$ are universal covers of $\Omega^\#$, so $h(z) = \psi^{-1}(\overline{\psi(\bar{z})})$ is a Möbius self map of $\mathbb{D}$ which fixes 0. In other words,

$$h(z) = e^{i\alpha}z, \quad \text{hence} \quad \psi(e^{i\alpha}z) = \overline{\psi(\bar{z})}.$$

Write $\ell(z) = \psi(e^{i\alpha/2}z)$, so that

$$\overline{\ell(\bar{z})} = \overline{\psi(\overline{e^{-i\alpha/2}z})} = \psi(e^{i\alpha}e^{-i\alpha/2}z) = \psi(e^{i\alpha/2}z) = \ell(z),$$

so we relabel with $M(r,\psi)$ unchanged and $h$ preserving the real axis. Thus $\psi$ preserves the real axis.

This reduction ensures that if $0 < r_0 < 1$, $\psi(r_0) = \sigma$ and $\gamma$ is the branch of $\psi^{-1}$ taking $[R_0, \sigma]$ to $[0, r_0]$, then

$$\int_{R_0}^{\sigma} \rho_{\Omega^\#}(t)\, dt = \int_0^{r_0} \frac{2}{(1-r^2)}\, dr = \log \frac{1+r_0}{1-r_0}. \tag{11.64}$$

To estimate $f$, let $r_1$ be the modulus with $M(r_1,f) = \sigma$, and by rotation of independent variable assume that $|f(r_1)| = \sigma$. If $\Gamma$ is the $f$-image of $[0, r_1]$, $f$ being a universal cover map ensures that

$$\log \frac{1+r_1}{1-r_1} = \int_0^{r_1} \frac{2}{1-|z|^2}\, |dz| = \int_\Gamma \frac{2|\eta'(\zeta)|}{1-|\eta(\zeta)|^2}\, |d\zeta| = \int_\Gamma \rho_\Omega(\zeta)\, |d\zeta|,$$

where $\eta$ is the inverse branch of $f$. If $\zeta = f(t) = R(t)e^{is(t)}$, then

$$|d\zeta| = |R'(t) + is'(t)R(t)|\, dt \geq |R'(t)|\, dt.$$

We recall (11.63) and (11.64) and find that

$$\log \frac{1+r_1}{1-r_1} \geq \int_{R_0}^{\sigma} \min_{|\zeta|=r} \rho_\Omega(\zeta)(t)\, dt \geq \int_{R_0}^{\sigma} \rho_{\Omega^\#}(t)\, dt = \log \frac{1+r_0}{1-r_0}.$$

Thus $r_1 \geq r_0$, which implies (ii).

## 11.10 Notes and Comments

1. The original definition of $T^\star$ in Baernstein (1973) had a multiplicative factor $(1/2\pi)$, but beginning with Baernstein (1974) the expression used here became standard. This choice does have one unfortunate consequence, namely the unusual-looking first relation of Proposition 11.2(ii).

2. That the ⋆-function preserves subharmonicity was shown in Corollary 9.10, and the account in Baernstein (2002) is in this spirit. The original proof in Baernstein (1973), augmented in Baernstein (1974) with contributions from Sjögren and Essén, has independent interest and explicitly relies on Jensen's formula, itself the linchpin of Nevanlinna's theory. An account centered on the 2-dimensional theory is in §9.1 of Hayman (1989).

3. One of the main remaining open problems in classical Nevanlinna theory is, given $\lambda > 1$, find the best bound for the deficiency sum $\sum \delta_j$ taken over all entire/meromorphic functions of (lower) order $\lambda$. This problem is settled for $\lambda \leq 1$ in Corollary 11.8; in this situation, entire functions are extremals among the class of meromorphic functions of order $\lambda$. This pattern is likely not true for general $\lambda > 1$, although examples suggest entire functions remain extremal over all functions of that order $\lambda$ when $\lambda$ is nearly an integer Drasin and Weitsman, 1975. When $\lambda > 1$ and this sum cannot reach 2, finding the precise upper bound is totally open. Further comments appear above in the discussion following (11.10). The ⋆-function no longer seem useful for

these investigations, since the presumed extremal functions $v$ do not have $v(re^{i\theta})$ symmetric decreasing; compare with the discussion of §11.1.

A similar issue is encountered when considering variants of Proposition 11.12. In that proposition $\alpha > 0$, but very little is known when $\alpha < 0$ other than Proposition 11.13 (the case $\alpha = 0$ is the spread relation, Proposition 11.7); the problem is that there is no natural way of continuing a trigonometrically convex function $h$ beyond an interval $I$ when $h < 0$ on $\partial I$.

4. Definition 11.3 in fact now refers to what are known as Pólya peaks of the first kind; if the inequality is reversed, we have peaks of the second kind. These were introduced in Shea (1966).

5. The definition of local indicator in §11.3 is due to Edrei, but the theory is far older, with Cartwright (1956) and Levin (1980) standard references. The subtrigonometric properties enumerated in Lemma 11.6 are always the main weapon, and they adapt naturally to the localization in Pólya peak intervals.

6. Our §11.8 was based on Baernstein (1978). However, §3 of that paper considers analytic functions $f$ which are Stieltjes integrals (11.43) of measures $\mu$, rather than only Poisson integrals of $L^1$ functions. A conjugate measure $\tilde{\mu}$ can be formally associated to $\mu$ using radial boundary values, and Theorem 11.47 has a formulation and proof in that setting. Note that the extremal function for $p = 1$ has boundary values which are in $L^1$. We have seen in the discussion surrounding (11.51) that when $p < 1$ that extremals are analytic functions $g$ which are Stieltjes integrals of measures, and this formulation was also used by Davis. By taking $g(re^{i\theta})$ with $r$ arbitrarily close to 1, $\Theta_p$ is thus the best possible constant for functions with absolutely continuous boundary values, but the bound is not achieved in this reduced class.

The precise range for which Theorem 11.28 is valid is not yet known.

7. The title of this chapter makes clear that only part of the well-developed theory of symmetrization in 2 dimensions is considered, specifically that part that depends on the $\star$-function. The subject is far more vast than this, with monographs dating back to at least the famous text of Pólya and Szegő (1951), with the treatise by Dubinin (2014) updating several areas. Several of the author's own contributions to symmetrization, some of which rely in part on the $\star$-function, are not discussed here; among them are Baernstein (1977, 1987a,b, 1989b, 1993, 1997), Baernstein and Brown (1982).

# References


Adams, D.R. 1988. A sharp inequality of J. Moser for higher order derivatives. *Ann. Math. (2)*, **128**(2), 385–398.

Ahlfors, L.V. 1954. On quasiconformal mappings. *J. Anal. Math.*, **3**, 1–58; correction, 207–208.

Ahlfors, L.V. 1966. *Lectures on Quasiconformal Mappings*. Manuscript prepared with the assistance of Clifford J. Earle, Jr. Van Nostrand Mathematical Studies, No. 10. Toronto, Ontario–New York–London: D. Van Nostrand Co., Inc.

Ahlfors, L.V. 1973. *Conformal Invariants: Topics in Geometric Function Theory*. McGraw-Hill Series in Higher Mathematics. New York: McGraw-Hill Book Co.

Ahlfors, L.V. 1978. *Complex Analysis: An Introduction to the Theory of Analytic Functions of One Complex Variable*. Third edn. International Series in Pure and Applied Mathematics. New York: McGraw-Hill Book Co.

Ahlfors, L.V. 1981. *Möbius Transformations in Several Dimensions*. Ordway Professorship Lectures in Mathematics. Minneapolis: University of Minnesota School of Mathematics.

Almgren, Jr., F.J., and Lieb, E.H. 1989. Symmetric decreasing rearrangement is sometimes continuous. *J. Amer. Math. Soc.*, **2**(4), 683–773.

Alvino, A., Trombetti, G., and Lions, P.-L. 1990. Comparison results for elliptic and parabolic equations via Schwarz symmetrization. *Ann. Inst. H. Poincaré Anal. Non Linéaire*, **7**(2), 37–65.

Alvino, A., Lions, P.-L., and Trombetti, G. 1991. Comparison results for elliptic and parabolic equations via symmetrization: a new approach. *Differ. Integr. Equ.*, **4**(1), 25–50.

Alvino, A., Trombetti, G., and Matarasso, S. 2002. Elliptic boundary value problems: comparison results via symmetrization. *Ricerche Mat.*, **51**(2), 341–355.

Anderson, G.D., Vamanamurthy, M.K., and Vuorinen, M.K. 1997. *Conformal Invariants, Inequalities, and Quasiconformal Maps*. Canadian Mathematical Society Series of Monographs and Advanced Texts. New York: John Wiley & Sons Inc.

Anderson, J.M., and Baernstein, II, A. 1978. The size of the set on which a meromorphic function is large. *Proc. London Math. Soc. (3)*, **36**(3), 518–539.

Andrews, G.E., Askey, R., and Roy, R. 1999. *Special Functions*. Encyclopedia of Mathematics and Its Applications, vol. 71. Cambridge: Cambridge University Press.

Ashbaugh, M.S., and Benguria, R.D. 1992. A sharp bound for the ratio of the first two eigenvalues of Dirichlet Laplacians and extensions. *Ann. Math. (2)*, **135**(3), 601–628.

Ashbaugh, M.S., and Benguria, R.D. 1994. Isoperimetric inequalities for eigenvalue ratios. Pages 1–36 of: *Partial Differential Equations of Elliptic Type (Cortona, 1992)*. Symposia Mathematica, XXXV. Cambridge: Cambridge University Press.

Ashbaugh, M.S., and Benguria, R.D. 1995. On Rayleigh's conjecture for the clamped plate and its generalization to three dimensions. *Duke Math. J.*, **78**(1), 1–17.

Ashbaugh, M.S., Benguria, R.D., and Laugesen, R.S. 1997. Inequalities for the first eigenvalues of the clamped plate and buckling problems. Pages 95–110 of: *General Inequalities, 7 (Oberwolfach, 1995)*. International Series of Numerical Mathematics, vol. 123. Basel: Birkhäuser.

Astala, K., Iwaniec, T., and Martin, G. 2009. *Elliptic Partial Differential Equations and Quasiconformal Mappings in the Plane*. Princeton Mathematical Series, vol. 48. Princeton, NJ: Princeton University Press.

Atkinson, K., and Han, W. 2012. *Spherical Harmonics and Approximations on the Unit Sphere: An Introduction*. Lecture Notes in Mathematics, vol. 2044. Heidelberg: Springer.

Aubin, T. 1976. Problèmes isopérimétriques et espaces de Sobolev. *J. Differential Geometry*, **11**(4), 573–598.

Baernstein, II, A. 1997. Comparison of $p$-harmonic measures of subsets of the unit circle. *Algebra i Analiz*, **9**(3), 141–149.

Baernstein, II, A. 1973. Proof of Edrei's spread conjecture. *Proc. London Math. Soc. (3)*, **26**, 418–434.

Baernstein, II, A. 1974. Integral means, univalent functions and circular symmetrization. *Acta Math.*, **133**, 139–169.

Baernstein, II, A. 1977. Regularity theorems associated with the spread relation. *J. Anal. Math.*, **31**, 76–111.

Baernstein, II, A. 1978. Some sharp inequalities for conjugate functions. *Indiana Univ. Math. J.*, **27**(5), 833–852.

Baernstein, II, A. 1980. How the $*$-function solves extremal problems. Pages 639–644 of: *Proceedings of the International Congress of Mathematicians (Helsinki, 1978)*. Helsinki: Acad. Sci. Fennica.

Baernstein, II, A. 1987a. Dubinin's symmetrization theorem. Pages 23–30 of: *Complex Analysis, I (College Park, Md., 1985–86)*. Lecture Notes in Mathematics, vol. 1275. Berlin: Springer.

Baernstein, II, A. 1987b. On the harmonic measure of slit domains. *Complex Variables Theory Appl.*, **9**(2–3), 131–142.

Baernstein, II, A. 1989a. Convolution and rearrangement on the circle. *Complex Variables Theory Appl.*, **12**(1–4), 33–37.

Baernstein, II, A. 1989b. Some topics in symmetrization. Pages 111–123 of: *Harmonic Analysis and Partial Differential Equations (El Escorial, 1987)*. Lecture Notes in Mathematics, vol. 1384. Berlin: Springer.

Baernstein, II, A. 1993. An extremal property of meromorphic functions with $n$-fold symmetry. *Complex Variables Theory Appl.*, **21**(3–4), 137–148.

Baernstein, II, A. 1994. A unified approach to symmetrization. Pages 47–91 of: *Partial Differential Equations of Elliptic Type (Cortona, 1992)*. Sympos. Math., XXXV. Cambridge: Cambridge University Press.

Baernstein, II, A. 2002. The $*$-function in complex analysis. Pages 229–271 of: *Handbook of Complex Analysis: Geometric Function Theory, Vol. 1*. Amsterdam: North-Holland.

Baernstein, II, A., and Brown, J.E. 1982. Integral means of derivatives of monotone slit mappings. *Comment. Math. Helv.*, **57**(2), 331–348.

Baernstein, II, A., and Loss, M. 1997. Some conjectures about $L^p$ norms of $k$-plane transforms. *Rend. Sem. Mat. Fis. Milano*, **67**, 9–26 (2000).

Baernstein, II, A., and Manfredi, J.J. 1983. Topics in quasiconformal mapping. Pages 819–862 of: *Topics in Modern Harmonic Analysis, Vol. I, II (Turin/Milan, 1982)*. Rome: Istituto Nazionale di Alta Matematica "Francesco Severi".

Baernstein, II, A., and Schober, G. 1980. Estimates for inverse coefficients of univalent functions from integral means. *Israel J. Math.*, **36**(1), 75–82.

Baernstein, II, A., and Taylor, B.A. 1976. Spherical rearrangements, subharmonic functions, and $*$-functions in $n$-space. *Duke Math. J.*, **43**(2), 245–268.

Baernstein, II, A., Laugesen, R.S., and Pritsker, I.E. 2011. Moment inequalities for equilibrium measures in the plane. *Pure Appl. Math. Q.*, **7**(1), 51–86.

Bandle, C. 1980. *Isoperimetric Inequalities and Applications*. Monographs and Studies in Mathematics, vol. 7. Boston: Pitman (Advanced Publishing Program).

Bañuelos, R., and Carroll, T. 1994. Brownian motion and the fundamental frequency of a drum. *Duke Math. J.*, **75**(3), 575–602.

Bañuelos, R., van den Berg, M., and Carroll, T. 2002. Torsional rigidity and expected lifetime of Brownian motion. *J. London Math. Soc. (2)*, **66**(2), 499–512.

Barthe, F. 1998. Optimal Young's inequality and its converse: a simple proof. *Geom. Funct. Anal.*, **8**(2), 234–242.

Barthe, F. 2003. Autour de l'inégalité de Brunn-Minkowski. *Ann. Fac. Sci. Toulouse Math. (6)*, **12**(2), 127–178.

Beckner, W. 1975. Inequalities in Fourier analysis. *Ann. Math. (2)*, **102**(1), 159–182.

Beckner, W. 1991. Moser-Trudinger inequality in higher dimensions. *Int. Math. Res. Notices*, no. 7, 83–91.

Beckner, W. 1992. Sobolev inequalities, the Poisson semigroup, and analysis on the sphere $S^n$. *Proc. Natl. Acad. Sci. U.S.A.*, **89**(11), 4816–4819.

Beckner, W. 1993. Sharp Sobolev inequalities on the sphere and the Moser-Trudinger inequality. *Ann. Math. (2)*, **138**(1), 213–242.

Beckner, W. 1995. Geometric inequalities in Fourier anaylsis. Pages 36–68 of: *Essays on Fourier Analysis in Honor of Elias M. Stein (Princeton, NJ, 1991)*. Princeton Math. Ser., vol. 42. Princeton, NJ: Princeton University Press.

Bennett, C., and Sharpley, R. 1988. *Interpolation of Operators*. Pure and Applied Mathematics, vol. 129. Boston: Academic Press Inc.

Bennett, J., Carbery, A., Christ, M., and Tao, T. 2008. The Brascamp–Lieb inequalities: finiteness, structure and extremals. *Geom. Funct. Anal.*, **17**(5), 1343–1415.

Bérard, P. 1986. *Analysis on Riemannian Manifolds and Geometric Applications: An Introduction*. Monografías de Matemática [Mathematical Monographs], vol. 42. Rio de Janeiro: Instituto de Matemática Pura e Aplicada (IMPA).

Betsakos, D. 2004. Symmetrization, symmetric stable processes, and Riesz capacities. *Trans. Amer. Math. Soc.*, **356**(2), 735–755.

Bhayo, B.A., and Vuorinen, M. 2011. On Mori's theorem for quasiconformal maps in the $n$-space. *Trans. Amer. Math. Soc.*, **363**(11), 5703–5719.

Blaschke, W. 1917. Über affine Geometrie XI: Lösung des "Vierpunktproblems" von Sylvester aus der Theorie der geometrischen Wahrscheinlichkeiten. *Leipziger Ber.*, **69**, 436–453.

Blaschke, W. 1923. *Vorlesungen über Differentialgeometrie und geometrische Grundlagen von Einsteins Relativitätstheorie. II. Affine Differentialgeometrie*. Berlin: Springer-Verlag.

Bliss, G.A. 1930. An integral inequality. *J. London Math. Soc.*, **5**(11), 40–46.

Blumenthal, R.M., and Getoor, R.K. 1968. *Markov Processes and Potential Theory*. Pure and Applied Mathematics, Vol. 29. New York: Academic Press.

Bobkov, S. 1996. A functional form of the isoperimetric inequality for the Gaussian measure. *J. Funct. Anal.*, **135**(1), 39–49.

Bobkov, S.G., and Houdré, C. 1997. Some connections between isoperimetric and Sobolev-type inequalities. *Mem. Amer. Math. Soc.*, **129**(616).

Bonnesen, T., and Fenchel, W. 1987. *Theory of Convex Bodies*. Translated from the German and edited by L. Boron, C. Christenson and B. Smith. Moscow, ID: BCS Associates.

Borell, C. 1975. The Brunn-Minkowski inequality in Gauss space. *Invent. Math.*, **30**(2), 207–216.

Borell, C. 1985. Geometric bounds on the Ornstein-Uhlenbeck velocity process. *Z. Wahrsch. Verw. Gebiete*, **70**(1), 1–13.

Brascamp, H.J., Lieb, E.H., and Luttinger, J.M. 1974. A general rearrangement inequality for multiple integrals. *J. Funct. Anal.*, **17**, 227–237.

Brascamp, H.J., and Lieb, E.H. 1976. On extensions of the Brunn–Minkowski and Prékopa–Leindler theorems, including inequalities for log concave functions, and with an application to the diffusion equation. *J. Funct. Anal.*, **22**(4), 366–389.

Brenier, Y. 1991. Polar factorization and monotone rearrangement of vector-valued functions. *Comm. Pure Appl. Math.*, **44**(4), 375–417.

Brock, F., and Solynin, A.Y. 2000. An approach to symmetrization via polarization. *Trans. Amer. Math. Soc.*, **352**(4), 1759–1796.

Brothers, J.E., and Ziemer, W.P. 1988. Minimal rearrangements of Sobolev functions. *J. Reine Angew. Math.*, **384**, 153–179.

Burago, Y.D., and Zalgaller, V.A. 1988. *Geometric Inequalities*. Grundlehren der Mathematischen Wissenschaften [Fundamental Principles of Mathematical Sciences], vol. 285. Translated from the Russian by A. B. Sosinskiĭ, Springer Series in Soviet Mathematics. Berlin: Springer-Verlag.

Burchard, A. 1997. Steiner symmetrization is continuous in $W^{1,p}$. *Geom. Funct. Anal.*, **7**(5), 823–860.

Burchard, A., and Schmuckenschläger, M. 2001. Comparison theorems for exit times. *Geom. Funct. Anal.*, **11**(4), 651–692.

Burchard, A. 1996. Cases of equality in the Riesz rearrangement inequality. *Ann. Math. (2)*, **143**(3), 499–527.

Burchard, A., and Ferone, A. 2015. On the extremals of the Pólya-Szegő inequality. *Indiana Univ. Math. J.*, **64**(5), 1447–1463.

Burchard, A., and Hajaiej, H. 2006. Rearrangement inequalities for functionals with monotone integrands. *J. Funct. Anal.*, **233**(2), 561–582.

Caccioppoli, R. 1953. Misura e integrazione degli insiemi dimensionalmente orientati. *Rend. Acc. Naz. Lincei*, **12**, 3–11.

Carleman, T. 1918. Über ein Minimalproblem der mathematischen Physik. *Math. Z.*, **1**(2–3), 208–212.

Carlen, E., and Loss, M. 1992. Competing symmetries, the logarithmic HLS inequality and Onofri's inequality on $S^n$. *Geom. Funct. Anal.*, **2**(1), 90–104.

Carleson, L., and Chang, S.-Y.A. 1986. On the existence of an extremal function for an inequality of J. Moser. *Bull. Sci. Math. (2)*, **110**(2), 113–127.

Cartwright, M.L. 1956. *Integral Functions*. Cambridge Tracts in Mathematics and Mathematical Physics, No. 44. Cambridge: Cambridge University Press.

Chang, S.-Y.A. 1996. The Moser-Trudinger inequality and applications to some problems in conformal geometry. Pages 65–125 of: *Nonlinear Partial Differential Equations in Differential Geometry (Park City, UT, 1992)*. IAS/Park City Math. Ser., vol. 2. Providence, RI: Amer. Math. Soc.

Chasman, L.M., and Langford, J.J. 2016. The clamped plate in Gauss space. *Ann. Mat. Pura Appl. (4)*, **195**(6), 1977–2005.

Chavel, I. 1993. *Riemannian Geometry—A Modern Introduction*. Cambridge Tracts in Mathematics, vol. 108. Cambridge: Cambridge University Press.

Chavel, I. 2001. *Isoperimetric Inequalities*. Cambridge Tracts in Mathematics, vol. 145. Cambridge: Cambridge University Press.

Chiti, G. 1979. Rearrangements of functions and convergence in Orlicz spaces. *Applicable Anal.*, **9**(1), 23–27.

Chong, K.M. 1975. Variation reducing properties of decreasing rearrangements. *Canad. J. Math.*, **27**, 330–336.

Christ, M. 1984. Estimates for the $k$-plane transform. *Indiana Univ. Math. J.*, **33**(6), 891–910.

Cianchi, A., and Fusco, N. 2006. Minimal rearrangements, strict convexity and critical points. *Appl. Anal.*, **85**(1–3), 67–85.

Cordero-Erausquin, D., Nazaret, B., and Villani, C. 2004. A mass-transportation approach to sharp Sobolev and Gagliardo–Nirenberg inequalities. *Adv. Math.*, **182**(2), 307–332.

Crandall, M.G., and Tartar, L. 1980. Some relations between nonexpansive and order preserving mappings. *Proc. Amer. Math. Soc.*, **78**(3), 385–390.

Crowe, J.A., Zweibel, J.A., and Rosenbloom, P.C. 1986. Rearrangements of functions. *J. Funct. Anal.*, **66**(3), 432–438.

Davies, E.B. 1989. *Heat Kernels and Spectral Theory*. Cambridge Tracts in Mathematics, vol. 92. Cambridge: Cambridge University Press.

Davis, B. 1974. On the weak type (1,1) inequality for conjugate functions. *Proc. Amer. Math. Soc.*, **44**, 307–311.

Davis, B. 1976. On Kolmogorov's inequalities $\tilde{f}_p \leq C_p f_1$, $0 < p < 1$. *Trans. Amer. Math. Soc.*, **222**, 179–192.

de Branges, L. 1985. A proof of the Bieberbach conjecture. *Acta Math.*, **154**(1–2), 137–152.

De Giorgi, E. 1954. Su una teoria generale della misura $(r-1)$-dimensionale in uno spazio ad $r$ dimensioni. *Ann. Mat. Pura Appl. (4)*, **36**, 191–213.

De Giorgi, E. 1955. Nuovi teoremi relativi alle misure $(r - 1)$-dimensionali in uno spazio ad $r$ dimensioni. *Ricerche Mat.*, **4**, 95–113.

Dieudonné, J. 1951. Sur les espaces de Köthe. *J. Anal. Math.*, **1**, 81–115.

Doob, J.L. 1953. *Stochastic Processes*. New York: John Wiley & Sons Inc.

Draghici, C. 2005. A general rearrangement inequality. *Proc. Amer. Math. Soc.*, **133**(3), 735–743.

Drasin, D. 1976. The inverse problem of the Nevanlinna theory. *Acta Math.*, **138**(1–2), 83–151.

Drasin, D. 2015. Albert Baernstein II 1941–2014. *Notices Amer. Math. Soc.*, **62**(7), 815–818.

Drasin, D., and Weitsman, A. 1975. Meromorphic functions with large sums of deficiencies. *Adv. Math.*, **15**, 93–126.

Dubinin, V.N. 1987. Transformation of condensers in space. *Dokl. Akad. Nauk SSSR*, **296**(1), 18–20.

Dubinin, V.N. 1993. Capacities and geometric transformations of subsets in $n$-space. *Geom. Funct. Anal.*, **3**(4), 342–369.

Dubinin, V.N. 2014. *Condenser Capacities and Symmetrization in Geometric Function Theory*. Translated from the Russian by N. G. Kruzhilin. Basel: Springer.

Dunford, N., and Schwartz, J.T. 1958. *Linear Operators. I. General Theory*. With the assistance of W.G. Bade and R.G. Bartle. Pure and Applied Mathematics, Vol. 7. New York: Interscience Publishers, Inc.

Duren, P.L. 1983. *Univalent Functions*. Grundlehren der Mathematischen Wissenschaften [Fundamental Principles of Mathematical Sciences], vol. 259. New York: Springer-Verlag.

Durrett, R. 1984. *Brownian Motion and Martingales in Analysis*. Wadsworth Mathematics Series. Belmont, CA: Wadsworth International Group.

Edrei, A. 1965. Sums of deficiencies of meromorphic functions. *J. Anal. Math.*, **14**, 79–107.

Edrei, A. 1970. A local form of the Phragmén–Lindelöf indicator. *Mathematika*, **17**, 149–172.

Edrei, A. 1973. Solution of the deficiency problem for functions of small lower order. *Proc. London Math. Soc. (3)*, **26**, 435–445.

Edrei, A., and Fuchs, W.H.J. 1960. The deficiencies of meromorphic functions of order less than one. *Duke Math. J.*, **27**, 233–249.

Ehrhard, A. 1983a. Symétrisation dans l'espace de Gauss. *Math. Scand.*, **53**(2), 281–301.

Ehrhard, A. 1983b. Un principe de symétrisation dans les espaces de Gauss. Pages 92–101 of: *Probability in Banach Spaces, IV (Oberwolfach, 1982)*. Lecture Notes in Math., vol. 990. Berlin: Springer.

Ehrhard, A. 1984. Inégalités isopérimétriques et intégrales de Dirichlet gaussiennes. *Ann. Sci. École Norm. Sup. (4)*, **17**(2), 317–332.

Epperson, J.B. 1990. A class of monotone decreasing rearrangements. *J. Math. Anal. Appl.*, **150**(1), 224–236.

Erdélyi, A., Magnus, W., Oberhettinger, F., and Tricomi, F.G. 1954. *Tables of Integral Transforms. Vol. II*. New York–Toronto–London: McGraw-Hill Book Company, Inc. Based, in part, on notes left by Harry Bateman.

Essén, M., Rossi, J., and Shea, D. 1993. A convolution inequality with applications to function theory. II. *J. Anal. Math.*, **61**, 339–366.

Evans, L.C. 1998. *Partial Differential Equations*. Graduate Studies in Mathematics, vol. 19. Providence, RI: American Mathematical Society.

Evans, L.C., and Gariepy, R.F. 1992. *Measure Theory and Fine Properties of Functions*. Studies in Advanced Mathematics. Boca Raton, FL: CRC Press.

Faber, G. 1923. Beweis, dass unter allen homogenen Membranen von gleicher Fläche und gleicher Spannung die kreisförmige den tiefsten Grundton gibt. *Sitzungsber. Bayer. Akad. Wiss. Mnchen, Math.-Phys. Kl.*, 169–172.

Federer, H. 1959. Curvature measures. *Trans. Amer. Math. Soc.*, **93**, 418–491.

Federer, H. 1969. *Geometric Measure Theory*. Die Grundlehren der mathematischen Wissenschaften, Band 153. New York: Springer-Verlag New York Inc.

Federer, H., and Fleming, W.H. 1960. Normal and integral currents. *Ann. of Math. (2)*, **72**, 458–520.

Folland, G.B. 1999. *Real Analysis: Modern Techniques and Their Applications*. Second edition. Pure and Applied Mathematics (New York). New York: John Wiley & Sons Inc. A Wiley-Interscience Publication.

Folland, G.B. 2002. *Advanced Calculus*. Upper Saddle River, NJ: Prentice Hall.

Frostman, O. 1935. Potentiel d'équilibre et capacité de ensembles avec quelques applications à la théorie des fonctions. *Med. Lund. Univ. Math. Sem.*, **3**, 1–118.

Fuchs, W.H.J. 1963. Proof of a conjecture of G. Pólya concerning gap series. *Illinois J. Math.*, **7**, 661–667.

Fuchs, W.H.J. 1967. Developments in the classical Nevanlinna theory of meromorphic functions. *Bull. Amer. Math. Soc.*, **73**, 275–291.

Fuchs, W.H.J. 1974. A theorem on $\min_{|z|} \log |f(z)|/T(r,f)$. Pages 69–72. London Math. Soc. Lecture Note Ser., No. 12 of: *Proceedings of the Symposium on Complex Analysis (Univ. Kent, Canterbury, 1973)*. London: Cambridge University Press.

Gardner, R.J. 2002. The Brunn–Minkowski inequality. *Bull. Amer. Math. Soc. (N.S.)*, **39**(3), 355–405.

Gehring, F.W. 1961. Symmetrization of rings in space. *Trans. Amer. Math. Soc.*, **101**, 499–519.

Gehring, F.W. 1962. Rings and quasiconformal mappings in space. *Trans. Amer. Math. Soc.*, **103**, 353–393.

Gilbarg, D., and Trudinger, N.S. 1983. *Elliptic Partial Differential Equations of Second Order*. Second edition. Grundlehren der Mathematischen Wissenschaften, vol. 224. Berlin: Springer-Verlag.

Giusti, E. 1984. *Minimal Surfaces and Functions of Bounded Variation*. Monographs in Mathematics, vol. 80. Basel: Birkhäuser Verlag.

Godula, J., and Nowak, M. 1987. Meromorphic univalent functions in an annulus. *Rend. Circ. Mat. Palermo (2)*, **36**(3), 474–481 (1988).

Goldberg, A.A., and Ostrovskii, I.V. 2008. *Value Distribution of Meromorphic Functions*. Translations of Mathematical Monographs, vol. 236. Translated from the 1970 Russian original by Mikhail Ostrovskii, With an appendix by Alexandre Eremenko and James K. Langley. Providence, RI: American Mathematical Society.

González, C. 2000. Differential inequalities associated with weighted symmetrization processes on the real line. *J. Anal. Math.*, **82**, 21–54.

Govorov, N.V. 1969. The Paley conjecture. *Funkcional. Anal. i Priložen.*, **3**(2), 41–45.

Gross, L. 1975. Logarithmic Sobolev inequalities. *Amer. J. Math.*, **97**(4), 1061–1083.

Hadwiger, H., and Ohmann, D. 1956. Brunn–Minkowskischer Satz und Isoperimetrie. *Math. Z.*, **66**, 1–8.

Haliste, K. 1965. Estimates of harmonic measures. *Ark. Mat.*, **6**, 1–31 (1965).

Hamel, F., Nadirashvili, N., and Russ, E. 2011. Rearrangement inequalities and applications to isoperimetric problems for eigenvalues. *Ann. Math. (2)*, **174**(2), 647–755.

Hardy, G.H., and Littlewood, J.E. 1930. A maximal theorem with function-theoretic applications. *Acta Math.*, **54**(1), 81–116.

Hardy, G.H., Littlewood, J.E., and Pólya, G. 1952. *Inequalities*. Second edition. Cambridge: Cambridge University Press.

Hayman, W.K. 1950. *Symmetrization in the Theory of Functions*. Tech. Rep. no. 11, Navy Contract N6-ori-106 Task Order 5. Stanford University.

Hayman, W.K. 1964. *Meromorphic Functions*. Oxford Mathematical Monographs. Oxford: Clarendon Press.

Hayman, W.K. 1967. *Research Problems in Function Theory*. London: The Athlone Press University of London.

Hayman, W.K. 1989. *Subharmonic Functions. Vol. 2*. London Mathematical Society Monographs, vol. 20. London: Academic Press Inc. [Harcourt Brace Jovanovich Publishers].

Hayman, W.K. 1994. *Multivalent Functions*. Second edition. Cambridge Tracts in Mathematics and Mathematical Physics, No. 48. Cambridge: Cambridge University Press.

Hayman, W.K., and Kennedy, P.B. 1976. *Subharmonic Functions. Vol. I*. London Mathematical Society Monographs, No. 9. London: Academic Press [Harcourt Brace Jovanovich Publishers].

Hazewinkel, M. (ed.). 1995. *Encyclopaedia of Mathematics*. Translated from the Russian, Reprint of the 1988–1994 English translation. Dordrecht: Kluwer Academic Publishers.

Heinonen, J., Kilpeläinen, T., and Martio, O. 1993. *Nonlinear Potential Theory of Degenerate Elliptic Equations*. Oxford Mathematical Monographs. New York: Clarendon Press.

Henrot, A. (ed). 2017. *Shape Optimization and Spectral Theory*. Warsaw: De Gruyter Open.

Herbst, I.W., and Zhao, Z.X. 1988. Sobolev spaces, Kac-regularity, and the Feynman–Kac formula. Pages 171–191 of: *Seminar on Stochastic Processes, 1987 (Princeton, NJ, 1987)*. Progr. Probab. Statist., vol. 15. Boston: Birkhäuser Boston.

Hildén, K. 1976. Symmetrization of functions in Sobolev spaces and the isoperimetric inequality. *Manuscripta Math.*, **18**(3), 215–235.

Hoel, P.G., Port, S.C., and Stone, C.J. 1972. *Introduction to Stochastic Processes*. The Houghton Mifflin Series in Statistics. Boston.: Houghton Mifflin Co.

Hoffman, K. 1962. *Banach Spaces of Analytic Functions*. Prentice-Hall Series in Modern Analysis. Englewood Cliffs, NJ: Prentice-Hall, Inc.

Hörmander, L. 1994. *Notions of Convexity*. Progress in Mathematics, vol. 127. Boston: Birkhäuser Boston Inc.

Iwaniec, T., and Manfredi, J.J. 1989. Regularity of $p$-harmonic functions on the plane. *Rev. Mat. Iberoamericana*, **5**(1-2), 1–19.

Iwaniec, T., and Martin, G. 2001. *Geometric Function Theory and Non-Linear Analysis*. Oxford Mathematical Monographs. New York: Clarendon Press.

Jenkins, J.A. 1955. On circularly symmetric functions. *Proc. Amer. Math. Soc.*, **6**, 620–624.

Kac, M. 1949. On distributions of certain Wiener functionals. *Trans. Amer. Math. Soc.*, **65**, 1–13.

Karatzas, I., and Shreve, S.E. 1991. *Brownian Motion and Stochastic Calculus*. Second edition. Graduate Texts in Mathematics, vol. 113. New York: Springer-Verlag.

Kato, T. 1978. Trotter's product formula for an arbitrary pair of self-adjoint contraction semigroups. Pages 185–195 of: *Topics in Functional Analysis (Essays Dedicated to M. G. Kreĭn on the Occasion of his 70th Birthday)*. Adv. in Math. Suppl. Stud., vol. 3. New York: Academic Press.

Kawohl, B. 1985. *Rearrangements and Convexity of Level Sets in PDE*. Lecture Notes in Mathematics, vol. 1150. Berlin: Springer-Verlag.

Kellogg, O.D. 1967. *Foundations of Potential Theory*. Reprint from the first edition of 1929. Die Grundlehren der Mathematischen Wissenschaften, Band 31. Berlin: Springer-Verlag.

Kesavan, S. 2006. *Symmetrization & Applications*. Series in Analysis, vol. 3. Hackensack, NJ: World Scientific Publishing Co. Pte. Ltd.

Kirwan, W.E., and Schober, G. 1976. Extremal problems for meromorphic univalent functions. *J. Anal. Math.*, **30**, 330–348.

Köthe, G. 1960. *Topologische lineare Räume. I.* Die Grundlehren der mathematischen Wissenschaften, Bd. 107. Berlin: Springer-Verlag.

Krahn, E. 1925. Über eine von Rayleigh formulierte Minimaleigenschaft des Kreises. *Math. Ann.*, **94**(1), 97–100.

Krzyż, J. 1959. Circular symmetrization and Green's function. *Bull. Acad. Polon. Sci. Sér. Sci. Math. Astr. Phys.*, **7**, 327–330 (unbound insert).

Landkof, N.S. 1972. *Foundations of Modern Potential Theory*. Translated from the Russian by A.P. Doohovskoy, Die Grundlehren der mathematischen Wissenschaften, Band 180. New York: Springer-Verlag.

Langford, J. J. 2015a. Neumann comparison theorems in elliptic PDEs. *Potential Anal.*, **43**(3), 415–459.

Langford, J.J. 2015b. Symmetrization of Poisson's equation with Neumann boundary conditions. *Ann. Sc. Norm. Super. Pisa Cl. Sci. (5)*, **14**(4), 1025–1063.

Laugesen, R. 1993. Extremal problems involving logarithmic and Green capacity. *Duke Math. J.*, **70**(2), 445–480.

Laugesen, R.S., and Morpurgo, C. 1998. Extremals for eigenvalues of Laplacians under conformal mapping. *J. Funct. Anal.*, **155**(1), 64–108.

Lebedev, N.N. 1972. *Special Functions and Their Applications*. Revised edition, translated from the Russian and edited by R.A. Silverman, Unabridged and corrected republication. New York: Dover Publications Inc.

Ledoux, M. 1994. Semigroup proofs of the isoperimetric inequality in Euclidean and Gauss space. *Bull. Sci. Math.*, **118**(6), 485–510.

Lehto, O. 1953. A majorant principle in the theory of functions. *Math. Scand.*, **1**, 5–17.

Lehto, O., and Virtanen, K.I. 1973. *Quasiconformal Mappings in the Plane*. Second edition. Translated from the German by K. W. Lucas, Die Grundlehren der mathematischen Wissenschaften, Band 126. New York: Springer-Verlag.

Leung, Y.J. 1979. Integral means of the derivatives of some univalent functions. *Bull. London Math. Soc.*, **11**(3), 289–294.

Levin, B.J. 1980. *Distribution of Zeros of Entire Functions*. Revised edition. Translations of Mathematical Monographs, vol. 5. Translated from the Russian by R.P. Boas, J.M. Danskin, F.M. Goodspeed, J. Korevaar, A.L. Shields and H.P. Thielman. Providence, RI: American Mathematical Society.

Lieb, E.H. 1976–1977. Existence and uniqueness of the minimizing solution of Choquard's nonlinear equation. *Studies Appl. Math.*, **57**(2), 93–105.

Lieb, E.H. 1983. Sharp constants in the Hardy–Littlewood–Sobolev and related inequalities. *Ann. Math. (2)*, **118**(2), 349–374.

Lieb, E.H. 1990. Gaussian kernels have only Gaussian maximizers. *Invent. Math.*, **102**(1), 179–208.

Lieb, E.H., and Loss, Michael. 1997. *Analysis*. Graduate Studies in Mathematics, vol. 14. Providence, RI: American Mathematical Society.

Littlewood, J.E. 1944. *Lectures on the Theory of Functions*. London: Oxford University Press.

Littman, W. 1959. A strong maximum principle for weakly $L$-subharmonic functions. *J. Math. Mech.*, **8**, 761–770.

Lumiste, Ü., and Peetre, J. (eds.). 1994. *Edgar Krahn, 1894–1961*. Amsterdam: IOS Press.

Luttinger, J.M. 1973a. Generalized isoperimetric inequalities. *J. Math. Phys.*, **14**, 586–593.

Luttinger, J.M. 1973b. Generalized isoperimetric inequalities. II, III. *J. Math. Phys.*, **14**, 1444–1447, 1448–1450.

Marshall, A.W., Olkin, I., and Arnold, B.C. 2011. *Inequalities: Theory of Majorization and Its Applications*. Second edition. Springer Series in Statistics. New York: Springer.

Marshall, D.E. 1989. A new proof of a sharp inequality concerning the Dirichlet integral. *Ark. Mat.*, **27**(1), 131–137.

Mattila, P. 1995. *Geometry of Sets and Measures in Euclidean Spaces*. Cambridge Studies in Advanced Mathematics, vol. 44. Cambridge: Cambridge University Press.

Maz'ja, V.G. 1985. *Sobolev Spaces*. Springer Series in Soviet Mathematics. Translated from the Russian by T. O. Shaposhnikova. Berlin: Springer-Verlag.

McCann, R.J. 1995. Existence and uniqueness of monotone measure-preserving maps. *Duke Math. J.*, **80**(2), 309–323.

McCann, R.J. 2001. Polar factorizat3ion of maps on Riemannian manifolds. *Geom. Funct. Anal.*, **11**(3), 589–608.

McKean, H.P. 1973. Geometry of differential space. *Ann. Prob.*, **1**, 197–206.

McKean, Jr., H.P., and Singer, I.M. 1967. Curvature and the eigenvalues of the Laplacian. *J. Differential Geometry*, **1**(1), 43–69.

Méndez-Hernández, P. J. 2006. An isoperimetric inequality for Riesz capacities. *Rocky Mountain J. Math.*, **36**(2), 675–682.

Mori, A. 1956. On an absolute constant in the theory of quasi-conformal mappings. *J. Math. Soc. Japan*, **8**, 156–166.

Morpurgo, C. 2002. Sharp inequalities for functional integrals and traces of conformally invariant operators. *Duke Math. J.*, **114**(3), 477–553.

Morrey, Jr., C.B. 1938. On the solutions of quasi-linear elliptic partial differential equations. *Trans. Amer. Math. Soc.*, **43**(1), 126–166.

Morrey, Jr., C.B. 1940. Existence and differentiability theorems for the solutions of variational problems for multiple integrals. *Bull. Amer. Math. Soc.*, **46**, 439–458.

Moser, J. 1970/71. A sharp form of an inequality by N. Trudinger. *Indiana Univ. Math. J.*, **20**, 1077–1092.

Mostow, G.D. 1968. Quasi-conformal mappings in $n$-space and the rigidity of hyperbolic space forms. *Inst. Hautes Études Sci. Publ. Math.*, 53–104.

Nadirashvili, N.S. 1993. An isoperimetric inequality for the main frequency of a clamped plate. *Dokl. Akad. Nauk*, **332**(4), 436–439.

Nelson, E. 1973. The free Markoff field. *J. Funct. Anal.*, **12**, 211–227.

Nevanlinna, R. 1970. *Analytic Functions*. Translated from the second German edition by Phillip Emig. Die Grundlehren der mathematischen Wissenschaften, Band 162. Springer-Verlag: New York-Berlin.

Onofri, E. 1982. On the positivity of the effective action in a theory of random surfaces. *Comm. Math. Phys.*, **86**(3), 321–326.

Osgood, B., Phillips, R., and Sarnak, P. 1988. Extremals of determinants of Laplacians. *J. Funct. Anal.*, **80**(1), 148–211.

Osserman, R. 1978. The isoperimetric inequality. *Bull. Amer. Math. Soc.*, **84**(6), 1182–1238.

Paouris, G., and Pivovarov, P. 2012. A probabilistic take on isoperimetric-type inequalities. *Adv. Math.*, **230**(3), 1402–1422.

Paouris, G., and Pivovarov, P. 2017. Random ball-polyhedra and inequalities for intrinsic volumes. *Monatsh. Math.*, **182**(3), 709–729.

Payne, L.E. 1967. Isoperimetric inequalities and their applications. *SIAM Rev.*, **9**, 453–488.

Payne, L.E., Pólya, G., and Weinberger, H.F. 1955. Sur le quotient de deux fréquences propres consécutives. *C. R. Acad. Sci. Paris*, **241**, 917–919.

Peretz, R. 2017. Applications of Steiner symmetrization to some extremal problems in geometric function theory. *WSEAS Trans. Math.*, **16**, 350–367.

Petrenko, V.P. 1969. Growth of meromorphic functions of finite lower order. *Izv. Akad. Nauk SSSR Ser. Mat.*, **33**, 414–454.

Pfiefer, R.E. 1990. Maximum and minimum sets for some geometric mean values. *J. Theoret. Probab.*, **3**(2), 169–179.

Pfluger, A. 1946. Zur Defektrelation ganzer Funktionen endlicher Ordnung. *Comment. Math. Helv.*, **19**, 91–104.

Pichorides, S.K. 1972. On the best values of the constants in the theorems of M. Riesz, Zygmund and Kolmogorov. *Studia Math.*, **44**, 165–179. (errata insert). Collection of articles honoring the completion by Antoni Zygmund of 50 years of scientific activity, II.

Pitt, L.D. 1977. A Gaussian correlation inequality for symmetric convex sets. *Ann. Prob.*, **5**(3), 470–474.

Poincaré, H. 1887. Sur un théorème de M. Liapounoff, relatif à l'équilibre d'une masse fluide. *Comptes Rendus hebdomadaires des séances de l'Académie des Sciences*, **104**, 622–625.

Pólya, G., and Szegő, G. 1945. Inequalities for the capacity of a condenser. *Amer. J. Math.*, **67**, 1–32.

Pólya, G., and Szegő, G. 1951. *Isoperimetric Inequalities in Mathematical Physics*. Annals of Mathematics Studies, no. 27. Princeton, NJ: Princeton University Press.

Pólya, G. 1948. Torsional rigidity, principal frequency, electrostatic capacity and symmetrization. *Quart. Appl. Math.*, **6**, 267–277.

Pólya, G. 1950. Sur la symétrisation circulaire. *C.R. Acad. Sci. Paris*, **230**, 25–27.

Pólya, G. 1984. *Collected Papers. Vol. III*. Mathematicians of Our Time, vol. 21. Analysis, Edited by J. Hersch, G.-C. Rota, M.C. Reynolds and R.M. Shortt. Cambridge, MA: MIT Press.

Pruss, A.R. 1996. Three counterexamples for a question concerning Green's functions and circular symmetrization. *Proc. Amer. Math. Soc.*, **124**(6), 1755–1761.

Ransford, T. 1995. *Potential Theory in the Complex Plane*. London Mathematical Society Student Texts, vol. 28. Cambridge: Cambridge University Press.

Rayleigh, Baron, J.W. 1945. *The Theory of Sound*. Second edition. New York: Dover Publications.

Reed, M., and Simon, B. 1972. *Methods of Modern Mathematical Physics. I. Functional Analysis*. New York: Academic Press.

Reed, M., and Simon, B. 1975. *Methods of Modern Mathematical Physics. II. Fourier Analysis, Self-Adjointness*. New York: Academic Press [Harcourt Brace Jovanovich Publishers].

Reshetnyak, Y.G. 1989. *Space Mappings with Bounded Distortion*. Translations of Mathematical Monographs, vol. 73. Translated from the Russian by H.H. McFaden. Providence, RI: American Mathematical Society.

Rickman, S. 1993. *Quasiregular Mappings*. Ergebnisse der Mathematik und ihrer Grenzgebiete (3), vol. 26. Berlin: Springer-Verlag.

Riesz, F. 1930. Sur une inégalité intégrale. *J. London Math. Soc.*, **5**(3), 162–168.

Riesz, M. 1928. Sur les fonctions conjuguées. *Math. Z.*, **27**(1), 218–244.

Roberts, A.W., and Varberg, D.E. 1973. *Convex Functions*. Pure and Applied Mathematics, Vol. 57. New York–London: Academic Press [A subsidiary of Harcourt Brace Jovanovich, Publishers].

Rogers, C.A. 1957. A single integral inequality. *J. London Math. Soc.*, **32**, 102–108.

Rossi, J., and Weitsman, A. 1983. A unified approach to certain questions in value distribution theory. *J. London Math. Soc. (2)*, **28**(2), 310–326.

Royen, T. 2014. A simple proof of the Gaussian correlation conjecture extended to some multivariate gamma distributions. *Far East J. Theor. Stat.*, **48**(2), 139–145.

Rudin, W. 1964. *Principles of Mathematical Analysis*. Second edition. New York: McGraw-Hill Book Co.

Rudin, W. 1966. *Real and Complex Analysis*. New York: McGraw-Hill Book Co.

Ryff, J.V. 1970. Measure preserving transformations and rearrangements. *J. Math. Anal. Appl.*, **31**, 449–458.

Sarvas, J. 1972. Symmetrization of condensers in $n$-space. *Ann. Acad. Sci. Fenn. Ser. A I*, 44.

Schaefer, H.H. 1971. *Topological Vector Spaces*. Third printing corrected, Graduate Texts in Mathematics, Vol. 3. New York: Springer-Verlag.

Schechtman, G., Schlumprecht, T., and Zinn, J. 1998. On the Gaussian measure of the intersection. *Ann. Prob.*, **26**(1), 346–357.

Schmidt, E. 1943. Beweis der isoperimetrischen Eigenschaft der Kugel im hyperbolischen und sphärischen Raum jeder Dimensionenzahl. *Math. Z.*, **49**, 1–109.

Schneider, R. 1993. *Convex Bodies: The Brunn–Minkowski Theory*. Encyclopedia of Mathematics and Its Applications, vol. 44. Cambridge: Cambridge University Press.

Schulz, F., and Vera de Serio, V. 1993. Symmetrization with respect to a measure. *Trans. Amer. Math. Soc.*, **337**(1), 195–210.

Schwarz, H.A. 1884. Beweis des Satzes, dass die Kugel kleinere Oberfläche besitzt, als jeder andere Körper gleichen Volumens. *Nachrichten Königlichen Gesellschaft Wissenschaften Göttingen*, 1–13.

Schwarz, H.A. 1890. *Gesammelte Mathematische Abhandlungen*. Berlin: Springer Verlag.

Shea, D. F. 1966. On the Valiron deficiencies of meromorphic functions of finite order. *Trans. Amer. Math. Soc.*, **124**, 201–227.

Sierpiński, W. 1920. Sur la question de la mesurabilité de la base de M. Hamel. *Fundamenta Mathematicae*, **1**(1), 105–111.

Simon, B. 1979. *Functional Integration and Quantum Physics*. Pure and Applied Mathematics, vol. 86. New York: Academic Press Inc. [Harcourt Brace Jovanovich Publishers].

Simon, B., and Høegh-Krohn, R. 1972. Hypercontractive semigroups and two dimensional self-coupled Bose fields. *J. Funct. Anal.*, **9**, 121–180.

Smith, K.T. 1983. *Primer of Modern Analysis*. Second edition. Undergraduate Texts in Mathematics. New York: Springer-Verlag.

Sobolev, S.L. 1938. On a theorem in functional analysis. *Sb. Math.*, **4**, 471–497.

Solynin, A.Y. 1996. Extremal configurations in some problems on capacity and harmonic measure. *Zap. Nauchn. Sem. S.-Peterburg. Otdel. Mat. Inst. Steklov. (POMI)*, **226** (Anal. Teor. Chisel i Teor. Funktsiĭ. 13), 170–195, 239.

Sperner, Jr., E. 1973. Zur Symmetrisierung von Funktionen auf Sphären. *Math. Z.*, **134**, 317–327.

Sperner, Jr., E. 1979. Symmetrization and currents. *Mathematika*, **26**(2), 269–289 (1980).

Spivak, M. 1965. *Calculus on Manifolds. A Modern Approach to Classical Theorems of Advanced Calculus*. New York-Amsterdam: W. A. Benjamin, Inc.

Stammbach, U. 2012. A letter of Hermann Amandus Schwarz on isoperimetric problems. *Math. Intelligencer*, **34**(1), 44–51. With an appendix containing German transcription of the letter.

Stanoyevitch, A. 1994. Integral inequalities and equalities for the rearrangement of Hardy and Littlewood. *J. Math. Anal. Appl.*, **183**(3), 509–517.

Stein, E.M. 1970. *Singular Integrals and Differentiability Properties of Functions*. Princeton Mathematical Series, No. 30. Princeton, NJ: Princeton University Press.

Stein, E.M., and Weiss, G. 1971. *Introduction to Fourier Analysis on Euclidean Spaces*. Princeton Mathematical Series, No. 32. Princeton, NJ: Princeton University Press.

Steiner, J. 1842. Sur le maximum et le minimum des figures dans le plan, sur la sphère et dans l'espace en géneral. *J. Reine Angew. Math.*, **24**, 93–152.

Steiner, J. 1882. *Gesammelte Werke Vol.2*. New York: Prussian Academy of Sciences.

Stroock, D.W. 1993. *Probability Theory, An Analytic View*. Cambridge: Cambridge University Press.

Szegő, G. 1930. Über einige Extremalaufgaben der Potentialtheorie. *Math. Z.*, **31**(1), 583–593.

Szegö, G. 1950. On membranes and plates. *Proc. Natl. Acad. Sci. U.S.A.*, **36**, 210–216.

Szegö, G. 1958. Note to my paper "On membranes and plates". *Proc. Natl. Acad. Sci. U.S.A.*, **44**, 314–316.

Talenti, G. 1976a. Best constant in Sobolev inequality. *Ann. Mat. Pura Appl. (4)*, **110**, 353–372.

Talenti, G. 1976b. Elliptic equations and rearrangements. *Ann. Scuola Norm. Sup. Pisa Cl. Sci. (4)*, **3**(4), 697–718.

Talenti, G. 1979. Nonlinear elliptic equations, rearrangements of functions and Orlicz spaces. *Ann. Mat. Pura Appl. (4)*, **120**, 160–184.

Talenti, G. 1993. The standard isoperimetric theorem. Pages 73–123 of: *Handbook of Convex Geometry, Vol. A, B*. Amsterdam: North-Holland.

Talenti, G. 1997. A weighted version of a rearrangement inequality. *Ann. Univ. Ferrara Sez. VII (N.S.)*, **43**, 121–133 (1998).

Talenti, G. 2016. The art of rearranging. *Milan J. Math.*, **84**(1), 105–157.

Teichmüller, O. 1939. Vermutungen und Sätze über die Wertverteilung gebrochener Funktionen endlicher Ordnung. *Deutsche Math.*, **4**, 163–190.

Tikhomirov, V.M. 1990. *Stories about Maxima and Minima*. Mathematical World, vol. 1. Translated from the 1986 Russian original by Abe Shenitzer. Providence, RI: American Mathematical Society.

Tsuji, M. 1975. *Potential Theory in Modern Function Theory*. Reprinting of the 1959 original. New York: Chelsea Publishing Co.

Väisälä, J. 1961. On quasiconformal mappings in space. *Ann. Acad. Sci. Fenn. Ser. A I No.*, **298**, 36.

Van Schaftingen, J. 2006. Approximation of symmetrizations and symmetry of critical points. *Topol. Methods Nonlinear Anal.*, **28**(1), 61–85.

Vilenkin, N.J. 1968. *Special Functions and the Theory of Group Representations*. Translated from the Russian by V.N. Singh. Translations of Mathematical Monographs, Vol. 22. Providence, RI: American Mathematical Society.

Vilenkin, N.J., and Klimyk, A.U. 1991. *Representation of Lie Groups and Special Functions. Vol. 1*. Mathematics and its Applications (Soviet Series), vol. 72. Simplest Lie groups, special functions and integral transforms, Translated from the Russian by V.A. Groza and A.A. Groza. Dordrecht: Kluwer Academic Publishers Group.

Villani, C. 2003. *Topics in Optimal Transportation*. Graduate Studies in Mathematics, vol. 58. Providence, RI: American Mathematical Society.

Watanabe, T. 1983. The isoperimetric inequality for isotropic unimodal Lévy processes. *Z. Wahrsch. Verw. Gebiete*, **63**(4), 487–499.

Weitsman, A. 1979. A symmetry property of the Poincaré metric. *Bull. London Math. Soc.*, **11**(3), 295–299.

Weitsman, A. 1986. Symmetrization and the Poincaré metric. *Ann. Math. (2)*, **124**(1), 159–169.

Wojtaszczyk, P. 1991. *Banach Spaces for Analysts*. Cambridge Studies in Advanced Mathematics, vol. 25. Cambridge: Cambridge University Press.

Wolontis, V. 1952. Properties of conformal invariants. *Amer. J. Math.*, **74**, 587–606.

Wu, J.-M. 2002. Harmonic measures for symmetric stable processes. *Studia Math.*, **149**(3), 281–293.

Yanagihara, H. 1993. An integral inequality for derivatives of equimeasurable rearrangements. *J. Math. Anal. Appl.*, **175**(2), 448–457.

Yang, S.S. 1994. Estimates for coefficients of univalent functions from integral means and Grunsky inequalities. *Israel J. Math.*, **87**(1–3), 129–142.

Ziemer, W.P. 1989. *Weakly Differentiable Functions*. Graduate Texts in Mathematics, vol. 120. New York: Springer-Verlag.

Zygmund, A. 1968. *Trigonometric Series: Vols. I, II*. Second edition, reprinted with corrections and some additions. London: Cambridge University Press.

# Index